COUNTABLE BOOLEAN ALGEBRAS AND DECIDABILITY

SIBERIAN SCHOOL OF ALGEBRA AND LOGIC

Series Editor: *Yuri L. Ershov*

Editorial Board: *Sergei S. Goncharov* (Associate Editor), *E. N. Kuz'min, V. D. Mazurov, V. K. Kharchenko, E. I. Khukhro, A. I. Ryaskin*

Editorial Council: *E. I. Zel'manov, O. H. Kegel, A. Macintyre, A. Nerode*

Managing Editor: *Tamara N. Rozhkovskaya*

Countable Boolean Algebras and Decidability • *Sergei S. Goncharov*

Definability and Computability • *Yuri L. Ershov*

Right-Ordered Groups • *Valeriĭ M. Kopytov and Nikolaĭ Ya. Medvedev*

COUNTABLE BOOLEAN ALGEBRAS AND DECIDABILITY

Sergei S. Goncharov
*Novosibirsk State University
and Institute of Mathematics
Novosibirsk, Russia*

Consultants Bureau • New York, London, and Moscow

Library of Congress Cataloging-in-Publication Data

On file

Siberian School of Algebra and Logic is a simultaneous translation
of the book series Sibirskaya Shkola Algebry i Logiki,
which is published in and translated by
Scientific Books (RIMIBE NGU)
2, Pirogova Street, Novosibirsk 630090, Russia

ISBN 0-306-11061-X

©1997 Consultants Bureau, New York
A Division of Plenum Publishing Corporation
233 Spring Street, New York, N.Y. 10013

http://www.plenum.com

10 9 8 7 6 5 4 3 2 1

All rights reserved

No part of this book may be reproduced, stored in a retrieval system, or transmitted in any
form or by any means, electronic, mechanical, photocopying, microfilming, recording, or
otherwise, without written permission from the Publisher

Printed in the United States of America

To the memory of Cris Ash

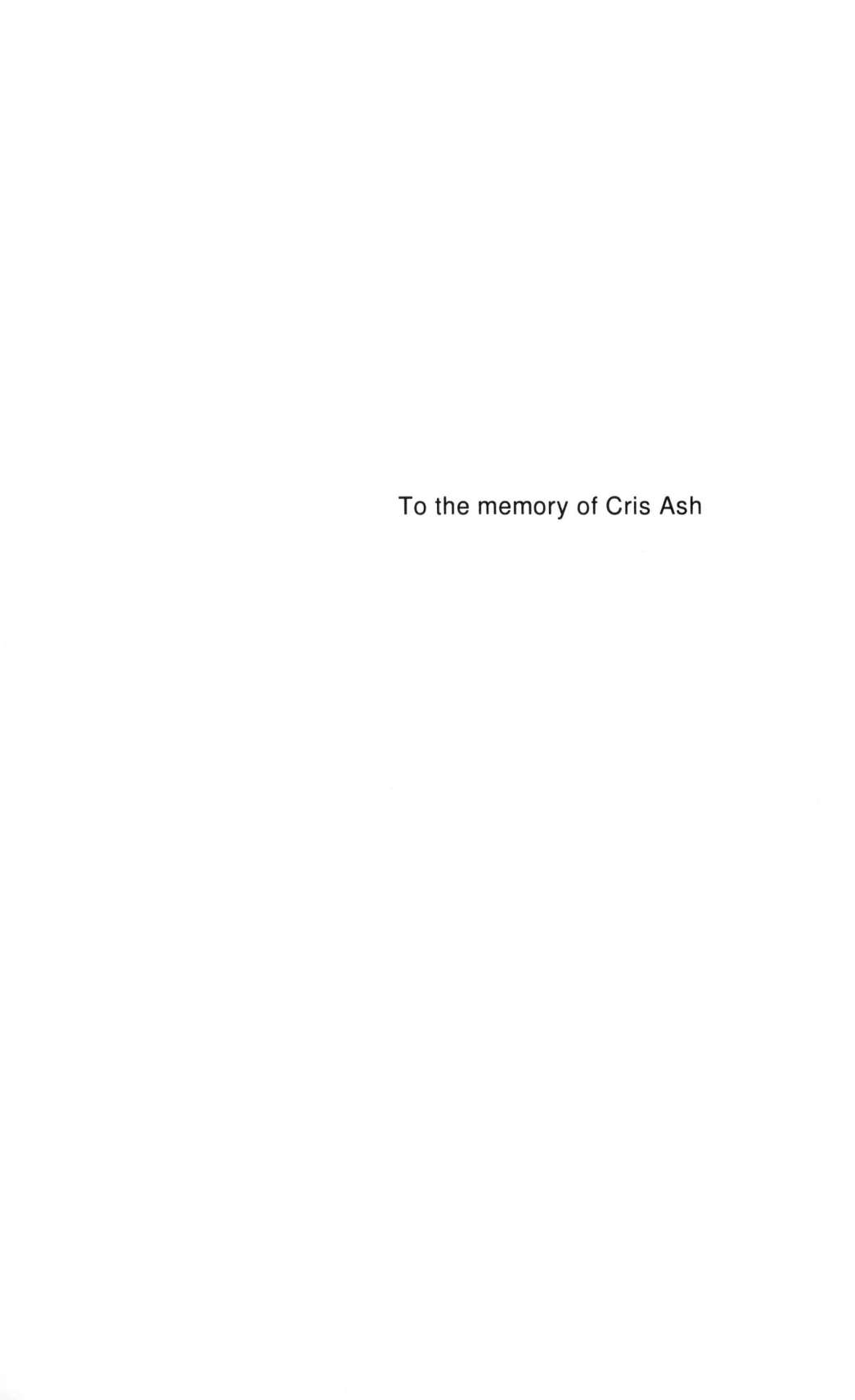

Preface

The theory of Boolean algebras studies algebras of subsets with respect to the operations of union, intersection, and complement. Such systems and their enrichments appear in various branches of mathematics and its applications. Due to this, the study of Boolean algebras has become a field of considerable interest to researchers.

Boolean algebras were introduced by Boole to analyze logical conclusions in accordance with logic laws. As was shown by Stone, such abstract algebras admit a natural set-theoretical representation as algebras of subsets. A classification of complete elementary theories of Boolean algebras was given by Tarski and Ershov and has become an important step in understanding the structure of Boolean algebras. The next important step in this direction was made by Ketonen, who characterized the types of isomorphisms of countable Boolean algebras. To obtain his result Ketonen used a topological approach. Using algebraic methods, Ershov described types of isomorphisms of countable distributive lattices with relative complements. The class considered by Ershov is wider than that studied by Ketonen, and the Ketonen characterization follows from the corresponding result of Ershov.

The use of Boolean algebras in various fields of mathematics leads to the necessity of studying additional structures on Boolean algebras such as measures, topologies, automorphisms, homomorphisms, ideals, etc. Questions on the algorithmic complexity of constructing Boolean algebras with given properties are also timely with respect to applications.

In this book, the author presents only the necessary algebraic and set-theoretic facts of the theory of Boolean algebras, referring, for details, to the recent handbook on Boolean algebras and other literature on the subject.

The main purpose is to present the theory of recursive and decidable Boolean algebras. In particular, the author aims to describe approaches and methods developed and used in his own works as well as in works of other Siberian algebraists and logicians.

The book focuses on the algebraic properties of Boolean algebras (Chapter 1), the classification of complete elementary theories of Boolean algebras and the description of first-order relations on Boolean algebras (Chapter 2), and the algorithmic properties of countable Boolean algebras which are studied on the basis of constructing effective representations of countable Boolean algebras on the set of natural numbers (Chapter 3). The book includes all sections of the author's book "Countable Boolean Algebras" published by the Siberian Branch of the Publishing House Nauka in 1988. Updated and expanded, the present book provides recent results. A new part of Chapter 1 is concerned with the class of Ershov algebras and their use in the algebraic classification of countable Boolean algebras. A description of the complexity of first-order relations constitutes new material in Chapter 2. The most significant revision was made in Chapter 3, which now includes a number of recent results on decidability problems, computable classes, automorphism groups. The results play an important role in understanding the model-theoretic properties of algebras in these enrichments and a variety of types of their elementary theories as well.

This book is based on lectures given by the author at Novosibirsk State University as well as at some universities in Kazakhstan, Uzbekistan (Tashkent University and Nukuss State University), and Australia (Monash University).

The book does not require a heavy mathematical background. The reader familiar with basic set-theoretic and model-theoretic notions and facts may omit the first sections of Chapters 1 and 2. However, to facilitate understanding of Chapter 3, it is desirable to know the elements of recursion theory, at least the possible formalizations of computability.

Preface

I would like to thank the many people who participated in the preparation of this book.

First of all, I would like to thank my teacher, Yu. L. Ershov, who directed my initial steps in science and who introduced me to this interesting field of research. His fundamental results were the basis of my work and of my students'. I am grateful to Yu. L. Ershov for his heavy editing of this book. His critical remarks, discussions, and careful editing significantly improved the book.

The author would like to acknowledge his colleagues and students, A. Morozov, S. Odintsov, V. Dzgoev, V. Selivanov, P. Alaev, S. Podzorov, S. Fedoryaev, and D. Pal'chunov, who work in the field and whose results are included in the book. In particular, Section 3.8 follows the results of A. Morozov, whose contribution to the study of recursive automorphisms of Boolean algebras is essential. The powerful methods for the study of recursive enrichments of Boolean algebras were developed by S. Odintsov. His results are also described here.

I express my gratitude to Professors A. Sorbi and A. Ursini for the wonderful opportunity (supported by PECO grant No. ERBCIPDCT 940615) of completing the work on this book at Siena University, Italy.

My investigations, the results of which are presented in this book, were partially supported by the Russian Fund for Fundamental Research (grant No. 96-01-01525).

I express warm and special thanks to my wife Lyuba Goncharova for her constant support of my scientific work although this activity does not allow me to dedicate much time to my family.

I am grateful to my editor Tamara Rozhkovskaya for her excellent editorial work on the manuscript, without whose efforts this work would not have been published.

Sergei S. Goncharov

Akademgorodok, Novosibirsk
May, 1996

Contents

Chapter 1. Algebraic Properties of Boolean Algebras.................. 1
  1.1 Algebraic Systems and Constructions 1
  1.2 Definitions and the Simplest Properties of Boolean Algebras .. 15
  1.3 Ideals and Quotient Algebras of Boolean Algebras 26
  1.4 The Stone Theorem on Representations of Boolean Algebras .. 39
  1.5 The Vaught Criterion 44
  1.6 Linearly Ordered Generating Sets 53
  1.7 Generating Trees 59
  1.8 Ershov Algebras and The Isomorphism Problem 71

Chapter 2. Elementary Classification of Boolean Algebras 91
  2.1 Basic Notions and Methods of Model Theory 91
  2.2 Definable Ershov–Tarski Ideals and Elementary Characteristics
      of Boolean Algebras 108
  2.3 Countably Saturated Boolean Algebras and Elementary Classification 114
  2.4 Model-Complete Theories of Boolean Algebras 127
  2.5 Consistent Complete Theories of Boolean Algebras 132
  2.6 Restricted Theories of Boolean Algebras 139

Chapter 3. Constructive Boolean Algebras		149
3.1	Basic Notions of the Theory of Algorithms and Constructive Models	150
3.2	Constructibility in Linear Orders and Boolean Algebras	174
3.3	Trees Generating Constructive Boolean Algebras	180
3.4	Decidable Boolean Algebras	190
3.5	Restricted Fragments of the Theory of Boolean Algebras and Decidable Algebras	214
3.6	Algorithmic Dimension of Boolean Algebras	249
3.7	Algorithmic Properties of Subalgebras and Quotient Algebras of Constructive Boolean Algebras	261
3.8	Automorphisms of Countable Boolean Algebras	290
References		303
Subject index		315

COUNTABLE BOOLEAN ALGEBRAS AND DECIDABILITY

Chapter 1

Algebraic Properties of Boolean Algebras

1.1. Algebraic Systems and Constructions

Boolean algebras form a special class of algebraic systems. There are different (but equivalent) representations of this class. In the present book, we apply the model-theoretic approach. For the definitions of models, algebraic systems, and constructions over them, we refer the reader to [18, 36, 71, 107, 108, 129]. Set-theoretic notions can be found in the above-mentioned books and in [84, 98] as well. As for topological notions, we follow [88, 99].

Algebraic systems can be characterized by basic operations and relations on a set. Indeed, a group is a system consisting of a set and operations on this set. Defining rings, we fix at least two operations. An order on a set means that a binary relation is defined.

To fix basic operations and relations, the notion of a signature is used. By a *signature* σ we mean disjoint sets $(\sigma_F, \sigma_P, \sigma_C)$ and a mapping $\rho\colon \sigma_F \cup \sigma_P \to \mathbb{N}^+$, where the *set of functional symbols* σ_F yields the names of basic operations, the *set of predicate symbols* σ_P gives the names of basic relations, and the *set of constant symbols* σ_C defines the names of elements. With every functional symbol or predicate symbol the mapping ρ associates its number of places. We use the notation $\sigma = ((\sigma_F, \sigma_P, \sigma_C), \rho)$.

By an *algebraic system* (model) $\mathfrak{A}$ of a signature $\sigma = ((\sigma_F, \sigma_P, \sigma_C), \rho)$ we mean a nonempty set A and a correspondence $\operatorname{int}_\sigma$ such that $\operatorname{int}_\sigma(f)$ is a mapping from $A^{\rho(f)}$ into A for every functional symbol $f \in \sigma_F$, $\operatorname{int}_\sigma(P)$ is a subset of the set $A^{\rho(P)}$ for every predicate symbol $P \in \sigma_P$, and $\operatorname{int}_\sigma(c)$ is an element of the set A for every constant symbol $c \in \sigma_C$. The set A is called the *universe* (sometimes denoted by $|\mathfrak{A}|$) and $\operatorname{int}_\sigma$ is called an *interpretation* of the algebraic system $\mathfrak{A}$. We use the notation $\mathfrak{A} = \langle A, \operatorname{int}_\sigma \rangle$. We often write $\sigma = \langle F_0^{n_0}, \ldots, F_k^{n_k}; P_0^{m_0}, \ldots, P_r^{m_r}; c_0, \ldots, c_s \rangle$ instead of $\sigma = ((\sigma_F, \sigma_P, \sigma_C), \rho)$ if the set σ_F consists of $F_0, \ldots, F_k$, the set σ_P consists of $P_0, \ldots, P_r$, the set σ_C consists of $c_0, \ldots, c_s$, $\rho(F_i) = n_i$, $i \leqslant k$, and $\rho(P_i) = m_i$, $i \leqslant r$.

If $\sigma = \langle F_0^{n_0}, \ldots, F_k^{n_k}; P_0^{m_0}, \ldots, P_r^{m_r}; c_0, \ldots, c_s \rangle$, then an algebraic system $\mathfrak{A}$ of the signature σ will be written in the form

$$\mathfrak{A} = \langle A; (F_0)_\mathfrak{A}, \ldots, (F_k)_\mathfrak{A}; (P_0)_\mathfrak{A}, \ldots, (P_r)_\mathfrak{A}; (c_0)_\mathfrak{A}, \ldots, (c_s)_\mathfrak{A} \rangle$$

where we regard $(F_i)_\mathfrak{A}$, $i = 1, \ldots, k$, as n_i-ary operations on A, $(P_i)_\mathfrak{A}$, $i = 1, \ldots, r$, as subsets of A^{m_i}, and $(c_i)_\mathfrak{A}$, $i = 1, \ldots, s$, as elements of A. We often omit the subscript $\mathfrak{A}$ and write simply $\mathfrak{A} = \langle A; F_0, \ldots, F_k; P_0, \ldots, P_r; c_0, \ldots, c_s \rangle$ when no confusion can arise.

For algebraic systems $\mathfrak{A}$ and $\mathfrak{B}$ of a fixed signature σ we introduce some standard algebraic constructions and notions.

A mapping φ from the universe $|\mathfrak{A}|$ of an algebraic system $\mathfrak{A}$ into the universe $|\mathfrak{B}|$ of an algebraic system $\mathfrak{B}$ is called a *homomorphism* from $\mathfrak{A}$ into $\mathfrak{B}$ (denoted by $\varphi \colon \mathfrak{A} \to \mathfrak{B}$) if the following conditions are fulfilled:

— for any F of σ_F and $a_1, \ldots, a_n$, $n = \rho(F)$, of $|\mathfrak{A}|$ the equality $\varphi((F)_\mathfrak{A}(a_1, \ldots, a_n)) = (F)_\mathfrak{B}(\varphi(a_1), \ldots, \varphi(a_n))$ holds,

— for any P of σ_P and $a_1, \ldots, a_n$, $n = \rho(P)$, of $|\mathfrak{A}|$ from $(a_1, \ldots, a_n) \in (P)_\mathfrak{A}$ it follows that $(\varphi(a_1), \ldots, \varphi(a_n)) \in (P)_\mathfrak{B}$,

— for any c of σ_C the equality $\varphi((c)_\mathfrak{A}) = (c)_\mathfrak{B}$ holds.

A homomorphism φ from $\mathfrak{A}$ into $\mathfrak{B}$ is called an *epimorphism* if φ maps $|\mathfrak{A}|$ onto $|\mathfrak{B}|$ and is called an *isomorphic embedding* if it is an injection, i.e., $\varphi(a) \neq \varphi(b)$ for different elements $a, b \in A$ and the following condition (regarded as the converse condition to the second one in the definition of a homomorphism) is fulfilled: if $(\varphi(a_1), \ldots, \varphi(a_n)) \in (P)_\mathfrak{B}$, then $(a_1, \ldots, a_n) \in (P)_\mathfrak{A}$.

1.1. Algebraic Systems and Constructions

For a mapping f from A into B we denote the domain of definition of f (i.e., the set A) by $\mathrm{dom}\,f$ and the range of values of f (i.e., the set $\{y \in B \mid \text{there exists } x \in A \text{ such that } f(x) = y\}$) by $\mathrm{range}\,f$.

If an isomorphic embedding of an algebraic system $\mathfrak{A}$ into an algebraic system $\mathfrak{B}$ is an epimorphism, then it is called an *isomorphism* of $\mathfrak{A}$ onto $\mathfrak{B}$ (or between $\mathfrak{A}$ and $\mathfrak{B}$). Two systems $\mathfrak{A}$ and $\mathfrak{B}$ are called *isomorphic* (denoted by $\mathfrak{A} \cong \mathfrak{B}$) if there exists an isomorphism of $\mathfrak{A}$ onto $\mathfrak{B}$. An isomorphism of $\mathfrak{A}$ onto itself is called an *automorphism* of $\mathfrak{A}$.

By an *equivalence relation* θ on a set A we mean a subset of A^2 such that

$(a, a) \in \theta$ for any $a \in A$ (reflexivity)

if $(a, b) \in \theta$, then $(b, a) \in \theta$ for any $a, b \in A$ (symmetry)

if $(a, b) \in \theta$ and $(b, c) \in \theta$, then $(a, c) \in \theta$ for any $a, b, c \in A$ (transitivity)

For a given equivalence relation θ we can divide the set A into classes of equivalent elements $a/\theta = \{x \in A \mid (x, a) \in \theta\}$, where $a \in A$. A class a/θ is called a *congruence class by* θ and the set of all congruence classes $\{a/\theta \mid a \in A\}$ is referred to as the *quotient set* (denoted by A/θ) of A by θ.

An equivalence relation θ on $|\mathfrak{A}|$ is called a *congruence on* $\mathfrak{A}$ if for any $a_i, b_i, i \leqslant n$, and functional symbol F the following condition holds: $(a_i, b_i) \in \theta$, $i \leqslant n$, imply $(F(a_1, \ldots, a_n), F(b_1, \ldots, b_n)) \in \theta$.

A congruence θ is called *strict* if for P of σ_P and $a_1, \ldots, a_n, b_1, \ldots, b_n$ of A we have $(b_1, \ldots, b_n) \in P$ if $(a_1, \ldots, a_n) \in P$ and $(a_i, b_i) \in \theta$, $i \leqslant n$.

For an algebraic system $\mathfrak{A}$ of a signature σ and a congruence θ on $\mathfrak{A}$ we can define the quotient system $\mathfrak{A}/\theta$ if take the quotient set A/θ as the universe and interpret the signature symbols as follows:

— $(F)_{\mathfrak{A}/\theta}(a_1/\theta, \ldots, a_n/\theta) = (F)_{\mathfrak{A}}(a_1, \ldots, a_n)/\theta$, where $F \in \sigma_F$ is an n-ary functional symbol and $a_1/\theta, \ldots, a_n/\theta$ are congruence classes of A/θ,

— for congruence classes $a_1/\theta, \ldots, a_n/\theta$ and an n-ary predicate symbol $P \in \sigma_P$ we set $(a_1/\theta, \ldots, a_n/\theta) \in (P)_{\mathfrak{A}/\theta}$ if there exist elements $b_1, \ldots, b_n$ such that $(a_i, b_i) \in \theta$, $i \leqslant n$, and $(b_1, \ldots, b_n) \in (P)_{\mathfrak{A}}$,

— for the value of a constant symbol $c \in \sigma_C$ we take the element $(c)_{\mathfrak{A}}/\theta$ of $\mathfrak{A}/\theta$.

Partially ordered sets are usually considered with the signature consisting of a single binary predicate $\langle \leqslant^2 \rangle$; moreover, the notation $x \leqslant y$ is often used instead of $\leqslant (x,y)$. Studying partially ordered sets possessing some additional properties (e.g., the presence of the least element and the greatest element, the least upper bound and the greatest lower bound, a complement to an element, etc.), we enrich the signature with new symbols. However, we may not include the order symbol into the signature if the partial order can be uniquely restored from the basic signature symbols.

Let $\mathfrak{A}$ be an algebraic system of a signature σ and let B be a subset of the universe $|\mathfrak{A}|$ such that B is closed under the operations of σ and contains all the values of constant symbols of σ, i.e., $(F)_\mathfrak{A}(a_1, \ldots, a_n) \in B$ if $a_1, \ldots, a_n \in B$ and F is a symbol of an n-ary operation of σ and $(c)_\mathfrak{A} \in B$ if c is a constant symbol of σ. In this case, we can consider restrictions on B of the interpretation of the signature symbols as follows:

$(P)_\mathfrak{B} \rightleftharpoons (P)_\mathfrak{A} \cap B^n$ is the restriction of the n-ary predicate corresponding to a predicate symbol $P \in \sigma$,

$(F)_\mathfrak{B} \rightleftharpoons (F)_\mathfrak{A} \cap B^{m+1}$ is the restriction of the m-ary function corresponding to a functional symbol $F \in \sigma$,

$(c)_\mathfrak{B} \rightleftharpoons (c)_\mathfrak{A}$ is the value of a constant symbol $c \in \sigma$.

Interpreting all the symbols of the signature σ on B in such a way, we obtain the algebraic system $\mathfrak{B} = \langle B; (P)_\mathfrak{B} : P \in \sigma_P; (F)_\mathfrak{B} : F \in \sigma_F; (c)_\mathfrak{B} : c \in \sigma_C \rangle$ which is a *subsystem* (of $\mathfrak{A}$) of the signature σ.

Let $\mathfrak{A}$ be an algebraic system of a signature σ and $X \subseteq A$. There exists the least subsystem of $\mathfrak{A}$ the universe of which contains all elements of X. We say that this subsystem *is generated* by X and denote it by $\mathrm{gr}_\mathfrak{A}(X)$. For a fixed $\mathfrak{A}$ we often omit the subscript $\mathfrak{A}$ to simplify the notation.

Let a signature σ be a part of a signature σ', i.e., $\sigma_P \subseteq \sigma'_P$, $\sigma_F \subseteq \sigma'_F$, $\sigma'_C \subseteq \sigma_C$. For any symbol of $\sigma_P \cup \sigma_F$ we suppose that its number of places with respect to σ coincides with its number of places with respect to σ'. We say that a *system $\mathfrak{A} \upharpoonright \sigma$ is obtained from a system $\mathfrak{A}$ of the signature σ' by restriction on the signature σ* if we restrict interpretations from the set of symbols of σ' onto the set of symbols of σ. In this case, we also say that $\mathfrak{A} \upharpoonright \sigma$ is a *restriction of $\mathfrak{A}$ on the signature σ* and $\mathfrak{A}$ is an *enrichment of $\mathfrak{A} \upharpoonright \sigma$ to the signature σ'*.

We introduce the *direct product* $\mathfrak{A} \times \mathfrak{B}$ of algebraic systems $\mathfrak{A}$ and $\mathfrak{B}$ of a signature σ. For the universe we take the Cartesian product $A \times B \rightleftharpoons \{(a,b) \mid a \in A, b \in B\}$ of the universes A and B of the systems $\mathfrak{A}$ and $\mathfrak{B}$. Further, we set

1.1. Algebraic Systems and Constructions

- $(P)_{\mathfrak{A}\times\mathfrak{B}} \rightleftharpoons \{((a_1,b_1),\ldots,(a_n,b_n)) \mid (a_i,b_i) \in A \times B,\ i \leqslant n, (a_1,\ldots,a_n) \in (P)_{\mathfrak{A}},\ (b_1,\ldots,b_n) \in (P)_{\mathfrak{B}}\}$ for an n-ary predicate symbol $P \in \sigma$,

- $(F)_{\mathfrak{A}\times\mathfrak{B}}((a_1,b_1),\ldots,(a_m,b_m)) \rightleftharpoons ((F)_{\mathfrak{A}}(a_1,\ldots,a_m),(F)_{\mathfrak{B}}(b_1,\ldots,b_m))$, where $(a_i,b_i) \in A \times B, i \leqslant m$, for an m-ary functional symbol $F \in \sigma$,

- $(c)_{\mathfrak{A}\times\mathfrak{B}} \rightleftharpoons ((c)_{\mathfrak{A}}, (c)_{\mathfrak{B}})$ for a constant symbol $c \in \sigma$.

We also introduce the *Cartesian product* $\prod_{i \in I} \mathfrak{A}_i$ for a family of algebraic systems $\mathfrak{A}_i$, $i \in I$, of a signature σ. As the universe we take the Cartesian product of the universes A_i, $i \in I$, of the systems $\mathfrak{A}_i$, $i \in I$:

$$\prod_{i \in I} A_i = \{f \colon I \to \cup A_i \mid f(i) \in A_i,\ i \in I\}$$

Predicates and operations are defined by coordinates as follows:

- $(f_1,\ldots,f_n) \in (P)_{\prod_{i \in I} \mathfrak{A}_i} \Leftrightarrow (f_1(i),\ldots,f_n(i)) \in (P)_{\mathfrak{A}_i}$ for all $i \in I$, where $f_j \in \prod_{i \in I} A_i$, $j \leqslant n$, for an n-ary predicate symbol $P \in \sigma$,

- $(F)_{\prod_{i \in I} \mathfrak{A}_i}(f_1,\ldots,f_n)(i) \rightleftharpoons (F)_{\mathfrak{A}_i}(f_1(i),\ldots,f_n(i))$ for any $f_j \in \prod_{i \in I} A_i$, $j \leqslant n$, for an m-ary functional symbol $F \in \sigma$,

- $(c)_{\prod_{i \in I} \mathfrak{A}_i} \rightleftharpoons f \in \prod_{i \in I} A_i$, where $f(i) \rightleftharpoons (c)_{\mathfrak{A}_i}$, $i \in I$, for a constant symbol $c \in \sigma$.

We emphasize that, in the case of infinite I, the cardinality of the Cartesian product $\prod_{i \in I} \mathfrak{A}_i$ regarded as an algebraic system may be as much as desired even if the systems $\mathfrak{A}_i$, $i \in I$, are finite.

For $\varnothing \neq C \subseteq \sigma_C$ we can introduce the *direct sum* $\sum_{i \in I} {}_C \mathfrak{A}_i$ as the subsystem of the Cartesian product $\prod_{i \in I} \mathfrak{A}_i$ generated by the following subset:

$\{f \in \prod_{i \in I} A_i \mid$ there exists a finite subset $K \subseteq I$ and a constant $c \in \sigma_C$

such that for any $i \in I \setminus K$ the value $f(i)$

is equal to the value of the constant c in $\mathfrak{A}_i\}$

It is easy to see that the universe of the system $\sum_{i \in I} c\mathfrak{A}_i$ consists of elements $f \in \prod_{i \in I} A_i$ such that there is a finite subset $K \subseteq I$ and a closed (without variables) term t with constants in C such that for any $i \in I \backslash K$ the value $f(i)$ is equal to the value of the term t in $\mathfrak{A}_i$. Let σ be a signature and let $V = \{v_0, v_1, \dots\}$ be a sequence of variables. The language describing the properties of algebraic systems of the signature σ consists of the set $\text{Term}_\sigma(V)$ constituted by terms of the signature σ in variables of V and the set $\text{Form}_\sigma(V)$ constituted by formulas of the signature σ in variables of V. For the sake of simplicity, we write simply Term and Form if σ and V are fixed.

The set Term can be inductively defined as the least set of words in the alphabet consisting of the symbols of $\sigma \cup V$ and auxiliary symbols (,) and , :

— variables of V and constant symbols of σ are terms,

— if $t_1, \dots, t_n$ are terms and F is an n-ary functional symbol of σ, the word $F(t_1, t_2, \dots, t_n)$ is a term.

Let $\overline{v}$ stand for a sequence of variables of the form $(v_{i_1}, \dots, v_{i_k})$. If all the variables of t are included in $\overline{v}$, then we write $t(\overline{v})$ or $t(v_{i_1}, \dots, v_{i_k})$ to denote the term t.

On an algebraic system $\mathfrak{A}$ of a signature σ, a term $t(v_{i_1}, \dots, v_{i_k})$ defines a function (called a termal operation) $\underline{t}$ from A^k into A. If $a_1, \dots, a_k$ is a sequence of elements of A, the value of the operation $\underline{t}$ on $a_1, \dots, a_k$ is denoted by $t_{\mathfrak{A}}(a_1, \dots, a_k)$. This value is inductively defined as follows:

— if a term t is a variable v_{i_n}, then $t_{\mathfrak{A}}(a_1, \dots, a_k) \leftrightharpoons a_n$; if a term t is a constant c, then the value $t_{\mathfrak{A}}(a_1, \dots, a_k)$ is equal to the value of the constant c in $\mathfrak{A}$,

— if a term t has the form $F(t_1, \dots, t_n)$, then the value $t_{\mathfrak{A}}(a_1, \dots, a_k)$ is equal to the value of the operation $(F)_{\mathfrak{A}}$ in $\mathfrak{A}$ on the sequence $(t_{1\mathfrak{A}}(a_1, \dots, a_k), \dots, t_{n\mathfrak{A}}(a_1, \dots, a_k))$.

It is easy to see that the termal operations are just operations obtained as compositions of basic operations of σ and the identity function $I(x) \equiv x$. In this case, we consider a constant as a 0-ary function on A.

It is not hard to see that the universe of the system $\text{gr}_{\mathfrak{A}}(X)$ consists of exactly the values of all terms of σ on sequences of elements of X. For a fixed system $\mathfrak{A}$ we write $t(a_1, \dots, a_k)$ instead of $t_{\mathfrak{A}}(a_1, \dots, a_k)$.

1.1. Algebraic Systems and Constructions 7

The language of the first-order predicate calculus of a signature σ can be inductively defined [36, 129]. This allows us to describe the properties of sequences of elements of an algebraic system $\mathfrak{A}$ in a precise way. To symbols of σ, variables of V, and the auxiliary symbols (,) and , we add the following *logical connectives*: & (the conjunction symbol), $\vee$ (the disjunction symbol), $\rightarrow$ (the implication symbol), $\neg$ (the negation symbol), the *quantifiers* $\exists$ (the existential quantifier), $\forall$ (the universal quantifier), and the *equality symbol* $=$.

By a *formula* we mean a word of Form such that

(1) if t and q are terms, then $t = q$ is a formula; if P is an n-ary predicate symbol of σ and $t_1, \ldots, t_n$ are terms, then $P(t_1, \ldots, t_n)$ is a formula; moreover, all the variables in such formulas are free; there are no bound occurrences of variables;

(2) if φ and ψ are formulas, then $(\varphi \& \psi)$, $(\varphi \vee \psi)$, $(\varphi \rightarrow \psi)$, and $\neg \varphi$ are formulas; moreover, the type of occurrences (free or bound) of variables remains unchanged.

(3) if φ is a formula and v_i is a variable of V, then $(\forall v_i)\varphi$ and $(\exists v_i)\varphi$ are formulas; moreover, the type of the occurrence of any variable different from v_i remains unchanged while all the occurrences of v_i are bound.

The language just introduced will be denoted by $\mathcal{L}_{\omega\omega}$. A formula satisfying (1) is called *atomic*.

We can extend the language $\mathcal{L}_{\omega\omega}$ by infinite formulas if we consider formulas of the form $\bigvee_{i \in \mathbb{N}} \varphi_i$ and $\bigwedge_{i \in \mathbb{N}} \varphi_i$ for a countable family of formulas φ_i, $i \in \mathbb{N}$, as admissible ones. The language obtained is denoted by $\mathcal{L}_{\omega\omega_1}$ [72].

We define a semantics of formulas of the language $\mathcal{L}_{\omega\omega}$, i.e., we assign the truth values of formulas in an algebraic system of a signature σ provided that the values of free variables are given.

Let all the variables occurring free in a formula φ belong to $\bar{v} = (v_{i_1}, \ldots, v_{i_k})$ and let $\bar{a} = (a_1, \ldots, a_n)$ be a sequence of elements of a system $\mathfrak{A}$ of a signature σ. We introduce the following relations:

$\mathfrak{A} \models [\varphi]_{\bar{a}}^{\bar{v}}$ a formula φ is true in $\mathfrak{A}$ after the substitution of the values of $\bar{a}$ for free occurrences of variables of $\bar{v}$

$\mathfrak{A} \not\models [\varphi]_{\bar{a}}^{\bar{v}}$ a formula φ is false in $\mathfrak{A}$ after the substitution of the values of $\bar{a}$ for free occurrences of variables of $\bar{v}$

The notation $\varphi(\overline{v})$ means that variables occurring free in φ belong to $\overline{v}$. We also write $\varphi(\overline{a})$ instead of $[\varphi]_{\overline{a}}^{\overline{v}}$ for brevity.

We define inductively:

for formulas of the form $t = q$

$\mathfrak{A} \vDash t = q(\overline{a})$ if $t_{\mathfrak{A}}(\overline{a}) = q_{\mathfrak{A}}(\overline{a})$

$\mathfrak{A} \nvDash t = q(\overline{a})$ otherwise

for formulas of the form $P(t_1, \ldots, t_n)$

$\mathfrak{A} \vDash P(t_1, \ldots, t_n)(\overline{a})$ if $((t_1)_{\mathfrak{A}}(\overline{a}), \ldots, (t_n)_{\mathfrak{A}}(\overline{a})) \in (P)_{\mathfrak{A}}$

$\mathfrak{A} \nvDash P(t_1, \ldots, t_n)(\overline{a})$ otherwise

for formulas of the form $(\varphi \,\&\, \psi)$

$\mathfrak{A} \vDash (\varphi \,\&\, \psi)(\overline{a})$ if $\mathfrak{A} \vDash \varphi(\overline{a})$ and $\mathfrak{A} \vDash \psi(\overline{a})$

$\mathfrak{A} \nvDash (\varphi \,\&\, \psi)(\overline{a})$ otherwise

for formulas of the form $(\varphi \vee \psi)$

$\mathfrak{A} \vDash (\varphi \vee \psi)(\overline{a})$ if $\mathfrak{A} \vDash \varphi(\overline{a})$ or $\mathfrak{A} \vDash \psi(\overline{a})$

$\mathfrak{A} \nvDash (\varphi \vee \psi)(\overline{a})$ otherwise

for formulas of the form $(\varphi \to \psi)$

$\mathfrak{A} \vDash (\varphi \to \psi)(\overline{a})$ if $\mathfrak{A} \vDash \psi(\overline{a})$ or $\mathfrak{A} \nvDash \varphi(\overline{a})$

$\mathfrak{A} \nvDash (\varphi \to \psi)(\overline{a})$ otherwise

for formulas of the form $\neg \varphi$

$\mathfrak{A} \vDash \neg\varphi(\overline{a})$ if $\mathfrak{A} \nvDash \varphi(\overline{a})$

$\mathfrak{A} \nvDash \neg\varphi(\overline{a})$ otherwise

for formulas of the forms $(\forall v_i)\varphi$ and $(\exists v_i)\varphi$ we define a new interpretation of the variable v_i as follows. We denote by $\overline{v}^{v_i}$ the sequence of variables obtained from $\overline{v}$ by eliminating v_i and denote by $\overline{a}^{v_i}$ the sequence of elements obtained from $\overline{a}$ by eliminating the element corresponding to v_i. We set

$\mathfrak{A} \vDash (\forall v_i)\varphi(\overline{a})$ if $\mathfrak{A} \vDash \varphi(\overline{a}^{v_i}, b)$ for any $b \in |\mathfrak{A}|$

$\mathfrak{A} \nvDash (\forall v_i)\varphi(\overline{a})$ otherwise

1.1. Algebraic Systems and Constructions

$\mathfrak{A} \vDash (\exists v_i)\varphi(\bar{a})$ if there is $b \in \mathfrak{A}$ such that $\mathfrak{A} \vDash \varphi(\bar{a}^{v_i}, b)$

$\mathfrak{A} \nvDash (\exists v_i)\varphi(\bar{a})$ otherwise

Let Δ be a set of formulas, φ a formula, and $\bar{v}$ a sequence of variables occurring free in formulas of $\Delta \cup \{\varphi\}$. Let $\bar{a}$ be an assignment of variables of $\bar{v}$. We say that

— Δ *holds in $\mathfrak{A}$ for the assignment $\bar{a}$* if all the formulas of Δ are true in $\mathfrak{A}$ for this assignment (denoted by $\mathfrak{A} \vDash \Delta(\bar{a})$ or $\mathfrak{A} \vDash [\Delta]\frac{\bar{v}}{\bar{a}}$ and $\mathfrak{A} \nvDash \Delta(\bar{a})$ otherwise),

— φ is a *semantic consequence* of Δ (denoted by $\Delta \vDash \varphi$) if for any system $\mathfrak{A}$ and assignment $\bar{a}$ the fact that Δ holds in $\mathfrak{A}$ for the assignment $\bar{a}$ implies that φ is true in $\mathfrak{A}$ for the assignment $\bar{a}$.

Two formulas φ and ψ are called *equivalent* (denoted by $\varphi \equiv \psi$) if $\{\varphi\} \vDash \psi$ and $\{\psi\} \vDash \varphi$. A set of formulas Δ is called *satisfiable* if there exists a system $\mathfrak{A}$ and an assignment $\bar{a}$ such that Δ holds in $\mathfrak{A}$ for the assignment $\bar{a}$ and is called *unsatisfiable* (denoted by $\Delta \vDash$) otherwise.

As was shown by Gödel, the semantic consequence relation can be syntactically expressed. We can introduce a system of axioms (the *first-order predicate calculus*) which defines the semantic consequence relation $\Gamma \vdash \alpha$ [36].

A set of formulas Γ is called *inconsistent* (denoted by $\Gamma \vdash$) if there exists a formula β such that $\Gamma \vdash \beta$, $\Gamma \vdash \neg \beta$ and *consistent* otherwise. A formula α is called *consistent (satisfiable)* if the set $\{\alpha\}$ is consistent (satisfiable).

Theorem 1.1.1 [the Gödel completeness theorem]. *A formula α is consistent if and only if α is satisfiable.*

Corollary 1.1.1. *Let Γ be a finite set of formulas and let φ be a formula. Then $\Gamma \vdash \varphi$ if and only if $\Gamma \vDash \varphi$, $\Gamma \vdash$ if and only if $\Gamma \vDash$.*

Mal'tsev [111] extended the completeness theorem to the case of an arbitrary set of formulas. He proved a theorem on local satisfiability known as the compactness theorem. This theorem is one of the important tools in the model theory and its applications.

Theorem 1.1.2 [the Mal'tsev compactness theorem]. *A set of formulas is satisfiable if any of its finite subsets is satisfiable.*

Corollary 1.1.2. *If a set of formulas is unsatisfiable, then it has a finite unsatisfiable subset.*

Corollary 1.1.3. *A set of formulas is satisfiable if and only if it is consistent.*

A class K of algebraic systems of a signature σ is called *axiomatizable* if there exists a set Δ of *sentences* (i.e., formulas having no free occurrences of variables) of the signature σ such that $K = \text{Mod}(\Delta)$, where $\text{Mod}(\Delta) \leftrightharpoons \{\mathfrak{A} \mid \mathfrak{A} \vDash \Delta\}$. By a *theory* $\text{Th}(K)$ of K we mean the set of sentences that are true in each system of K. By an *elementary theory* (or simply *theory*) (in a signature σ) we mean any set of sentences (of the signature σ) closed under the deduction relation. The Mal'tsev compactness theorem shows that T is a theory if and only if there is a class of systems K such that $\text{Th}(K) = T$. A theory is called *complete* if $\varphi \in T$ or $\neg\varphi \in T$ for any sentence φ. It is easy to see that $K \subseteq \text{Mod}(\text{Th}(K))$. However, an arbitrary class of systems is not necessarily axiomatizable.

We note that the axiom method is the basic tool of constructing a class of algebraic systems. In accordance with this method, we first define basic relations, operations, and constants and then define their basic properties by means of axioms. Taking different signatures, we arrive at different possibilities in studying classes of systems. Thus, we can choose a signature depending on the properties of a class we wish to examine.

We consider a class K of algebraic systems of a signature σ and a class K' of algebraic systems of a signature σ'. To introduce the notion of the semantic definability of K in K' we fix a first-order definable description Φ of the signature symbols of σ, i.e., for an n-ary predicate symbol $P \in \sigma$, an m-ary functional symbol $F \in \sigma$, and a constant symbol $c \in \sigma$ there exist formulas $\Phi_P(x_1, \ldots, x_n)$, $\Phi_F(x_1, \ldots, x_m)$, and $\Phi_c(x)$ of the signature σ'. For an arbitrary system $\mathfrak{A}' \in K'$, on the universe A', we interpret the signature symbols P, F, and c of σ as follows:

— as the interpretation of P on A' we take the set $\{(a_1, \ldots, a_n) \mid \mathfrak{A}' \vDash \Phi_P(a_1, \ldots, a_n)\}$,

— as the interpretation of F on A' we take the mapping from A'^m into A' such that the formula $\Phi_F(x_1, \ldots, x_{m+1})$ defines on A' its graph,

— as the interpretation of c on A' we take the single element on which the formula $\Phi_c(x)$ is true.

Thus, on $|\mathfrak{A}|$ we have defined a model of the signature σ. Denote it by $\text{int}_\Phi(\mathfrak{A}')$. We say that a *class K of algebraic systems of a signature σ is semantically definable in a class K' of algebraic systems of a signature σ' by a first-order definable description* Φ if for any $\mathfrak{A}' \in K'$ the system

1.1. Algebraic Systems and Constructions

int$_\Phi(\mathfrak{A}')$ belongs to K and for any $\mathfrak{A} \in K$ there exists an algebraic system $\mathfrak{A}' \in K'$ such that the systems int$_\Phi(\mathfrak{A}')$ and $\mathfrak{A}'$ are isomorphic.

Classes K and K' of algebraic systems of signatures σ and σ' respectively are called *semantically equivalent* if K is semantically definable in K' by a first-order definable description Φ, K' is semantically definable in K by a first-order definable description Φ', and int$_{\Phi'}(\text{int}_\Phi(\mathfrak{A}')) = \mathfrak{A}'$, int$_\Phi(\text{int}_{\Phi'}(\mathfrak{A})) = \mathfrak{A}$ for any algebraic systems $\mathfrak{A} \in K$ and $\mathfrak{A}' \in K'$.

To study semantically equivalent classes various methods can be applied. We note that many properties of a class are transferred to semantically equivalent classes. However, as for algebraic properties, they turn out to be different for semantically equivalent classes because of the difference in basic structures and the complexity of formulas of Φ and Φ'. In the sequel, we omit external universal quantifiers in variables for the sake of simplicity.

The complexity of a formula can be characterized by the presence of quantifiers and their mutual location in the formula. Any formula φ can be transformed into a formula of the form $Q_1x_1\ldots Q_nx_n\psi$ up to an equivalence, where ψ is a quantifier-free formula and $Q_i \in \{\forall, \exists\}$, $1 \leqslant i \leqslant n$. However, the most important characteristic of the complexity of a formula is not the number of quantifiers but the number of their alternations.

A formula $Q_1x_1\ldots Q_nx_n\psi$ is called an $\exists$-*formula* if $Q_i = \exists$, $i = 1,\ldots,n$, and is called a $\forall$-*formula* if $Q_i = \forall$, $i = 1,\ldots,n$. We define Σ_n and Π_n by induction. The class Σ_1 consists of $\exists$-formulas and the class Π_1 is formed by $\forall$-formulas. The class Σ_{n+1} contains only formulas of the form $\exists x_1\ldots \exists x_m\psi$, where $\psi \in \Pi_n$, and the class Π_{n+1} consists of formulas of the form $\forall x_1\ldots \forall x_m\psi$, where $\psi \in \Sigma_n$. We note that the "prefix" $(\exists x_1)\ldots(\exists x_m)$ can be empty for $m = 0$. Thus, we have obtained the following classification of formulas by the complexity of quantifier blocks:

$$\Pi_0 \subseteq \Pi_1 \subseteq \Pi_2 \subseteq \ldots \subseteq \Pi_n \subseteq \Pi_{n+1} \subseteq \ldots$$
$$\Sigma_0 \subseteq \Sigma_1 \subseteq \Sigma_2 \subseteq \ldots \subseteq \Sigma_n \subseteq \Sigma_{n+1} \subseteq \ldots$$

We note that $\exists$-formulas preserve the truth value under taking extensions of models and $\forall$-formulas have the same property under taking submodels; moreover, these properties can be regarded as characteristics of $\exists$-formulas and $\forall$-formulas. The classes Σ_n and Π_n, $n > 1$, can be characterized in terms of the sandwich language [18].

A quantifier-free formula of the form $\bigvee_{i=1}^{n} \&_{j=1}^{m_i} \Phi_{ij}$, where Φ_{ij} is an atomic formula or the negation of an atomic formula, is said to be in *disjunctive normal form*. Using the simple associativity and distributivity laws for logical connectives as well as equivalences of the form $\varphi \to \psi \equiv \neg \varphi \vee \psi$, $\neg(\varphi \vee \psi) \equiv \neg \varphi \& \neg \psi$, $\neg(\varphi \& \psi) \equiv \neg \varphi \vee \neg \psi$, we can easily establish that any quantifier-free formula is equivalent to a formula in disjunctive normal form.

In the case of a finite signature σ, a formula $\varphi(x_1, \ldots, x_n)$ of the form $\underset{i=1}{\overset{k}{\&}} \Phi_i$, where Φ_i is an atomic formula or the negation of an atomic formula, is called *perfect* if the following conditions hold:

— for any sequence $y_1, \ldots, y_m$ constituted by variables of $\{x_1, \ldots, x_n\}$ and constant symbols of σ as well and for any m-ary predicate symbol $P \in \sigma$, the formula $P(y_1, \ldots, y_m)$ or the formula $\neg P(y_1, \ldots, y_m)$ is one of the formulas $\Phi_1, \ldots, \Phi_k$,

— for any sequence (y_1, y_2) constituted by variables of $\{x_1, \ldots, x_n\}$ and constant symbols of σ as well, the formula $y_1 = y_2$ or the formula $y_1 \neq y_2$ is one of the formulas $\Phi_1, \ldots, \Phi_k$,

— there are no i and j such that $\Phi_i \equiv \neg \Phi_j$.

A disjunctive normal form $\overset{n}{\underset{i=1}{\vee}} \overset{m_i}{\underset{j=1}{\&}} \Phi_{ij}$ is called *perfect* if $\overset{m}{\underset{j=1}{\&}} \Phi_{ij}$ is perfect for any $1 \leqslant i \leqslant n$.

Using the provability of a formula of the form $\Phi \vee \neg \Phi$, we can add $\Phi \vee \neg \Phi$, where Φ is the required formula, to any conjunction and reduce a result to a disjunctive normal form by the same transformations. We can exclude multiple occurrences with the help of the equivalence $\Phi \& \Phi \equiv \Phi$. Since the formula $\Phi \& \neg \Phi$ is identically false, we can eliminate all the identically false disjunctive members. As a result, we either obtain a formula in perfect disjunctive normal form or establish that the formula is identically false, i.e., it is inconsistent. Thus, any consistent formula $\varphi(x_1, \ldots, x_n)$ of a finite signature σ is equivalent (with respect to σ) to some formula in perfect disjunctive normal form.

If the signature σ contains functional symbols, then we define an *atomical formula* as a formula of the form

$$x = F(y_1, \ldots, y_n), \quad x = c, \quad P(x_1, \ldots, x_n) \qquad (1.1.1)$$

1.1. Algebraic Systems and Constructions

where $x, y_1, \ldots, y_n, x_1, \ldots, x_n$ are variables of the language $L_{\omega\omega}$. Using the equivalences

$$t = F(t_1, \ldots, t_n) \Leftrightarrow (\forall x \forall y_1 \ldots \forall y_n)$$
$$\left(x = t \,\&\, \underset{i=1}{\overset{n}{\&}}\, y_i = t_i \to x = F(y_1, \ldots, y_n) \right)$$

$$P(t_1, \ldots, t_n) \Leftrightarrow (\forall x \forall y_1 \ldots \forall y_n) \left(\underset{i=1}{\overset{n}{\&}}\, y_i = t_i \to P(y_1, \ldots, y_n) \right)$$

$$t = F(t_1, \ldots, t_n) \Leftrightarrow (\exists x \exists y_1 \ldots \exists y_n)$$
$$\left(x = t \,\&\, \underset{i=1}{\overset{n}{\&}}\, y_i = t_i \,\&\, x = F(y_1, \ldots, y_n) \right)$$

$$P(t_1, \ldots, t_n) \Leftrightarrow (\exists x \exists y_1 \ldots \exists y_n) \left(\underset{i=1}{\overset{n}{\&}}\, y_i = t_i \,\&\, P(y_1, \ldots, y_n) \right)$$

one can show that any atomic formula (i.e., a formula of the form $t = q$ or $P(t_1, \ldots, t_n)$, where t, q, and $t_1, \ldots, t_n$ are terms) is equivalent to an $\exists$-formula φ_0 and a $\forall$-formula ψ_0 as well that have atomical subformulas of the form (1.1.1). Since a quantifier-free formula φ is equivalent to a formula in disjunctive normal form φ', we can replace each of the positive occurrences of a complex atomic subformula by an equivalent $\exists$-formula ($\forall$-formula) and each of the negative occurrences of such a subformula by $\forall$-formula ($\exists$-formula) of the required form. As a result, we conclude that a quantifier-free formula is equivalent to an $\exists$-formula and a $\forall$-formula that have atomical subformulas of the form (1.1.1). Therefore, any Σ_{n+1}-formula (Π_{n+1}-formula) is equivalent to a Σ_{n+1}-formula (Π_{n+1}-formula) having atomical subformulas of the form (1.1.1). Thus, examining Σ_{n+1}-formulas or Π_{n+1}-formulas, we can require that their quantifier-free subformulas be in perfect disjunctive normal form and contain atomical subformulas only of the form (1.1.1). Therefore, studying perfect disjunctive normal forms appearing as quantifier-free subformulas of a formula with quantifiers, we always assume that each of their atomical subformulas have the form (1.1.1).

The notion of a partially ordered set is fundamental in mathematics. For partially ordered sets we usually consider the signature consisting of the single binary predicate $\langle \leqslant^2 \rangle$. We often write xPy instead of $P(x, y)$ for a binary predicate P and $(x\,f\,y)$ instead of $f(x, y)$ for a binary function f.

By a *partially ordered set* $\langle A, \leqslant \rangle$ [8] we mean a system of the signature $\langle \leqslant^2 \rangle$ subject to the following axioms:

$(\forall x)(x \leqslant x)$ \hfill (reflexivity)

$(\forall x)(\forall y)(x \leqslant y \,\&\, y \leqslant x \to x = y)$ (antisymmetry)

$(\forall x)(\forall y)(\forall z)(x \leqslant y \,\&\, y \leqslant z \to x \leqslant z)$ (transitivity)

EXAMPLE 1.1.1. Consider the set $\mathcal{P}(A)$ of all subsets of a set A, a nonempty subset X of $\mathcal{P}(A)$, and the inclusion relation $\subseteq$. Then $\langle X; \subseteq \rangle$ is a partially ordered set.

Let $\mathfrak{A} = \langle A; \leqslant \rangle$ be a partially ordered set and let $X \subseteq A$. An element $a \in A$ is called

— an *upper (lower) bound* of X if $x \leqslant a$ $(a \leqslant x)$ for any $x \in X$,

— the *least (greatest) element* of X if $a \in X$ and $x \geqslant a$ $(a \geqslant x)$ for any $x \in X$,

— the *least upper (greatest lower) bound* of X in $\mathfrak{A}$ if a is the least (greatest) element among all upper (lower) bounds of X.

We denote the least upper (greatest lower) bound of X in $\mathfrak{A}$ by $\vee_\mathfrak{A} X$ ($\wedge_\mathfrak{A} X$) or $\sup_\mathfrak{A}(X)$ ($\inf_\mathfrak{A}(X)$). If X consists of two elements a and b, then their least upper bound (if it exists) is also denoted by $a \vee_\mathfrak{A} b$ and their greatest lower bound is denoted by $a \wedge_\mathfrak{A} b$.

In the case of arbitrary partial orders, the least upper bounds cannot exist for arbitrary subsets and even for pairs of elements.

Let $a, b \in A$ and let $\langle A, \leqslant \rangle$ be a partially ordered set. We define $\hat{a} = \{x \mid x \leqslant a\}$ and $\check{a} = \{x \mid a \leqslant x\}$. A subset $X \subseteq A$ is called an *initial segment* of $\langle A, \leqslant \rangle$ if for any $x \in X$ and $a \in A$ from $a \leqslant x$ it follows that $a \in X$.

The subsets

(1) $[a, b] \rightleftharpoons \{x \in A \mid a \leqslant x \leqslant b\}$,

(2) $]a, b] \rightleftharpoons \{x \in A \mid a < x \leqslant b\}$,

(3) $[a, b[\rightleftharpoons \{x \in A \mid a \leqslant x < b\}$,

(4) $]a, b[\rightleftharpoons \{x \in A \mid a < x < b\}$

are called *intervals* of $\langle A, \leqslant \rangle$. They are also called *closed intervals* in the case (1), *half-open from the left intervals* in the case (2), *half-open from the right intervals* in the case (3), and *open intervals* in the case (4).

Any partially ordered set can be completed by elements ∞ and $-\infty$ provided that ∞ is regarded as the greatest element and $-\infty$ as the least element of the set $A \cup \{-\infty, \infty\}$. We always mean such a completion in the cases in which these symbols are used.

Subsets of A of the form $]-\infty, a]$, $]-\infty, a[$, $[a, \infty[$, and $]a, \infty[$ are also referred to as *intervals* of $\langle A, \leqslant \rangle$. Intervals of the form $]-\infty, a]$ and $]-\infty, a[$ are called *initial intervals* of $\langle A, \leqslant \rangle$ and intervals of the form $[a, \infty[$ and $]a, \infty[$ are called *end intervals* of $\langle A, \leqslant \rangle$.

A partially ordered set $\langle A, \leqslant \rangle$ is called an *upper semilattice* if for any two elements $x, y \in A$ there exists the least upper bound $x \vee y$. If a partially ordered set is an upper semilattice, then we often consider it with the signature $\langle \vee^2 \rangle$, where $x \vee y$ is the operation of taking the least upper bound. With the help of this operation, the order is determined by the relation $x \leqslant y \Leftrightarrow x \vee y = y$. We introduce the dual notion of a *lower semilattice*: for any two elements x and y there exists the greatest lower bound $x \wedge y$. Lower semilattices can also be considered with the signature $\langle \wedge^2 \rangle$. It is possible to define an order on it by the equivalence $x \leqslant y \Leftrightarrow x \wedge y = x$.

In a partially ordered set, the least element is called the *zero* and the greatest element is called the *unit*.

A lattice $\langle A, \leqslant \rangle$ is called *distributive* if $a \wedge (b \vee c) = (a \wedge b) \vee (a \wedge c)$ and $a \vee (b \wedge c) = (a \vee b) \wedge (a \vee c)$ for any $a, b, c \in A$.

By an *ideal* I of an upper semilattice $\langle A, \leqslant \rangle$ with zero we mean a subset $I \subseteq A$ such that $0 \in I$ and the following condition holds:

if $x \leqslant y$ and $y \in I$, then $x \in I$;   if $x, y \in I$, then $x \vee y \in I$

By a *filter* $\mathcal{D}$ of a lower semilattice $\langle A, \leqslant \rangle$ with unit we mean a subset $\mathcal{D} \subseteq A$ such that $1 \in \mathcal{D}$ and the following condition holds:

if $x \leqslant y$ and $x \in \mathcal{D}$, then $y \in \mathcal{D}$;   if $x, y \in \mathcal{D}$, then $x \wedge y \in \mathcal{D}$

An ideal (filter) is called *proper* if it is not equal to A.

1.2. Definitions and the Simplest Properties of Boolean Algebras

Boolean algebras appear in different branches of mathematics for different purposes. Therefore, there are several equivalent approaches to their definition. Here, we give the most popular definitions of a Boolean algebra and prove that they are equivalent. Considering a Boolean algebra as an algebraic system, we introduce various algebraic constructions on it and study its algebraic structure.

- An algebraic system $\mathfrak{A} = \langle A, \vee, \wedge, C \rangle$ of the signature $\langle \vee^2, \wedge^2, C^1 \rangle$ is called a *Boolean algebra* if for any $a, b \in A$ the following conditions hold:

    (1) $a \vee b = b \vee a$ (commutativity),

    (2) $(a \vee b) \vee c = a \vee (b \vee c)$ (associativity),

    (3) $a \vee (b \wedge c) = (a \vee b) \wedge (a \vee c)$ (distributivity),

    (4) $C(a \vee b) = C(a) \wedge C(b)$,

    (5) $a \vee a = a$,

    (6) $(a \wedge C(a)) \vee b = b$,

    (7) $CC(a) = a$.

EXAMPLE 1.2.1. Let $\mathcal{P}(A)$ be the set of all subsets of a set A and let $\cup$, $\cap$, and C be the ordinary operations of union, intersection, and complement for subsets of A. The system $\langle \mathcal{P}(A); \cup, \cap, C \rangle$ is a Boolean algebra because all the relations (1)–(7) hold. It is called the *algebra of all subsets of the set A*. Let a nonempty subset X of $\mathcal{P}(A)$ be closed under the operations $\cup$, $\cap$, and C. For the subalgebra $\langle X; \cup, \cap, C \rangle$ all the relations (1)–(7) also hold. Such Boolean algebras are called *algebras of sets*.

EXAMPLE 1.2.2. Let (A, τ) be a topological space and let $\boldsymbol{O}_\tau$ be the set of all clopen, i.e., open and closed, in the topology τ subsets of the set A. Then $\boldsymbol{O}_\tau$ is closed under the ordinary set-theoretic operations $\cup$, $\cap$, and C, hence $O_\tau = \langle \boldsymbol{O}_\tau; \cup, \cap, C \rangle$ is a Boolean algebra. It is called the *algebra of clopen sets of the space* (A, τ).

EXAMPLE 1.2.3. Let $\mathfrak{L}_n$ be the set of all formulas of a signature σ in free variables of the set $\{x_1, \cdots, x_n\}$ and let T be a theory in the signature σ. On $\mathfrak{L}_n$, we introduce the T-equivalence relation $\sim_T$ as follows:

$$\varphi \sim_T \psi \rightleftharpoons T \vdash (\varphi \to \psi) \,\&\, (\psi \to \varphi)$$

It is easy to check that this relation possesses the following property:

if $\varphi \sim_T \varphi'$ and $\psi \sim_T \psi'$, then $\varphi \vee \psi \sim_T \varphi' \vee \psi'$, $\varphi \wedge \psi \sim_T \varphi' \wedge \psi'$,
$$\varphi \to \psi \sim_T \varphi' \to \psi', \; \neg \varphi \sim_T \neg \varphi' \qquad (1.2.1)$$

1.2. Definitions and the Simplest Properties 17

We consider the quotient set $B_n(T) \rightleftharpoons \mathfrak{L}_n/{\sim_T}$ and introduce the operations $\cup$, $\cap$, and C by setting

$$\varphi/{\sim_T} \vee \psi/{\sim_T} \rightleftharpoons \varphi \vee \psi/{\sim_T}$$
$$\varphi/{\sim_T} \wedge \psi/{\sim_T} \rightleftharpoons \varphi \wedge \psi/{\sim_T}$$
$$C(\varphi/{\sim_T}) \rightleftharpoons \neg\varphi/{\sim_T}$$

where $\varphi/{\sim_T}$ denotes a congruence class by $\sim_T$ containing φ. In view of (1.2.1), these operations are independent of the choice of representatives of congruence classes. By the properties of the equivalence relation, the algebra $\mathfrak{B}_n(T) \rightleftharpoons \langle B_n(T); \cup, \cap, C \rangle$ is a Boolean one. The algebra $\mathfrak{B}_n(T)$ is called the *Lindenbaum algebra* of formulas in n variables of the theory T.

Proposition 1.2.1. *In any Boolean algebra $\langle A, \vee, \wedge, C \rangle$, for any elements $a, b, c \in A$ the following relations, which are dual to (1)–(6), hold:*

(1') $a \wedge b = b \wedge a$,

(2') $(a \wedge b) \wedge c = a \wedge (b \wedge c)$,

(3') $a \wedge (b \vee c) = (a \wedge b) \vee (a \wedge c)$,

(4') $C(a \wedge b) = C(a) \vee C(b)$,

(5') $a \wedge a = a$,

(6') $(a \vee C(a)) \wedge b = b$.

PROOF. From (7) and (4) we obtain (4'):

$$C(a \wedge b) = C(CC(a) \wedge CC(b)) = CC(C(a) \vee C(b)) = C(a) \vee C(b)$$

From (4') and (1)–(7) it is easy to derive the remaining relations. For instance, let us prove (1') and (2'). The rest of the relations are left to the reader. We have

$$a \wedge b = CC(a \wedge b) = C(C(a) \vee C(b)) = C(C(b) \vee C(a))$$
$$= CC(b \wedge a) = b \wedge a$$
$$(a \wedge b) \wedge c = CC((a \wedge b) \wedge c) = C(C(a \wedge b) \vee C(c))$$
$$= C((C(a) \vee C(b)) \vee C(c)) = C(C(a) \vee (C(b) \vee C(c)))$$
$$= C(C(a) \vee C(b \wedge c)) = CC(a \wedge (b \wedge c)) = a \wedge (b \wedge c)$$

The proposition is proved. □

For any Boolean algebra $\mathfrak{A}$ we define the dual system $\mathfrak{A}^*$ as follows: $A^* \leftrightharpoons A$, $a \vee^* b \leftrightharpoons a \wedge b$, $a \wedge^* b \leftrightharpoons a \vee b$, and $C^*(a) = C(a)$.

REMARK 1.2.1. For a Boolean algebra $\mathfrak{A}$ its dual system $\mathfrak{A}^*$ is isomorphic to $\mathfrak{A}$ and $\mathfrak{A}^{**} = \mathfrak{A}$. Indeed, consider the mapping $C : A \to A$ and show that C is an isomorphism of $\mathfrak{A}$ onto $\mathfrak{A}^*$. From (7) we have $a = CC(a)$, and from $C(a) = C(b)$ the equality $a = b$ follows. Therefore, C is a one-to-one mapping of A onto A. It remains to check that C preserves the operations. In view of (4) and (4'),

$$C(a \vee b) = C(a) \wedge C(b) = C(a) \vee^* C(b)$$
$$C(a \wedge b) = C(a) \vee C(b) = C(a) \wedge^* C(b)$$

moreover, $CC(a) = C^*C(a)$ since $C^* = C$. Hence C preserves the operations. Consequently, it is an isomorphism. The equality $\mathfrak{A}^{**} = \mathfrak{A}$ follows from the definition.

Proposition 1.2.2. *For any elements a and b of a Boolean algebra $\langle A, \vee, \wedge, C \rangle$ the following equalities hold:*

(a) $a \vee C(a) = b \vee C(b)$,

(b) $a \wedge C(a) = b \wedge C(b)$.

PROOF. In view of (1') and (6'),

$$b \vee C(b) = (a \vee C(a)) \wedge (b \vee C(b)) = (b \vee C(b)) \wedge (a \vee C(a)) = a \vee C(a)$$

Similarly, in view of (1) and (6),

$$a \wedge C(a) = (b \wedge C(b)) \vee (a \wedge C(a)) = (a \wedge C(a)) \vee (b \wedge C(b)) = b \wedge C(b)$$

Thus, the values of $a \wedge C(a)$ and $a \vee C(a)$ are independent of the choice of a. These values are denoted by $\mathbf{0}_{\mathfrak{A}}$ and $\mathbf{1}_{\mathfrak{A}}$ and are referred to as the "null element" and the "unit" of the Boolean algebra $\mathfrak{A}$. We omit the subscript $\mathfrak{A}$ when no confusion can arise.

Proposition 1.2.3. *For any element $a \in A$ of a Boolean algebra $\langle A, \vee, \wedge, C \rangle$ the following relations hold:*

(a) $a \vee \mathbf{0} = a$,   (a') $a \wedge \mathbf{0} = \mathbf{0}$

(b) $a \vee \mathbf{1} = \mathbf{1}$,   (b') $a \wedge \mathbf{1} = a$

PROOF. The conditions (a) and (b') are an immediate consequence of the relations (6) and (6'), the definition of $\mathbf{0}$ and $\mathbf{1}$, and the commutativity

1.2. Definitions and the Simplest Properties

of the operations $\vee$ and $\wedge$. Let us prove (b) and (a'). Since $1 = a \vee C(a)$, from (2) and (5) it follows that

$$a \vee 1 = a \vee (a \vee C(a)) = (a \vee a) \vee C(a) = a \vee C(a) = 1$$

Since $0 = a \wedge C(a)$, from (2') and (5') we obtain

$$a \wedge 0 = a \wedge (a \wedge C(a)) = (a \wedge a) \wedge C(a) = a \wedge C(a) = 0$$

The proposition is proved. □

For Boolean algebras we consider the standard signature containing two supplementary constants $\mathbf{0}$, $\mathbf{1}$ and subject to the axioms $(\forall x)(x \wedge C(x) = \mathbf{0})$ and $(\forall x)(x \vee C(x) = \mathbf{1})$ defining constants. Unless otherwise indicated, by the signature of Boolean algebras we mean $\langle \wedge^2, \vee^2, C^1, \mathbf{0}, \mathbf{1} \rangle$. It is clear that the class of Boolean algebras of the signature $\langle \wedge^2, \vee^2, C^1 \rangle$ and that of the signature $\langle \wedge^2, \vee^2, C^1, \mathbf{0}, \mathbf{1} \rangle$ are semantically equivalent; moreover, the enrichment of $\langle \wedge^2, \vee^2, C^1 \rangle$ to $\langle \wedge^2, \vee^2, C^1, \mathbf{0}, \mathbf{1} \rangle$ can be defined by the formulas $\Phi_0(x) = (\forall a)(a \wedge C(a) = x)$, $\Phi_1(x) = (\forall a)(a \vee C(a) = x)$.

A binary relation U on a set A is called a *partial order on A* if $\langle A, U \rangle$ is a partially ordered set with the signature consisting of a single binary predicate U. A partial order U is said to be *linear* if it satisfies the connectivity condition: $(a,b) \in U$ or $(b,a) \in U$ for any $a, b \in A$. It is obvious that the restriction on $B \subseteq A$ of a partial (linear) order defined on A is a partial (linear) order on B.

We regard a partially ordered set $\langle A, U \rangle$ as a system of the signature $\langle \leqslant^2 \rangle$ with a single binary predicate. We write $x \leqslant_U y$ instead of $(x, y) \in U$.

Let $\mathfrak{A} = \langle A, U \rangle$ be a partially ordered set. For $a, b \in A$ let $a \vee_\mathfrak{A} b$ denote the least upper bound and let $a \wedge_\mathfrak{A} b$ stand for the greatest lower bound (if they exist) of the set $\{a, b\}$ in $\mathfrak{A}$.

If for any a, b of a partially ordered set $\mathfrak{A}$ there exist $a \vee_\mathfrak{A} b$ and $a \wedge_\mathfrak{A} b$, then the algebra $\langle A, \vee_\mathfrak{A}, \wedge_\mathfrak{A} \rangle$ can be considered with the signature including two binary operations $\vee_\mathfrak{A}$ and $\wedge_\mathfrak{A}$. In this case, the partially ordered set $\langle A, U \rangle$ is called a *lattice*. A lattice $\mathfrak{A}$ is called *distributive* if for any $a, b, c \in A$ the following distributivity laws hold:

(D) $$a \vee_\mathfrak{A} (b \wedge_\mathfrak{A} c) = (a \vee_\mathfrak{A} b) \wedge_\mathfrak{A} (a \vee_\mathfrak{A} c)$$

(D') $$a \wedge_\mathfrak{A} (b \vee_\mathfrak{A} c) = (a \wedge_\mathfrak{A} b) \vee_\mathfrak{A} (a \wedge_\mathfrak{A} c)$$

A distributive lattice $\mathfrak{A} = \langle A, U \rangle$ is called a *Boolean lattice* if it has the largest element $\mathbf{1}$, the least element $\mathbf{0}$, and for any $a \in A$ there exists $b \in A$ such that $a \wedge_\mathfrak{A} b = \mathbf{0}$ and $a \vee_\mathfrak{A} b = \mathbf{1}$. In this case, the element b is called a *complement to a* in $\mathfrak{A}$.

Lemma 1.2.1. *A complement to any element of a Boolean lattice is unique.*

PROOF. Let b_1 and b_2 be two complements to an element a. Then
$$b_1 = b_1 \vee_{\mathfrak{A}} 0 = b_1 \vee_{\mathfrak{A}} (b_2 \wedge_{\mathfrak{A}} a) = (b_1 \vee_{\mathfrak{A}} b_2) \wedge_{\mathfrak{A}} (b_1 \vee_{\mathfrak{A}} a)$$
$$= (b_1 \vee_{\mathfrak{A}} b_2) \wedge \mathbf{1} = b_1 \vee_{\mathfrak{A}} b_2 = b_2 \vee_{\mathfrak{A}} b_1$$

On the other hand,
$$b_2 = b_2 \vee_{\mathfrak{A}} 0 = b_2 \vee_{\mathfrak{A}} (b_1 \wedge_{\mathfrak{A}} a) = (b_2 \vee_{\mathfrak{A}} b_1) \wedge_{\mathfrak{A}} (b_2 \vee_{\mathfrak{A}} a)$$
$$= (b_2 \vee_{\mathfrak{A}} b_1) \wedge \mathbf{1} = b_2 \vee_{\mathfrak{A}} b_1$$

Therefore, $b_1 = b_2$. □

On the universe A of a Boolean lattice $\mathfrak{A} = \langle A, U \rangle$, we introduce a unary operation $C_{\mathfrak{A}}$ as follows. Let $C_{\mathfrak{A}}(a)$ be a unique complement to $a \in A$.

The algebraic system $\langle A, \vee_{\mathfrak{A}}, \wedge_{\mathfrak{A}}, C_{\mathfrak{A}} \rangle$ constructed from a Boolean lattice $\mathfrak{A}$ will be denoted by $A(\mathfrak{A})$.

Proposition 1.2.4. *If $\mathfrak{A}$ is a Boolean lattice, then $A(\mathfrak{A})$ is a Boolean algebra.*

PROOF. We verify only the condition (2) of the definition of a Boolean algebra. The remaining conditions immediately follow from the definition of $\vee_{\mathfrak{A}}$, $\wedge_{\mathfrak{A}}$, and $C_{\mathfrak{A}}$. We need to show that $\sup_{\mathfrak{A}}(\{\sup_{\mathfrak{A}}(\{a,b\}), c\}) = \sup_{\mathfrak{A}}(\{a, \sup_{\mathfrak{A}}(\{b,c\})\})$. It suffices to show that the sets A_0 and B_0 of upper bounds of the sets $\{\sup_{\mathfrak{A}}(\{a,b\}), c\}$ and $\{a, \sup_{\mathfrak{A}}(\{b,c\})\}$ coincide. Let $d \in A_0$. Then $d \geqslant c$ and $d \geqslant \sup_{\mathfrak{A}}(\{a,b\})$. Hence $d \geqslant a$ and $d \geqslant b$. Consequently, $d \geqslant \sup_{\mathfrak{A}}(\{b,c\})$ and $d \in B_0$. The inverse inclusion is proved in a similar way. □

On a Boolean algebra $\mathfrak{A}$, we introduce the relation $\leqslant$ by setting
$$a \leqslant b \leftrightharpoons a \wedge b = a$$

Lemma 1.2.2. *The relation $\leqslant$ is a partial order on A, and for any $a, b \in A$ the following relations hold:*

(a) $a \leqslant \mathbf{1}$, $\mathbf{0} \leqslant a$,

(b) $a \vee b = b \Leftrightarrow a \leqslant b$,

(c) $\inf_{\mathfrak{A}}(\{a,b\}) = a \wedge b$,

(d) $\sup_{\mathfrak{A}}(\{a,b\}) = a \vee b$.

1.2. Definitions and the Simplest Properties

PROOF. By the definition of a partial order, we need to show that the relation $\leqslant$ is reflexive, antisymmetric, and transitive. Since $a \wedge a = a$, we have $a \leqslant a$. If $a \leqslant b$ and $b \leqslant a$, then $a \wedge b = a$ and $a \wedge b = b$. Consequently, $a = b$. If $a \leqslant b$ and $b \leqslant c$, then $a \wedge b = a$ and $b \wedge c = b$. But in this case, $a = a \wedge b = a \wedge (b \wedge c) = (a \wedge b) \wedge c = a \wedge c$, i.e., $a \leqslant c$. In view of Proposition 1.2.3, $a \wedge 0 = 0$ and $a \wedge 1 = a$. Therefore, $0 \leqslant a$ and $a \leqslant 1$ for any $a \in B$, i.e., (a) is valid.

We now prove (b). Let $a \vee b = b$. Then $a \wedge b = b \wedge a = (b \wedge a) \vee (b \wedge C(b)) = b \wedge (a \vee C(b)) = (a \vee b) \wedge (a \vee C(b)) = a \vee (b \wedge C(b))$ in view of (6'), (3'), (3), and (1'). Consequently, $a \leqslant b$. Let $a \leqslant b$. Then $a \wedge b = a$. We obtain $a \vee b = b$ in a similar way.

We now prove (c). The relation (d) is proved in a similar way and is left to the reader as an exercise. We first note that $a \wedge b$ is a lower bound of the set $\{a, b\}$. Indeed, since $a \wedge (a \wedge b) = (a \wedge a) \wedge b = a \wedge b$ and $b \wedge (a \wedge b) = b \wedge (b \wedge a) = (b \wedge b) \wedge a = b \wedge a = a \wedge b$, we have $a \wedge b \leqslant a$ and $a \wedge b \leqslant b$ by the definition of an order. It remains to note that $a \wedge b$ is the greatest lower bound. Let c be a lower bound of the set $\{a, b\}$. Then $c \leqslant a$ and $c \leqslant b$. By the definition of an order, $c = c \wedge a$ and $c = c \wedge b$. By associativity, $c \wedge (a \wedge b) = (c \wedge a) \wedge b = c \wedge b = c$. Hence $c \leqslant a \wedge b$. □

The partially ordered set constructed from a Boolean algebra $\mathfrak{B}$ will be denoted by $P(\mathfrak{B}) = \langle B; \leqslant \rangle$.

Proposition 1.2.5. *If $\mathfrak{B}$ is a Boolean algebra, then $P(\mathfrak{B})$ is a Boolean lattice.*

PROOF. By Lemma 1.2.2, for any $a, b \in B$ there exist $\sup_{\mathfrak{A}} (\{a, b\})$ and $\inf_{\mathfrak{A}} (\{a, b\})$ in $\langle B; \leqslant \rangle$. Therefore, $P(\mathfrak{B})$ is a lattice. The distributivity condition follows from the relations (c) and (d) of Lemma 1.2.2 and the distributivity properties (3) and (3') of Boolean algebras. It remains to check the existence of the greatest element and the least element of $P(\mathfrak{B})$ and the existence of complements. By Lemma 1.2.2, the unit $\mathbf{1}$ is the greatest element and the null element $\mathbf{0}$ is the least element of $P(\mathfrak{B})$. For a complement to an element a we can take the element $C(a)$ since, by the definition of the null element and the unit in a Boolean algebra, $a \vee C(a) = \mathbf{1}$ and $a \wedge C(a) = \mathbf{0}$. □

Theorem 1.2.1. *If $\mathfrak{A}$ is a Boolean algebra and $\mathfrak{R}$ is a Boolean lattice, then $A(\mathfrak{R})$ is a Boolean algebra and $P(\mathfrak{A})$ is a Boolean lattice; moreover, $AP(\mathfrak{A}) = \mathfrak{A}$ and $PA(\mathfrak{R}) = \mathfrak{R}$.*

PROOF. The first assertion of the theorem follows from Propositions 1.2.4 and 1.2.5. Let us prove the equalities $AP(\mathfrak{A}) = \mathfrak{A}$ and $PA(\mathfrak{R}) = \mathfrak{R}$. We note that the universe of a system remains unchanged under the operations P and A. Therefore, it suffices to show that the operations and relations coincide in $PA(\mathfrak{R})$, $\mathfrak{R}$ and $AP(\mathfrak{A})$, $\mathfrak{A}$. By definitions,

$$a \leqslant_{PA(\mathfrak{R})} b \Leftrightarrow a \wedge_{A(\mathfrak{R})} b = a \Leftrightarrow \inf{}_{\mathfrak{R}}\{a,b\} = a \Leftrightarrow a \leqslant b$$

Thus, $PA(\mathfrak{R}) = \mathfrak{R}$. By the definition of A, we have $a \vee_{A(\mathfrak{R})} b = \sup{}_{\mathfrak{R}}\{a,b\}$. However, $\sup_{P(\mathfrak{A})}\{a,b\} = a \vee b$ by Lemma 1.2.2. Consequently, $a \vee_{AP(\mathfrak{A})} b = a \vee b$. The relation $a \wedge_{AP(\mathfrak{A})} b = a \wedge b$ can be established in a similar way.

We now prove the equality $C_{AP(\mathfrak{A})}(a) = C(a)$. Since **1** is the greatest element and **0** is the least element of $P(\mathfrak{A})$ and

$$a \wedge_{AP(\mathfrak{A})} C(a) = a \wedge C(a) = \mathbf{0}, \quad a \vee_{AP(\mathfrak{A})} C(a) = a \vee C(a) = \mathbf{1}$$

by Lemma 1.2.1, we conclude that $C_{AP(\mathfrak{A})}(a) = C(a)$. □

One of natural classes of commutative rings is presented by Boolean rings which are closely connected with Boolean algebras. A *Boolean ring with unit* is a commutative and associative ring $\mathfrak{K} = \langle K; +, \cdot, \mathbf{0}, \mathbf{1} \rangle$ with unit such that $x \cdot x = x$. Starting from a Boolean ring $\mathfrak{K}$, we define a Boolean lattice $L(\mathfrak{K})$ by setting $|L(\mathfrak{K})| = K$ and $a \leqslant_{L(\mathfrak{K})} b$ if $a \cdot b = a$.

Lemma 1.2.3. *In a Boolean ring $\mathfrak{K}$, for any element $a \in K$ the equality $a + a = \mathbf{0}$ holds.*

PROOF. For an arbitrary element $a \in K$ we have $a + a = (a+a)(a+a) = a^2 + a^2 + a^2 + a^2 = a + a + a + a$. Since $\langle K; + \rangle$ is an Abelian group with $\mathbf{0}$, the above equality implies $a + a = \mathbf{0}$. □

Lemma 1.2.4. *For a Boolean ring $\mathfrak{K}$, the relation $\leqslant_{L(\mathfrak{K})}$ defines a Boolean lattice on the universe of $\mathfrak{K}$.*

PROOF. We first check that the relation $\leqslant_{L(\mathfrak{K})}$ defines a partial order on K. We omit $L(\mathfrak{K})$ in the notation $\leqslant_{L(\mathfrak{K})}$ when no confusion can arise. Since $a \cdot a = a$, the relation $\leqslant$ is reflexive. We show that it is antisymmetric. Let $a \leqslant b$ and $b \leqslant a$. By definition, $a = a \cdot b$ and $b = b \cdot a$. Since the ring $\mathfrak{K}$ is commutative, the elements $a \cdot b$ and $b \cdot a$ are equal. Consequently, $a = b$. The transitivity property follows from the associativity of $\mathfrak{K}$. Indeed, let $a \leqslant b$ and $b \leqslant c$. Then $a = a \cdot b$ and $b = b \cdot c$. Therefore, $a = a \cdot b = a \cdot (b \cdot c) = (a \cdot b) \cdot c = a \cdot c$. Consequently, $a \leqslant c$.

To prove that $\langle K; \leqslant \rangle$ is a lattice, for an arbitrary pair of elements a and b we find their least upper bound and greatest lower bound. We

1.2. Definitions and the Simplest Properties

show that $a \cdot b$ is the greatest lower bound and $a + b + a \cdot b$ is the least upper bound. Indeed, since $a \cdot (a \cdot b) = (a \cdot a) \cdot b = a \cdot b$ in K, we have $a \cdot b \leqslant a$. Similarly, $a \cdot b \leqslant b$. Let c be a lower bound of $\{a, b\}$. Then $c \cdot a = c$ and $c \cdot b = c$. But in this case, $c \cdot (a \cdot b) = (c \cdot a) \cdot b = c \cdot b = c$. Consequently, $c \leqslant a \cdot b$. Thus, we have proved that $a \cdot b$ is the greatest lower bound. Since $a \cdot (a + b + a \cdot b) = a + a \cdot b + a \cdot b = a + (a \cdot b + a \cdot b) = a + 0 = a$, we obtain $a \leqslant a + b + a \cdot b$. Similarly, $b \leqslant a + b + a \cdot b$. Let c be an upper bound of $\{a, b\}$. Then $a \cdot c = a$ and $b \cdot c = b$. In this case, $(a + b + a \cdot b) \cdot c = a \cdot c + b \cdot c + (a \cdot b) \cdot c = a + b + a \cdot (b \cdot c) = a + b + a \cdot b$. Hence $a + b + a \cdot b$ is the least upper bound of $\{a, b\}$. Thus, for $\{a, b\}$ we have defined the greatest lower bound $a \cdot b$ and the least upper bound $a + b + a \cdot b$, which will be denoted by $a \wedge b$ and $a \vee b$ respectively (cf. Sec. 1.1). We show that the lattice obtained is distributive. By definition, from Lemma 1.2.3 we find

$$a \vee (b \wedge c) = a + (b \wedge c) + a \cdot (b \wedge c) = a + b \cdot c + a \cdot b \cdot c$$
$$(a \vee b) \wedge (a \vee c) = (a + b + a \cdot b) \wedge (a + c + a \cdot c)$$
$$= (a + b + a \cdot b) \cdot (a + c + a \cdot c) = a^2 + a \cdot c + a^2 \cdot c$$
$$+ a \cdot b + b \cdot c + a \cdot b \cdot c + a^2 \cdot b + a \cdot b \cdot c + a^2 \cdot b \cdot c$$
$$= a + (a \cdot c + a \cdot c) + (a \cdot b + a \cdot b) + b \cdot c + a \cdot b \cdot c$$
$$+ (a \cdot b \cdot c + a \cdot b \cdot c) = a + b \cdot c + a \cdot b \cdot c$$

Thus, $a \vee (b \wedge c) = (a \vee b) \wedge (a \vee c)$, i.e., the condition (D) is fulfilled. The validity of (D′) is established in a similar way. By Lemma 1.2.3, $a \cdot 0 = 0$. Hence **0** is the least element of $\langle K; \leqslant \rangle$. The ring $\mathfrak{K}$ has the unit **1**, hence $x \cdot 1 = x$. Consequently, **1** is the greatest element of $\langle K; \leqslant \rangle$. It remains to check that in $\langle K; \leqslant \rangle$ for any element there exists its complement. Let a be an arbitrary element of K. Let us show that $1 + a$ is a complement to a in $\langle K; \leqslant \rangle$. By the definition of a Boolean ring with unit, from Lemma 1.2.3 we find

$$a \wedge (1 + a) = a \cdot (1 + a) = a \cdot 1 + a \cdot a = a + a = 0$$
$$a \vee (1 + a) = a + (1 + a) + a \cdot (1 + a) = a + 1 + a = (a + a) + 1 = 1$$

The lemma is proved. □

Let $\mathfrak{L} = \langle L; \leqslant \rangle$ be a Boolean lattice. We define the Boolean ring R($\mathfrak{L}$). For the universe we take L and set $a \cdot_{R(\mathfrak{L})} b \rightleftharpoons a \wedge_{\mathfrak{L}} b$, $a +_{R(\mathfrak{L})} b \rightleftharpoons (a \wedge_{\mathfrak{L}} C(b)) \vee_{\mathfrak{L}} (b \wedge_{\mathfrak{L}} C(a))$. For the null element and the unit we take the least element and the greatest element of R respectively.

Lemma 1.2.5. *If $\mathfrak{L}$ is a Boolean lattice, then R($\mathfrak{L}$) is a Boolean ring with unit.*

PROOF. The operations $+$ and $\cdot$ on $\mathrm{R}(\mathfrak{L})$ are commutative, associative, and distributive, because the operations $\wedge$ and $\vee$ possess the same properties. In $\langle L; + \rangle$, for any $a \in L$ there exists its complement, and $\mathbf{0}$ is the null element since

$$0 + a = (\mathbf{0} \wedge C(a)) \vee (a \wedge C(\mathbf{0})) = \mathbf{0} \vee (a \wedge \mathbf{1}) = \mathbf{0} \vee a = a$$
$$a + a = (a \wedge C(a)) \vee (a \wedge C(a)) = a \wedge C(a) = \mathbf{0}$$

In $\langle L; \cdot \rangle$ the element $\mathbf{1}$ is the unit since $a \cdot \mathbf{1} = a \wedge \mathbf{1} = a$. The identity $a \cdot a = a$ follows from the definition of multiplication. □

Theorem 1.2.2. *For any Boolean ring* $\mathfrak{K} = \langle K; +, \cdot, 0, 1 \rangle$ *and Boolean lattice* $\mathfrak{L} = \langle L; \leqslant \rangle$ *the equality* $\mathrm{R}(\mathfrak{L}) = \mathfrak{K}$ *holds if and only if* $\mathrm{L}(\mathfrak{K}) = \mathfrak{L}$.

PROOF. Let $\mathrm{L}(\mathfrak{K}) = \mathfrak{L}$. Then $a \leqslant b \Leftrightarrow a \cdot b = a$ and, by Lemma 1.2.4, $a \vee b = a + b + a \cdot b$, $a \wedge b = a \cdot b$, and $C(a) = 1 + a$. However,

$$a +_{\mathrm{R}(\mathfrak{L})} b = (a \wedge C(b)) \vee (b \wedge C(a)) = (a \cdot (1 + b)) \vee (b + a \cdot b)$$
$$= (a + a \cdot b) + (b + a \cdot b) + (a \cdot (1 + b))(b + a \cdot b)$$
$$= a + b + (a \cdot b + a \cdot b) + a \cdot b + a^2 \cdot b + a \cdot b^2 + (a \cdot b)^2$$
$$= a + b + (a \cdot b + a \cdot b) + (a \cdot b + a \cdot b) = a + b$$
$$a \cdot_{\mathrm{R}(\mathfrak{L})} b \Leftrightarrow a \wedge b = a \cdot b$$

The least element $\mathbf{0}$ and the greatest element $\mathbf{1}$ of $\mathfrak{L}$ are equal to the elements 0 and 1 of K respectively since $0 \cdot a = 0$ and $1 \cdot a = a$ for any a. Consequently, $\mathrm{R}(\mathfrak{L}) = \mathfrak{K}$.

To prove the converse assertion we put $\mathrm{R}(\mathfrak{L}) = \mathfrak{K}$. Then $a \wedge b = a \cdot_{\mathrm{R}(\mathfrak{L})} b = a \cdot b$. However, $a \leqslant b \Leftrightarrow a = a \wedge b \Leftrightarrow a \leqslant_{\mathfrak{K}} b$. Thus, we can conclude that $\mathrm{R}(\mathfrak{L}) = \mathfrak{K}$. □

Corollary 1.2.1. *Let $\mathfrak{K}$ be a Boolean ring with unit and let $\mathfrak{B}$ be a Boolean algebra. Then* $\mathrm{AL}(\mathfrak{K}) = \mathfrak{B}$ *if and only if* $\mathfrak{K} = \mathrm{RP}(\mathfrak{B})$.

Exercises

1. Prove the properties $(3')$, $(5')$, and $(6')$ of Boolean algebras (see Proposition 1.2.1).

2. Prove the relation (b) of Lemma 1.2.2.

1.2. Definitions and the Simplest Properties

3. Below we use the abbreviations $\bigvee_{i=1}^{n} a_i$, $\bigwedge_{i=1}^{n} a_i$ for the expressions
$$(\ldots((a_1 \vee a_2) \vee a_3) \vee \ldots \vee a_n), \quad (\ldots((a_1 \wedge a_2) \wedge a_3) \wedge \ldots \wedge a_n)$$
Prove that for any elements $a_1, \ldots, a_n, b_1, \ldots, b_m, a_{n+1}, \ldots, a_{n+m}$, $n \geqslant 2$, of a Boolean algebra $\mathfrak{B}$ the following relations hold:

$$\left(\bigvee_{i=1}^{n} a_i\right) \vee \left(\bigvee_{i=n+1}^{m} a_i\right) = \bigvee_{i=1}^{m} a_i,$$

$$\left(\bigwedge_{i=1}^{n} a_i\right) \wedge \left(\bigwedge_{i=n+1}^{m} a_i\right) = \bigwedge_{i=1}^{m} a_i,$$

$$\left(\bigvee_{i=1}^{n} a_i\right) = \left(\bigvee_{i=1}^{n} a_{s(i)}\right), \quad \left(\bigwedge_{i=1}^{n} a_i\right) = \left(\bigwedge_{i=1}^{n} a_{s(i)}\right), \text{ where } s \text{ denotes a permutation of } \{1, \ldots, n\},$$

$$\left(\bigvee_{i=1}^{n} a_i\right) \wedge \left(\bigvee_{j=1}^{m} b_j\right) = \bigvee_{\substack{1 \leqslant i \leqslant n \\ 1 \leqslant j \leqslant m}} (a_i \wedge b_i),$$

$$\left(\bigwedge_{i=1}^{n} a_i\right) \vee \left(\bigwedge_{j=1}^{m} b_j\right) = \bigwedge_{\substack{1 \leqslant i \leqslant n \\ 1 \leqslant j \leqslant m}} (a_i \vee b_i).$$

HINT: Apply induction on n and m.

4. Define the structure of a Boolean ring with unity for a Boolean algebra $\langle \mathcal{P}(A); \cup, \cap, C \rangle$.

5. Let $\mathfrak{B}^* = \langle B; \wedge^*, \vee^*, C^* \rangle$ be the dual algebra for a Boolean algebra $\mathfrak{B} = \langle B; \wedge, \vee, C \rangle$.

   (a) Prove that $\mathfrak{B}^*$ is a Boolean algebra.
   (b) Prove that $x \leqslant_{P(\mathfrak{B}^*)} y \Leftrightarrow y \leqslant_{P(\mathfrak{B})} x$.
   (c) How are the operations connected in $\text{RP}(\mathfrak{B})$ and $\text{RP}(\mathfrak{B}^*)$?

6. Show that if $\mathfrak{A}$ is a Boolean algebra, then the subsystem generated by $X \subseteq A$ consists of the elements
$$\{\mathbf{0}, \mathbf{1}\} \cup \{\bigvee_{i=1}^{n} (x_{i1}^{\varepsilon_{i1}} \wedge x_{i2}^{\varepsilon_{i2}} \wedge \ldots \wedge x_{ik_i}^{\varepsilon_{ik_i}}) \mid x_{ij} \in X, \varepsilon_{ij} \in \{\mathbf{0}, \mathbf{1}\}\}$$
where $x^0 \rightleftharpoons C(x)$ and $x^1 \rightleftharpoons x$.

7. Prove that a subalgebra of a Boolean algebra is a Boolean algebra.

8. A nonzero element a is called an *atom* if only zero is under a, i.e., $(\forall b)(b < a \to b = 0)$. Prove that under any element $x \neq 0$ of a finite Boolean algebra there is an atom.

9. Let $\{a_i \mid 1 \leq i \leq n\}$ and $\{b_i \mid 1 \leq i \leq n\}$ be two sets of pairwise disjoint atoms. Show that

$$\{a_i \mid 1 \leq i \leq n\} = \{b_i \mid 1 \leq i \leq n\} \Leftrightarrow \bigvee_{i=1}^{n} a_i = \bigvee_{i=1}^{n} b_i$$

10. Show that the classes of Boolean algebras, Boolean lattices, and Boolean rings are semantically equivalent.

1.3. Ideals and Quotient Algebras of Boolean Algebras

The identification of different elements is often used in mathematical constructions. In the class of Boolean algebras, such an identification can be realized with the help of ideals.

A subset $I \subseteq B$ is called an *ideal* of a Boolean algebra $\mathfrak{B}$ if

(1) $\mathbf{0} \in I$,

(2) $(\forall x \in I)(\forall y)(y \leq x \to y \in I)$,

(3) $(\forall x \in I)(\forall y \in I)(x \vee y \in I)$.

An ideal I is called *principal* if there exists the greatest element of I.

The notion of a filter of a Boolean algebra $\mathfrak{B}$ is the dual notion of an ideal. A subset $F \subseteq B$ is called a *filter* of a Boolean algebra $\mathfrak{B}$ if

(1) $\mathbf{1} \in F$,

(2) $(\forall x \in F)(\forall y)(x \leq y \Rightarrow y \in F)$,

(3) $(\forall x \in F)(\forall y \in F)(x \wedge y \in F)$.

A filter F is called *principal* if there exists the least element of F.

It is easy to see that I is a proper ideal of a Boolean algebra $\mathfrak{B}$ if and only if $C(I)$ is a proper filter of the Boolean algebra $\mathfrak{B}$.

1.3. Ideals and Quotient Algebras of Boolean Algebras

A proper filter F is called an *ultrafilter* if $x \in F$ or $C(x) \in F$ for any $x \in \mathfrak{B}$.

For a given ideal I we introduce the binary relation $x \sim_I y$ by setting $x \sim_I y \Leftrightarrow x\Delta y \rightleftharpoons (x\backslash y) \vee (y\backslash x) \in I$, where $a\backslash b \rightleftharpoons a \wedge C(b)$.

Lemma 1.3.1. *The relation $\sim_I$ is an equivalence.*

PROOF. It is required to check that the relation $\sim_I$ is reflexive, symmetric, and transitive. We have $x\Delta x \in I$ and $x \sim_I x$ (reflexivity) because of the relation $x\Delta x = x\backslash x \vee x\backslash x = 0$. Since $x\Delta y = y\Delta x$, we obtain the symmetry property: $x \sim_I y$ if $y \sim_I x$. To prove the transitivity property, we note that $x \sim_I y$ and $y \sim_I z$ imply $x\Delta y \in I$ and $y\Delta z \in I$. However, $x\backslash z \leqslant x\backslash y \vee y\backslash z$ since

$$\begin{aligned}x\backslash z &= x\backslash z \wedge 1 = (x \wedge C(z)) \wedge (y \vee C(y)) \\ &= (x \wedge y \wedge C(z)) \vee (x \wedge C(y) \wedge C(z)) \\ &\leqslant (y \wedge C(z)) \vee (x \wedge C(y)) = x\backslash y \vee y\backslash z\end{aligned}$$

Similarly, $z\backslash x \leqslant y\backslash x \vee z\backslash y$. In this case, $x\backslash z \vee z\backslash x \leqslant (x\Delta y) \vee (y\Delta z) \in I$. Taking into account the properties of ideals, we conclude that $x\Delta z \in I$. □

Lemma 1.3.2. *The relation $\sim_I$ is a congruence.*

PROOF. By the definition of a congruence, we need to show that $f(a_1, \ldots, a_n) \sim_I f(b_1, \ldots, b_n)$ for any basic operations $f(x_1, \ldots, x_n)$ and arbitrary sequences of equivalent elements $a_i \sim_I b_i$, $1 \leqslant i \leqslant n$. Let $a_1 \sim_I b_1$ and $a_2 \sim_I b_2$. Then

$$\begin{aligned}(a_1 \vee a_2)\Delta(b_1 \vee b_2) &= a_1\backslash(b_1 \vee b_2) \vee a_2\backslash(b_1 \vee b_2) \vee b_1\backslash(a_1 \vee a_2) \\ \vee\, b_2\backslash(a_1 \vee a_2) &\leqslant a_1\backslash b_1 \vee a_2\backslash b_2 \vee b_1\backslash a_1 \vee b_2\backslash a_2 = (a_1\Delta b_1) \vee (a_2\Delta b_2)\end{aligned}$$

By the definition of an ideal, $(a_1 \vee a_2)\Delta(b_1 \vee b_2) \in I$ and $a_1 \vee a_2 \sim_I b_1 \vee b_2$. For the operations $\wedge$ and C the proof is similar. □

We consider the quotient set $A/I \rightleftharpoons \{a/I \mid a \in I\}$, where $a/I \rightleftharpoons \{x \mid x \sim_I a\}$. It is easy to check that $a/I = b/I \Leftrightarrow a/I \cap b/I \neq \emptyset \Leftrightarrow a \sim_I b$. On A/I, we introduce the structure of Boolean algebra that is induced from $\mathfrak{A}$ by setting $a/I \vee b/I \rightleftharpoons a \vee b/I$, $a/I \wedge b/I \rightleftharpoons a \wedge b/I$, and $C(a/I) \rightleftharpoons C(a)/I$. By Lemma 1.3.2, the constructed quotient algebra (denoted by $\mathfrak{A}/I$) is independent of the choice of a representative a of the congruence class a/I. We note that $\mathfrak{A}/I$ is a Boolean algebra.

We state without proof the following simple but useful claim.

Lemma 1.3.3. *For any elements a/I, b/I, x_i/I, $i \leqslant n$, of the quotient algebra $\mathfrak{A}/I$ the following conditions hold:*

(a) *if $a/I \leqslant b/I$, then there exists $x \leqslant b$ such that $x/I = a/I$,*

(b) *if $\bigvee_{i=0}^{n} x_i/I = a/I$, $i \neq j$, imply $x_i/I \wedge x_j/I = \mathbf{0}/I$, then there exists $y_i \leqslant a$ such that $\bigvee_{i=0}^{n} y_i = a$, $y_i \wedge y_j = \mathbf{0}$ for $i \neq j$ and $x_i/I = y_i/I$ for $i \leqslant n$.*

Let $\varphi(a) \leftrightharpoons a/I$. It is easy to check that φ is a homomorphism of the Boolean algebra $\mathfrak{A}$ onto the quotient algebra $\mathfrak{A}/I$. This homomorphism is called *canonical* and is denoted by φ_I.

We now study the converse problem. Let φ be a homomorphism of a Boolean algebra $\mathfrak{A}$ onto a Boolean algebra $\mathfrak{B}$. Then $I_\varphi = \{x \mid \varphi(x) = \mathbf{0}\}$ is an ideal of $\mathfrak{A}$. This ideal is called the *kernel* of the homomorphism φ and is denoted by $\ker \varphi$. It is easy to check that $\varphi(x) = \varphi(y) \Leftrightarrow x \Delta y \in I_\varphi$. Now, we are in a position to introduce a mapping $\widehat{\varphi}$ of the quotient algebra $\mathfrak{A}/I_\varphi$ onto the Boolean algebra $\mathfrak{B}$ by setting $\widehat{\varphi}(a/I_\varphi) \leftrightharpoons \varphi(a)$. A direct verification shows that $\widehat{\varphi}$ is an isomorphism between $\mathfrak{A}/I_\varphi$ and $\mathfrak{B}$.

Let I be an ideal of a Boolean algebra $\mathfrak{A}$ and let J be an ideal of the quotient algebra $\mathfrak{A}/I$. We introduce a new ideal $I \circ J$ of the Boolean algebra $\mathfrak{A}$ (called the *composition of I and J*) by setting $I \circ J \leftrightharpoons \{x \in A \mid x/I \in J\}$. It is clear that $I \circ J \supseteq I$, $I \circ J$ is the kernel of the composition of the homomorphisms $\varphi_J \circ \varphi_I$, and $\widehat{\varphi_J \circ \varphi_I}$ is an isomorphism between the Boolean algebras $\mathfrak{A}/I \circ J$ and $(\mathfrak{A}/I)/J$.

The composition allows us to iterate the quotient algebras by some ideals. We consider one such sequence of iterated ideals. Let $\mathfrak{A}$ be a Boolean algebra. We define the *Frechét ideal $F(\mathfrak{A})$* of $\mathfrak{A}$ as follows:

$$F(\mathfrak{A}) \leftrightharpoons \{x \in A \mid x = \mathbf{0} \text{ or there exists a finite number of atoms}$$

$$a_1, \ldots, a_n \text{ of } \mathfrak{A} \text{ such that } x = \bigvee_{i=1}^{n} a_i\}$$

First of all we check that $F(\mathfrak{A})$ is an ideal. If $x \in F(\mathfrak{A})$ and $y \leqslant x$, then $x = \bigvee_{i=0}^{n} a_i$ and a_i, $i \leqslant n$, are atoms. Therefore, $y = \bigvee_{i=0}^{n} (a_i \wedge y)$ and $a_i \wedge y = a_i$ or $a_i \wedge y = \mathbf{0}$, $i \leqslant n$. Consequently, $y \in F(\mathfrak{A})$. If $x = \bigvee_{i=0}^{n} a_i$ and

1.3. Ideals and Quotient Algebras of Boolean Algebras

$y = \bigvee_{i=n+1}^{m} a_i$, where a_i, $i \leqslant m$, are atoms of $\mathfrak{A}$, then $x \vee y = \bigvee_{i=0}^{m} a_i$. Hence $x \vee y$ also belongs to $F(\mathfrak{A})$.

A Boolean algebra $\mathfrak{A}$ is called *atomic* if there is an atom under any of its nonzero elements and is called *atomless* if there is no atom in it.

For an ordinal α we define the ideal $F_\alpha(\mathfrak{A})$ as follows:

$$F_0(\mathfrak{A}) \doteqdot \{0\}, \quad F_{\alpha+1}(\mathfrak{A}) \doteqdot F_\alpha(\mathfrak{A}) \circ F\left(\mathfrak{A}/F_\alpha(\mathfrak{A})\right), \quad F_\gamma(\mathfrak{A}) \doteqdot \bigcup_{\alpha < \gamma} F_\alpha(\mathfrak{A})$$

where γ is a limit ordinal. Using the transfinite induction on α, we easily verify that $F_\alpha(\mathfrak{A})$ is an ideal.

In the sequel, we need some definitions. A sequence of ideals $\{F_\alpha(\mathfrak{A}) \mid \alpha \text{ is an ordinal}\}$ is called a *sequence of iterated Frechét ideals*. A Boolean algebra $\mathfrak{A}$ is called α-*atomic* if the quotient algebras $\mathfrak{A}/F_\gamma(\mathfrak{A})$, $\gamma < \alpha$, are atomic. The least ordinal α such that $F_\alpha(\mathfrak{A}) = F_{\alpha+1}(\mathfrak{A})$ is called the *ordinal type* of the Boolean algebra $\mathfrak{A}$ and is denoted by $o(\mathfrak{A})$.

Lemma 1.3.4. *If $\alpha = o(\mathfrak{A})$ and $F_\alpha(\mathfrak{A}) = A$, then α is not a limit ordinal.*

PROOF. We assume that α is a limit ordinal. Since $F_\alpha(\mathfrak{A}) = A$, we have $\mathbf{1} \in F_\alpha(\mathfrak{A})$. Consequently, $\mathbf{1} \in \bigcup_{\gamma < \alpha} F_\alpha(\mathfrak{A})$. In this case, $\mathbf{1} \in F_\gamma(\mathfrak{A})$ for some $\gamma < \alpha$; therefore, $A = F_\gamma(\mathfrak{A})$. Since $A = F_\gamma(\mathfrak{A}) \subseteq F_{\gamma+1}(\mathfrak{A}) \subseteq A$, we obtain $F_\gamma(\mathfrak{A}) = F_{\gamma+1}(\mathfrak{A})$, which contradicts the condition $\alpha = o(\mathfrak{A})$. □

Lemma 1.3.5. *The ordinal type $o(\mathfrak{A})$ of a countable Boolean algebra $\mathfrak{A}$ is countable.*

PROOF. The assertion follows from the axiom of choice. Let $F_\alpha(\mathfrak{A}) \neq F_{\alpha+1}(\mathfrak{A})$ for all countable ordinals α. Using the choice function, it is possible to choose an element $a_\alpha \in F_{\alpha+1}(\mathfrak{A}) \setminus F_\alpha(\mathfrak{A})$ for every countable ordinal α. But in this case, all of these elements are pairwise disjoint. Thus, we have constructed an uncountable subset of $\mathfrak{A}$. □

Let $\alpha = o(\mathfrak{A})$ and $F_\alpha(\mathfrak{A}) = A$. By Lemma 1.3.4, α is not a limit ordinal. Consequently, it is equal to $\beta + 1$. Hence the unit of the algebra $\mathfrak{A}/F_\beta(\mathfrak{A})$ belongs to the Frechét ideal of this algebra and is equal to a union of a finite number of atoms. Therefore, the Boolean algebra $\mathfrak{A}/F_\beta(\mathfrak{A})$ is finite.

Let m be the number of atoms of $\mathfrak{A}/F_\beta(\mathfrak{A})$. The pair (β, m) is called a *type* of the Boolean algebra $\mathfrak{A}$ and is denoted by type$(\mathfrak{A})$. By Lemma 1.3.5, the ordinal β is countable provided that the Boolean algebra $\mathfrak{A}$ is countable. For an element $a \in \mathfrak{A}$ we define type (a) as the pair (γ, k) such that $a \in F_{\gamma+1}(\mathfrak{A}) \setminus F_\gamma(\mathfrak{A})$ and k is equal to the number of different atoms of $\mathfrak{A}/F_\gamma(\mathfrak{A})$ located under $a/F_\gamma(\mathfrak{A})$. If type $(x) = (\alpha, m)$, then type $_1(x)$ stands for the ordinal α, and type $_2(x)$ denotes the number m. Since $a \in F_{\gamma+1}(\mathfrak{A})$, there is a finite number of atoms located under $a/F_\gamma(\mathfrak{A})$.

We emphasize that ideals of Boolean algebras are not Boolean algebras. A comprehensive study of systems that are ideals of Boolean algebras was carried out by Ershov.

We consider an algebraic system $\mathfrak{A} = \langle A, \vee, \wedge, \mathbf{0} \rangle$ with two binary operations $\vee, \wedge$ and null element $\mathbf{0}$. The system $\mathfrak{A}$ is called a *distributive lattice with null* if the following axioms hold:

(1) $a \vee a = a, a \wedge a = a$,

(2) $a \vee b = b \vee a, a \wedge b = b \wedge a$,

(3) $(a \vee b) \vee c = a \vee (b \vee c), (a \wedge b) \wedge c = a \wedge (b \wedge c)$,

(4) $a \vee (b \wedge c) = (a \vee b) \wedge (a \vee c), a \wedge (b \vee c) = (a \wedge b) \vee (a \wedge c)$,

(5) $a \vee \mathbf{0} = a, a \wedge \mathbf{0} = \mathbf{0}$.

As for Boolean lattices, we can define the relation $\leqslant$ on distributive lattices by setting $a \leqslant b$ if $a \wedge b = a$. This relation is a partial order on A; moreover, $a \wedge b$ is the greatest lower bound and $a \vee b$ is the least upper bound of the set $\{a, b\}$ with respect to this order. By (5), $\mathbf{0}$ is the least element of A with respect to this order.

In the case $a \leqslant b$, an element c is called a *relative complement* to a with respect to b if $c \wedge a = \mathbf{0}$ and $a \vee c = b$. We emphasize that relative complements cannot exist in an arbitrary distributive lattice with null element. However, if a relative complement exists, then it is unique.

A distributive lattice $\mathfrak{A} = \langle A, \vee, \wedge, \mathbf{0} \rangle$ with null element is called an *Ershov algebra* if for any two elements $a, b \in A$ such that $a \leqslant b$ there exists a relative complement to a with respect to b.

Since a relative complement is unique, we can extend an Ershov algebra by the relative complement operation $\setminus$. A relative complement to $a \wedge b$ with respect to b is denoted by $b \setminus a$. This operation satisfies the following identities:

(1) $y \wedge (x \setminus y) = \mathbf{0}, \quad y \vee (x \setminus y) = x \vee y$,

(2) $x \setminus y = x \wedge (x \setminus y)$,

(3) $(x \vee y) \setminus z = (x \setminus z) \vee (y \setminus z)$,

1.3. Ideals and Quotient Algebras of Boolean Algebras

(4) $(x \wedge y) \backslash z = (x \backslash z) \wedge (y \backslash z) = x \wedge (y \backslash z)$,

(5) $x \backslash (y \vee z) = (x \backslash y) \wedge (x \backslash z)$,

(6) $x \backslash (y \wedge z) = x \backslash y \vee x \backslash z$,

(7) $(x \wedge y) \vee (x \backslash y) = x$.

It is obvious that the enrichment of a distributive lattice with null element by the operation $\backslash$ satisfies (1)–(6) if and only if it is an Ershov algebra and the operation $a \backslash b$ defines exactly a relative complement to $a \wedge b$ with respect to a. It is easy to see that any homomorphism φ from an Ershov algebra $\mathfrak{A}$ into an Ershov algebra $\mathfrak{B}$ preserves the operation $\backslash$, i.e., $\varphi(a \backslash b) = \varphi(a) \backslash \varphi(b)$ for any $a, b \in \mathfrak{A}$. Thus, from the model theory point of view, as well as from the universal algebra standpoint, Ershov algebras can be considered with this additional operation.

Boolean algebras form a subclass of the class of Ershov algebras, which is precisely the class of Ershov algebras with greatest elements.

Let $\mathfrak{A}$ be a Boolean algebra and let I be ideal of $\mathfrak{A}$. We consider the restrictions on I of the operations $\wedge$, $\vee$ defined on $\mathfrak{A}$. By the definition of an ideal, $\mathbf{0} \in I$. Therefore, it is possible to induce on I the structure $\langle I; \wedge, \vee, \mathbf{0} \rangle$. By the properties of Boolean algebras, $\langle I; \wedge, \vee, \mathbf{0} \rangle$ is a distributive lattice with null element. We define the binary operation $x \backslash y \rightleftharpoons x \wedge C(y)$. For any $x, y \in I$ we have $x \backslash y \in I$; moreover, the axioms (1)–(7) hold. The kernel of any homomorphism φ mapping an Ershov algebra $\mathfrak{A}$ onto an Ershov algebra $\mathfrak{B}$, i.e., the set $I = \ker \varphi \rightleftharpoons \{x \mid \varphi(x) = 0\}$, forms an Ershov algebra on which the operations $\wedge$ and $\vee$ are induced from $\mathfrak{A}$ onto I and the unique null element $\mathbf{0} \in I$ is that of $\mathfrak{A}$. Consequently, the class of Ershov algebras is closed under not only epimorphisms, but also under taking kernels of homomorphisms.

We show that any Ershov algebra E is an ideal of some Boolean algebra $\mathfrak{A}$ and any given Boolean algebra can be obtained as some quotient algebra. Following Ershov [38, 39], we consider this problem in a more general statement.

Let two Ershov algebras $\mathfrak{A}$ and $\mathfrak{B}$ be fixed. Is there an Ershov algebra $\mathfrak{C}$ such that an isomorphic embedding $\varphi \colon \mathfrak{A} \to \mathfrak{C}$ as an ideal exists and there is an epimorphism $\psi \colon \mathfrak{C} \to \mathfrak{B}$ whose kernel coincides with the φ-image of $\mathfrak{A}$?

For the sequence $\mathfrak{A} \xrightarrow{\varphi} \mathfrak{C} \xrightarrow{\psi} \mathfrak{B}$, where φ is an isomorphic embedding as an ideal coinciding with the kernel of the epimorphism ψ, we adopt the notation $\mathbf{0} \to \mathfrak{A} \xrightarrow{\varphi} \mathfrak{C} \xrightarrow{\psi} \mathfrak{B} \to \mathbf{0}$ and call this sequence *exact*. Here

0 stands for the Ershov algebra containing only the null element, $\mathbf{0} \to \mathfrak{A}$ denotes the natural embedding, and $\mathfrak{B} \to \mathbf{0}$ is an epimorphism of $\mathfrak{B}$ onto the null element. By definition, the exactness of $\mathfrak{A} \xrightarrow{\varphi} \mathfrak{C} \xrightarrow{\psi} \mathfrak{B}$ means that $\ker \psi = \operatorname{range} \varphi$ and the exactness of

$$\mathfrak{A}_1 \xrightarrow{\varphi_1} \mathfrak{A}_2 \xrightarrow{\varphi_2} \mathfrak{A}_3 \xrightarrow{\varphi_3} \mathfrak{A}_4 \xrightarrow{\varphi_4} \ldots \xrightarrow{\varphi_{n-1}} \mathfrak{A}_n$$

means that any of its subsequences $\mathfrak{A}_i \xrightarrow{\varphi_i} \mathfrak{A}_{i+1} \xrightarrow{\varphi_{i+1}} \mathfrak{A}_{i+2}$ is exact. Hence $\mathbf{0} \to \mathfrak{A} \xrightarrow{\varphi} \mathfrak{C} \xrightarrow{\psi} \mathfrak{B} \to \mathbf{0}$ is exact if and only if φ is an isomorphic embedding, ψ is an epimorphism, and $\ker \psi = \operatorname{range} \varphi$.

Two exact sequences $\mathbf{0} \to \mathfrak{A} \xrightarrow{\varphi} \mathfrak{C} \xrightarrow{\psi} \mathfrak{B} \to \mathbf{0}$ and $\mathbf{0} \to \mathfrak{A} \xrightarrow{\varphi'} \mathfrak{C}' \xrightarrow{\psi'} \mathfrak{B} \to \mathbf{0}$ are called *equivalent* if there exists a homomorphism $\varepsilon: \mathfrak{C} \to \mathfrak{C}'$ such that the diagram

$$\begin{array}{ccccccccc} \mathbf{0} & \to & \mathfrak{A} & \xrightarrow{\varphi} & \mathfrak{C} & \xrightarrow{\psi} & \mathfrak{B} & \to & \mathbf{0} \\ \downarrow \operatorname{id} & & \downarrow \operatorname{id} & & \downarrow \varepsilon & & \downarrow \operatorname{id} & & \downarrow \operatorname{id} \\ \mathbf{0} & \to & \mathfrak{A} & \xrightarrow{\varphi'} & \mathfrak{C}' & \xrightarrow{\psi'} & \mathfrak{B} & \to & \mathbf{0} \end{array} \quad (1.3.1)$$

is commutative. In this case, ε is an isomorphism between $\mathfrak{C}$ and $\mathfrak{C}'$, where id denotes the identity mapping of a set onto itself.

An exact sequence $\mathbf{0} \to \mathfrak{A} \xrightarrow{\varphi} \mathfrak{C} \xrightarrow{\psi} \mathfrak{B} \to \mathbf{0}$ defines up to an equivalence an *extension of* $\mathfrak{A}$ *by* $\mathfrak{B}$. Let $\operatorname{Ext}(\mathfrak{A}, \mathfrak{B})$ denote the family of all extensions of $\mathfrak{A}$ by $\mathfrak{B}$ up to an equivalence. In what follows, we describe all such extensions.

Let an algebra $\mathfrak{D}$ be an ideal of an Ershov algebra $\mathfrak{D}'$. We say that $\mathfrak{D}'$ is an *ideal completion* of $\mathfrak{D}$ if the following condition holds:

for any Ershov algebra $\mathfrak{D}''$ and embedding $\varphi: \mathfrak{D} \to \mathfrak{D}''$ as an ideal, there exists a unique homomorphism $\varphi: \mathfrak{D}'' \to \mathfrak{D}'$ such that the following diagram is commutative (i.e., the composition $\varphi \circ \varphi'$ is the identity embedding id of the algebra $\mathfrak{D}$ into the algebra $\mathfrak{D}'$):

$$\begin{array}{ccc} & \xrightarrow{\varphi} & \mathfrak{D}'' \\ \mathfrak{D} & & \downarrow \varphi' \\ & \xrightarrow{\operatorname{id}} & \mathfrak{D}' \end{array}$$

Let $\operatorname{Hom}(\mathfrak{A}, \mathfrak{B})$ denote the set of homomorphisms from $\mathfrak{A}$ into $\mathfrak{B}$.

1.3. Ideals and Quotient Algebras of Boolean Algebras 33

Theorem 1.3.1. *If $\mathfrak{A}'$ is an ideal completion of an Ershov algebra $\mathfrak{A}$, then for the quotient algebra $\mathfrak{A}^* \rightleftharpoons \mathfrak{A}'/\mathfrak{A}$ there exists a one-to-one correspondence between homomorphisms $\mathrm{Hom}\,(\mathfrak{B}, \mathfrak{A}^*)$ and extensions of $\mathfrak{A}$ by $\mathfrak{B}$ up to an equivalence $\mathrm{Ext}\,(\mathfrak{A}, \mathfrak{B})$.*

PROOF. Consider the extension of $\mathfrak{A}$ by $\mathfrak{B}$: $0 \to \mathfrak{A} \xrightarrow{\varphi} \mathfrak{C} \xrightarrow{\psi} \mathfrak{B} \to 0$. Let $\mathfrak{A}'$ be an ideal completion of $\mathfrak{A}$. By definition, there exists a unique homomorphism $\varphi': \mathfrak{C} \to \mathfrak{A}'$ such that the diagram

$$\begin{array}{ccc} & \varphi \nearrow & \mathfrak{C} \\ \mathfrak{A} & & \downarrow \varphi' \\ & \searrow_{\mathrm{id}} & \mathfrak{A}' \end{array} \qquad (1.3.2)$$

is commutative, where id is the identity embedding of $\mathfrak{A}$ into $\mathfrak{A}'$. We consider the quotient algebras $\mathfrak{C}/\varphi(\mathfrak{A})$, $\mathfrak{A}^* \rightleftharpoons \mathfrak{A}'/\mathfrak{A}$ and the homomorphism φ'' from $\mathfrak{C}/\varphi(\mathfrak{A})$ into $\mathfrak{A}'/\mathfrak{A}$ that is induced by φ'. We note that the mapping ψ induces the homomorphism ψ^* from $\mathfrak{C}/\varphi(\mathfrak{A})$ into $\mathfrak{B}$ which turns out to be an isomorphism. Consequently, there exists a unique homomorphism φ^* from $\mathfrak{B}$ into $\mathfrak{A}^*$ such that $\varphi^* = (\psi^*)^{-1} \circ \varphi''$. It is clear that $\varphi^* \in \mathrm{Hom}\,(\mathfrak{B}, \mathfrak{A}^*)$. Consequently, we have defined a mapping from $\mathrm{Ext}\,(\mathfrak{A}, \mathfrak{B})$ into $\mathrm{Hom}\,(\mathfrak{B}, \mathfrak{A}^*)$. Since the sequences $0 \to \mathfrak{A} \xrightarrow{\varphi} \mathfrak{C} \xrightarrow{\psi} \mathfrak{B} \to 0$ and $0 \to \mathfrak{A} \xrightarrow{\varphi'} \mathfrak{C}' \xrightarrow{\psi'} \mathfrak{B} \to 0$ are equivalent, the same homomorphism is associated with them. Hence this mapping is well defined.

With any homomorphism $\varphi^{\#} \in \mathrm{Hom}\,(\mathfrak{B}, \mathfrak{A}^*)$ we can associate some extension of $\mathfrak{A}$ by $\mathfrak{B}$. Indeed, consider the direct product $\mathfrak{B} \times \mathfrak{A}'$ of the Ershov algebras $\mathfrak{B}$, $\mathfrak{A}'$ and the semilattice $\mathfrak{C} \rightleftharpoons \{\langle b, a \rangle \mid \varphi^{\#} b = p_{\mathfrak{A}} a\}$, where $p_{\mathfrak{A}}: \mathfrak{A}' \to \mathfrak{A}^*$ gives the quotient algebra $\mathfrak{A}^*$ of $\mathfrak{A}'$ by an ideal of $\mathfrak{A}$. It is easy to check that $\mathfrak{C}$ is an Ershov algebra. It is obvious that $I = \{\langle 0, a \rangle \mid a \in \mathfrak{A}\}$ is an ideal of $\mathfrak{C}$ and the mapping φ from $\mathfrak{A}$ onto I such that $\varphi(a) = \langle 0, a \rangle$ is an isomorphic embedding of $\mathfrak{A}$ into $\mathfrak{C}$. For ψ we take the projection of $\mathfrak{C}$ onto the first coordinate. We note that ψ is a homomorphism from $\mathfrak{C}$ into $\mathfrak{B}$. It is obvious that for any $b \in \mathfrak{B}$ there exists $a \in \mathfrak{A}$ such that $\langle b, a \rangle \in \mathfrak{B}$. Consequently, ψ is an epimorphism from $\mathfrak{C}$ onto $\mathfrak{B}$. Since φ is an isomorphic embedding, ψ is an epimorphism, and range $\varphi = \ker \psi$, we conclude that the sequence $0 \to \mathfrak{A} \xrightarrow{\varphi} \mathfrak{C} \xrightarrow{\psi} \mathfrak{B} \to 0$ is exact and defines an extension of $\mathfrak{A}$ by $\mathfrak{B}$. We consider the corresponding homomorphism

φ^* from $\mathfrak{B}$ into $\mathfrak{A}^*$ just constructed. We have $\varphi^* = (\psi^*)^{-1} \circ \varphi''$, where $\varphi'' \colon \mathfrak{C}/\varphi(\mathfrak{A}) \to \mathfrak{A}^*$ and ψ^* is the isomorphism between $\mathfrak{C}/\varphi(\mathfrak{A})$ and $\mathfrak{B}$ that is obtained by taking quotient homomorphism of $\psi \colon \mathfrak{C} \to \mathfrak{B}$ by $\varphi(\mathfrak{A})$. In this case, $\varphi^*(b) = \varphi''((\psi^*)^{-1}(b)) = \varphi''(\langle b, a\rangle/\varphi(\mathfrak{A})) = \varphi'(\langle b, a\rangle)/\mathfrak{A}$, where φ' is a unique homomorphism such that the diagram (1.3.2) is commutative. By the definition of $\mathfrak{C}$, we obtain $\varphi^\#(b) = p_\mathfrak{A}(a)$ for $\langle b, a\rangle \in \mathfrak{C}$. Consider the following mapping $\widetilde{\varphi}$ from $\mathfrak{C}$ into $\mathfrak{A}'$: $\widetilde{\varphi}(\langle b, a\rangle) = a$ for any pair $\langle b, a\rangle \in \mathfrak{C}$. It is clear that the composition $\varphi \circ \widetilde{\varphi}$ is identical on elements of $\mathfrak{A}$. Consequently, the diagram

$$\mathfrak{A} \xrightarrow{\varphi} \mathfrak{C} \xrightarrow{\widetilde{\varphi}} \mathfrak{A}'$$
(with id from $\mathfrak{A}$ to $\mathfrak{A}'$)

is commutative. Therefore, $\widetilde{\varphi} = \varphi'$. Thus, $\varphi'(\langle b, a\rangle)/\mathfrak{A} = a/\mathfrak{A} = p_\mathfrak{A}(a) = \varphi^\#(b)$. Consequently, $\varphi^*(b) = \varphi^\#(b)$, which is required. $\square$

To complete the description of extensions of $\mathfrak{A}$ by $\mathfrak{B}$, it remains to prove the existence of an ideal completion of an arbitrary Ershov algebra.

Let $\mathfrak{A}$ be an Ershov algebra. We consider the family $I(\mathfrak{A})$ of all ideals of $\mathfrak{A}$, including the improper ideal $I = A$. On $I(\mathfrak{A})$, we define the operations $\wedge$ and $\vee$ by setting $j \wedge j' = \{x \mid x \in j \,\&\, x \in j'\}$ and $j \vee j' = \{x \vee y \mid x \in j \,\&\, y \in j'\}$. For the null element we take the ideal $\boldsymbol{O} \rightleftharpoons \{0\}$.

Lemma 1.3.6. *If $\mathfrak{A}$ is an Ershov algebra, then $\langle I(\mathfrak{A}); \wedge, \vee, \boldsymbol{O}\rangle$ is a distributive lattice.*

The proof is a direct verification of the corresponding set-theoretic relations and is left to the reader.

There exists the natural embedding (denoted by $\widehat{}$) of $\mathfrak{A}$ into $I(\mathfrak{A})$, where $\widehat{a} \rightleftharpoons \{x \in |\mathfrak{A}| \mid x \leqslant a\}$.

Lemma 1.3.7. *If $\mathfrak{A}$ is an Ershov algebra, then the mapping $\widehat{}$ is an isomorphic embedding of $\mathfrak{A}$ into $\langle I(\mathfrak{A}); \wedge, \vee, \boldsymbol{O}\rangle$.*

The proof is a direct verification of the conditions of an isomorphic embedding and is left to the reader.

1.3. Ideals and Quotient Algebras of Boolean Algebras

An ideal I of an Ershov algebra $\mathfrak{A}$ is called *locally principal* if for any $a \in \mathfrak{A}$ the intersection $I \cap \widehat{a}$ is a principal ideal, i.e., there exists $b \in \mathfrak{A}$ such that $I \cap \widehat{a} = \widehat{b}$. We note that the improper ideal A of an Ershov algebra $\mathfrak{A}$, as well as any principal ideal, including $O = \{0\} = \widehat{0}$, is a locally principal ideal. We also note that locally principal ideals form a subalgebra of $\langle I(\mathfrak{A}); \wedge, \vee, O \rangle$. Denote by $C_E(\mathfrak{A})$ the sublattice $\langle I(\mathfrak{A}); \wedge, \vee, 0 \rangle$ such that its universe is formed by locally principal ideals of $\mathfrak{A}$. Thus, $\mathfrak{A}$ can be regarded as a sublattice $C_E(\mathfrak{A})$ if a is identified with $\widehat{a}$ for any $a \in \mathfrak{A}$.

Proposition 1.3.1. *For any Ershov algebra $\mathfrak{A}$ the distributive lattice $C_E(\mathfrak{A})$ is an Ershov algebra with the greatest element and an ideal completion of $\mathfrak{A}$ with respect to the isomorphic embedding $\widehat{}$.*

PROOF. We first show that $C_E(\mathfrak{A})$ is an Ershov algebra. Furthermore, we show that $C_E(\mathfrak{A})$ is a Boolean algebra. The improper ideal A is a locally principal ideal. Consequently, $A \in C_E(\mathfrak{A})$. Therefore, A is the greatest element of $C_E(\mathfrak{A})$. For any $j \in C_E(\mathfrak{A})$ we define a complement to the greatest element, i.e., an element $C(j) \in C_E(\mathfrak{A})$ such that $C(j) \vee j = A$ and $C(j) \wedge j = O$. In this case, $C_E(\mathfrak{A})$ is a Boolean lattice. Consequently, it is an Ershov algebra with the greatest element.

Let $C(j) \rightleftharpoons \{a \in \mathfrak{A} \mid a \wedge b = 0 \text{ for any } b \in j\}$. It is obvious that all the properties of an ideal are fulfilled:

(1) $0 \in C(j)$,

(2) if $x \leqslant y$ and $y \in C(j)$, then $x \in C(j)$,

(3) if $x, y \in C(j)$, then $x \vee y \in C(j)$.

To verify that $C(j)$ is a locally principal ideal we consider an arbitrary element $a \in \mathfrak{A}$. Since the ideal j is a locally principal one, from definitions it follows that there exists $b \in j$ such that $j \cap \widehat{a} = \widehat{b}$. Let $c \rightleftharpoons a \backslash b$. We show that $C(j) \wedge \widehat{a} = \widehat{c}$. If $d \in C(j) \cap \widehat{a}$, then $d \leqslant a$ and $d \wedge x = 0$ for all $x \in j$. Consequently, $d \wedge b = 0$ and $d = d \wedge a = d \wedge (a \backslash b \vee b) = d \wedge (a \backslash b) = d \wedge c$, i.e., $d \leqslant c$ and $d \in \widehat{c}$.

To prove the converse assertion, we note that $d \leqslant c$ implies $d \leqslant a$ since $c \leqslant a$ by the definition of relative complement. Let us show $d \in C(j)$. On the contrary, we assume that $d \notin C(j)$. By definition, there is $x \in j$ such that $d \wedge x \neq 0$. In this case, $d \wedge x \in j$ and $d \wedge x \leqslant c = a \backslash b \leqslant a$. Thus, $d \wedge x \in j \cap \widehat{a}$. But then $d \wedge x \leqslant b$ by the definition of b. Consequently, $d \wedge x \leqslant a \backslash b \wedge b = 0$. Therefore, we conclude that $d \wedge x = 0$, which contradicts the choice of x. Thus, $C(j) \in C_E(\mathfrak{A})$.

The condition $j \wedge C(j) = O$ follows from the definition of $C(j)$. We show that $j \vee C(j) = A$. It suffices to show that $x \in j \vee C(j)$ for any

$x \in \mathfrak{A}$. Consider the intersection $j \cap \widehat{x}$. By the definition of a locally principal ideal, there exists an element y such that $j \cap \widehat{x} = \widehat{y}$. For $x \backslash y$ we have $x = y \vee x \backslash y$. In view of the choice of y, we have $y \in j$. It remains to show that $x \backslash y \in C(j)$. In the opposite case, let there exist $z \in j$ such that $x \backslash y \wedge z \neq 0$. However, $x \backslash y \leqslant x$, hence $x \backslash y \wedge z \leqslant x$ belongs to j. In this case, $x \backslash y \wedge z \leqslant y$ by the choice of y. However, $x \backslash y \wedge y = 0$; moreover, $x \backslash y \wedge z \wedge y = 0$. But $(x \backslash y \wedge z) \wedge y = y$. Consequently, $x \backslash y \wedge z = 0$, which contradicts the choice of z. Thus, $C_E(\mathfrak{A})$ is a Boolean algebra and $\mathfrak{A}$ is embedded into $C_E(\mathfrak{A})$ (as a subalgebra) by means of the isomorphic embedding $\widehat{}$.

Let us show that the sublattice $\mathfrak{A}$ of $C_E(\mathfrak{A})$ is an ideal. It suffices to note that for any $j \in C_E(\mathfrak{A})$ and $a \in \mathfrak{A}$ the following condition holds: $j \leqslant \widehat{a}$ implies $j = \widehat{b}$ for some $b \in \mathfrak{A}$. Since j is a locally principal ideal, there exists $b \in \mathfrak{A}$ such that $j \cap \widehat{a} = \widehat{b}$. On the other hand, $j \cap \widehat{a} = j$.

It remains to verify the main condition of an ideal completion. Let $\mathfrak{B}$ be an Ershov algebra and let φ be an isomorphic embedding of $\mathfrak{A}$ into $\mathfrak{B}$ as an ideal. We define a mapping φ' from $\mathfrak{B}$ into $C_E(\mathfrak{A})$ by setting $\varphi'(x) \rightleftharpoons \{a \in \mathfrak{A} \mid \varphi(a) \leqslant x\}$. It is obvious that $\varphi'(x)$ is an ideal of $\mathfrak{A}$. We show that this ideal is locally principal. Let z be some element of $\mathfrak{A}$. Since $\varphi(z) \wedge x \leqslant \varphi(z)$ and φ is an embedding of $\mathfrak{A}$ into $\mathfrak{B}$ as an ideal, there exists an element $a \in \mathfrak{A}$ such that $\varphi(a) = \varphi(z) \wedge x$. We show that $\varphi'(x) \cap \widehat{z} = \widehat{a}$. If $y \in \varphi'(x) \cap \widehat{z}$, then $\varphi(y) \leqslant x$ and $y \leqslant z$. But then $\varphi(y) \leqslant \varphi(z)$ and $\varphi(y) \leqslant \varphi(a)$. Therefore, $y \leqslant a$ and $y \in \widehat{a}$. In the case $y \in \widehat{a}$, we have $y \leqslant a$ and $\varphi(y) \leqslant \varphi(a) = \varphi(z) \wedge x$, which implies $\varphi(y) \leqslant \varphi(z)$ and $\varphi(y) \leqslant x$. Consequently, $y \leqslant z$ and $y \in \varphi'(x) \cap \widehat{z}$. Thus, we have proved that $C_E(\mathfrak{A})$ is an ideal completion of $\mathfrak{A}$ with respect to an isomorphic embedding $\widehat{}$ of $\mathfrak{A}$ into $C_E(\mathfrak{A})$ as an ideal. □

Corollary 1.3.1. *An Ershov algebra $\mathfrak{C}$ extending $\mathfrak{A}$ by $\mathfrak{B}$ is a Boolean algebra if and only if $\mathfrak{B}$ is a Boolean algebra and the corresponding homomorphism φ^* is a homomorphism from the Boolean algebra $\mathfrak{B}$ into the Boolean algebra $C_E(\mathfrak{A})$.*

PROOF. NECESSITY. It suffices to verify that the greatest element of $\mathfrak{B}$ goes into the greatest element of $C_E(\mathfrak{A})$ under the corresponding homomorphism φ^*. But if φ is an embedding of $\mathfrak{A}$ into $\mathfrak{C}$ as an ideal, then, in accordance with the definition of φ', for $C_E(\mathfrak{A})$ we have the relation

$$\varphi'(1) = \{x \in \mathfrak{A} \mid \varphi(x) \leqslant 1_{\mathfrak{C}}\} = A$$

which implies the required assertion.

1.3. Ideals and Quotient Algebras of Boolean Algebras

SUFFICIENCY. The assertion immediately follows from the existence of the greatest element of $\mathfrak{C}$. By Theorem 1.3.1, we can consider $\mathfrak{B}$, up to an equivalence, as a subalgebra of $\mathfrak{B} \times C_E(\mathfrak{A})$ consisting of elements of the form $\langle b, a \rangle$, where $\varphi^*(b) = p_\mathfrak{A}(a)$. It is clear that $\langle 1_\mathfrak{A}, 1_{C_E(\mathfrak{A})} \rangle$ satisfies all these conditions and defines the greatest element of $\mathfrak{C}$, where $1_\mathfrak{B}$ and $1_{C_E(\mathfrak{A})}$ are the greatest elements of $\mathfrak{B}$ and of $C_E(\mathfrak{A})$ respectively. $\square$

Corollary 1.3.2. *If $\mathfrak{A}$ is an Ershov algebra and $\mathfrak{B}_2$ is a Boolean algebra containing only two elements, then there exists a unique extension $B(\mathfrak{A})$ of $\mathfrak{A}$ by $\mathfrak{B}_2$ which is a Boolean algebra.*

The proof follows from the uniqueness of a Boolean homomorphism of $\mathfrak{B}_2$ into $C_E(\mathfrak{A})$.

The Boolean algebra $B(\mathfrak{A})$ is called the *minimal Boolean extension* of the Ershov algebra $\mathfrak{A}$.

Corollary 1.3.3. *In the minimal Boolean extension $B(\mathfrak{A})$ of an Ershov algebra $\mathfrak{A}$, the ideal $\mathfrak{A}$ is a maximal proper ideal.*

Exercises

1. Let $\mathcal{P}(A) = \langle \mathcal{P}(A); \cup, \cap, C \rangle$ be the Boolean algebra of all subsets of a set A and let $F \rightleftharpoons \{X \mid X \subseteq A, X \text{ is finite}\}$.

   (a) Prove that F is an ideal of $\mathcal{P}(A)$.

   (b) Prove that in $\mathcal{P}(A)/F \setminus \{\emptyset/F\}$ there are no minimal elements if the set A is infinite.

   (c) Define the ordinal type of $\mathcal{P}(A)$.

2. Prove that if a is an element of a Boolean algebra $\mathfrak{A}$, then $\widehat{a} \rightleftharpoons \{x \mid x \leqslant a\}$ is an ideal.

3. Let $\mathfrak{B}$ be a Boolean algebra. Prove that an ideal of the ring $K(\mathfrak{B})$ is an ideal of $\mathfrak{B}$. Prove the converse assertion.

4. Prove that if I is an ideal and $\mathfrak{B}$ is a subalgebra of a Boolean algebra $\mathfrak{A}$, then $F_\gamma(\mathfrak{A}) \subseteq I \circ F_\gamma(\mathfrak{A}/I)$ and $F_\gamma(\mathfrak{A}) \cap \mathfrak{B} \subseteq F_\gamma(\mathfrak{B})$ for any ordinal γ.

5. Prove that if $a \neq 0$ is an atomless element, then $a \notin F_\gamma(\mathfrak{A})$ for any ordinal γ. An element a is called *atomless* if there is no atom under a.

6. Prove that if I is an ideal of a Boolean algebra $\mathfrak{A}$ and $\mathbf{1} \in I$, then $I = A$.

7. Show that any ideal of a finite Boolean algebra is a principal ideal.

8. Let $\mathfrak{B}$ be a Boolean algebra and let $a \in \mathfrak{B}$. Prove that $\widehat{C(a)}$ is a maximal ideal of $\mathfrak{B}$ if and only if a is an atom.

9. Let $\mathfrak{B}$ be a finite Boolean algebra. Show that any of its elements $a \neq \mathbf{0}$ can be represented as a finite union of atoms, i.e., $a = \bigvee_{i=1}^{n} a_i$, where a_i, $1 \leqslant i \leqslant n$, are atoms of $\mathfrak{B}$.

10. Prove that $I \subseteq A$ is a proper ideal of a Boolean algebra $\mathfrak{A}$ if and only if $C(I)$ is a filter of $\mathfrak{A}$.

11. Prove that the set of all atomless elements of a Boolean algebra is its ideal. An element a is called *atomic* if there is an atom under any $x \leqslant a$ different from the null element.

12. Let $\mathfrak{B}$ be a Boolean algebra and let I be the set of all those elements of $\mathfrak{B}$ that admit the representation in the form of a union of atomic elements and atomless ones. Prove that I is an ideal of $\mathfrak{B}$.

13. Prove that the kernel of any homomorphism is an ideal.

14. Let L be a linearly ordered set and let $\{I_\xi\}_{\xi \in L}$ be a family of ideals such that $I_\xi \subseteq I_\eta$, $\xi \leqslant \eta$. Prove that $\bigcup_{\xi \in L} I_\xi$ is an ideal.

15. Show that a principal ultrafilter of a Boolean algebra contains an atom.

16. Prove that a homomorphism ε of an Ershov algebra $\mathfrak{C}$ into an Ershov algebra $\mathfrak{C}'$ for the exact sequences $\mathfrak{A} \xrightarrow{\varphi} \mathfrak{C} \xrightarrow{\psi} \mathfrak{B}$ and $\mathfrak{A} \xrightarrow{\varphi'} \mathfrak{C}' \xrightarrow{\psi'} \mathfrak{B}$ such that the diagram (1.3.1) commutative, is an isomorphism between $\mathfrak{C}$ and $\mathfrak{C}'$.

17. Prove the uniqueness of an ideal completion of an Ershov algebra.

18. Prove that a homomorphism ψ^* from the quotient algebra $\mathfrak{C}/\mathfrak{A}$ into $\mathfrak{B}$ that is induced by an epimorphism ψ (we assume that the subsystem $\mathfrak{A}$ of the Ershov algebra $\mathfrak{C}$ coincides with the kernel of the homomorphism ψ^*) is an isomorphism between $\mathfrak{C}/\mathfrak{A}$ and $\mathfrak{B}$.

1.4. The Stone Theorem

19. Show the correctness of the definition of the mapping from Ext $(\mathfrak{A}, \mathfrak{B})$ into Hom $(\mathfrak{B}, \mathfrak{A}^*)$ constructed in Theorem 1.3.1.

20. Prove that the ideal of $\mathfrak{A}$ in the minimal Boolean extension $B(\mathfrak{A})$ of an Ershov algebra $\mathfrak{A}$ is a maximal proper ideal with respect to inclusion.

21. Let $\langle \mathcal{P}(A); \cup, \cap, C \rangle$ be the Boolean algebra of all subsets of a set A, $F \rightleftharpoons \{X \mid X \subseteq A,\ X \text{ is finite}\}$ the Frechét ideal of $\langle \mathcal{P}(A); \cup, \cap, C \rangle$, and $\mathcal{F}(A) \rightleftharpoons \langle F; \cup, \cap, \varnothing \rangle$. Prove the following claims:

   (a) any ideal of $\mathcal{F}(A)$ is locally principal,
   (b) if $A_0 \subseteq A$ and $I_{A_0} \rightleftharpoons \{x \in F \mid x \subseteq A_0\}$, then I_{A_0} is an ideal of $\mathcal{F}(A)$,
   (c) any ideal I of $\mathcal{F}(A)$ has the form I_{A_0} with a suitable $A_0 \subseteq A$,
   (d) the algebra $\langle \mathcal{P}(A); \cup, \cap, \varnothing \rangle$ is an ideal completion of $\mathcal{F}(A)$.

1.4. The Stone Theorem on Representations of Boolean Algebras

In this section, we describe Boolean algebras in terms of algebras of subsets. This enables us, without special stipulations, to use various set-theoretic relations.

A proper ideal I of a Boolean algebra $\mathfrak{A}$ is called

— *maximal* if there exists no proper ideal J extending I,

— *prime* if $x \in I$ or $C(x) \in I$ for any $x \in A$.

It is obvious that a prime ideal is maximal. The converse assertion is also true.

Proposition 1.4.1. *A maximal ideal I of a Boolean algebra $\mathfrak{A}$ is a prime ideal.*

PROOF. On the contrary, let I be not a prime ideal. Then there exists $x \in A$ such that $x \notin I$ and $C(x) \notin I$. It is easy to check that the sets

$$I_x \rightleftharpoons \{y \in A \mid (\exists a \in I) y \leqslant x \vee a\}$$
$$I_{C(x)} \rightleftharpoons \{y \in A \mid (\exists a \in I) y \leqslant C(x) \vee a\}$$

are ideals. Since the ideal I is maximal, $I \subseteq I_x$, $I \subseteq I_{C(x)}$, $x \in I_x \setminus I$, and $C(x) \in I_{C(x)} \setminus I$, we conclude that $I_x = A$ and $I_{C(x)} = A$. Consequently, there exist elements $a \in I$ and $b \in I$ such that $x \vee a = 1$ and $C(x) \vee b = 1$. Therefore,

$$(a \vee b) \vee (x \wedge C(x)) = ((a \vee b) \vee x) \wedge ((a \vee b) \vee C(x)) = 1$$
$$a \vee b = (a \vee b) \vee 0, \quad a \vee b = 1$$

However, the ideal I is proper, i.e., $1 \notin I$, which yields a contradiction. □

Proposition 1.4.2. *For any element $a \in A$ of a Boolean algebra $\mathfrak{A}$ different from the null element there exists a maximal ideal I which does not contain a.*

PROOF. We consider the set $\mathcal{J}$ of all proper ideals I such that $C(a) \in I$ (thereby, $a \notin I$). It is clear that $\mathcal{J} \neq \varnothing$ since the ideal of $\widehat{C(a)}$ belongs to $\mathcal{J}$. The set $\langle \mathcal{J}; \subseteq \rangle$ equipped with the inclusion relation is inductive since the union of a directed family of ideals containing $C(a)$ is an ideal containing $C(a)$. By the Zorn lemma [36, 72], in $\langle \mathcal{J}; \subseteq \rangle$ there exists a maximal element, say I. We show that I is a maximal ideal. We assume the contrary. Then there exists a proper ideal J that extends I and is not equal to I. However, $C(a) \in I \subseteq J$ and J is a proper ideal. Consequently, $J \in \mathcal{J}$, which contradicts the maximality of I in J. □

We define the set $\mathcal{J}(\mathfrak{A})$ of all maximal ideals of a Boolean algebra $\mathfrak{A}$ and consider the Boolean algebra $\mathcal{P}(\mathcal{J}(\mathfrak{A}))$ of all subsets of $\mathcal{J}(\mathfrak{A})$. Introduce a mapping φ from A into $\mathcal{P}(\mathcal{J}(\mathfrak{A}))$ by setting $\varphi(a) \rightleftharpoons \{I \in \mathcal{J}(\mathfrak{A}) \mid a \notin I\}$.

Lemma 1.4.1. *The mapping φ is an isomorphic embedding of $\mathfrak{A}$ into $\mathcal{P}(\mathcal{J}(\mathfrak{A}))$.*

PROOF. The proof proceeds in several steps.

⟨1⟩ *The mapping φ is an injection.* Assume that there are two different elements a and b of A such that $\varphi(a) = \varphi(b)$. Since $a \neq b$, we have $a \not\leq b$ or $b \not\leq a$. We consider only the first case because the second case is reduced to the first one by renaming a and b. Let $a \not\leq b$. Then $a \wedge C(b) \neq 0$; otherwise, $a \wedge b = a$ since $a = a \wedge 1 = a \wedge (b \vee C(b)) = (a \wedge b) \vee (a \wedge C(b))$. By Proposition 1.4.2, there exists a maximal ideal I such that $a \wedge C(b) \notin I$. Therefore, $a \notin I$ and $C(b) \notin I$. By Proposition 1.4.1, $b \in I$. Consequently, $I \in \varphi(a)$ and $I \notin \varphi(b)$, which contradicts the assumption $\varphi(a) = \varphi(b)$.

1.4. The Stone Theorem

⟨2⟩ *The mapping φ preserves the operation $\vee$*, i.e., $\varphi(a \vee b) = \varphi(a) \cup \varphi(b)$ for any $a, b \in A$. In other words, prove that the following set-theoretic inclusions hold:

$$\varphi(a \vee b) \subseteq \varphi(a) \cup \varphi(b), \quad \varphi(a) \cup \varphi(b) \subseteq \varphi(a \vee b)$$

We prove the first inclusion. Let $I \in \varphi(a \vee b)$. We need to show that $I \in \varphi(a)$ or $I \in \varphi(b)$. On the contrary, assume that $a \in I$ and $b \in I$. By the maximality of I and Proposition 1.4.1, we have $C(a \vee b) \in I$. Since $a \in I$ and $b \in I$, we conclude that $a \vee b \in I$. But then $1 = (a \vee b) \vee C(a \vee b)$ is an element of I, which contradicts I being maximal. To prove the second inclusion we assume that $I \in \varphi(a) \cup \varphi(b)$ and $I \notin \varphi(a \vee b)$. Then $a \vee b \in I$. By the definition of an ideal, $a \in I$ and $b \in I$. Consequently, $I \notin \varphi(a)$ and $I \notin \varphi(b)$, i.e., $I \notin \varphi(a) \cup \varphi(b)$, which yields a contradiction.

⟨3⟩ *The mapping φ preserves the operation $\wedge$*, i.e., $\varphi(a \wedge b) = \varphi(a) \cap \varphi(b)$ for any $a, b \in A$. The proof of this assertion is similar to that of ⟨2⟩.

⟨4⟩ *The mapping φ preserves the operation C*, i.e., $C(\varphi(a)) = \varphi(C(a))$. If $I \in C(\varphi(a))$, then $I \notin \varphi(a)$ and $a \in I$. Since a maximal ideal is proper, $C(a) \notin I$. Consequently, $I \in \varphi(C(a))$. Thus, $C(\varphi(a)) \subseteq \varphi(C(a))$. Let $I \in \varphi(C(a))$. Then $C(a) \notin I$. Since a maximal ideal is prime, $a \in I$, which means $I \notin \varphi(a)$ and $I \in C(\varphi(a))$.

Thus, the lemma is proved. □

Lemma 1.4.2. *If a Boolean algebra $\mathfrak{A}$ is isomorphically embedded into a Boolean algebra $\mathfrak{B}$, then $\mathfrak{A}$ is isomorphic to a subalgebra of $\mathfrak{B}$.*

PROOF. Let φ be an isomorphic embedding of $\mathfrak{A}$ into $\mathfrak{B}$. As the universe of a desired subalgebra, we take $B_0 \rightleftharpoons \varphi(A)$. It is easy to check that B_0 is closed under the basic operations on $\mathfrak{B}$. As the required isomorphism, we take a mapping φ from $\mathfrak{A}$ onto $\mathfrak{B}_0$, where $\mathfrak{B}_0$ is a subalgebra of $\mathfrak{B}$ with the universe B_0. □

Theorem 1.4.1 [the Stone theorem]. *Every Boolean algebra is isomorphic to some algebra of subsets.*

PROOF. By Lemma 1.4.1, any Boolean algebra $\mathfrak{A}$ is isomorphically embedded into the Boolean algebra of all subsets of some set. By Lemma 1.4.2, $\mathfrak{A}$ is isomorphic to some of its subalgebras. □

Using Theorem 1.4.1, it is easy to describe all finite Boolean algebras. We denote by $\mathfrak{B}_n$ the Boolean algebra of all subsets of some set containing n elements. The Boolean algebras $\mathfrak{B}_n$ have precisely n atoms which are defined by single-element sets and, consequently, the algebras $\mathfrak{B}_n$, $n \in \mathbb{N}$, are not pairwise isomorphic, i.e., $\mathfrak{B}_n \neq \mathfrak{B}_m$ if $n \neq m$.

In a finite Boolean algebra $\mathfrak{B}$, under any element $x \neq 0$ there is an atom. We consider the set Atom$(\mathfrak{B})$ of all atoms of $\mathfrak{B}$ and note that their intersection is exactly the null element and any element $x \neq 0$ is a union of atoms. It is obvious that $\mathfrak{B}$ is isomorphic to $\mathfrak{B}_n$, where n is the number of atoms of $\mathfrak{B}$.

Corollary 1.4.1. *The isomorphism type of a finite Boolean algebra is determined by the number of its atoms.*

We study finitely generated subalgebras of Boolean algebras.

Corollary 1.4.2. *Any finitely generated subalgebra of a Boolean algebra is finite.*

Let $\mathfrak{A}$ be a Boolean algebra and let $a_1, \ldots, a_n$ be a finite sequence of its elements. We consider the subalgebra $\operatorname{gr}_{\mathfrak{A}}(a_1, \ldots, a_n) \subseteq \mathfrak{A}$. We set $a^0 \rightleftharpoons C(a)$ and $a^1 \rightleftharpoons a$. We define the term $t_{\bar{\varepsilon}}(x_1, \ldots, x_n) \rightleftharpoons x_1^{\varepsilon_1} \wedge x_2^{\varepsilon_2} \wedge \ldots \wedge x_n^{\varepsilon_n}$ from the sequence $\bar{\varepsilon} = (\varepsilon_1, \ldots, \varepsilon_n) \in \{0,1\}^n$. It is obvious that $t_{\bar{\varepsilon}}(a_1, \ldots, a_n) \in \operatorname{gr}_{\mathfrak{A}}(a_1, \ldots, a_n)$. Starting from an arbitrary subset $K \subseteq \{0,1\}^n$, we define the term

$$t_K(x_1, \ldots, x_n) \rightleftharpoons \bigvee_{\bar{\varepsilon} \in K} t_{\bar{\varepsilon}}(x_1, \ldots, x_n)$$

We now consider the set

$$T(a_1, \ldots, a_n) \rightleftharpoons \{t_K(a_1, \ldots, a_n) \mid K \subseteq \{0,1\}^n\} \cup \{0\}$$

It is easy to see that $\{a_1, \ldots, a_n\} \subseteq T(a_1, \ldots, a_n) \subseteq \operatorname{gr}_{\mathfrak{A}}(a_1, \ldots, a_n)$. Furthermore, $T(a_1, \ldots, a_n)$ forms a subalgebra. Hence $T(a_1, \ldots, a_n) = \operatorname{gr}_{\mathfrak{A}}(a_1, \ldots, a_n)$. Thus, the finiteness of $\operatorname{gr}_{\mathfrak{A}}(a_1, \ldots, a_n)$ is proved. We note that the values of terms $t_{\bar{\varepsilon}}(a_1, \ldots, a_n)$ are equal to the null element 0 or to an atom of the subalgebra $\operatorname{gr}_{\mathfrak{A}}(a_1, \ldots, a_n)$.

We denote the set of all ultrafilters of a Boolean algebra $\mathfrak{B}$ by $\mathcal{F}(\mathfrak{B})$ and define the topology $\tau_{\mathfrak{B}}$ on $\mathcal{F}(\mathfrak{B})$ taking the family $\{v_a \mid a \in \mathfrak{B}\}$, where $v_a = \{F \in \mathcal{F}(\mathfrak{B}) \mid a \in F\}$, as the basis of open sets. This topology is called the *Stone topology* and the pair $(\mathcal{F}(\mathfrak{B}), \tau_{\mathfrak{B}})$ is called the *topological Stone space*. We note that it is a compact topological space and v_a are simply its clopen sets.

1.4. The Stone Theorem

Exercises

1. Show that any finite Boolean algebra is isomorphic to the Boolean algebra of all subsets of the set of maximal ideals and is isomorphic to the algebra of subsets of the set of its atoms.

2. Using Theorem 1.4.1, for an arbitrary Boolean algebra establish the following identities:

$$(a\backslash b)\backslash c = a\backslash(b \vee c), \quad a\backslash(b\backslash c) = a\backslash b \vee (a \wedge c),$$

$$\left(\bigwedge_{i=1}^{n} x_i\right) \vee y = \bigwedge_{i=1}^{n} (x_i \vee y), \quad \left(\bigvee_{i=1}^{n} x_i\right) \wedge y = \bigvee_{i=1}^{n} (x_i \wedge y).$$

3. A Boolean algebra $\mathfrak{A}$ is called *complete* if any of its nonempty subsets has the least upper bound and the greatest lower bound in $\mathfrak{A}$. Prove that the algebra of all subsets of some set is complete.

4. Prove that any Boolean algebra is isomorphically embedded into a complete Boolean algebra.

5. Prove that any finite Boolean algebra is generated by the set of all atoms.

6. Prove that every finitely generated subalgebra of an Ershov algebra is finite.

7. Let $\mathfrak{A}$ and $\mathfrak{B}$ be finite Boolean algebras and let φ be a one-to-one mapping of $\text{Atom}(\mathfrak{A})$ onto $\text{Atom}(\mathfrak{B})$. Prove that the mapping $\overline{\varphi}(x) \leftrightharpoons \vee\{\varphi(a) \mid a \leqslant x, a \in \text{Atom}(\mathfrak{A})\}$ is an isomorphism between $\mathfrak{A}$ and $\mathfrak{B}$.

8. Let $x_1, \ldots, x_n$ be elements of a Boolean algebra $\mathfrak{A}$ such that $x_i \wedge x_j = 0$ for any $i \neq j$ and $\bigvee_{i=1}^{n} x_i = 1$. Show that the subalgebra generated by $\{x_1, \ldots, x_n\}$ is exactly $\{0\} \cup \{\bigvee_{i \in K} x_i \mid K \subseteq \{1, \ldots, n\}\}$.

9. Let a sequence $x_1, \ldots, x_n$ be as in Exercise 8 and let a mapping $\varphi \colon \{x_1, \ldots, x_n\} \to \mathfrak{B}$ be such that $\varphi(x_i) = 0$ if $x_i = 0$, $1 \leqslant i \leqslant n$, $\bigvee_{i=1}^{n} \varphi(x_i) = 1$ and $\varphi(x_i) \wedge \varphi(x_j) = 0$, $i \neq j$. Prove that a mapping $\overline{\varphi} \colon \text{gr}\{x_1, \ldots, x_n\} \to \mathfrak{B}$ such that $\overline{\varphi}(x) \leftrightharpoons \vee\{\varphi(x_i) \mid 1 \leqslant i \leqslant n, x_i \leqslant x\}$ is a homomorphism from $\text{gr}\{x_1, \ldots, x_n\}$ into $\mathfrak{B}$.

10. Let A be a subset of a Boolean algebra $\mathfrak{A}$. Prove that the subalgebra $\mathfrak{B}$ generated by A consists of the elements $\bigvee_{l=1}^{s} a_{i_{1l}}^{\varepsilon_{1l}} \wedge \ldots \wedge a_{i_{kl}}^{\varepsilon_{kl}}$, where $a_{i_{1l}}, \ldots, a_{i_{kl}}$ are elements of A, $\varepsilon_{1l}, \ldots, \varepsilon_{kl}$ are elements of $\{0,1\}$, $a^0 \rightleftharpoons C(a)$, and $a^1 \rightleftharpoons a$.

11. Prove that the subalgebra of $\mathfrak{A}$ generated by atoms of $\mathfrak{A}$ is an atomic Boolean algebra.

12. Let $\mathfrak{A}$ be a finite Boolean algebra, let $a_1, \ldots, a_n$ be all the atoms of $\mathfrak{A}$, and let $b_1, \ldots, b_n$ be pairwise disjoint elements (i.e., $b_i \wedge b_j = \mathbf{0}$ for $i \neq j$) of $\mathfrak{B}$ different from the null element and $\bigvee_{i \in I} b_i = \mathbf{1}$. Show that a mapping $\varphi \colon \mathfrak{A} \to \mathfrak{B}$ such that $\varphi\left(\bigvee_{i \in I} a_i\right) \rightleftharpoons \bigvee_{i \in I} b_i$ for all subsets $I \subseteq \{1, \ldots, n\}$ is an isomorphic embedding of $\mathfrak{A}$ into $\mathfrak{B}$.

13. Construct an example of an infinite Boolean algebra without atoms.

14. Prove that an atomless Boolean algebra is infinite.

15. How many atoms can an n-generated Boolean algebra have? Describe all such algebras.

1.5. The Vaught Criterion

Isomorphisms and descriptions of algebraic systems in terms of their properties provide a good base for solving many algebraic problems. Therefore, the problem of finding various isomorphism criteria is very important in algebra. For countable Boolean algebras such a criterion was found by Vaught [201]. Due to his criterion, it is possible to obtain numerous results by direct verification of some conditions.

In this section, we present various generalizations and modifications of the Vaught criterion. In particular, we consider the criteria for embeddings, homomorphic embeddings, isomorphisms, and epimorphisms of countable Boolean algebras from a common point of view.

Let $\mathfrak{A}$ and $\mathfrak{B}$ be Boolean algebras and let $S \subseteq A \times B$, where A and B denote the universes of $\mathfrak{A}$ and $\mathfrak{B}$ respectively. We introduce the following axioms about S:

Ax1: $\langle \mathbf{0}, \mathbf{0} \rangle \in S \,\&\, \langle \mathbf{1}, \mathbf{1} \rangle \in S$,

1.5. The Vaught Criterion

Ax2: $(\langle x, y \rangle \in S \,\&\, x = \mathbf{0}) \Rightarrow y = \mathbf{0}$,

Ax3: $(\langle x, y \rangle \in S \,\&\, y = \mathbf{0}) \Rightarrow x = \mathbf{0}$,

Ax4: $\langle x, y \rangle \in S \,\&\, a \leqslant x \Rightarrow (\exists b)(b \leqslant y \,\&\, \langle a, b \rangle \in S \,\&\, \langle x \backslash a, y \backslash b \rangle \in S)$,

Ax5: $\langle x, y \rangle \in S \,\&\, b \leqslant y \Rightarrow (\exists a)(a \leqslant x \,\&\, \langle a, b \rangle \in S \,\&\, \langle x \backslash a, y \backslash b \rangle \in S)$,

Ax6: $\left(\left(\bigvee_{i=1}^{n} x_i = \mathbf{1} \right) \,\&\, \left(\bigvee_{i=1}^{n} y_i = \mathbf{1} \right) \,\&\, \left(\underset{i \neq j}{\&} (x_i \wedge x_j = \mathbf{0} \,\&\, y_i \wedge y_j = \mathbf{0}) \right) \right.$
$\left. \,\&\, \left(\underset{i=1}{\overset{n}{\&}} \langle x_i, y_i \rangle \in S \right) \right) \Rightarrow \langle \bigvee_{i \in K} x_i, \bigvee_{i \in K} y_i \rangle \in S$,

where $K \subseteq \{1, 2, \ldots, n\}$.

- A relation $S \subseteq A \times B$ is called

— a *condition of homomorphic embedding* of $\mathfrak{A}$ into $\mathfrak{B}$ if Ax1, Ax2, and Ax4 hold,

— a *condition of epimorphism* of $\mathfrak{A}$ onto $\mathfrak{B}$ if S is a condition of homomorphic embedding and Ax5 holds,

— a *condition of embedding* of $\mathfrak{A}$ into $\mathfrak{B}$ if S is a condition of homomorphic embedding and Ax3 holds,

— a *condition of isomorphism* of $\mathfrak{A}$ onto $\mathfrak{B}$ if Ax1–Ax5 hold.

A homomorphism φ of a Boolean algebra $\mathfrak{A}$ into a Boolean algebra $\mathfrak{B}$ is called *S-definite* if for any $x \in A$ there exist pairwise disjoint elements $a_0, \ldots, a_m$ of A such that $\bigvee_{i=1}^{m} a_i = x$ and the pairs $\langle a_i, \varphi(a_i) \rangle$, $i \leqslant m$, belong to S.

Our purpose is to find conditions providing for the existence of an S-definite homomorphism extending a finite partial mapping. We prescribe a partial mapping φ by indicating its graph Γ, i.e., by enumerating its elements: $\Gamma \rightleftharpoons \{\langle x_1, y_1 \rangle, \ldots, \langle x_n, y_n \rangle\}$.

A sequence $\langle x_i, y_i \rangle$, $i \leqslant m$, is called *disjunctive* if $\bigvee_{i=0}^{m} x_i = \mathbf{1}$, $\bigvee_{i=0}^{m} y_i = \mathbf{1}$, and $x_i \wedge x_j = \mathbf{0} \,\&\, y_i \wedge y_j = \mathbf{0}$, $0 \leqslant i < j \leqslant m$. We say that a homomorphism φ *continues* the sequence $\langle x_i, y_i \rangle$, $i \leqslant m$, if $\varphi(x_i) = y_i$, $i \leqslant m$.

Theorem 1.5.1 [on homomorphic embedding]. *Let S be a condition of homomorphic embedding of a countable Boolean algebra $\mathfrak{A}$*

into a Boolean algebra $\mathfrak{B}$. Then for any finite disjunctive sequences of S there exists an S-definite homomorphism φ from $\mathfrak{A}$ into $\mathfrak{B}$ that continues this sequence; moreover, $\varphi \subseteq S$ if Ax6 holds for S.

PROOF. We fix some enumeration of elements of the universe A of a Boolean algebra $\mathfrak{A}$: $A = \{d_0, d_1, \ldots, d_k, \ldots\}$. Let $\{\langle x_i, y_i \rangle \mid i \leqslant n\}$ be a disjunctive sequence. We construct step-by-step a sequence of finite subalgebras A_k of $\mathfrak{A}$ and homomorphisms $\varphi_k : A_k \to \mathfrak{B}$ subject to the following conditions:

(1) $x_0, x_1, \ldots, x_n \in A_0$ and $\varphi_0(x_i) = y_i$ for $i \leqslant n$,

(2) $A_{k+1} \supseteq A_k$ and $\varphi_{k+1} \supseteq \varphi_k$,

(3) $d_k \in A_{k+1}$,

(4) for any atom a of the subalgebra A_k the pair $\langle a, \varphi_k(a) \rangle$ belongs to S.

STEP 0. Let A_0 be the subalgebra of $\mathfrak{A}$ generated by elements $x_0, \ldots, x_n$ and $\varphi_0(\bigvee_{i \in K} x_i) \rightleftharpoons \bigvee_{i \in K} y_i$ for $K \subseteq \{0, 1, \ldots, n\}$. It is clear that the constructed mapping is a homomorphism and satisfies the required conditions.

STEP $t+1$. Let $a_0, \ldots, a_k$ be all atoms of the Boolean algebra A_t. Let $\widetilde{a}_{2i} \rightleftharpoons a_i \wedge d_t$, $\widetilde{a}_{2i+1} \rightleftharpoons a_i \wedge C(d_t)$. By the induction hypothesis, $\langle a_i, \varphi_t(a_i) \rangle \in S$. Therefore, there is $\widetilde{b}_{2i}$ such that $\widetilde{b}_{2i} \leqslant \varphi_t(a_i)$, $\langle \widetilde{a}_{2i}, \widetilde{b}_{2i} \rangle \in S$, and $\langle \widetilde{a}_{2i+1}, \varphi_t(a_i) \backslash \widetilde{b}_{2i} \rangle \in S$. We put $\widetilde{b}_{2i+1} \rightleftharpoons \varphi_t(a_i) \backslash \widetilde{b}_{2i}$ and define the subalgebra A_{t+1} of $\mathfrak{A}$ generated by $\widetilde{a}_i$, $i \leqslant 2k+1$. We set $\varphi_{t+1}(\bigvee_{i \in K} \widetilde{a}_i) \rightleftharpoons \bigvee_{i \in K} \widetilde{b}_i$, where $K \subseteq \{0, 1, 2, \ldots, 2k+1\}$. The construction is complete.

It is easy to check that A_{t+1} and φ_{t+1} just constructed possess the required properties (cf. Sec. 1.4, Exercise 9).

We have $d_t \in A_{t+1}$ and $\bigcup_{t \geqslant 0} A_t = A$ by (3). From (2) it follows that $\varphi = \bigcup_{t \geqslant 0} \varphi_t$ is a homomorphism of $\mathfrak{A}$ into $\mathfrak{B}$. Since $\varphi \supseteq \varphi_0$, the homomorphism φ continues the sequence $\{\langle x_i, y_i \rangle \mid i \leqslant n\}$. The condition (3) implies that the homomorphism φ is S-definite. If Ax6 holds for S, then $\varphi \subseteq S$ by definition. □

REMARK 1.5.1. If a homomorphism φ is S-definite and Ax3 holds for S, then φ is an isomorphic embedding.

1.5. The Vaught Criterion

Corollary 1.5.1. *If S is a condition of embedding of a countable Boolean algebra $\mathfrak{A}$ into a Boolean algebra $\mathfrak{B}$, then there exists an S-definite isomorphic embedding of $\mathfrak{A}$ into $\mathfrak{B}$ that continues a given disjunctive sequence of elements of S.*

Corollary 1.5.2. *If S is a condition of embedding of a countable Boolean algebra $\mathfrak{A}$ into a Boolean algebra $\mathfrak{B}$, then there exists an S-definite isomorphic embedding of $\mathfrak{A}$ into $\mathfrak{B}$.*

Theorem 1.5.2 [on covering]. *If S is a condition of epimorphism of a countable Boolean algebra $\mathfrak{A}$ onto a Boolean algebra $\mathfrak{B}$, then for any disjunctive sequence of elements of S there exists an S-definite epimorphism extending this sequence.*

PROOF. Let
$$A = \{a_0, a_1, a_2, \ldots, a_n, \ldots\}, \quad B = \{b_0, b_1, b_2, \ldots, b_n, \ldots\}$$
Proceeding by induction, we construct an extending sequence of subalgebras $\{A_n \mid n \in \mathbb{N}\}$ and homomorphisms $\{\varphi_n \mid n \in \mathbb{N}\}$ such that

$A_0 = \operatorname{gr}\{x_0, \ldots, x_m\}$, $\varphi_0 : A_0 \to B$, and $\varphi(x_i) = y_i$, $i \leqslant m$, where $\Gamma = \{\langle x_i, y_i \rangle \mid i \leqslant m\}$ is a fixed disjunctive sequence,

$A_n \subseteq A_{n+1}$ and $\varphi_n \subseteq \varphi_{n+1} : A_{n+1} \to B$ for any $n \in \mathbb{N}$,

$a_n \in A_{2n+1}$ and $b_n \in \varphi_{2n+2}(A_{2n+2})$ for any $n \in \mathbb{N}$,

for any atom a of A_n the pair $\langle a, \varphi_n(a) \rangle$ belongs to S.

We proceed as in Theorem 1.5.1: we add b_n to $\varphi_{2n+1}(A_{2n+1})$ at even steps and a_n to A_{2n} at odd steps. At even steps, the construction is perfectly similar to the case of Theorem 1.5.1 since Ax5 holds for S. Setting $\varphi = \bigcup_{t \geqslant 0} \varphi_t$, we obtain the required epimorphism. □

Corollary 1.5.3 [the Vaught theorem on isomorphism]. *If S is a condition of isomorphism between countable Boolean algebras $\mathfrak{A}$ and $\mathfrak{B}$, then for any disjunctive sequence of elements of S there exists an S-definite isomorphism between $\mathfrak{A}$ and $\mathfrak{B}$ that extends this sequence.*

REMARK 1.5.2. If φ is an S-definite homomorphism and Ax6 holds for S, then $\varphi \subseteq S$.

Applying Corollary 1.5.3, we can characterize some Boolean algebras.

Proposition 1.5.1. *Any two countable atomless Boolean algebras are isomorphic.*

PROOF. Let $\mathfrak{A}$ and $\mathfrak{B}$ be countable atomless Boolean algebras. We define

$$S = \{\langle a, b\rangle \mid a \in A,\ b \in B,\ (a = \mathbf{1} \Leftrightarrow b = \mathbf{1})\ \&\ (a = \mathbf{0} \Leftrightarrow b = \mathbf{0})\}$$

and show that S is a condition of isomorphism. The validity of Ax1–Ax3 is obvious. Since Ax4 and Ax5 are symmetric, it suffices to verify only one of them. Let $\langle x, y\rangle \in S$ and $a \leqslant x$. If $a \neq \mathbf{0}$ and $x\backslash a \neq \mathbf{0}$, then $y \neq \mathbf{0}$. Since $\mathfrak{B}$ is atomless, there is b such that $\mathbf{0} < b < y$. Then the pairs $\langle a, b\rangle$ and $\langle x\backslash a, y\backslash b\rangle$ belong to S. In the case $a = \mathbf{0}$ or $x\backslash a = \mathbf{0}$, the validity of Ax4 is obvious. □

We note that any finite Boolean algebra, as well as the algebra of all subsets of some nonempty set, is atomic. A Boolean algebra is atomic provided that it contains no atomless element $x \neq \mathbf{0}$.

Proposition 1.5.2. *If $\mathfrak{A}$ and $\mathfrak{B}$ are countable atomic Boolean algebras and the quotient algebras by the Frechét ideals $\mathfrak{A}/F(\mathfrak{A})$ and $\mathfrak{B}/F(\mathfrak{B})$ are isomorphic, then $\mathfrak{A}$ and $\mathfrak{B}$ are isomorphic.*

PROOF. Let ψ be an isomorphism between $\mathfrak{A}/F(\mathfrak{A})$ and $\mathfrak{B}/F(\mathfrak{B})$. We set

$$S = \{\langle a, b\rangle \mid a \in A\ \&\ b \in B\ \&\ \big(a/F(\mathfrak{A}) = \mathbf{0} \Rightarrow \mathrm{At}(a) = \mathrm{At}(b)\big)$$
$$\&\ \big(a/F(\mathfrak{A}) = \mathbf{1} \Rightarrow \mathrm{At}(\mathbf{1}\backslash a) = \mathrm{At}(\mathbf{1}\backslash b)\big)\ \&\ \psi\big(a/F(\mathfrak{A})\big) = b/F(\mathfrak{B})\},$$

where $\mathrm{At}(a)$ denotes the cardinality of the set of atoms located under a. As in Proposition 1.5.1, we check only Ax4. If $a/F(\mathfrak{A}) \neq \mathbf{0}$ and $x\backslash a/F(\mathfrak{A}) \neq \mathbf{0}$, then for any b such that $\varphi\big(a/F(\mathfrak{A})\big) = b/F(\mathfrak{B})$ all the conditions are obviously fulfilled. If $a/F(\mathfrak{A}) = \mathbf{0}$ or $x\backslash a/F(\mathfrak{A}) = \mathbf{0}$, then, choosing the required number of different atoms located under y, we obtain the desired b as the union of these atoms or as the difference of y and the union of these atoms. □

From the proof of Proposition 1.5.2 we immediately obtain the following claim.

Corollary 1.5.4. *If $\mathfrak{A}$ and $\mathfrak{B}$ are countable atomic Boolean algebras and $\psi: \mathfrak{A}/F(\mathfrak{A}) \to \mathfrak{B}/F(\mathfrak{B})$ is an isomorphism, then there exists an isomorphism $\varphi: \mathfrak{A} \overset{\mathrm{onto}}{\to} \mathfrak{B}$ such that $\varphi\big(x/F(\mathfrak{A})\big) = \psi(x)/F(\mathfrak{B})$.*

1.5. The Vaught Criterion

The following Remmel proposition yields an analogous result for arbitrary algebras.

Proposition 1.5.3. *Let $\mathfrak{A}$ be a subalgebra of a countable Boolean algebra $\mathfrak{B}$. If for any atom $a \in \mathfrak{A}$ there exists a finite set A_a of atoms of $\mathfrak{B}$ such that $a = \vee A_a$, the set of atoms of $\mathfrak{A}$ is infinite, and $\mathfrak{B}$ is generated by the union of the set A and the set of atoms of $\mathfrak{B}$, then $\mathfrak{A}$ and $\mathfrak{B}$ are isomorphic.*

PROOF. We construct a condition of isomorphism by setting

$$S \rightleftharpoons \{(a,b) \mid (a = b \text{ and } a \text{ is not equal to a union of a finite number of atoms) or } (a \text{ and } b \text{ are equal to unions of the same finite number of different atoms})\}$$

As above, it is easy to verify that S is a condition of isomorphism. By the Vaught theorem on isomorphism (cf. Corollary 1.5.3), $\mathfrak{A}$ and $\mathfrak{B}$ are isomorphic. □

Proposition 1.5.4. *If φ is a homomorphism of a Boolean algebra $\mathfrak{B}$ onto a countable Boolean algebra $\mathfrak{A}$, then there exists a subalgebra $\mathfrak{A}_0$ of $\mathfrak{B}$ such that $\varphi \restriction \mathfrak{A}_0$ is an isomorphism between $\mathfrak{A}_0$ and $\mathfrak{A}$.*

PROOF. The set $S = \{(a,b) \mid \varphi(b) = a \,\&\, (a = \mathbf{0} \Leftrightarrow b = \mathbf{0})\}$ is closed under taking disjunction. It is a condition of embedding of $\mathfrak{A}$ into $\mathfrak{B}$. Therefore, there exists an isomorphic embedding ψ of $\mathfrak{A}$ into $\mathfrak{B}$ such that $\psi \subseteq S$. Setting $\mathfrak{A}_0 = \psi(\mathfrak{A})$, we obtain the required subalgebra. □

Proposition 1.5.5. *If $\mathfrak{A}$ and $\mathfrak{B}$ are countable Boolean algebras and $\mathfrak{A}$ is atomless, then there exists a homomorphism φ of $\mathfrak{A}$ onto $\mathfrak{E}$.*

PROOF. It is easy to see that $S \rightleftharpoons \{(a,b) \mid a \in A \,\&\, b \in B \,\&\, (a = \mathbf{0} \Rightarrow b = \mathbf{0})\}$ is a condition of epimorphism of $\mathfrak{A}$ onto $\mathfrak{B}$. □

Corollary 1.5.5. *Every countable Boolean algebra is isomorphically embedded into a countable atomless Boolean algebra.*

We apply the results obtained to characterize countable superatomic Boolean algebras. A Boolean algebra is called *superatomic* if each of its subalgebras is atomic.

Proposition 1.5.6. *If $\mathfrak{A}$ is an atomless subalgebra of a countable Boolean algebra $\mathfrak{B}$, then there exists a homomorphism φ of $\mathfrak{B}$ onto $\mathfrak{A}$.*

PROOF. Since the Boolean algebra $\mathfrak{A}$ is atomless, the set
$$S \leftrightharpoons \{(a,b) \mid a \in B \,\&\, b \in A \,\&\, b \neq 0 \Rightarrow (\exists x)(x \in A \,\&\, x \neq 0 \,\&\, x \leqslant a)\}$$
is a condition of epimorphism and there exists an epimorphism φ of $\mathfrak{B}$ onto $\mathfrak{A}$. □

Corollary 1.5.6. *If in a countable Boolean algebra $\mathfrak{A}$ there is a countable atomless subalgebra, then there exists an ideal I such that $\mathfrak{A}/I$ is a countable atomless quotient algebra.*

Remark 1.5.3. If a Boolean algebra has an atomless subalgebra, then it has a countable atomless subalgebra.

Remark 1.5.4. If a Boolean algebra has a nonatomic subalgebra, then it has a countable nonatomic subalgebra.

We consider an element a of a Boolean algebra $\mathfrak{A}$. On the set $\hat{a} \leftrightharpoons \{x \mid x \leqslant a\}$, we introduce the operations $\vee_a$ and $\wedge_a$ as the restrictions on $\hat{a}$ of the corresponding operations defined on $\mathfrak{A}$. We also introduce the operation of complement C_a on $\hat{a}$ by setting $C_a(x) \leftrightharpoons C(x) \wedge a$. Let $\mathbf{0}_a \leftrightharpoons \mathbf{0}$ and $\mathbf{1} \leftrightharpoons a$. As a result, we obtain a Boolean algebra $\langle \hat{a}, \vee_a, \wedge_a, C_a \rangle$ called the *restriction* of the Boolean algebra $\mathfrak{A}$ onto a. It is denoted by $\hat{a}$.

Proposition 1.5.7. *For any Boolean algebra $\mathfrak{A}$ the following conditions are equivalent:*

(a) *$\mathfrak{A}$ is a superatomic Boolean algebra,*

(b) *$\mathfrak{A}$ has no atomless subalgebras,*

(c) *any Boolean quotient algebra of $\mathfrak{A}$ is atomic,*

(d) *there is no atomless quotient algebra of $\mathfrak{A}$.*

(e) *there is an ordinal α such that $F_\alpha(\mathfrak{A}) = A$.*

PROOF. We show (a)⇒(b)⇒(c)⇒(d) ⇒(e)⇒(a). The implications (a)⇒(b), (c)⇒(d), and (d)⇒(e) follow immediately from definitions. Hence it requires to prove (b)⇒ (c) and (e)⇒(a).

(b)⇒(c) We assume that $\mathfrak{A}$ satisfies (b), but I is an ideal of the Boolean algebra $\mathfrak{A}$ such that the algebra $\mathfrak{A}/I$ is not atomic. There exists a such that $a/I \neq \mathbf{0}$ and there is no atom located under a/I in $\mathfrak{A}/I$. We consider the principal ideal J of $\mathfrak{A}/I$ that is defined by $C(a)/I$. We have $(\mathfrak{A}/I)/J \cong \hat{a}/I \cap \hat{a}$. Taking the composition I_0 of the ideals I and

1.5. The Vaught Criterion

J, we conclude that $\mathfrak{A}/I_0$ is an atomless Boolean algebra. We consider a countable subalgebra $\mathfrak{A}_0$ of the algebra $\mathfrak{A}$ such that $\mathfrak{A}_0/I$ is a countable atomless subalgebra of the quotient algebra $\mathfrak{A}/I$. We also consider the natural homomorphism φ of $\mathfrak{A}_0$ onto $\mathfrak{A}_0/I_0$. By Proposition 1.5.4, there exists a subalgebra $\mathfrak{B} \leqslant \mathfrak{A}_0$ such that $\varphi \upharpoonright \mathfrak{B}$ is an isomorphism between $\mathfrak{B}$ and $\mathfrak{A}_0/I_0$. Consequently, $\mathfrak{B}$ is an atomless subalgebra of $\mathfrak{A}$, which contradicts our assumption.

(e)$\Rightarrow$(a) We assume that $F_\alpha(\mathfrak{A}) = \mathfrak{A}$ for some α but there is a nonatomic subalgebra $\mathfrak{B} \subseteq \mathfrak{A}$. Let b be an atomless element of $\mathfrak{B}$. In view of $F_\alpha(\mathfrak{A}) = \mathfrak{A}$, from Exercise 4 of Sec. 1.3 it follows that $F_\alpha(\mathfrak{B}) = \mathfrak{B}$. From Exercise 5 of Sec. 1.4 it follows that $b \notin F_\alpha(\mathfrak{B}) = \mathfrak{B}$. We arrive at a contradiction. □

Proposition 1.5.8. *Let $\mathfrak{A}$ be a Boolean algebra and let $a \in \mathfrak{A}$. Then $\mathfrak{A}$ is isomorphic to the direct product of the Boolean algebras $\widehat{a}$ and $\widehat{C(a)}$.*

PROOF. Define a mapping φ from $\widehat{a} \times \widehat{C(a)}$ into $\mathfrak{A}$ as follows: $\varphi((x, y)) = x \vee y$ for any $x \leqslant a$ and $y \leqslant C(a)$. It is easy to check that φ is an isomorphism. □

Let $\mathfrak{B}$ be a superatomic Boolean algebra. For an arbitrary ordinal α we consider an α-ideal $F_\alpha(\mathfrak{B})$ from the sequence of iterated Frechét ideals. Since the Boolean algebra $\mathfrak{B}$ is superatomic, for any ordinal α the quotient algebra $\mathfrak{B}/F_\alpha(\mathfrak{B})$ is atomic or trivial. In view of cardinality reasoning and Lemma 1.3.5, we can assert that there exists an ordinal γ such that $F_\gamma(\mathfrak{B}) = B$. However, if α is equal to the ordinal type of $\mathfrak{B}$, then the same equality $F_\alpha(\mathfrak{B}) = B$ is fulfilled. In this case, the type type$(\mathfrak{B}) = (\beta, m)$ of $\mathfrak{B}$, being a superatomic Boolean algebra, and the types type(a) of its elements $a \in B$ are defined.

If type$(\mathfrak{B}) = (\beta, m)$ is defined, then for any $a, b \in \mathfrak{B}$ the following conditions hold:

(1) type$(\mathbf{1}) =$ type$(\mathfrak{B})$,

(2) if $a \leqslant b$, type$(a) = (\delta, k)$, and type$(b) = (\gamma, n)$, then $\delta < \gamma$ or $\delta = \gamma$, but $k \leqslant n$,

(3) if $a = b \vee c$ and $b \wedge c = \mathbf{0}$, then
$$\text{type}_1(a) = \max\{\text{type}_1(b), \text{type}_1(c)\}$$

$$\text{type}_1(a) = \max\{\text{type}_1(b), \text{type}_1(c)\}$$
$$\text{type}_2(a) = \text{type}_2(b) \text{ if } \text{type}_1(c) < \text{type}_1(b)$$
$$\text{type}_2(a) = \text{type}_2(b) + \text{type}_2(c) \text{ if } \text{type}_1(b) = \text{type}_1(c)$$

(4) if $\text{type}(a) = (\alpha, n)$, $\gamma < \alpha$, $0 < m \in \mathbb{N}$, then there exists $b < a$ such that $\text{type}(b) = (\gamma, m)$,

(5) if $\text{type}(a) = (\alpha, n)$ and $0 < k < n$, then there exists $b < a$ such that $\text{type}(b) = (\alpha, k)$.

Proposition 1.5.9. *If the types of countable Boolean algebras $\mathfrak{A}$ and $\mathfrak{B}$ are defined, then $\mathfrak{A}$ and $\mathfrak{B}$ are isomorphic if and only if their types coincide*: $\text{type}(\mathfrak{A}) = \text{type}(\mathfrak{B})$.

PROOF. The necessity is obvious. Let us prove the sufficiency. By superatomicity, the types of Boolean algebras $\mathfrak{A}$ and $\mathfrak{B}$, as well as the types of all their elements, are defined. We introduce a set $S \subseteq A \times B$ as follows:

$$S \rightleftharpoons \{(a,b) \mid a \in A, b \in B, \text{type}(a) = \text{type}(b), \text{type}(C(a)) = \text{type}(C(b))\}$$

In view of the properties of the types of elements, S is a condition of isomorphism. Consequently, $\mathfrak{A}$ and $\mathfrak{B}$ are isomorphic. □

Exercises

1. Prove that two countable atomic Boolean algebras are isomorphic if the following condition holds: for any element a the set $\text{at}(a)$ or the set $\text{at}(C(a))$ is finite.

2. Prove that any two countable Boolean algebras are isomorphic if they have the following properties:

   — for any element a the set $\text{at}(a)$ or the set $\text{at}(C(a))$ is finite,
   — there exists no atomless element a such that $C(a)$ is atomic (x is called *atomic* if there is no atomless element that is located under it).

3. Prove that two countable α-atomic Boolean algebras $\mathfrak{A}$ and $\mathfrak{B}$ are isomorphic if the quotient algebras $\mathfrak{A}/F_\alpha(\mathfrak{A})$ and $\mathfrak{B}/F_\alpha(\mathfrak{B})$ are isomorphic and each of them contains more than one element, where $\{F_\gamma(\mathfrak{A})\}$ denotes the sequence of Frechét ideals.

1.6. Linearly Ordered Generating Sets

4. Prove that two countable atomic Boolean algebras $\mathfrak{A}$ and $\mathfrak{B}$ are isomorphic if they have the following property: for any element a such that the set $\operatorname{at}(a)$ is infinite, there exists an element b such that the sets $\operatorname{at}(b)$ and $\operatorname{at}(a\backslash b)$ are infinite. Construct an example of a Boolean algebra with such a property.

5. Prove that for an element a of a Boolean algebra $\mathfrak{A}$ the system $\hat{a}$ is a Boolean algebra; moreover, the order induced by this system on the universe of $\hat{a}$ coincides with the restriction onto $\hat{a}$ of the natural order defined on $\mathfrak{A}$.

6. Prove that the Cartesian product $\mathfrak{A} \times \mathfrak{B}$ of Boolean algebras $\mathfrak{A}$ and $\mathfrak{B}$ is a Boolean algebra.

7. Let $\mathfrak{A}$ be a countable atomless Boolean algebra and let $a \neq \mathbf{0}$ be an element of $\mathfrak{A}$. Prove that $\mathfrak{A}$ and $\hat{a}$ are isomorphic.

8. Prove that for any subalgebra $\mathfrak{A}$ of a countable Boolean algebra $\mathfrak{B}$ there exists an epimorphism of $\mathfrak{B}$ onto $\mathfrak{A}$.

    HINT: Consider the cases in which the algebra $\mathfrak{B}$ is superatomic and nonsuperatomic.

9. Prove that a Boolean algebra possessing an atomless subalgebra has a countable atomless subalgebra.

10. Prove that an algebra possessing a nonatomic subalgebra has a countable nonatomic subalgebra.

1.6. Linearly Ordered Generating Sets

In order to construct Boolean algebras with prescribed properties and characterize them, it is useful to know about generating sets whose properties are closely connected with those of Boolean algebras. Linearly ordered generating sets form an important class of generating sets.

Let $\mathfrak{B}$ be a Boolean algebra. A subset L of B is called *linearly ordered in* $\mathfrak{B}$ if $a \leqslant b$ or $b \leqslant a$ for any $a, b \in L$. If L is a linearly ordered set and generates $\mathfrak{B}$, then L is called a *linearly ordered generating set*.

REMARK 1.6.1. If L is a linearly ordered set generating a Boolean algebra $\mathfrak{B}$, then $a \vee b \in L$ and $a \wedge b \in L$ for any $a, b \in L$.

A linearly ordered set L generating a Boolean algebra $\mathfrak{B}$ is called a *linear basis* of $\mathfrak{B}$ if $\mathbf{1} \notin L$, but $\mathbf{0} \in L$.

Theorem 1.6.1. *For any countable Boolean algebra $\mathfrak{B}$ there exists a linear basis.*

PROOF. Let $B = \{b_0, b_1, b_2, \ldots\}$ be an enumeration of all elements of B. We construct a linear basis L step-by-step. At the step n, we construct a linearly ordered finite set L_n such that $b_n \in \operatorname{gr}(L_{n+1})$.

STEP 0. We set $L_0 \rightleftharpoons \{\mathbf{0}, \mathbf{1}\}$.

STEP $n+1$. We assume that the set $L_n = \{a_0 < a_1 < a_2 < \ldots < a_k < a_{k+1}\}$ has been constructed and $a_0 = \mathbf{0}$, $a_{k+1} = \mathbf{1}$. Setting $c_{i+1} \rightleftharpoons a_i \vee (a_{i+1} \wedge b_n)$, $0 \leqslant i \leqslant k$, we obtain elements of $\mathfrak{B}$ which are ordered as follows: $a_0 \leqslant c_1 \leqslant a_1 \leqslant c_2 \leqslant a_2 \leqslant \ldots \leqslant a_k \leqslant c_{k+1} \leqslant a_{k+1}$. We include them into L_{n+1}, i.e., we put $L_{n+1} = L_n \cup \{c_i \mid 1 \leqslant i \leqslant k+1\}$. Then the set L_{n+1} is linearly ordered and $b_n = \bigvee_{i=0}^{k} (c_{i+1} \setminus a_i)$.

Setting $L = \bigcup_{n \geqslant 0} L_n$, we obtain a linearly ordered generating set, and setting $L^* = L \setminus \{\mathbf{1}\}$, we obtain a linearly ordered basis. □

We study the converse problem: starting from an arbitrary linearly ordered set $\boldsymbol{L} = \langle L; \leqslant \rangle$, construct a Boolean algebra $\mathfrak{B}_{\boldsymbol{L}}$ with a linear basis of the form $\langle L; \leqslant \rangle$ provided that L has the least element. The desired Boolean algebra is a subalgebra of the Boolean algebra $\langle \mathcal{P}(L); \cup, \cap, C \rangle$ of all subsets of the set L.

For $a, b \in L$ we introduce the subsets

$$[a, b[\rightleftharpoons \{c \mid c \in L \,\&\, a \leqslant c < b\}, \quad [-\infty, b[\rightleftharpoons \{c \mid c \in L \,\&\, c < b\}$$
$$[a, \infty[\rightleftharpoons \{c \mid c \in L \,\&\, a \leqslant c\}, \quad [-\infty, \infty[\rightleftharpoons L$$

which will be called *half-open intervals* in $\mathcal{P}(L)$. The subalgebra of the Boolean algebra $\langle \mathcal{P}(L); \cup, \cap, C \rangle$ generated by half-open intervals is the desired algebra $\mathfrak{B}_{\boldsymbol{L}}$.

Proposition 1.6.1. *The linearly ordered set $\widehat{L} \rightleftharpoons \langle \{[-\infty, a[\mid a \in L\}, \subseteq \rangle$ is ordered in the same way as L, generates a Boolean algebra $\mathfrak{B}_{\boldsymbol{L}}$, and forms a linearly ordered basis of $\mathfrak{B}_{\boldsymbol{L}}$ provided that L has the least element.*

PROOF. It suffices to show that any half-open interval is generated by elements of $\widehat{L}$. It is not hard to check the equalities $[a, b[= C([-\infty, a[) \cap [-\infty, b[$ and $[a, +\infty[= C([-\infty, a[)$. □

1.6. Linearly Ordered Generating Sets

Proposition 1.6.2. *If a linearly ordered set L generates a Boolean algebra $\mathfrak{B}$, then any element of B can be represented as $\bigvee_{i=0}^{n}(a_i\setminus b_i)$, where $a_i, b_i \in L \cup \{0, 1\}$.*

PROOF. The set of elements of the above-mentioned form is closed under the operations of union and complement. Indeed, using set-theoretic relations and the Stone theorem, we arrive at the following relations:

$$\left(\bigvee_{i=0}^{n} a_i\setminus b_i\right)\bigwedge\left(\bigvee_{i=0}^{m} a'_i\setminus b'_i\right) = \bigvee_{i=0}^{n}\bigvee_{j=0}^{m}(a_i \wedge a'_j) \wedge C(b_i \vee b'_j)$$

$$\bigvee_{i=0}^{n} a_i\setminus b_i \vee \bigvee_{i=n+1}^{m} a_i\setminus b_i = \bigvee_{i=0}^{m} a_i\setminus b_i$$

The fact that the above set is closed under complement is proved by induction. The basis of induction is obvious. Assume that the assertion holds for $C(\bigvee_{i=0}^{n} a_i\setminus b_i)$. Then

$$C\left(\bigvee_{i=0}^{n+1}(a_i\setminus b_i)\right) = C\left(\bigvee_{i=0}^{n} a_i\setminus b_i\right)\bigwedge C(a_{n+1}\setminus b_{n+1})$$

$$= \left(\bigvee_{i=0}^{m} a'_i\setminus b'_i\right)\bigwedge(C(a_{n+1}) \vee b_{n+1})$$

$$= \bigvee_{i=0}^{m} a'_i\setminus(b'_i \vee a_{n+1})\bigvee \bigvee_{i=0}^{m}(a'_i \vee b_{n+1})\setminus b'_i$$

The proposition is proved. □

Proposition 1.6.3. *If $\mathbf{L} = \langle L; \leqslant \rangle$ is a linearly ordered set with the least element, then there exists a one-to-one correspondence between ultrafilters of the Boolean algebra $\mathfrak{B}_L$ and nonempty initial segments of the linearly ordered set $\mathbf{L}$.*

PROOF. Let $\emptyset \neq X \subseteq L$ be an initial segment of $\mathbf{L}$, i.e., for any $x, y \in L$ from $x \leqslant y$ and $y \in X$ it follows that $x \in X$. We set

$$\mathcal{D}_X \rightleftharpoons \{a \in \mathfrak{B}_L \mid \text{ for any } y \in X \text{ there exists } x \in a \cap X \text{ and } y \leqslant x\}$$

We prove that $\mathcal{D}_X$ is a filter of the Boolean algebra $\mathfrak{B}_L$. It suffices to verify that for any $a, b \in \mathcal{D}_X$ the element $a \wedge b$ belongs to $\mathcal{D}_X$ because the rest of the properties are obvious. Let

$$a = \bigcup_{i=0}^{n} [x_{2i}, x_{2i+1}[, \quad b = \bigcup_{i=1}^{m} [y_{2i}, y_{2i+1}[$$

$$x_0 < x_1 < \ldots < x_{2n+1}, \quad y_0 < y_1 < y_2 < \ldots < y_{2m+1}$$

and let $y \in X$. If there is no $y' \in X \cap a \wedge b$ such that $y \leqslant y'$, then for any $y' \in X \cap a$ and $y \leqslant y'$ there is $y'' \in X \cap b$, $y' < y''$, and for any $y' \in X \cap b$ there is $y'' \in X \cap a$ such that $y' < y''$ and $y'' \in X \cap a$. Hence we can construct an infinite increasing sequence of elements $z_0 < z_1 < \ldots < z_n \ldots$ such that $z_0 = y$, $z_{2i+1} \in a \cap X$, and $z_{2i+2} \in b \cap X$, which contradicts the fact that a and b consist of a finite number of intervals. Since any element $a \in \mathfrak{B}_L$ is a union of half-open intervals of the form $[x, y[$, $x, y \in L \cup \{\infty\}$, we have $a \in \mathcal{D}_X$ or $C(a) \in \mathcal{D}_X$. Consequently, $\mathcal{D}_X$ is an ultrafilter of $\mathfrak{B}_L$. It is obvious that for different initial segments we obtain different ultrafilters. To complete the proof it suffices to show that any ultrafilter $\mathcal{D}$ can be represented as $\mathcal{D}_X$. Let $\mathcal{D}$ be an arbitrary ultrafilter. We set

$$X \rightleftharpoons \{l \in L \mid \text{ for any } a \in \mathcal{D} \text{ there exists } l' \in a \text{ such that } l \leqslant l'\}$$

Since the least element l_0 belongs to X, we have $X \neq \emptyset$. By definition, X is an initial segment and $\mathcal{D} \subseteq \mathcal{D}_X$. Since an ultrafilter is a maximal filter, we conclude that $\mathcal{D} = \mathcal{D}_X$. □

Proposition 1.6.4. *If L is a linearly ordered set generating a Boolean algebra $\mathfrak{B}$ and for every element (which is not the lagrest element) of L there is successive element (i.e., for any $x, w \in L$, $x < w$, there exists $y \in L$ such that $x < y \& \neg(\exists z)(z \in L \& x < z < y))$, then $\mathfrak{B}_L$ is an atomic Boolean algebra.*

PROOF. We consider an element $a \neq \mathbf{0}$ of the Boolean algebra $\mathfrak{B}$. By Proposition 1.6.2, $a = \bigvee_{i=0}^{n} (a_i \backslash b_i)$, where $a_i, b_i \in L \cup \{\mathbf{0}, \mathbf{1}\}$. Since $a \neq \mathbf{0}$, there exists $i \leqslant n$ such that $a_i \backslash b_i \neq \mathbf{0}$. This is possible only if $b_i < a_i$ because $\langle L \cup \{\mathbf{0}, \mathbf{1}\}; \leqslant \rangle$ is a linearly ordered set. For $b_i \in L$ we consider an element $c \in L$ such that $b_i < c \& \neg(\exists z)(z \in L \& b_i < z < c)$. This means that $c \backslash b_i \leqslant a_i \backslash b_i$ and under $c \backslash b_i$ there is only the null element, i.e., $c \backslash b_i$ is an atom located under a. In the case $b_i \notin L$ and $b_i = \mathbf{0}$, we consider an element $d \in L$ such that $b_i < d < a_i$. Taking the element of L following after d, we again obtain an atom $c \backslash d$. If such an element is absent, then the element $a_i \backslash b_i$ is an atom. □

1.6. Linearly Ordered Generating Sets

Proposition 1.6.5. *If L is a linearly ordered basis of a Boolean algebra $\mathfrak{A}$ and L is a dense linear order without right end, then $\mathfrak{A}$ is atomless.*

PROOF. If $a \neq \mathbf{0}$ is an element of a Boolean algebra $\mathfrak{A}$, then there exist elements $c, d \in L$ such that $c < d$ and $d \backslash c \leqslant a$. By density, there exists an element e such that $c < e < d$. Therefore, $e \backslash c < d \backslash c \leqslant a$ and $e \backslash c \neq 0$. Consequently, a is not an atom. □

We describe countable superatomic Boolean algebras in terms of well-ordered sets.

Lemma 1.6.1. *If $A = \langle A, \leqslant \rangle$ is a well-ordered set, then a Boolean algebra $\mathfrak{B}_A$ is superatomic.*

PROOF. If the Boolean algebra $\mathfrak{B}_A$ is nonsuperatomic, then, by Proposition 1.5.7, there exists its atomless subalgebra. For an arbitrary element a of $\mathfrak{B}_A$ we define the element end $(a) \in A \cup \{\infty\}$ such that end (a) is equal to the least element among elements greater than all elements of the set $\{x \mid x \in a\}$. Since for $a \neq 0$ we have $a = \bigcup_{i=1}^{n} [x_i, y_i[$, where $x_i, y_i \in A \cup \{\infty\}$ and $x_i < y_i$, we find that end $(a) = \max\{y_i \mid 1 \leqslant i \leqslant n\}$. Let us define a sequence of elements a_i, $i \in \mathbb{N}$, of a subalgebra of the algebra $\mathfrak{B}$ such that end $(a_{i+1}) <$ end (a_i), $i \in \mathbb{N}$. Let $a_0 = \mathbf{1}$. Then end $(a_0) = \infty$. If $a_n < a_{n-1} < \ldots < a_0$ of $\mathfrak{B}$ are already defined and end $(a_n) <$ end $(a_{n-1}) < \ldots <$ end (a_0), then a_n is not an atom because $\mathfrak{B}$ is atomless. Consequently, there exists an element b of a subalgebra of the algebra $\mathfrak{B}$ such that $0 < b < a_n$. We consider a pair of elements b and $a_n \backslash b$. Since $b \subseteq a_n$ and $a_n \backslash b \subseteq a_n$, we have end $(b) \leqslant$ end (a_n) and end $(a_n \backslash b) \leqslant$ end (a_n). Both elements are unions of intervals. Therefore, for some $x \in \{b, a_n \backslash b\}$ we have end $(x) <$ end (a_n). Let a_{n+1} be equal to this element. Then we obtain the strictly decreasing sequence of elements end $(a_1) >$ end $(a_2) > \ldots >$ end $(a_n) > \ldots$ of A, which contradicts the fact that the set A is well-ordered. □

Proposition 1.6.6. *For any ordinal β and number $m > 0$, the type of the Boolean algebra $\mathfrak{B}_{\omega^\beta \times m}$ is (β, m).*

PROOF. By Lemma 1.6.1, the Boolean algebra $\mathfrak{B}_{\omega^\beta \times m}$ is superatomic and its type is defined. We consider the sequence of iterated Frechét ideals $F_\gamma(\mathfrak{B}_{\omega^\beta \times m})$ of this algebra. Each element a of the Boolean algebra $\mathfrak{B}_{\omega^\beta \times m}$ is a subset of the ordinal $\omega^\beta \times m$. Therefore, it is a well-ordered set with respect to the induced order and, thereby, defines some ordinal ord $(a) \leqslant \omega^\beta \times m$. By induction on ordinals γ, it is easy to show that $F_\gamma(\mathfrak{B}_{\omega^\beta \times m}) = \{a \in \mathfrak{B}_{\omega^\beta \times m} \mid \text{ord}(a) < \omega^\gamma\}$. Hence $\mathfrak{B}_{\omega^\beta \times m}/F_\beta(\mathfrak{B}_{\omega^\beta \times m})$ has exactly m atoms. Thus, type $(\mathfrak{B}_{\omega^\beta \times m}) = (\beta, m)$, which is required. □

Corollary 1.6.1. *For a countable Boolean algebra $\mathfrak{B}$ the following conditions are equivalent:*

(a) *$\mathfrak{B}$ is a superatomic Boolean algebra,*

(b) *$\mathfrak{B}$ is isomorphic to a Boolean algebra of the form $\mathfrak{B}_{\omega^\beta \times m}$, where β is an ordinal and m is a nonzero number.*

Exercises

1. Prove that $\mathfrak{B}_\omega$ is an atomic Boolean algebra such that there is only a finite number of atoms located under any element or its complement (ω is ordered as natural numbers).

2. Prove that $\mathfrak{B}_\eta$ is an atomless Boolean algebra (η is ordered as rational numbers).

3. Prove that in $\mathfrak{B}_{\omega+\eta}$ there is only a finite number of atoms located under any element or its complement and there is no atomless element a such that the complement to a is atomic.

4. Prove that in the Boolean algebra $\mathfrak{B}_{\omega \times \eta}$ for any element a under which there are infinitely many atoms, there is an element $b \leqslant a$ such that the sets at (b) and at $(a \setminus b)$ are infinite ($\boldsymbol{L}_1 \times \boldsymbol{L}_2$ stands for the linear order on the product $L_1 \times L_2$ such that $(a,b) \leqslant (c,d) \Leftrightarrow b <_{L_2} d \vee (b = d \,\&\, a \leqslant_{L_1} c)$).

5. Prove that

    (a) $F_1(\mathfrak{B}_\omega) \leftrightharpoons \{\bigcup_{i=1}^{n}[a_i, b_i[\mid n, a_i, b_i \in \mathbb{N}\}$,

    (b) $F_1(\mathfrak{B}_{\omega^2}) \leftrightharpoons \{\bigcup_{i=1}^{n}[a_i, b_i[\mid b_i = (n_i, k_i) \,\&\, a_i = (m_i, k_i) \,\&\, n_i, k_i, m_i,$
    $n \in \mathbb{N}$ for all $1 \leqslant i \leqslant n\}$,
    where $\omega^2 \leftrightharpoons \langle \mathbb{N}^2; \leqslant \rangle$, $(a,b) \leqslant (a', b') \Leftrightarrow b < b' \vee (b = b' \,\&\, a \leqslant a')$.

6. Prove that $\mathfrak{B}_L$ is a countable Boolean algebra provided that $\boldsymbol{L}$ is a countable linear order.

1.7. Generating Trees

7. Prove that $\mathfrak{B}_{n+1}$ is a Boolean algebra with $n+1$ atoms provided that n is a linearly ordered set containing n elements.

8. For an ordinal α we define the degree β^α by setting $\beta^0 = 1$, $\beta^\gamma = \beta^\alpha \times \beta$ if $\gamma = \alpha + 1$ and $\beta^\gamma = \sup\{\beta^\alpha \mid \alpha < \gamma\}$ if γ is a limit ordinal. Prove that the sequence of Frechét ideals $F_\gamma(\mathfrak{B}_{\omega^\alpha})$ of the Boolean algebra $\mathfrak{B}_{\omega^\alpha}$, where $\gamma \leqslant \alpha$ and α is an ordinal, is equal to the set

$$\{[a_1, b_1[\cup \ldots \cup [a_n, b_n[\in \mathfrak{B}_{\omega^\alpha} \mid \text{the order type of the well-ordered set } \langle [a_i, b_i[, \leqslant \rangle, \text{ obtained by restriction of the order from } \omega^\alpha \text{ strictly less than } \omega^\gamma\}.$$

9. Prove that the Boolean algebra $\mathfrak{B}_{\omega^\alpha}/F_\alpha(\mathfrak{B}_{\omega^\alpha})$, where α is an ordinal, consists of exactly two elements.

10. Prove that $\mathfrak{B}_{\omega^\alpha}$ is a superatomic Boolean algebra.

11. Prove that the direct product of superatomic Boolean algebras is a superatomic Boolean algebra.

12. Prove that a countable Boolean algebra is superatomic if and only if it has a well-ordered linear basis.

13. Construct a countable Boolean algebra $\mathfrak{B}$ and its countable subalgebra $\mathfrak{A}$ so that there is no epimorphism φ from $\mathfrak{B}$ onto $\mathfrak{A}$ such that it is identical on elements of $\mathfrak{A}$.

1.7. Generating Trees

In this section, we show how to construct a countable Boolean algebra $\mathfrak{B}_D$ for any tree D (Fig. 1.7.1). It turns out that the construction presented below exhausts all countable Boolean algebras. On the set $\mathbb{N}$, we introduce the following functions and subsets [158]:

$$R(n) \rightleftharpoons 2n+2, \quad L(n) \rightleftharpoons 2n+1$$
$$H(n) = 0 \text{ if } n = 0, \quad H(n) = [(n-1)/2] \text{ if } n > 0,$$
$$\text{where } [x] \text{ denotes the integer part of } x$$

$$S(n) \leftrightharpoons \begin{cases} n-1, & n \text{ is even, } n > 0 \\ n+1, & n \text{ is odd, } n > 0 \\ 0, & n = 0, \end{cases}$$

$$h(0) = 0, \quad h(n+1) = h(H(n+1)) + 1$$
$$E_n \leftrightharpoons \{x \mid h(x) = n\}$$
$$H(x,0) = x, \quad H(x, n+1) = H(H(x,n))$$
$$x \preccurlyeq y \Leftrightarrow \prod_{n=0}^{h(x)} |H(x,n) - y| = 0$$

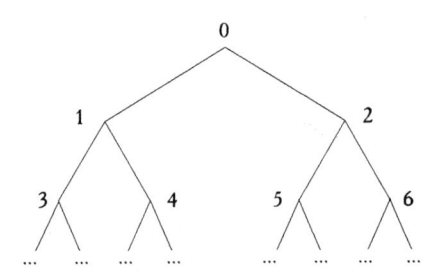

Fig. 1.7.1

It is easy to observe the correspondence between vertices whose numbers are defined by the functions $R(n)$, $L(n)$, $S(n)$, $H(n)$ and the vertex n with the number n:

$R(n)$ is the number of the right-hand vertex located below n,

$L(n)$ is the number of the left-hand vertex located below n,

$H(n)$ is the number of the nearest vertex located above n, $H(0) = 0$,

$S(n)$ is the number of the neighboring with respect to n vertex located under $H(n)$,

Further, $h(n)$ indicates the distance between n and 0, i.e., the number of preceding vertices of the branch on which n is situated.

By definition, the set E_n consists of all the vertices of the same level.

The notation $x \preccurlyeq y$ means that x and y belong to the same branch and x is located under y.

We recall that a subset $D \subseteq \mathbb{N}$ is called a *tree* if for any $n \in D$ the elements $H(n)$ and $S(n)$ belong to D.

REMARK 1.7.1. If D is a tree, $n \prec m$, and $n \in D$, then $m \in D$.

1.7. Generating Trees

By a *tree generating* a Boolean algebra $\mathfrak{A}$ we mean a pair $\langle D; \varphi \rangle$, where D is a tree and φ is an injection of D into A such that

(1) $\varphi(0) = \mathbf{1}$,

(2) $(\forall n \in D)(\varphi(n) \vee \varphi(S(n)) = \varphi(H(n)) \,\&\, \varphi(n) \wedge \varphi(S(n)) = \mathbf{0} \,\&\, \varphi(n) \neq \mathbf{0})$,

(3) $(\forall a \neq \mathbf{0})(\exists n_1, \ldots, n_k \in D)(\varphi(n_1) \vee \varphi(n_2) \vee \ldots \vee \varphi(n_k) = a)$.

REMARK 1.7.2. If $\langle D; \varphi \rangle$ is a generating tree, then $n \preccurlyeq m \Leftrightarrow \varphi(n) \leqslant \varphi(m)$. If n, m are incomparable with respect to $\preccurlyeq$, then $\varphi(n) \wedge \varphi(m) = \mathbf{0}$.

Let $\langle D, \varphi \rangle$ be a tree generating a Boolean algebra $\mathfrak{A}$.

- A subset $H \subseteq D$ is called a *canonical representation* of an element $a \in \mathfrak{A}$, if the following conditions hold:

  (1) $(\forall n \in H)(S(n) \notin H)$,
  (2) $(\forall n \in H)(\forall m \in H)(n \neq m \Rightarrow \neg\, (n \preccurlyeq m))$,
  (3) $a = \mathbf{0} \Leftrightarrow H = \varnothing$ and $a \neq \mathbf{0} \Rightarrow a = \bigvee_{n \in H} \varphi(n)$.

Lemma 1.7.1. *For any $a \in A$ there exists a unique canonical representation.*

PROOF. For $a = \mathbf{0}$ the assertion is obvious. For $a \neq \mathbf{0}$ we consider a finite subset $H \subseteq D$ such that $\bigvee_{n \in H} \varphi(n) = a$. We set

$$H^* \rightleftharpoons \{n \in H \mid \text{there is no } m \in H \text{ such that } n \preccurlyeq m \text{ and } n \neq m\}$$

It is obvious that $\bigvee_{n \in H} \varphi(n) = \bigvee_{n \in H^*} \varphi(n)$. Consider the subsets

$$H_0 = H^*$$
$$H_{n+1} = \{m \mid S(m) \notin H_n \,\&\, m \in H_n\} \cup \{H(m) \mid m, S(m) \in H_n\}$$

It is easy to see that $\bigvee_{m \in H_{n+1}} \varphi(m) = \bigvee_{m \in H_n} \varphi(m)$. In the case $H_n \neq H_{n+1}$, we have $|H_n| > |H_{n+1}|$. Therefore, beginning from some n_0, we do not change H_n. Consequently, H_{n_0} is a canonical representation of a.

To prove the uniqueness of a canonical representation it suffices to consider the case $a \neq \mathbf{0}$. Let H and G be two canonical representations of a: $a = \bigvee_{n \in H} \varphi(n) = \bigvee_{n \in G} \varphi(n)$. Then $\varphi(n) \leqslant \bigvee_{m \in G} \varphi(m)$ for any $n \in H$.

By Remark 1.7.2, for $\varphi(n)$ there exists $m \in G$ such that $\varphi(n) \leqslant \varphi(m)$ or $\varphi(m) \leqslant \varphi(n)$. We show that $\varphi(n) = \varphi(m)$. Assume that $\varphi(n) < \varphi(m)$. Then $n \prec m$. Consequently, $\varphi(S(n)) < \varphi(m)$, which means $\varphi(S(n)) \leqslant \varphi(m)$ and $\varphi(S(n)) \leqslant \bigvee_{i \in H} \varphi(i)$. Since H is a canonical representation, $S(n) \preccurlyeq k$ fails for any $k \in H$.

We consider a minimal element $k \in H$ (with respect to $\preccurlyeq$) such that $k \preccurlyeq S(n)$ or $k = n$ and there is no element of H under $S(k)$. Then $k \in H$, $S(k) \notin H$, and there is no element under $S(k)$. Such an element k exists because H is finite. By condition, under $S(k)$ and over $S(k)$ there are no elements of H. Therefore, $\varphi(S(k)) \wedge \varphi(m) = \mathbf{0}$ for any $m \in H$. However, $\varphi(S(k)) \leqslant \varphi(S(n)) \leqslant \bigvee_{m \in H} \varphi(m)$. Consequently,

$$\varphi(S(k)) \wedge \left(\bigvee_{m \in H} \varphi(m) \right) = \bigvee_{m \in H} (\varphi(S(k)) \wedge \varphi(m)) = \mathbf{0}$$

$$\varphi(S(k)) \leqslant \bigvee_{m \in H} \varphi(m)$$

Therefore, $\varphi(S(k)) = \varphi(S(k)) \wedge (\bigvee_{m \in H} \varphi(m))$. We obtain $\varphi(S(k)) = \mathbf{0}$, which contradicts the condition on $\langle D; \varphi \rangle$. The contradiction obtained shows that the case $\varphi(n) < \varphi(m)$ is impossible. The case $\varphi(m) < \varphi(n)$ can be handled in a similar way. Thus, $H = G$. □

A vertex n of a tree D is called an *end vertex* if $n \in D$, but $R(n) \notin D$ and $L(n) \notin D$.

Lemma 1.7.2. *If $\langle D; \varphi \rangle$ is a tree generating a Boolean algebra $\mathfrak{B}$ and $a \in B$, then a is an atom if and only if there exists an end vertex n of the tree D such that $\varphi(n) = a$.*

PROOF. Let a be an atom. Then $a \neq \mathbf{0}$ and there exist $n_1, \ldots, n_k \in D$ such that $a = \bigvee_{i=1}^{k} \varphi(n_i)$. By Lemma 1.7.1, we can assume that $\{n_1, \ldots, n_k\}$ is a canonical representation of a. Hence $\varphi(n_i) \wedge \varphi(n_j) = \mathbf{0}$, $i \neq j$. Since a is an atom, we have $k = 1$, $a = \varphi(n_1)$, and n_1 is an end vertex. Conversely, if $a = \varphi(n)$ and n is an end vertex, then $a \neq \mathbf{0}$. We assume that there is $b = \bigvee_{i=1}^{k} \varphi(n_i)$, $b < a$. Then $\bigvee_{i=1}^{k} \varphi(n_i) \leqslant \varphi(n)$. Therefore, $\varphi(n_1) \leqslant \varphi(n)$, which yields $n_1 \preccurlyeq n$. If $n_1 = n$, then $a = b$, which contradicts the hypothesis. If $n_1 \prec n$, then $n_1 \preccurlyeq R(n)$ or $n_1 \preccurlyeq L(n)$ and, consequently, $R(n) \in D$ or

1.7. Generating Trees

$L(n) \in D$, which also contradicts the hypothesis. Consequently, a is an atom. □

Theorem 1.7.1. *For any countable Boolean algebra there exists a generating tree.*

PROOF. Consider a countable Boolean algebra $\mathfrak{B}$. Let

$$B = \{b_0, b_1, \ldots, b_n, \ldots\}$$

We construct a generating tree step-by-step. At the step n, we construct a finite tree D_n, a subalgebra B_n of $\mathfrak{B}$, and a mapping $\varphi_n: D_n \to B_n$ such that $\langle D_n; \varphi_n \rangle$ is a tree generating B_n, and $b_n \in B_{n+1}$ for any n.

STEP 0. We set $D_0 = \{0\}$ and $\varphi_0(0) = \mathbf{1}$.

STEP $n+1$. Let D_n be the tree constructed at the step n, let $m_1, \ldots, m_k$ be all the end vertices of D_n, and let $\varphi_n: D_n \to B_n$ be the mapping constructed at the step n. By Lemma 1.7.2, $\{\varphi_n(m_i) \mid 1 \leqslant i \leqslant k\}$ is the set of all atoms of the Boolean algebra B_n. Let $a_i^0 \rightleftharpoons b_n \wedge \varphi_n(m_i)$, $a_i^1 \rightleftharpoons C(b_n) \wedge \varphi_n(m_i)$, $D_{n+1} \rightleftharpoons D_n \cup \{R(m_i), L(m_i) \mid a_i^0 \neq \mathbf{0}, a_i^1 \neq \mathbf{0}\}$, and $\varphi_{n+1}(R(m_i)) \rightleftharpoons a_i^0$, $\varphi_{n+1}(L(m_i)) \rightleftharpoons a_i^1$ if $a_i^0 \neq \mathbf{0} \,\&\, a_i^1 \neq \mathbf{0}$ and $\varphi_{n+1}(k) \rightleftharpoons \varphi_n(k)$ if $k \in D_n$. Let B_{n+1} be the subalgebra of $\mathfrak{B}$ generated by $\{\varphi_{n+1}(k) \mid k \in D_{n+1}\}$. It is easy to check that $B_n \subseteq B_{n+1}$, $\langle D_{n+1}; \varphi_{n+1} \rangle$ is a tree generating B_{n+1} and $b_n \in B_{n+1}$. □

Lemma 1.7.3. *If $\langle D; \varphi_1 \rangle$ and $\langle D; \varphi_2 \rangle$ are trees generating Boolean algebras $\mathfrak{A}$ and $\mathfrak{B}$ respectively, then there exists an isomorphism ψ of $\mathfrak{A}$ onto $\mathfrak{B}$ such that $\psi(\varphi_1(n)) = \varphi_2(n)$ for any $n \in D$.*

PROOF. We define ψ as follows: for $a \neq \mathbf{0}$ we set $\psi(a) = \bigvee_{n \in H_a} \varphi_2(n)$, where H_a is a canonical representation and $\psi(\mathbf{0}) = \mathbf{0}$. Since H_a is a canonical representation of a, the mapping ψ is an injection. For any $b \neq \mathbf{0}$ there exists a canonical representation H such that $b = \bigvee_{n \in H} \varphi_2(n)$. Setting $a = \bigvee_{n \in H} \varphi_1(n)$, we obtain $\psi(a) = b$ in view of the uniqueness of a canonical representation. Thus, ψ is a one-to-one mapping of A onto B. It remains to check that ψ preserves the basic operations. But this fact follows from the uniqueness of a canonical representation and the properties of a generating tree. □

Corollary 1.7.1. *If $\langle D_0; \varphi_0 \rangle$ and $\langle D_0; \varphi_1 \rangle$ are two trees generating a Boolean algebra $\mathfrak{A}$, then there exists an automorphism ψ of $\mathfrak{A}$ such that $\psi \varphi_0 = \varphi_1$.*

Thus, we may not distinguish, up to an automorphism of a Boolean algebra, the mappings φ from D into A for generating trees $\langle D; \varphi \rangle$. Therefore we often omit φ in the notation $\langle D; \varphi \rangle$ and by a generating tree we mean only the tree D in this pair.

We leave as an exercise the proof of the following lemma.

Lemma 1.7.4. *If $\langle D; \varphi \rangle$ is a tree generating a Boolean algebra $\mathfrak{A}$ and $a = \bigvee_{n \in H_a} \varphi(n)$, $b = \bigvee_{n \in H_b} \varphi(n)$, where H_a and H_b are canonical representations of a and b respectively, then*

(a) $C(a) = \bigvee_{n \in H} \varphi(n)$, *where* $H = \{n \mid (\exists m)(m \in H_a \,\&\, m \preccurlyeq H(n)) \,\&\, \neg(\exists m \in H_a)(m \preccurlyeq n)\}$ *is a canonical representations of $C(a)$,*

(b) $a \vee b = \bigvee_{n \in H_a \cup H_b} \varphi(n)$,

(c) $a \wedge b = \bigvee_{n \in H} \varphi(n)$, *where* $H = \{n \in H_a \mid \text{there exists } m \in H_b \text{ such that } n \preccurlyeq m\} \cup \{n \in H_b \mid \text{there exists } m \in H_a \text{ such that } n \preccurlyeq m\}$.

We now consider the converse problem: starting from a tree D, construct a Boolean algebra $\mathfrak{B}_D$ and a mapping φ_D such that $\langle D; \varphi_D \rangle$ is a tree generating $\mathfrak{B}_D$. With each element $n \in \mathbb{N}$ we associate an infinite subset $A_n \subseteq \mathbb{N}$ so that $A_0 = \mathbb{N}$, $A_{R(n)} \cup A_{L(n)} = A_n$, and $A_{R(n)} \cap A_{L(n)} = \varnothing$ for any $n \in \mathbb{N}$. It is easy to see that for any k

$$(\forall n \forall m \in E_k)(n \neq m \Rightarrow A_n \cap A_m = \varnothing), \quad \bigcup_{n \in E_k} A_n = \mathbb{N}$$

The set $B_{\mathbb{N}} \rightleftharpoons \{ \bigcup_{n \in K} A_n \mid K \text{ is a finite subset of } \mathbb{N}\}$ is closed under union, intersection, and complement in $\mathbb{N}$. Therefore, there is a subalgebra (denoted by $\mathfrak{B}_{\mathbb{N}}$) of the Boolean algebra $\langle \mathcal{P}(\mathbb{N}); \cup, \cap, C \rangle$ whose universe is $B_{\mathbb{N}}$. It is clear that $\langle \mathbb{N}; \varphi \rangle$, where $\varphi(n) \rightleftharpoons A_n$, is a tree generating the Boolean algebra $\mathfrak{B}_{\mathbb{N}}$.

For an arbitrary tree $D \subseteq \mathbb{N}$ we define $\mathfrak{B}_D$ as the subalgebra of the algebra $\mathfrak{B}_{\mathbb{N}}$ generated by the set $\{\varphi(n) \mid n \in D\}$.

Lemma 1.7.5. *If $\langle D; \varphi \rangle$ is a generating tree and $D_0 \subseteq D$ is a tree, then $\langle D_0; \varphi \upharpoonright D_0 \rangle$ is a tree generating the subalgebra that is generated by the set $\{\varphi(n) \mid n \in D_0\}$.*

PROOF. By Lemma 1.7.4, the set

$$B_0 \rightleftharpoons \{ \bigcup_{n \in K} \varphi(n) \mid K \text{ is a finite subset of } D_0 \}$$

1.7. Generating Trees

is closed under union, intersection, and complement. Therefore, the universe of the subalgebra generated by $\{\varphi(n) \mid n \in D_0\}$ coincides with B_0. Thus, $\langle D_0, \varphi \upharpoonright D_0 \rangle$ is the tree generating this subalgebra. □

For a subalgebra $\mathfrak{A}$ of an atomless Boolean algebra $\mathfrak{B}$ and the tree $\langle D; \varphi \rangle$ generating $\mathfrak{A}$, there is a tree $\langle D'; \psi \rangle$ generating $\mathfrak{B}$; moreover, $D' = \mathbb{N}$. Therefore, $D \subseteq D'$. However, the requirement $\psi \supseteq \varphi$ is not always satisfied. It suffices to take a proper atomless subalgebra $\mathfrak{A}$ of $\mathfrak{B}$. In this case, any tree $\langle D; \varphi \rangle$ generating $\mathfrak{A}$ assumes the form $\langle \mathbb{N}; \varphi \rangle$ and does not admit an extension.

By Lemma 1.7.5, for the subalgebra $\mathfrak{B}_D$ of the algebra $\mathfrak{B}_\mathbb{N}$ the pair $\langle D; \varphi \upharpoonright D \rangle$ is a generating tree. Thus, we have proved the following theorem.

Theorem 1.7.2. *For any tree D there exists a unique up to an isomorphism Boolean algebra $\mathfrak{B}$ and a mapping $\varphi \colon D \to B$ such that $\langle D; \varphi \rangle$ is a tree generating $\mathfrak{B}$.*

We now clarify relationships between the properties of a tree D and those of the Boolean algebra $\mathfrak{B}_D$.

Proposition 1.7.1. *For any tree D the following conditions hold:*

(a) *$\mathfrak{B}_D$ is an atomic Boolean algebra if and only if for any $n \in D$ there exists $m \preccurlyeq n$ such that m is an end vertex of D,*

(b) *$\mathfrak{B}_D$ is an atomless Boolean algebra if and only if $D = \mathbb{N}$.*

For any tree D we define a subtree $D^{(n)}$ that starts in $\mathbb{N}$ from an element n and preserves the structure of D. Let

$$\varphi^0(k) = k, \quad \varphi^{n+1}(0) = n+1$$
$$\varphi^{n+1}(m+1) = R\varphi^{n+1}H(m+1), \text{ if } RH(m+1) = m+1$$
$$\varphi^{n+1}(m+1) = L\varphi^{n+1}H(m+1), \text{ if } LH(m+1) = m+1$$
$$D^{(n)} \rightleftharpoons \{\varphi^n(k) \mid k \in D\}$$

We denote by $T(D)$ the set of end vertices of a tree D. For a set of trees $I = \{D_i \mid i \in T(D)\}$ we introduce the new tree $D^I = D \cup \bigcup_{i \in T(D)} D_i^{(i)}$

Proposition 1.7.2. *If for any $i \in T(D)$ the tree D_i is finite and there are infinitely many end vertices of D, then the Boolean algebras $\mathfrak{B}_D$ and $\mathfrak{B}_{D^I}$ are isomorphic.*

PROOF. It is clear that $\mathfrak{B}_D \subseteq \mathfrak{B}_{D^I}$. Let

$$S \rightleftharpoons \{(a,b) \mid a \in \mathfrak{B}_D \& b \in \mathfrak{B}_{D^I} \& \big((a/F(\mathfrak{B}_D) \neq 0 \& C(a)/F(\mathfrak{B}_D) \neq 0$$
$$\& a = b\big) \vee \big(a/F(\mathfrak{B}_D) = b/F(\mathfrak{B}_D) = 0$$
$$\& \operatorname{At}(a) = \operatorname{At}(b)\big) \vee \big(C(a)/F(\mathfrak{B}_D) = C(b)/F(\mathfrak{B}_D) = 0$$
$$\& \operatorname{At}(C(a)) = \operatorname{At}(C(b))\big)\big)\}$$

It is easy to check that S is a condition of isomorphism on $(\mathfrak{B}_D, \mathfrak{B}_{D^I})$. By the Vaught theorem (cf. Corollary 1.5.3), these Boolean algebras are isomorphic. □

By a (maximal) *branch* of a tree we mean any (maximal) linearly ordered subset of the tree. For the sake of simplicity, a maximal branch is referred to as a branch.

For an element of a tree D, instead of the natural number $n+1$ we can consider its binary form $1\varepsilon_0\ldots\varepsilon_m$. With the vertex $n+1$ we associate the sequence $\langle \varepsilon_0, \ldots, \varepsilon_m \rangle$ consisting of 0 and 1. Moreover, 0 can be regarded as the empty sequence. Then all the basic functions on the tree $\langle \mathbb{N}; \preccurlyeq \rangle$ have a simple description. Denote by α_n a sequence of 0 and 1 associated with a natural number n. If $\alpha_n = \langle \varepsilon_1, \ldots, \varepsilon_m \rangle$, then $n+1 = 1\varepsilon_1\ldots\varepsilon_m$ provided that $n \neq 0$, where instead of n its binary representation is taken. Then

$$L(1\varepsilon_1,\ldots,\varepsilon_m) = 1\varepsilon_1\ldots\varepsilon_m 0, \quad R(1\varepsilon_1,\ldots,\varepsilon_m) = 1\varepsilon_1\ldots\varepsilon_m 1$$

$$H(1\varepsilon_1,\ldots,\varepsilon_m) = \begin{cases} 1\varepsilon_1\varepsilon_2\ldots\varepsilon_{m-1}, & m \neq \varnothing \\ 0 & \text{otherwise} \end{cases}$$

$$h(1\varepsilon_1,\ldots,\varepsilon_m) = m, \quad 1\varepsilon_1\ldots\varepsilon_m \preccurlyeq 1\delta_1\ldots\delta_k \Leftrightarrow k \leqslant m, \forall i \leqslant k\; \varepsilon_i = \delta_i$$

From the tree $\langle \mathbb{N}; \preccurlyeq \rangle$ we induce the order on the set of sequences and keep the notation $\preccurlyeq$ for the induced order. With this notation, the tree can be represented as in Fig. 1.7.2.

Using the above notation, we can give a rather simple notion of a tree. A set $D \subseteq 2^{<\omega} \rightleftharpoons \bigcup_n \{0,1\}^n$ is a tree if for any element $\langle \varepsilon_1, \ldots, \varepsilon_{m+1} \rangle \in D$ the element $\langle \varepsilon_1, \ldots, \varepsilon_m \rangle$ and the elements $\langle \varepsilon_1, \ldots, \varepsilon_m, \mathbf{0} \rangle$, $\langle \varepsilon_1, \ldots, \varepsilon_m, \mathbf{1} \rangle$ belong to D. We will use lower-case Greek letters $\alpha, \beta, \gamma, \ldots$ to denote finite sequences consisting of 0 and 1. Let $\sigma * \tau$ stand for the concatenation of sequences. Then $\sigma \preccurlyeq \tau$ with respect to the order defined on a tree if and only if there exists ρ such that $\tau = \sigma * \rho$. We denote by $\mathbf{0}^k$ ($\mathbf{1}^k$) the sequence consisting of k zeros (units). Let lh designate the length of a

1.7. Generating Trees

sequence. It defines levels in the tree, i.e., all the elements of the length n have the level n in the tree.

Any sequence σ consisting of 0 and 1 can be regarded as some characteristic function on the set $\{i \mid i < \mathrm{lh}\,(\sigma)\}$. In this interpretation, branches determine functions from $\mathbb{N}$ into $\{0,1\}$.

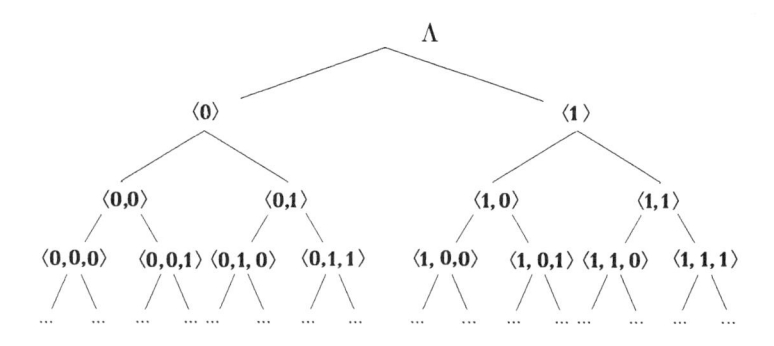

Fig. 1.7.2

The notion of a (2-branching) tree can also be introduced in terms of partially ordered sets. By a *tree* we mean a partial order which is a lower semilattice with least element such that each of its elements either has exactly two heirs or is maximal and every initial segment is a finite linearly ordered set. A tree is called *complete* if it has no maximal element. By a *segment* we mean any subset $P \subseteq 2^{<\omega}$ such that $\tau \in P$ and $\sigma \preccurlyeq \tau$ imply $\sigma \in P$. If a segment of a complete tree of $2^{<\omega}$ is a tree itself, then it is called a *subtree*. We note that any countable tree is isomorphic (with respect to a partial order) to a subtree of $2^{<\omega}$.

A mapping φ from a tree $T \subset 2^\omega$ into a Boolean algebra $\mathfrak{B}$ is called *admissible* if the following conditions hold:

(1) $\varphi(\varnothing) = \mathbf{1}$,

(2) $\varphi(\tau) = \varphi(\tau * \mathbf{1}) \vee \varphi(\tau * \mathbf{0})$, $\tau * \mathbf{0} \in T$,

(3) if τ are σ incomparable, then $\varphi(\tau) \wedge \varphi(\sigma) = \mathbf{0}$, $\tau, \sigma \in T$.

We say that a *tree T generates a Boolean algebra* $\mathfrak{B}$ if there exists an admissible mapping $\varphi \colon T \to \mathfrak{B}$ satisfying the following additional conditions:

(4) $\varphi(\tau) \neq \mathbf{0}$, $\tau \in T$,

(5) any element of $\mathfrak{B}$ can be represented as a finite union of elements of the form $\varphi(\tau)$.

Let $\mathfrak{A}$ and $\mathfrak{B}$ be Boolean algebras. Suppose that a tree $T \subset 2^{<\omega}$ generates $\mathfrak{B}$ and φ is the corresponding admissible mapping. Let $\psi : T \to \mathfrak{A}$ be an admissible mapping. It is easy to see that there exists a unique homomorphism ξ from $\mathfrak{B}$ into $\mathfrak{A}$ such that the diagram

$$\begin{array}{ccc} T & \xrightarrow{\varphi} & \mathfrak{B} \\ & \searrow_{\psi} \quad \swarrow_{\xi} & \\ & \mathfrak{A} & \end{array}$$

is commutative. If ψ satisfies (4), then ξ is a monomorphism. If ψ satisfies (5), then ξ is an epimorphism. Consequently, a Boolean algebra generated by a given tree is uniquely defined up to an isomorphism. From a tree T one can define the Boolean algebra satisfying the above conditions. We now explain how to construct it. To this end, we consider an example. Let $\mathcal{P}^{<\omega}(T)$ denote the set of all finite subsets of a tree $T \subset 2^{<\omega}$. We set

$$F \leqslant_{BA} G \Leftrightarrow \forall \rho \in T (\exists \sigma \in F (\sigma \preccurlyeq \rho \vee \rho \preccurlyeq \sigma) \Rightarrow \exists \tau \in G (\tau \preccurlyeq \rho \vee \rho \preccurlyeq \tau))$$

The relation $\leqslant_{BA}$ is a preorder. Let $\sim_{BA}$ be the corresponding equivalence relation. The induced partial order defines the structure of the Boolean algebras on $\mathcal{P}^{<\omega}(T)/{\sim_{BA}}$. Denote the corresponding Boolean algebra by $BA(T)$. It satisfies the above-mentioned conditions.

Using the tree method, we give one more characterization of countable superatomic Boolean algebras.

Theorem 1.7.3. *A countable Boolean algebra $\mathfrak{B}$ is superatomic if and only if the set of its ultrafilters $\mathcal{F}(\mathfrak{B})$ is at most countable.*

PROOF. To prove the sufficiency, consider a nonsuperatomic Boolean algebra $\mathfrak{B}$. By Proposition 1.5.7, there exists an isomorphic embedding ψ of the countable atomless Boolean algebra $\mathfrak{B}_{\mathbb{N}}$ into the Boolean algebra $\mathfrak{B}$. Starting from an arbitrary maximal branch L of the tree $\mathbb{N}$, we define a filter

$$F_L^0 = \{x \in \mathfrak{B} \mid \text{ there exists } l \in L \text{ such that } \psi(\varphi(l)) \leqslant x\}$$

where the tree $\langle \mathbb{N}; \varphi \rangle$ generates $\mathfrak{B}_{\mathbb{N}}$. By the Zorn lemma, any filter F_L^0 can be extended to an ultrafilter F_L. We note that for any two different branches L and L' there exist elements $n \in L$ and $m \in L'$ such that

1.7. Generating Trees

$n \ne m$ and $H(n) = H(m)$. In this case, $\varphi(n) \wedge \varphi(m) = \varnothing$ and $\psi(\varphi(n)) \wedge \psi(\varphi(m)) = \mathbf{0}$. On the other hand, $\psi(\varphi(n)) \in F_L^0$ and $\psi(\varphi(m)) \in F_{L'}^0$. Consequently, $F_{L'} \ne F_L$; otherwise, $\mathbf{0} = \psi(\varphi(m)) \wedge \psi(\varphi(n)) \in F_L = F_{L'}$, which is impossible in view of the definition of a filter. Thus, $\mathcal{F}(\mathfrak{B})$ is more than countable.

We now prove the necessity. Let a countable Boolean algebra $\mathfrak{B}$ be superatomic. By Corollary 1.6.1, there exists a countable ordinal β and a number $m > 0$ such that $\mathfrak{B} \cong \mathfrak{B}_{\omega^\beta \times m}$. By Proposition 1.6.3, the set of ultrafilters of the Boolean algebra $\mathfrak{B}_{\omega^\beta \times m}$ and the set of initial segments of the linearly ordered set $\omega^\beta \times m$ are of the same cardinality, but $\omega^\beta \times m$ is well-ordered and for any initial segment $X \subseteq \omega^\beta \times m$ there exists $\gamma \leqslant \omega^\beta \times m + 1$ such that $X = \{\delta \mid \delta < \gamma\}$. Consequently, the set of ultrafilters is countable. □

We wish to describe all ultrafilters of a countable superatomic Boolean algebra $\mathfrak{A}$. By Lemma 1.3.5, the ordinal type $o(\mathfrak{A})$ of $\mathfrak{A}$ is countable. By Lemma 1.3.4, $o(\mathfrak{A}) = \beta + 1$. But in this case, $\mathfrak{A}/F_\beta(\mathfrak{A})$ is a nontrivial finite Boolean algebra. For $\alpha \leqslant \beta$ and an atom $a/F_\alpha(\mathfrak{A}) \in \mathfrak{A}/F_\alpha(\mathfrak{A})$ we define the set $F_{\alpha,a} \rightleftharpoons \{b \mid a/F_\alpha(\mathfrak{A}) \leqslant b/F_\alpha(\mathfrak{A})\}$. It is obvious that $\mathbf{0} \notin F_{\alpha,a}$. If $x \leqslant y$ and $x \in F_{\alpha,a}$, then $y \in F_{\alpha,a}$, and if $x, y \in F_{\alpha,a}$, then $x \wedge y \in F_{\alpha,a}$, i.e., $F_{\alpha,a}$ is a filter. For an arbitrary element $x \in A$ we have $a/F_\alpha(\mathfrak{A}) \leqslant x/F_\alpha(\mathfrak{A})$ or $a/F_\alpha(\mathfrak{A}) \leqslant C(x)/F_\alpha(\mathfrak{A})$ since $a/F_\alpha(\mathfrak{A})$ is an atom. Consequently, $x \in F_{\alpha,a}$ or $C(x) \in F_{\alpha,a}$. Hence $F_{\alpha,a}$ is an ultrafilter. Let us show that any ultrafilter F of $\mathfrak{A}$ has the same form. If the set $F/F_\beta(\mathfrak{A}) \rightleftharpoons \{b/F_\beta(\mathfrak{A}) \mid b \in F\}$ does not contain $\mathbf{0}/F_\beta(\mathfrak{A})$, then $F/F_\beta(\mathfrak{A})$ is an ultrafilter of the finite Boolean algebra $\mathfrak{A}/F_\beta(\mathfrak{A})$. In this case, $F/F_\beta(\mathfrak{A})$ contains the least nonzero element $a/F_\beta(\mathfrak{A})$ which is an atom of $\mathfrak{A}/F_\beta(\mathfrak{A})$ since $F/F_\beta(\mathfrak{A})$ is an ultrafilter. We consider the ultrafilter $F_{a,\beta}$. Since $a/F_\beta(\mathfrak{A})$ is the least element of the ultrafilter $F/F_\beta(\mathfrak{A})$, any element $b \in F$ belongs to $F_{a,\beta}$. Since F is an ultrafilter, it is maximal and $F = F_{a,\beta}$. In the case $\mathbf{0}/F_\beta(\mathfrak{A}) \in F/F_\beta(\mathfrak{A})$, we consider the least ordinal $\alpha \leqslant \beta$ such that $\mathbf{0}/F_\alpha(\mathfrak{A}) \in F/F_\alpha(\mathfrak{A})$. There exists $a \in F$ such that $a \in F_\alpha(\mathfrak{A})$. Since $\mathbf{0}/F_\gamma(\mathfrak{A}) \notin F/F_\gamma(\mathfrak{A})$ for $\gamma < \alpha$, we have $a \notin F_\gamma(\mathfrak{A})$ for $\gamma < \alpha$. If α is a limit ordinal, then $F_\alpha(\mathfrak{A}) = \bigcup_{\gamma < \alpha} F_\gamma(\mathfrak{A})$ and hence $a \notin F_\alpha(\mathfrak{A})$. Consequently, $\alpha = \gamma + 1$. By the choice of $\gamma < \alpha$, we have $\mathbf{0}/F_\gamma(\mathfrak{A}) \notin F/F_\gamma(\mathfrak{A})$. Consequently, $F/F_\gamma(\mathfrak{A})$ is an ultrafilter of $\mathfrak{A}/F_\gamma(\mathfrak{A})$. However, $a/F_\gamma(\mathfrak{A}) \in F(\mathfrak{A}/F_\gamma(\mathfrak{A}))$ because $a \in F_{\gamma+1}(\mathfrak{A})$. Consequently, $a/F_\gamma(\mathfrak{A})$ is equal to the union of a finite number of atoms $b_i/F_\gamma(\mathfrak{A})$, $i \leqslant k$, of the Boolean algebra $\mathfrak{A}/F_\gamma(\mathfrak{A})$. It is easy to choose representatives b_i so that $\vee b_i = a$ and $b_i \wedge b_j = \mathbf{0}$, $i \neq j$. Since F is an ultrafilter, there is i such that $b_i \in F$. Then $F/F_\gamma(\mathfrak{A})$ is a principal filter containing an atom. Since the ultrafilter $F/F_\gamma(\mathfrak{A})$ is principal, $F \subseteq F_{b_i,\gamma}$. Since F is an ultrafilter, $F = F_{b_i,\gamma}$. Thus, we have shown that any ultrafilter of a superatomic Boolean algebra is a principal ultrafilter of a suitable quotient algebra and is uniquely defined by the corresponding ordinal and atom. The aforesaid proves the following claim.

Corollary 1.7.2. *If a countable Boolean algebra $\mathfrak{B}$ is superatomic, then for any ultrafilter D there exists an ordinal $\alpha < o(\mathfrak{B})$ and an element $a \in \mathfrak{B}$ such that $a/F_\alpha(\mathfrak{B})$ is an atom of $\mathfrak{B}/F_\alpha(\mathfrak{B})$ and $D = \{b \mid a/F_\alpha(\mathfrak{B}) \leqslant b/F_\alpha(\mathfrak{B})\}$.*

Exercises

1. Construct trees generating the Boolean algebras $\mathfrak{B}_\omega$ and $\mathfrak{B}_{\omega^2}$.

2. Prove that a tree generating an atomless Boolean algebra coincides with $\mathbb{N}$.

3. Let D_1 and D_2 be trees generating Boolean algebras $\mathfrak{B}_1$ and $\mathfrak{B}_2$ respectively. Construct a tree generating the Boolean algebra $\mathfrak{B}_1 \times \mathfrak{B}_2$.

4. Let D be a tree generating a Boolean algebra $\mathfrak{B}$. Construct a generating tree D_0 such that the Boolean quotient algebra of $\mathfrak{B}_{D_0}$ by the Frechét ideal is isomorphic to $\mathfrak{B}$.

5. Describe atomic and atomless Boolean algebras $\mathfrak{B}_D$, and its atoms as well, in terms of the properties of elements of a tree D.

6. Describe elements of the Frechét ideal of the Boolean algebra $\mathfrak{B}_D$ in terms of the properties of elements of a tree D.

7. Prove that the Boolean algebra $\mathfrak{B}_{\omega \cdot n}$ has exactly n infinite branches in any generating tree.

8. Define the rank on branches of a tree D as follows: for a maximal branch γ of a tree D, we set

    $\operatorname{rank}(\gamma) = 0$ if γ is finite,

    $\operatorname{rank}(\gamma) = \alpha + 1$ if there exists an extension of any initial segment of γ in D to a maximal branch of rank α, but there exists an initial segment of γ such that any of its extensions either coincides with γ or is of rank $\leqslant \alpha$,

rank$(\gamma) = \alpha$, where α is a limit ordinal, if for any $\delta < \alpha$ there exists an extension of any initial segment of γ of rank at least δ, but there exists an initial segment of γ such that each of its extensions coincides with γ or is of rank less than α.

Prove the following assertions:

(a) $\mathfrak{B}_D$ is a superatomic Boolean algebra if any branch of D has rank,

(b) type $\mathfrak{B}_D = (\alpha, n)$ if and only if each of the branches of D has rank at most α and there exists exactly n branches of rank α.

9. Construct trees generating the Boolean algebras $\mathfrak{B}_{\omega \times \eta}$ and $\mathfrak{B}_{\omega+\eta}$.

10. Describe trees generating the Boolean algebras $\mathfrak{B}_\omega$ and $\mathfrak{B}_{\omega+\eta}$.

11. Give an algebraic description of the Boolean algebras $\mathfrak{B}_\omega$, $\mathfrak{B}_{\omega^2}$, $\mathfrak{B}_{\omega \times \eta}$, $\mathfrak{B}_{\omega+\eta}$, and $\mathfrak{B}_{(\omega+\eta) \times \eta}$ [65, 66].

12. Indicate relationships between maximal branches of a tree D and ultrafilters of the Boolean algebra $\mathfrak{B}_D$ generated by D.

1.8. Ershov Algebras and The Isomorphism Problem

In this section, we give a full characterization of the types of isomorphism of countable Ershov algebras. As a corollary, we obtain the Ketonen characterization of countable Boolean algebras. We follow the definitions and the proof used in [38].

We now expand the characterization of the types of isomorphism for superatomic Boolean algebras to the class of superatomic Ershov algebras. An Ershov algebra $\mathfrak{A}$ is called *superatomic* if for any $a \in \mathfrak{A}$ the system $\hat{a} = \langle \hat{a}, \vee, \wedge, C \rangle$ is a superatomic Boolean algebra, where $C(x) \rightleftharpoons a \setminus x$, $\hat{a} = \{x \mid x \leqslant a\}$, and $\vee$ and $\wedge$ are the restrictions of the corresponding operations defined on $\mathfrak{A}$. An element a of an Ershov algebra $\mathfrak{A}$ is called an *atom* if $a > 0$ and there is no element $x \in \mathfrak{A}$ such that $0 < x < a$, where 0 is the least element of $\mathfrak{A}$. As in the case of Boolean algebras, we introduce the *Frechét ideal* $F(\mathfrak{A})$ of an Ershov algebra $\mathfrak{A}$ as follows:

$$F(\mathfrak{A}) \rightleftharpoons \{x \in \mathfrak{A} \mid \text{there exists a finite number of atoms} \\ x_1, \ldots, x_n \text{ of } \mathfrak{A} \text{ such that } x = \vee x_i\}$$

By the definition of the least upper bound, the union of the empty set of atoms is equal to the least element $\mathbf{0}$ of the algebra $\mathfrak{A}$. It is easy to check that the Frechét ideal of $\mathfrak{A}$ is an ideal of $\mathfrak{A}$.

We will iterate the construction of the Frechét ideal by ordinals:

$$F_0(\mathfrak{A}) = \{\mathbf{0}\}$$
$$F_\alpha(\mathfrak{A}) = \bigcup_{\gamma < \alpha} F_\gamma(\mathfrak{A}) \text{ if } \alpha \text{ is a limit ordinal}$$
$$F_\alpha(\mathfrak{A}) = \{x \mid x/F_\beta(\mathfrak{A}) \in F(\mathfrak{A}/F_\beta(\mathfrak{A}))\} \text{ if } \alpha = \beta + 1$$

It is easy to see that for any ordinal α the set $F_\alpha(\mathfrak{A})$ is an ideal of an Ershov algebra $\mathfrak{A}$ and for any element $a \in \mathfrak{A}$ the equality $F_\alpha(\widehat{a}) = F_\alpha(\mathfrak{A}) \cap \widehat{a}$ holds.

Corollary 1.8.1 *An Ershov algebra $\mathfrak{A}$ is superatomic if and only if there exists an ordinal α such that $F_\alpha(\mathfrak{A}) = |\mathfrak{A}|$.*

As in the case of Boolean algebras, for an Ershov algebra $\mathfrak{A}$ the least ordinal α such that $F_\alpha(\mathfrak{A}) = F_{\alpha+1}(\mathfrak{A})$ is called the *ordinal type* of $\mathfrak{A}$ and is denoted by $o(\mathfrak{A})$. We note that for superatomic Boolean algebras the ordinal type is not a limit ordinal while for superatomic Ershov algebras it may be a limit one.

We consider a superatomic Boolean algebra $\mathfrak{B}_n$ of the ordinal type n. It is an Ershov algebra with respect to the signature $\sigma_e = \langle \vee, \wedge, \backslash, \mathbf{0} \rangle$. We redefine the operations on $\mathfrak{B}$ and introduce the structure of Ershov algebras $\mathfrak{B}^e$ of the signature $\langle \vee, \wedge, \backslash, \mathbf{0} \rangle$, where $a \backslash b = a \wedge C(b)$. We consider the direct sum $\mathfrak{B} = \sum_{\{0\}} \mathfrak{B}_n^e$ of Ershov algebras over the set consisting of the single constant symbol $\mathbf{0}$. It is easy to see that $o(\mathfrak{B}) = \omega$ and $\mathfrak{B}$ is superatomic.

Let an Ershov algebra $\mathfrak{A}$ be superatomic. If there is no $\gamma < o(\mathfrak{A})$ such that $\mathfrak{A}/F_\gamma(\mathfrak{A})$ has the greatest element and $\alpha = o(\mathfrak{A})$, then we take $(\alpha, \alpha, 0)$ as the *superatomicity type* of $\mathfrak{A}$. If $\mathfrak{A}/F_\gamma(\mathfrak{A})$ has the greatest element for some $\gamma < o(\mathfrak{A})$, then the quotient algebra $\mathfrak{A}/F_\gamma(\mathfrak{A})$ is a Boolean algebra and the ordinal type $o(\mathfrak{A})$ is not a limit ordinal, i.e., $o(\mathfrak{A}) = \alpha^* + 1$ and $\gamma \leq \alpha^*$. Let n be the number of atoms of the Boolean algebra $\mathfrak{A}/F_{\alpha^*}(\mathfrak{A})$ and let β be the least ordinal such that $\mathfrak{A}/F_\beta(\mathfrak{A})$ has the greatest element. As the *superatomicity type* we take (α^*, β, n). Denote the superatomicity type of the Ershov algebra $\mathfrak{A}$ by $\tau(\mathfrak{A})$. By the *superatomicity type of an element* $a \in \mathfrak{A}$ we mean the superatomicity type of the subalgebra $\widehat{a} = \{x \mid x \leq a\}$ of the Ershov algebra $\mathfrak{A}$. Denote it by $\tau(a)$.

1.8. Ershov Algebras and The Isomorphism Problem

REMARK 1.8.1. If $\mathfrak{B}$ is a superatomic Boolean algebra and type($\mathfrak{B}$) = $\langle \alpha, n \rangle$, then the superatomicity type $\tau(\mathfrak{B})$ of $\mathfrak{B}$ as an Ershov algebra is $\langle \alpha, 0, n \rangle$.

An Ershov algebra $\mathfrak{A}$ is called *atomic* if for any $b \in \mathfrak{A}$, $b \neq \mathbf{0}$, there exists an atom $a \in \mathfrak{A}$ such that $a \leqslant b$. It is easy to see that an Ershov algebra $\mathfrak{A}$ is atomic if and only if for any $a \in \mathfrak{A}$ the Boolean algebra $\hat{a}$ is atomic. Using the characterization of superatomic Boolean algebras, we easily establish the following claim.

Proposition 1.8.1. *For an Ershov algebra $\mathfrak{A}$ the following conditions are equivalent*:

(1) *the algebra $\mathfrak{A}$ is superatomic,*

(2) *there exists an ordinal α such that $F_\alpha(\mathfrak{A}) = |\mathfrak{A}|$,*

(3) *any subalgebra of $\mathfrak{A}$ is atomic,*

(4) *any subalgebra of $\mathfrak{A}$ is superatomic,*

(5) *any epimorphic image of $\mathfrak{A}$ is an atomic algebra,*

(6) *any epimorphic image of $\mathfrak{A}$ is a superatomic algebra.*

The Vaught criterion for isomorphism of Boolean algebras can be adapted to the case of Ershov algebras in an obvious way. Let $\mathfrak{A}$ and $\mathfrak{B}$ be two Ershov algebras. We define a condition of isomorphism S on $(\mathfrak{A}, \mathfrak{B})$ as a subset $S \subseteq A \times B$ subject to the following conditions:

(1) $\langle \mathbf{0}, \mathbf{0} \rangle \in S$,

(2) if $\langle x, y \rangle \in S$, then $x = \mathbf{0} \Leftrightarrow y = \mathbf{0}$,

(3) if $\langle a, b \rangle \in S$ and $c \in A$, then there exist $d_0, d_1 \in B$ such that $\langle a \wedge c, b \wedge d_0 \rangle$, $\langle a \backslash c, b \backslash d_0 \rangle$, $\langle a \vee c, b \vee d_1 \rangle$, and $\langle c \backslash a, d_1 \backslash b \rangle$ belong to S,

(4) if $\langle a, b \rangle \in S$ and $d \in B$, then there exist $d_0, d_1 \in A$ such that $\langle a \wedge d_0, b \wedge d \rangle$, $\langle a \backslash d_0, b \backslash d \rangle$, $\langle a \vee d_1, b \vee d \rangle$, $\langle d_1 \backslash a, d \backslash b \rangle$ belong to S.

Proposition 1.8.2. *If $\mathfrak{A}$ and $\mathfrak{B}$ are countable Ershov algebras, S is a condition of isomorphism on $(\mathfrak{A}, \mathfrak{B})$, $a_0 < a_1 < \ldots < a_n$ is an increasing chain of elements of $\mathfrak{A}$, and $b_0 < b_1 < \ldots < b_n$ is an increasing chain of elements of $\mathfrak{B}$, where $a_0 = \mathbf{0}$, $b_0 = \mathbf{0}$, and $\langle a_{i+1} \backslash a_i, b_{i+1} \backslash b_i \rangle \in S$, $i < n$, $\langle a_n, b_n \rangle \in S$, then there exists an isomorphism φ between $\mathfrak{A}$ and $\mathfrak{B}$ such that $\varphi(a_i) = b_i$, $0 \leqslant i \leqslant n$. Furthermore, the following condition holds*:

1. Algebraic Properties of Boolean Algebras

(∗) *for any element $a \in \mathfrak{A}$ there exist elements $a'_0 < a'_1 < \ldots < a'_m$ of $\mathfrak{A}$ such that the pairs $\langle a'_{i+1} \backslash a'_i, \varphi(a'_{i+1}) \backslash \varphi(a'_i) \rangle$, $i < m$, and $\langle a'_m, \varphi(a'_m) \rangle$ belong to S and $\bigvee_{i \in K}(a'_{i+1} \backslash a'_i) = a$ for some $K \subseteq \{0, \ldots, m\}$.*

A homomorphism φ possessing the property (∗) is called *S-definite*.

PROOF OF PROPOSITION 1.8.2. We first note that a homomorphism φ such that $\varphi(a_i) = b_i$, $i \leqslant n$, is S-definite on elements of the finite subalgebra generated by elements $a_0, a_1, \ldots, a_n$. The proof is based on the "back and forth" method. We construct finite subalgebras $\mathfrak{A}^t$ and $\mathfrak{B}^t$ of the algebras $\mathfrak{A}$ and $\mathfrak{B}$ and an S-definite partial isomorphism φ^t between $\mathfrak{A}$ and $\mathfrak{B}$ with domain of definition dom $\varphi^t = \mathfrak{A}^t$ and range range $\varphi^t = \mathfrak{B}^t$ so as to satisfy the following conditions:

(1) $\mathfrak{A}^t$ is generated by $\{a_0^t < a_1^t < \ldots < a_{n_t}^t\}$, $\mathfrak{B}^t$ is generated by $\{b_0^t < \ldots < b_{n_t}^t\}$, $\varphi^t(a_i^t) = b_i^t$, $i \leqslant n_t$, $a_0^t = \mathbf{0}$, and $b_0^t = \mathbf{0}$,

(2) $\langle a_{i+1}^t \backslash a_i^t, b_{i+1}^t \backslash b_i^t \rangle \in S$, $0 \leqslant i < n_t$, and $\langle a_{n_t}^t, b_{n_t}^t \rangle \in S$,

(3) $\varphi^t \subseteq \varphi^{t+1}$,

(4) for any n the element $c_n \in \mathfrak{A}$ belongs to $\mathfrak{A}^{2n+1}$ and the element $d_n \in \mathfrak{B}$ belongs to $\mathfrak{B}^{2n+2}$, where the lists $\mathfrak{A} = \{c_0, c_1, \ldots, c_n, \ldots\}$ and $\mathfrak{B} = \{d_0, d_1, \ldots, d_n, \ldots\}$ of elements of $\mathfrak{A}$ and $\mathfrak{B}$ are fixed.

Such a construction is possible. Indeed, any element of the subalgebra $\mathfrak{C}_0$ of the Ershov algebra $\mathfrak{C}$ generated by the linearly ordered sequence $\{h_0, \ldots, h_n\}$, where $h_0 = \mathbf{0}$ and $h_0 < h_1 < \ldots < h_n$, can be uniquely represented in the form $\bigvee_{i \in K}(h_{i+1} \backslash h_i)$ for some $K \subseteq \{0, 1, \ldots, n-1\}$. Let $a_0 < a_1 < \ldots < a_n$ and $b_0 < b_1 < \ldots < b_n$ be two linearly ordered sequences such that $a_0 = \mathbf{0}$, $b_0 = \mathbf{0}$, $\langle a_{i+1} \backslash a_i, b_{i+1} \backslash b_i \rangle \in S$, $i < n$, and $\langle a_n, b_n \rangle \in S$. For any $c \in \mathfrak{B}$ we consider the elements $c_i = a_i \vee (a_{i+1} \wedge c)$, $i < n$, and $c_n = a_n \vee c$. It is obvious that $a_i \leqslant c_i \leqslant a_{i+1}$, $a_n \leqslant c_n$, and $c = \bigvee_{i=1}^{n}(c_i \backslash a_i)$. Using the condition of isomorphism S, we can find elements d_i, $0 \leqslant i \leqslant n$, such that the pairs $\langle c_i \backslash a_i, d_i \backslash b_i \rangle$ and $\langle a_{i+1} \backslash c_i, b_{i+1} \backslash d_i \rangle$, $i < n$, belong to S and the following condition holds: $\langle c_n, d_n \rangle \in S$; moreover, they are linearly ordered: $b_i \leqslant d_i \leqslant b_{i+1}$, $i < n$, and $b_n \leqslant d_n$. In view of the condition (2) imposed on S, these inequalities hold consistently for both sequences:

$$a_i < c_i \Leftrightarrow b_i < d_i, \quad c_i < a_{i+1} \Leftrightarrow d_i < b_{i+1}, \quad a_n < c_n \Leftrightarrow b_n < d_n$$

1.8. Ershov Algebras and The Isomorphism Problem

Since $\mathfrak{A}$ and $\mathfrak{B}$ play symmetric roles in the condition of isomorphism S, for any $d \in \mathfrak{B}$ we can choose its preimage in $\mathfrak{A}$. Thus, the "back and forth" method allows us to construct the desired isomorphism. □

We now apply the criterion obtained to describe the atomless and superatomic Ershov algebras.

An Ershov algebra is called *atomless* if it contains no atom.

Proposition 1.8.3. *There are exactly two types of isomorphism for countable atomless Ershov algebras which differ from each other in the presence or absence of the greatest element.*

PROOF. If there is the greatest element of a countable Boolean algebra, then the latter is atomless. But an atomless countable Boolean algebra is unique up to an isomorphism. Thus, we need to prove that two atomless countable Ershov algebras $\mathfrak{A}$ and $\mathfrak{B}$ without greatest elements are isomorphic. For a condition of isomorphism we take the set

$$S = \{(a,b) \mid a \in \mathfrak{A},\ b \in \mathfrak{B},\ a = \mathbf{0} \Leftrightarrow b = \mathbf{0}\}$$

The verification of the required conditions on S is a trivial exercise. □

We now prove that the superatomicity types characterize countable superatomic Ershov algebras in a unique way up to an isomorphism. It is clear that a finite Ershov algebra is a Boolean algebra and the type of isomorphism is uniquely determined by the number of atoms. Therefore, a nontrivial part of the proof is to establish an isomorphism between countable Ershov algebras.

Proposition 1.8.4. *Any two countable superatomic Ershov algebras of the same superatomicity type are isomorphic.*

PROOF. It suffices to define a condition of isomorphism. Let $\mathfrak{A}$ and $\mathfrak{B}$ be countable superatomic Ershov algebras and let $\tau(\mathfrak{A}) = \tau(\mathfrak{B})$. For an arbitrary element a we introduce two subalgebras of the algebra $\mathfrak{A}$ as follows: $\widehat{a} \rightleftharpoons \{x \mid x \leqslant a\}$ and $a^\perp \rightleftharpoons \{x \mid x \wedge a = \mathbf{0}\}$. It is easy to check that $a^\perp$ also forms a subalgebra of $\mathfrak{A}$ which will be called the *algebra of orthogonal elements* to a, and elements of $a^\perp$ will be called elements *orthogonal* to a. We set

$$S \rightleftharpoons \{(a,b) \mid a \in \mathfrak{A},\ b \in \mathfrak{B},\ \tau(a) = \tau(b),\ \tau(a^\perp) = \tau(b^\perp)\}$$

A nontrivial part of the proof is to verify the conditions (3) and (4). Since these conditions are symmetric and the algebras $\mathfrak{A}$, $\mathfrak{B}$ are arbitrary, it suffices to check only the condition (3).

Let a pair (a,b) belong to S and let c be an element of $\mathfrak{A}$. Since $\widehat{a}$ and $\widehat{b}$ are Boolean algebras, from their characterization we immediately conclude that there exists an element $d_0 \leqslant b$ such that $\tau(a \wedge c) = \tau(d_0)$, $\tau(c\backslash a) = \tau(b\backslash d_0)$. Note that $(a\wedge c)^{\perp} = \widehat{a\backslash c} + a^{\perp}$ and $(b\wedge d_0)^{\perp} = \widehat{(b\backslash d_0)} + b^{\perp}$; moreover, the superatomicity type of the direct sum can be computed from the superatomicity types of the summands. Therefore, $\tau((a\wedge c)^{\perp}) = \tau((b\wedge d_0)^{\perp})$ and $\langle a \wedge c, b \wedge d_0 \rangle \in S$. Similarly, $\tau((a\backslash c)^{\perp}) = \tau((b\backslash d_0)^{\perp})$. Consequently, $(a\backslash c, b\backslash d_0) \in S$.

Since $\tau(a^{\perp}) = \tau(b^{\perp})$, there exists an element d_1 of $b^{\perp}$ such that $\tau(c\backslash a) = \tau(d_1)$ and $\tau((c\backslash a)^{\perp} \cap a^{\perp}) = \tau(d_1^{\perp} \cap b^{\perp})$. The existence of d_1 follows from the definition of τ and the characterization of superatomic Boolean algebras.

Since $d_1 \wedge b = 0$, we have $\tau(c\backslash a) = \tau(d_1) = \tau(d_1\backslash b)$ and $\tau(a \vee c) = \tau(a \vee (c\backslash a)) = \tau(b \vee (d_1\backslash b)) = \tau(b \vee d_1)$ in view of equalities $a \vee (c\backslash a) = \widehat{a} + \widehat{c\backslash a}$ and $b \vee (d_1\backslash b) = \widehat{b} + \widehat{d_1\backslash b}$. Taking into account the decompositions of $(c\backslash a)^{\perp}$ and $(d_1\backslash b)^{\perp}$ into the direct sums of subalgebras: $(c\backslash a)^{\perp} = ((c\backslash a)^{\perp} \cap a^{\perp}) + \widehat{a}$ and $(d_1\backslash b)^{\perp} = (d_1^{\perp} \cap b^{\perp}) + \widehat{b}$ of the same superatomicity type, we conclude that $\tau((c\backslash a)^{\perp}) = \tau((d_1\backslash b)^{\perp})$. Therefore, $\langle c\backslash a, d_1\backslash b \rangle \in S$. Similarly, $\tau((a \vee c)^{\perp}) = \tau((b \vee d_1)^{\perp})$ and $\langle a \vee c, b \vee d_1 \rangle \in S$. In view of Proposition 1.8.2, the algebras $\mathfrak{A}$ and $\mathfrak{B}$ are isomorphic. □

We now define the superatomicity type of an arbitrary Ershov algebra $\mathfrak{A}$. We denote by $F_*(\mathfrak{A})$ the set of all superatomic elements of $\mathfrak{A}$. It is obvious that $F_\alpha(\mathfrak{A}) = F_*(\mathfrak{A})$ for an ordinal α such that $F_{\alpha+1}(\mathfrak{A}) = F_\alpha(\mathfrak{A})$. Let the superatomicity type $\tau(\mathfrak{A})$ be equal to that of the superatomic subalgebra $F_*(\mathfrak{A})$ of the algebra $\mathfrak{A}$, i.e., $\tau(\mathfrak{A}) \rightleftharpoons \tau(F_*(\mathfrak{A}))$. For an arbitrary element $a \in \mathfrak{A}$ we define $\tau(a) = \tau(\widehat{a} \cap F_*(\mathfrak{A}))$.

If $\tau(\mathfrak{A}) = \langle \alpha, \beta, k \rangle$, then β is called an *E-rank of an algebra* $\mathfrak{A}$ and is denoted by $\sigma(\mathfrak{A})$. We also define an *E-rank of an element* $a \in \mathfrak{A}$ by setting $\sigma(a) \rightleftharpoons \sigma(\widehat{a})$ and call σ the *E*-rank.

Proposition 1.8.5. *If* $\mathfrak{A}$ *is an Ershov algebra of E-rank* $\sigma(\mathfrak{A}) = \alpha$, *then the function* σ *maps* $\mathfrak{A}$ *into the ordinal* $\alpha + 1$ *and* $\sigma(0) = 0$, $\sigma(a \vee b) = \max\{\sigma(a), \sigma(b)\}$ *for any* $a, b \in \mathfrak{A}$.

PROOF. The reasoning consists in checking the fact that, by the equality $F_*(\mathfrak{A}) = F_\alpha(\mathfrak{A})$, the ideal $\widehat{a \vee b} \cap F_*(\mathfrak{A})$ of the quotient algebra by $F_\gamma(\mathfrak{A})$, where $\gamma = \max\{\sigma(a), \sigma(b)\}$, is a principal ideal of $\mathfrak{A}/F_\gamma(\mathfrak{A})$. □

A function $r\colon \mathfrak{A} \to \alpha + 1$ acting from an Ershov algebra $\mathfrak{A}$ of E-rank d into an ordinal $\alpha + 1$ is called *additive* if $r(0) = 0$ and $r(a \vee b) =$

1.8. Ershov Algebras and The Isomorphism Problem

$\max\{r(a), r(b)\}$ for any $a, b \in \mathfrak{A}$. A superatomic Ershov algebra $\mathfrak{A}$ is called *special* if $\tau(\mathfrak{A}) = \langle \sigma(\mathfrak{A}), \sigma(\mathfrak{A}), 0 \rangle$.

Proposition 1.8.6 [on decomposition of superatomic Ershov algebras]. *If $\mathfrak{A}$ is a superatomic Ershov algebra, then $\mathfrak{A}$ either is a special algebra or is represented as the direct sum $\mathfrak{A}_0 + \mathfrak{A}_1$ of Ershov algebras $\mathfrak{A}_0$ and $\mathfrak{A}_1$ such that*

(1) *$\mathfrak{A}_0$ is a Boolean algebra of the superatomicity type $\langle \alpha, 0, n \rangle$,*

(2) *$\mathfrak{A}_1$ is a special Ershov algebra,*

(3) *$\sigma(\mathfrak{A}) = \sigma(\mathfrak{A}_1) \leqslant \alpha$.*

PROOF. We consider an element d defining the greatest element of the quotient algebra $F_*(\mathfrak{A})/F_{\sigma(\mathfrak{A})}(\mathfrak{A})$. The algebra $\mathfrak{A}$ can be represented as the direct sum $\widehat{d} + d^\perp$. Taking $\mathfrak{A}_0 = \widehat{d}$ and $\mathfrak{A}_1 = d^\perp$, we easily check that all the requirements on $\mathfrak{A}_0$ and $\mathfrak{A}_1$ are fulfilled. □

Proposition 1.8.7. *Let $\mathfrak{A}$ be a special Ershov algebra and let $\sigma(\mathfrak{A}) = \beta$. Then*

(a) *if $\beta = \alpha + 1$, then the algebra $\mathfrak{A}$ is isomorphic to the direct sum $\sum_n {}_{\{0\}} \mathfrak{A}_n$ of Boolean algebras $\mathfrak{A}_n$ of the superatomicity type $\langle \alpha, 0, 1 \rangle$,*

(b) *if β is a limit ordinal and $\alpha_0 < \alpha_1 < \ldots < \alpha_n < \ldots$ is a sequence of ordinals such that $\sup_n \alpha_n = \beta$, then $\mathfrak{A}$ is isomorphic to the direct sum $\sum_n {}_{\{0\}} \mathfrak{A}_n$ of superatomic Boolean algebras $\mathfrak{A}_n$ of the superatomicity type $\langle \alpha_n, 0, 1 \rangle$.*

PROOF. The assertion follows from the fact that the superatomicity types of such Ershov algebras coincide. □

Proposition 1.8.8. *If an Ershov algebra $\mathfrak{A}$ is isomorphic to the direct sum $\sum_n {}_{\{0\}} \mathfrak{A}_n$ of Ershov algebras $\mathfrak{A}_n$, then an ideal completion $\mathfrak{A}'$ of $\mathfrak{A}$ is isomorphic to the direct product $\prod_n \mathfrak{A}'_n$ of ideal completions $\mathfrak{A}'_n$ of $\mathfrak{A}_n$.*

PROOF. The assertion is an immediate consequence of the description of an ideal completion as the lattice of locally principal ideals. □

We consider the decomposition of a special Boolean algebra $\mathfrak{A}$ into the direct sum of Boolean algebras (regarded as Ershov algebras) from Proposition 1.8.7: $\mathfrak{A} \cong \sum_{n} {}_{\{0\}}\mathfrak{A}_n$. Since a locally principal ideal of a Boolean algebra is principal, we have $\mathfrak{A}'_n = \mathfrak{A}_n$. Consequently, $\mathfrak{A}' \cong \prod_{n} \mathfrak{A}_n$. Starting from an arbitrary ordinal β, we define an ideal

$$\Phi_\beta(\mathfrak{A}') \rightleftharpoons \{f \in \prod_{n} \mathfrak{A}_n \mid f(n) \in F_\beta(\mathfrak{A}_n) \text{ for any } n\}$$

of $\mathfrak{A}'$. Let $\mathfrak{A}^* \rightleftharpoons \mathfrak{A}'/\mathfrak{A} = \prod_{n} \mathfrak{A}_n / \sum_{n} {}_{\{0\}}\mathfrak{A}_n$. For $\beta \leqslant \sigma(\mathfrak{A})$ we set $\Psi_\beta(\mathfrak{A}^*) \rightleftharpoons \Phi_\beta(\mathfrak{A}')/\mathfrak{A}$. We now define a mapping ρ from $\mathfrak{A}^*$ into $\sigma(\mathfrak{A}) + 1$ by setting $\rho(d) = \min\{\beta \mid d \in \Psi_\beta(\mathfrak{A}^*)\}$ for $d \in \mathfrak{A}^*$.

REMARK 1.8.2. If j is a locally principal ideal of $\mathfrak{A}$, i.e., $j \in \mathfrak{A}'$, then $\rho(j/\mathfrak{A}) = \sigma(j)$.

REMARK 1.8.3. If j is a locally principal ideal of $\mathfrak{A}$ and $d \in \mathfrak{A}$, then $\widehat{j/\mathfrak{A}} = j + \widehat{d}/\mathfrak{A}$.

Proposition 1.8.9. *The mapping $\rho\colon \mathfrak{A}^* \to \sigma(\mathfrak{A}) + 1$ possesses the following properties*:

(a) $\rho(a \vee b) = \max\{\rho(a), \rho(b)\}$ *and* $\rho(d) = \mathbf{0} \Leftrightarrow d = \mathbf{0}$,

(b) *for any $\beta \leqslant \sigma(\mathfrak{A})$ there exists $d \in \mathfrak{A}^*$ such that $\rho(d) = \beta$*,

(c) *if $\beta \leqslant \rho(d)$, then there exists $d_0 \in \mathfrak{A}^*$ such that $d_0 \leqslant d$, $\rho(d_0) = \beta$, and $\rho(d \backslash d_0) = \rho(d)$.*

PROOF. Property (a) follows from the definition. Property (b) takes place in view of (c) and the existence of the greatest element of $A = |\mathfrak{A}|$ in $\mathfrak{A}'$ such that $\rho(A/\mathfrak{A}) = \sigma(\mathfrak{A})$.

We now prove (c). Let $\rho(d) = \gamma$ and let an element $f \in \prod_{n} \mathfrak{B}_n$ be such that d is the image of f under the natural epimorphism. Without loss of generality, we assume that $f(n) \in F_\gamma(\mathfrak{B}_n)$ for any Boolean algebra $\mathfrak{B}_n$. On the other hand, under the assumption that $\rho(d) = \gamma$, the set $\{n \mid f(n) \notin F_\delta(\mathfrak{B}_n)\}$ is infinite for any $\delta < \gamma$. We consider a sequence of ordinals β_i that are not limit ones and $\beta_0 \leqslant \beta_1 \leqslant \ldots \leqslant \beta_n \leqslant \ldots \leqslant \beta$, $\sup_n \beta_n = \beta$. It is possible to choose a sequence $n_0 \leqslant n_1 \leqslant \ldots \leqslant n_k \leqslant \ldots$ such that $\tau(f(n_i)) = \langle \xi, 0, m \rangle$ and $\xi \geqslant \beta_i$ for all $i \in \mathbb{N}$. Therefore, there exists an element d_i of the algebra $\mathfrak{B}_{n_i}$ such that $\tau(d_i) = \langle \beta_i, 0, 1 \rangle$. We

1.8. Ershov Algebras and The Isomorphism Problem

define

$$f'(n) \rightleftharpoons \begin{cases} d_{2i}, & n = n_{2i} \\ 0, & n \notin \{n_0, n_2, n_4, \ldots\} \end{cases}$$

We have $f' \leqslant f$. Consider the image d' of f' under the natural epimorphism. By definition, $\rho(d') = \beta$. We now prove the second equality in (c). If $\beta < \gamma$, then $\rho(d\backslash d') = \rho(d)$ in view of (a). If $\beta = \gamma$, then we consider the function

$$f^*(n) = \begin{cases} d_{2i+1}, & n = n_{2i+1} \\ 0, & n \notin \{n_1, n_3, n_5, \ldots\} \end{cases}$$

As in the case of f', we have $\rho(d^*) = \beta = \gamma$, where d^* is the image of f^* under the natural epimorphism. However, $d^* \leqslant d\backslash d'$. Consequently, $\rho(d\backslash d') = \gamma$. □

We now consider an additive function $r: \mathfrak{A} \to \alpha + 1$, where $\mathfrak{A}$ is an Ershov algebra. A function $r': \mathfrak{A} \to \alpha + 1$ is called *complementary* to the function r if the following conditions hold:

(1) $\alpha = \max\{ra, r'a\}$ for any $a \in \mathfrak{A}$,

(2) if $a \leqslant b$, then $r'(a) = \max\{r(b\backslash a), r'(b)\}$.

For a given function, there are many complementary functions. As a trivial example, we indicate the function r' such that $r'(d) = \alpha$ for all $a \in \mathfrak{A}$. One more example of a complementary function to $\sigma: \mathfrak{A} \to \alpha + 1$, where $\alpha = \sigma(\mathfrak{A})$, is presented by $\sigma'(a) \rightleftharpoons \sigma(a^\perp)$, i.e., the E-rank of the orthogonal complement to a.

We now describe all functions complementary to an additive function $r: \mathfrak{A} \to \alpha + 1$, where $\mathfrak{A}$ is an Ershov algebra.

Proposition 1.8.10. *The following assertions hold:*

(a) *if r' is a nontrivial function complementary to r and $\gamma_{r'}$ is the least ordinal from the values of the function r, then $\gamma_{r'} < \alpha$, the set $\mathcal{D}_{r'} = \{a \in |\mathfrak{A}| \mid r'(a) = \gamma_{r'}\}$ is a filter of $\mathfrak{A}$, $\mathcal{D}_{r'} \subseteq \{a \mid r(a) = \alpha\}$, and $r(a\backslash b) \leqslant \gamma_{r'} \Leftrightarrow b \in \mathcal{D}_{r'}$ for any $a \in \mathcal{D}_{r'}$, $b \in \mathfrak{A}$,*

(b) *for a filter $\mathcal{D}$ of $\mathfrak{A}$ such that $\mathcal{D} \subseteq \{a \in |\mathfrak{A}| \mid r(a) = \alpha\}$ and an ordinal $\gamma < \alpha$ such that $r(a\backslash b) \leqslant \gamma \Leftrightarrow b \in \mathcal{D}$ for any $a \in \mathcal{D}$ and $b \in \mathfrak{A}$ there exists a unique function r' complementary to r such that $\mathcal{D} = \mathcal{D}_{r'}$ and $\gamma = \gamma_{r'}$.*

PROOF. For (a) it suffices to prove that $\mathcal{D}_{r'}$ is closed under intersections. All the remaining properties follow from definitions.

Let a and b be two elements of $\mathcal{D}_{r'}$, i.e., $r'(a) = \gamma_{r'}$ and $r'(b) = \gamma_{r'}$. We show that $r'(a \wedge b) = \gamma_{r'}$. We assume that it is not so. By the minimality of $\gamma_{r'}$, we have $r'(a \wedge b) > \gamma_{r'}$. By the condition (2) in the definition of a complementary function r', we obtain $r'(a \wedge b) = \max\{r(a \backslash (a \wedge b)), r'(a)\}$. Consequently, $\gamma = r'(a \wedge b) = r(a \backslash (a \wedge b))$. However, $(a \vee b) \backslash b = a \backslash (a \wedge b)$ and $r((a \vee b) \backslash b) = \gamma$. Therefore, $r'(b) = \max\{r((a \vee b) \backslash b), r'(a \vee b)\}$, and $r'(b) \geqslant \gamma > \gamma_{r'}$, which contradicts the above assumption.

We now prove (b). Let a filter $\mathcal{D}$ and an ordinal γ be subject to the hypotheses of (b). We construct a complementary function r' such that $\mathcal{D}_{r'} = \mathcal{D}$, $\gamma_{r'} = \gamma$ and prove that it is unique. We consider an arbitrary element $a \in \mathcal{D}$ and define a function r'_a as follows: $r'_a(b) = \max\{r(a \backslash b), \gamma\}$, $b \in \mathfrak{A}$. We note that r'_a is complementary to r. Indeed, let $r'_a(b) = \delta$. If $\delta = \alpha$, then (1) is obvious. Let $\delta < \alpha$. In view of the relations $r(a \backslash b) \leqslant r'_a(b)$, $\mathcal{D} \subseteq \{a \mid r(a) = \alpha\}$, and the additivity of r, we can conclude that $\alpha = r(a \vee b) = \max\{r(a \backslash b), r(a)\} = r(a)$ Hence (1) is valid. We now verify (2). Let $b_0 \leqslant b_1$. We show that $r'_a(b_0) = \max\{r(b_1 \backslash b_0), r'_a(b_1)\}$. By definition, $r'_a(b_0) = \max\{r(a \backslash b_0), \gamma\}$. However, $b_0 \leqslant b_1$. In view of additivity, $r((b_1 \vee a) \backslash a) = \max\{r(b_1 \backslash b_0), r(a \backslash b_1)\}$ and $r((a \vee b_1) \backslash b_0) = \max\{r(a \backslash b_0), r((a \vee b_1) \backslash (a \vee b_0))\}$. Therefore,

$$r'_a(b_0) = r'_{a \vee b_1}(b_0) = \max\{r(b_1 \backslash b_0), r(a \backslash b_1), \gamma\}$$
$$= \max\{\max\{r(b_1 \backslash a), \gamma\}, r(b_1 \backslash b_0)\}$$
$$= \max\{r'_a(b_1), r(b_1 \backslash b_0)\}$$

Thus, r'_a is complementary to r for any $a \in \mathcal{D}$. By definition, $\gamma_{r'_a} = \gamma$.

We now check the equality $\mathcal{D} = \mathcal{D}_{r'_a}$. Assume that $b \in \mathcal{D}$ and $r'_a(b) = \delta > \gamma$. Then $r(a \backslash b) = \delta$. But in view of the condition on $\mathcal{D}$, we have $r(a \backslash b) \leqslant \gamma$ for any $a, b \in \mathcal{D}$. Consequently, $b \in \mathcal{D}_{r'_a}$. If $b \in \mathcal{D}_{r'_a}$, then $r'_a(b) = \gamma$. Therefore, $r(a \backslash b) \leqslant \gamma$. By the condition on $\mathcal{D}$, the element b belongs to $\mathcal{D}$.

It remains to prove the uniqueness. Let both $r' \colon \mathfrak{A}_0 \to \alpha + 1$ and $r'' \colon \mathfrak{A}_0 \to \alpha + 1$ be complementary to r under the condition that $\mathcal{D} = \mathcal{D}_{r'} = \mathcal{D}_{r''}$ and $\gamma = \gamma_{r'} = \gamma_{r''}$. Let $a \in \mathcal{D}$ and $b \in \mathfrak{A}$. Then $r'(b) = \max\{r(a \backslash b), r'(a \vee b)\}$ and $r''(b) = \max\{r(a \backslash b), r''(a \vee b)\}$. But $a \vee b \in \mathcal{D}$ since $\mathcal{D}$ is a filter. From $\mathcal{D}_{r'} = \mathcal{D} = \mathcal{D}_{r''}$ we obtain $r'(a \vee b) = r''(a \vee b) = \gamma$. Consequently, $r'(b) = r''(b)$. The proof of the uniqueness completes the proof of the proposition. □

1.8. Ershov Algebras and The Isomorphism Problem

Proposition 1.8.11. *If $\mathfrak{A}$ is a countable special Ershov algebra, $\alpha = \sigma(\mathfrak{A})$, and $\mathfrak{A}^* = \mathfrak{A}'/\mathfrak{A}$, where $\mathfrak{A}'$ is an ideal completion of $\mathfrak{A}$ and $\rho\colon \mathfrak{A}^* \to \sigma(\mathfrak{A}) + 1$ is an additive function constructed before Proposition 1.8.9, then for any at most countable Ershov algebra $\mathfrak{A}_0$, an additive function $r\colon \mathfrak{A}_0 \to \alpha + 1$, and its complementary function $r'\colon \mathfrak{A}_0 \to \alpha + 1$, there exists a homomorphism $\psi\colon \mathfrak{A}_0 \to \mathfrak{A}^*$ such that $ra = \rho\psi a$ and $r'a = \rho(C\psi(a))$ for any $a \in \mathfrak{A}_0$.*

PROOF. As in the case of Boolean algebras, in order to construct a homomorphism, we weaken a condition of isomorphism: we eliminate (4) and, instead of (2), require only the following implication:

(2') if $\langle \mathbf{0}, b \rangle \in S$, then $b = \mathbf{0}$.

From the same construction it is easy to see that there exists an S-definite homomorphism of a countable Ershov algebra $\mathfrak{A}$ into an arbitrary Ershov algebra $\mathfrak{B}$ provided that S is a condition of homomorphism from $\mathfrak{A}_0$ into $\mathfrak{B}$. We now define the required condition of homomorphism from $\mathfrak{A}$ into $\mathfrak{A}^*$ by setting

$$S = \{\langle a, b \rangle \mid a \in \mathfrak{A}_0,\ b \in \mathfrak{A}^*,\ ra = \rho b,\ r'a = \rho C(b)\}$$

where C is the operation of taking the complement in $\mathfrak{A}^*$. As was emphasized in constructing the ideal completion of $\mathfrak{A}'$, it is a Boolean algebra. Conditions (1) and (2') are obvious.

We verify (3). Let $\langle a, b \rangle \in S$, $c \in \mathfrak{A}_0$, $ra = \beta$, $r(a \wedge c) = \gamma$, and $r(a \backslash c) = \delta$. In view of the relations $\beta = ra = \rho b$, $\max\{\gamma, \delta\} = \beta$, and Proposition 1.8.9, there exists an element $d_0 \leqslant b$ such that $\rho d_0 = \gamma$ and $\rho(b \backslash d_0) = \delta$. Since

$$\rho(C(d_0)) = \rho(C(b) \vee (b \backslash d_0)) = \max\{\rho(b \backslash d_0), \rho(C(b))\}$$
$$= \max\{\delta, r'(a)\} = \max\{r(a \backslash (a \wedge c)), r'a\} = r'(a \wedge c)$$
$$\rho(C(b \backslash d_0)) = \rho(C(b) \vee d_0) = \max\{\rho d_0, \rho(C(b))\}$$
$$= \max\{\gamma, r'a\} = \max\{r(a \wedge c), r'a\}$$
$$= \max\{r(a \backslash (a \backslash c)), r'a\} = r'(a \backslash c)$$

$\langle a \wedge c, b \wedge d_0 \rangle$ and $\langle a \backslash c, b \backslash d_0 \rangle$ belong to S. We now construct an element d_1 required by the condition (2). To this end, we consider $r'(a)$, $r'(c \backslash a)$, and $r'(a \vee c)$. In view of the relations $r'a = \max\{r(c \backslash a), r'(a \vee c)\}$, $\rho(C(b)) = r'(a)$, and Proposition 1.8.9, there exists an element $d_1 \leqslant C(b)$ such that $\rho(d_1) = r(c \backslash a)$ and $\rho(C(b) \backslash d_1) = r'(a \vee c)$. Consequently, $\langle c \backslash a, d_1 \backslash b \rangle$ and $\langle a \vee c, b \vee d_1 \rangle$ belong to S, all the conditions of homomorphism from $\mathfrak{A}_0$

82   1. Algebraic Properties of Boolean Algebras

into $\mathfrak{A}^*$ hold, and the desired homomorphism exists. Since it is S-definite, it satisfies the needed equalities. □

We now proceed to the problem of describing the isomorphism types of countable Ershov algebras. If an Ershov algebra $\mathfrak{A}$ is not superatomic, then the quotient algebra $\mathfrak{A}/F_*(\mathfrak{A})$, where $F_*(\mathfrak{A})$ is the set of all superatomic elements of $\mathfrak{A}$, is already an atomless Ershov algebra; moreover, it is nontrivial. Hence it is an extension of a superatomic Ershov algebra $F_*(\mathfrak{A})$ by an atomless nontrivial Ershov algebra $\mathfrak{A}/F_*(\mathfrak{A})$. In other words, we have an exact sequence $\mathbf{0} \to F_*(\mathfrak{A}) \to \mathfrak{A} \to \mathfrak{A}/F_*(\mathfrak{A}) \to \mathbf{0}$. For a superatomic Ershov algebra we have described its isomorphism type in terms of the superatomicity type τ. But there are only two atomless countable Ershov algebras: $\mathfrak{B}_\eta$ and $\mathfrak{B}_\eta^\omega$, where $\mathfrak{B}_\eta$ is an atomless Boolean algebra and $\mathfrak{B}_\eta^\omega$ is obtained as the direct sum of a countable number of copies of $\mathfrak{B}_\eta$ for the set of constants that consists of the single element $\mathbf{0}$.

Thus, the characterization of isomorphism types of countable Ershov algebras is reduced to that for those Ershov algebras $\mathfrak{A}$ that are extensions of superatomic countable Ershov algebras by $\mathfrak{B}_\eta$ and $\mathfrak{B}_\eta^\omega$.

A countable Ershov algebra $\mathfrak{A}$ is called *normal* if $F_*(\mathfrak{A})$ is a special Ershov algebra, $\mathfrak{A}/F_*(\mathfrak{A})$ is a nontrivial algebra, and $\mathfrak{A}$ is a Boolean algebra when $\mathfrak{A}/F_*(\mathfrak{A})$ is a Boolean algebra.

We introduce the *Boolean rank* of an Ershov algebra $\mathfrak{A}$ as follows: $R_{\text{Bool}}(\mathfrak{A}) = \alpha$, where α is an ordinal such that $\mathfrak{A}/F_\alpha(\mathfrak{A})$ is a Boolean algebra and for any $\gamma < \alpha$ the quotient algebra $\mathfrak{A}/F_\alpha(\mathfrak{A})$ has no largest element. If there is no ordinal α such that $\mathfrak{A}/F_\alpha(\mathfrak{A})$ is a Boolean algebra, we set $R_{\text{Bool}}(\mathfrak{A}) = \infty$. We note that $R_{\text{Bool}}(\mathfrak{A}) = \infty$ if and only if $\mathfrak{A}/F_*(\mathfrak{A})$ is not a Boolean algebra, where $F_*(\mathfrak{A})$ denotes the set of superatomic elements of $\mathfrak{A}$.

Let $\mathfrak{A}$ be a nonsuperatomic countable Ershov algebra and $\tau(\mathfrak{A}) = (\alpha, \beta, n)$. We say that $\mathfrak{A}$ *has a normal decomposition into the direct sum of subalgebras* $\mathfrak{A}_0$ *and* $\mathfrak{A}_1$ if $\mathfrak{A}_0 \cap \mathfrak{A}_1 = \{\mathbf{0}\}$ and $\mathfrak{A}_0 + \mathfrak{A}_1 = \mathfrak{A}$ and the following conditions hold:

(1) $\mathfrak{A}_0$ is a normal Ershov algebra and $\mathfrak{A}_1$ is a superatomic Ershov algebra,

(2) if $R_{\text{Bool}}(\mathfrak{A}) = \gamma < \infty$, then $\mathfrak{A}_0$ is a Boolean algebra and $\mathfrak{A}_1$ is a superatomic Ershov algebra of type (α, β, n), where $\alpha \geqslant \sigma(\mathfrak{A}_0) = \sigma(\mathfrak{A})$,

1.8. Ershov Algebras and The Isomorphism Problem

(3) if $R_{\text{Bool}}(\mathfrak{A}) = \infty$, then $\mathfrak{A}_1$ is a Boolean algebra and for $\mathfrak{A}_1$ one of the following conditions holds:

$$\tau(\mathfrak{A}_1) = (\alpha, 0, n) \& \sigma(\mathfrak{A}) = \sigma(\mathfrak{A}_0) \leqslant \alpha \,\&\, \tau(\mathfrak{A}) = (\alpha, \beta, n)$$
$$\tau(\mathfrak{A}) = \tau(\mathfrak{A}_0) = (\alpha, \alpha, 0) \& \mathfrak{A}_1 = \{\mathbf{0}\}$$

The pair $(\mathfrak{A}_0, \mathfrak{A}_1)$ is called a *normal decomposition* of the Ershov algebra $\mathfrak{A}$.

Proposition 1.8.12. *For any countable not superatomic Ershov algebra $\mathfrak{A}$ there exists a normal decomposition $(\mathfrak{A}_0, \mathfrak{A}_1)$.*

PROOF. Consider two cases.

CASE 1: $\mathfrak{A}/F_*(\mathfrak{A})$ is a Boolean algebra, i.e., $R_{\text{Bool}}(\mathfrak{A}) < \infty$. We consider the least ordinal γ such that $\mathfrak{A}^\gamma = \mathfrak{A}/F_\gamma(\mathfrak{A})$ is a Boolean algebra. Let $a \in \mathfrak{A}$ be such that $a/F_\gamma(\mathfrak{A})$ is the greatest element of the quotient algebra $\mathfrak{A}/F_\gamma(\mathfrak{A})$. Then $\mathfrak{A}$ can be presented as the direct sum of the ideal $\widehat{a}$ and its orthogonal complement $a^\perp$. In this case, $a^\perp$ is a special algebra. We consider the superatomicity type $\langle \alpha, \beta, n \rangle$ of a. If $n = 0$, i.e., the algebra $\widehat{a} \cap F_*(\mathfrak{A})$ is special, we set $\mathfrak{A}_0 = \widehat{a}$ and $\mathfrak{A}_1 = a^\perp$, which yields the required decomposition. For $n \neq 0$, by the definition of type, there exists an element $a_0 \leqslant a$ such that $a_0 \in F_*(\mathfrak{A})$ and $\tau(a_0) = \langle \alpha, \beta, n \rangle$. Then $F_*(\mathfrak{A})$ can be represented as the sum $\widehat{a_0} + j$, where j is a special algebra such that $\sigma(j) = \beta$. Consequently, by setting $\mathfrak{A}_0 = \widehat{a \setminus a_0}$ and $\mathfrak{A}_1 = \widehat{a_0} + a^\perp$, we arrive at the desired decomposition.

CASE 2: $\mathfrak{A}/F_*(\mathfrak{A})$ is not a Boolean algebra and the superatomicity type of $F_*(\mathfrak{A})$ is (α, β, n). If $n = 0$, then $F_*(\mathfrak{A})$ is a special algebra and $\mathfrak{A}$ is a normal algebra. In this case, we set $\mathfrak{A}_0 = \mathfrak{A}$ and $\mathfrak{A}_1 = \{\mathbf{0}\}$. For $n \neq 0$ we consider an element $a \in F_*(\mathfrak{A})$ of the superatomicity type $\langle \alpha, 0, n \rangle$. Then $F_*(\mathfrak{A}) \cap a^\perp$ is a special algebra such that $\sigma(F_*(\mathfrak{A}) \cap a^\perp) = \beta$. We put $\mathfrak{A}_0 = a^\perp$ and $\mathfrak{A}_1 = \widehat{a}$ and obtain the required decomposition.

The proposition is proved. □

We now study extensions of a countable special Ershov algebra $\mathfrak{A}_0$. Our purpose is to describe the isomorphism types in the case of a countable Ershov algebra $\mathfrak{A}$ such that the sequence $0 \to \mathfrak{A}_0 \to \mathfrak{A} \xrightarrow{\varphi} \Lambda \to 0$ is exact; here $\Lambda \in \{\mathfrak{B}_\eta, \mathfrak{B}_\eta^\omega\}$. We consider the function $\sigma \colon \mathfrak{A} \to \alpha + 1$ defining the E-rank of elements of $\mathfrak{A}$, where $\tau(\mathfrak{A}_0) = (\alpha, \alpha, 0)$. It is obvious that $\sigma(a) = 0$ for $a \in F_*(\mathfrak{A})$. An epimorphism φ from $\mathfrak{A}$ onto Λ induces on Λ the function $r_0^\varphi \colon \Lambda \to \alpha + 1$ such that $r_0^\varphi(\varphi(a)) = \sigma(a)$ for $a \in \mathfrak{A}$. The additivity of r_φ is obvious. The mapping $\sigma'(a) = \sigma(a^\perp)$ is complementary to σ. Since $\sigma'(a) = \sigma(a^\perp) = \sigma((a \vee b)^\perp) = \sigma'(a \vee b)$ for any $a \in \mathfrak{A}$ and $b \in F_*(\mathfrak{A})$, the function σ' also induces on Λ the mapping $r_1^\varphi \colon \Lambda \to \alpha + 1$, where $r_1^\varphi(\varphi(a)) = \sigma'(a)$. It is obvious that r_1^φ is complementary to r_0^φ.

Proposition 1.8.13. *Let* $\Lambda \in \{\mathfrak{B}_\eta, \mathfrak{B}_\eta^\omega\}$. *Two extensions*

$$0 \to \mathfrak{A}_0 \to \mathfrak{A} \xrightarrow{\varphi} \Lambda \to 0, \quad 0 \to \mathfrak{A}_0 \to \mathfrak{A}' \xrightarrow{\varphi'} \Lambda \to 0$$

of a special algebra $\mathfrak{A}_0$ *of E-rank* α *by the algebra* Λ *are isomorphic if and only if the triples* $\langle \Lambda, r_0^\varphi, r_1^\varphi \rangle$ *and* $\langle \Lambda, r_0^{\varphi'}, r_1^{\varphi'} \rangle$ *are isomorphic, i.e., there exists an automorphism* μ *of the algebra* Λ *such that* $r_0^\varphi = r_0^{\varphi'} \mu$ *and* $r_1^\varphi = r_1^{\varphi'} \mu$.

PROOF. We consider the case $\Lambda \cong \mathfrak{B}_\eta^\omega$. Let λ be an isomorphism between the algebras $\mathfrak{A}$ and $\mathfrak{A}'$. Then $\mathfrak{A}_0 = F_*(\mathfrak{A}) = F_*(\mathfrak{A}') = \mathfrak{A}_0$. Consequently, $\lambda(\mathfrak{A}_0) = \mathfrak{A}_0$ and λ induces an automorphism $\mu \colon \mathfrak{B}_\eta^\omega \to \mathfrak{B}_\eta^\omega$, where $\mu(\varphi(a)) = \varphi'\lambda(a)$. Since λ is an isomorphism, $\lambda \upharpoonright \widehat{d} \colon \widehat{d} \to \widehat{\lambda(d)}$ and $\lambda \upharpoonright d^\perp \colon d^\perp \to \lambda(d)^\perp$ are isomorphisms for any $d \in \mathfrak{A}$. But $\sigma(\widehat{d}) = \sigma(\lambda(\widehat{d}))$, $\sigma(d^\perp) = \sigma(\lambda(d)^\perp)$, $r_0^\varphi c = r_0^{\varphi'} \mu c$, and $r_1^\varphi c = r_1^{\varphi'} \mu c$ for any $c \in \mathfrak{B}_\eta^\omega$. Consequently, μ is an isomorphism between $\langle \mathfrak{B}_\eta^\omega, r_0^\varphi, r_1^\varphi \rangle$ and $\langle \mathfrak{B}_\eta^\omega, r_0^{\varphi'}, r_1^{\varphi'} \rangle$.

We now prove the converse assertion. Let μ be an automorphism of $\mathfrak{B}_\eta^\omega$ such that it establishes an isomorphism between $\langle \mathfrak{B}_\eta^\omega, r_0^\varphi, r_1^\varphi \rangle$ and $\langle \mathfrak{B}_\eta^\omega, r_0^{\varphi'}, r_1^{\varphi'} \rangle$. We define a condition of isomorphism S on $(\mathfrak{A} \times \mathfrak{A}')$ by setting

$$S = \{\langle a, b \rangle \mid a \in \mathfrak{A},\ b \in \mathfrak{A}',\ \mu\varphi(a) = \varphi'(b),\ \tau(a) = \tau(b),\ \tau(a^\perp) = \tau(b^\perp)\}$$

The validity of (1) and (2) is obvious. In view of symmetricity, it suffices to check only (3).

Let $\langle a, b \rangle \in S$ and $c \in \mathfrak{A}$. Consider $\widetilde{d}_0 \rightleftharpoons \mu\varphi(a \wedge c)$. We have $r_0^{\varphi'}(\widetilde{d}_0) = r_0^{\varphi'} \mu\varphi(a \wedge c) = r_0^\varphi(\varphi(a \wedge c))$ and $r_0^\varphi \varphi(a \backslash c) = r_1(\psi b \backslash \widetilde{d}_0)$ since $\mu\varphi(a) = \psi(b)$. We choose an element $d' \in \widehat{b}$ such that $\psi(d') = \widetilde{d}_0$. Since $\sigma d' = r_1 \widetilde{d}_0 = \sigma(a \wedge c)$ and $\sigma(b \backslash d') = r_1(\psi b \backslash \widetilde{d}_0) = r_0(\varphi(a \backslash c)) = \sigma(a \backslash c)$, by the definition of the superatomicity type, there exists an element $d_0 \leqslant b$ such that $\tau(d_0) = \tau(a \wedge c)$, $\tau(b \backslash d_0) = \tau(a \backslash c)$, and $\psi(d_0) = \psi d' = \widetilde{d}_0$. Therefore, $\langle a \wedge c, b \wedge d_0 \rangle$ and $\langle a \backslash c, b \backslash d_0 \rangle$ belong to S.

We now find the second element of d_1. Let $\widetilde{d}_1 = \mu\varphi(c \backslash a)$. We choose $d' \in b^\perp$ such that $\psi d' = \widetilde{d}_1$. From the equalities $\tau(a) = \tau(b)$, $\tau(a^\perp) = \tau(b^\perp)$, and definitions, it follows that there is an element $d' \in b^\perp$

1.8. Ershov Algebras and The Isomorphism Problem 85

such that $\tau(d_1) = \tau(c\backslash a)$, $\tau(d_1^\perp \cap b^\perp) = \tau((a \vee c)^\perp)$, and $\psi d_1 = \psi d' = \tilde{d}_1$. Therefore, $\langle c\backslash a, d_1\backslash b\rangle$ and $\langle a \vee c, b \vee d_1\rangle$ belong to S. Consequently, there exists an isomorphism between $\mathfrak{A}$ and $\mathfrak{A}'$.

The case $\Lambda = \mathfrak{B}_\eta$ is handled in a similar way. $\square$

As a consequence of the characterization of functions r' complementary to r via the ordinal $\gamma_{r_1^\varphi}$ and the filter $P_{r_1^\varphi}$, we obtain the description of types of isomorphism which is formulated in the following claim.

Corollary 1.8.2. *Let the hypotheses of Proposition 1.8.13 be fulfilled. Ershov algebras $\mathfrak{A}$ and $\mathfrak{A}'$ are isomorphic if and only if there exists an automorphism μ of the algebra Λ such that $r_0^\varphi = r_0^{\varphi'}\mu$, $P_{r_1^{\varphi'}} = \mu P_{r_1^\varphi}$, and $\gamma_{r_1^{\varphi'}} = \gamma_{r_1^\varphi}$.*

Proposition 1.8.14. *Let $\Lambda \in \{\mathfrak{B}_\eta, \mathfrak{B}_\eta^\omega\}$. If $\mathfrak{A}_0$ is a countable special algebra with $\sigma(\mathfrak{A}_0) = \alpha$, then for any additive function $r\colon \Lambda \to \alpha + 1$ and its complementary function $r'\colon \Lambda \to \alpha + 1$ there exists an extension $\mathfrak{A}$ of $\mathfrak{A}_0$ by Λ such that $r_0^\varphi = r_0$ and $r_1^\varphi = r_1$.*

PROOF. By Proposition 1.8.11, there exists a homomorphism $\psi\colon \Lambda \to \mathfrak{A}_0^*$ such that $r(a) = \rho\psi(a)$ and $r'(a) = \rho(C(\psi(a)))$ for any $a \in \mathfrak{B}_\eta^\omega$. By Theorem 1.3.1, we obtain the conclusion of the proposition. $\square$

We now consider the characterization of types of isomorphisms of countable normal Boolean algebras. Let $\mathfrak{A}_0$ be a countable special Ershov algebra of E-rank α. From any exact sequence $0 \to \mathfrak{A}_0 \to \mathfrak{B} \xrightarrow{\varphi} \mathfrak{B}_\eta \to 0$, i.e., from an extension $\mathfrak{A}_0$ by a Boolean algebra $\mathfrak{B}_\eta$, where $\mathfrak{B}$ is a Boolean algebra, an additive function of E-rank $\sigma\colon \mathfrak{B} \to \alpha+1$ we define an additive function $r_\mathfrak{B}\colon \mathfrak{B}_\eta \to \alpha + 1$ induced by σ, i.e., $r_\mathfrak{B}(\varphi(a)) = \sigma(a)$. As already mentioned, this function is well defined. In the case of Boolean algebras, complementary functions are already not necessary.

Proposition 1.8.15. *The following assertions hold:*

(a) *A Boolean algebra $\mathfrak{A}$ is isomorphic to a Boolean algebra $\mathfrak{B}$, being an extension of $\mathfrak{A}_0$ by $\mathfrak{B}_\eta$, if and only if there exists an automorphism μ of $\mathfrak{B}_\eta$ such that $r_\mathfrak{A}(a) = r_\mathfrak{B}(\varphi(a))$ for any $a \in \mathfrak{B}_\eta$,*

(b) *for any additive function $r\colon \mathfrak{B}_\eta \to \alpha + 1$ such that $r(\mathbf{1}) = \alpha$ there exists a Boolean algebra $\mathfrak{A}$ such that it is an extension of $\mathfrak{A}_0$ by $\mathfrak{B}_\eta$ and $r_\mathfrak{A} = r$.*

PROOF. To establish (a) we consider two exact sequences $0 \to \mathfrak{A}_0 \to \mathfrak{A} \xrightarrow{\varphi} \mathfrak{B}_\eta \to 0$ and $0 \to \mathfrak{A}_0 \to \mathfrak{B} \xrightarrow{\psi} \mathfrak{B}_\eta \to 0$. Let λ be an isomorphism between the Boolean algebras $\mathfrak{A}$ and $\mathfrak{B}$. Then $\lambda(F_*(\mathfrak{A})) = F_*(\mathfrak{B})$. Consequently, λ induces an isomorphism μ between $\mathfrak{A}/F_*(\mathfrak{A})$ and $\mathfrak{B}/F_*(\mathfrak{B})$. Since for any $a \in F_*(\mathfrak{A})$ the algebras $\widehat{a}$ and $\widehat{\lambda(a)}$ are isomorphic, for any $b \in \mathfrak{A}$ we have $\sigma(b) = \sigma(\lambda(b))$ and $r_\mathfrak{A}(b/F_*(\mathfrak{A})) = r_\mathfrak{B}(\lambda(b)/F_*(\mathfrak{A}))$. We now prove the converse assertion. Let us show that

$$S = \{(a,b) \mid \mu\varphi(a) = \psi(b),\ \tau(a) = \tau(b),\ \tau(C(a)) = \tau(C(b))\}$$

is a condition of isomorphism of the algebras $\mathfrak{A}$ and $\mathfrak{B}$ (cf. Sec. 1.5). Indeed, Ax1–Ax3 are obvious and Ax4 and Ax5 are symmetric. Therefore, it suffices to establish Ax4.

Let $\langle a,b \rangle \in S$ and $c \in \mathfrak{A}$. We consider $a \wedge c$ and $d \leqslant b$ such that $\mu\varphi(a \wedge c) = \psi(d)$. Let $\tau(a) = \langle \alpha, \beta, n \rangle = \tau(b)$ and $\tau(a \wedge c) = \langle \alpha', \beta', n' \rangle$. Since $a \wedge c \leqslant a$, we have $\alpha' \leqslant \alpha$ and $\beta' \leqslant \beta$. If $\alpha' = \alpha$ and $\beta' = \beta$, then $n' \leqslant n$. Since $\mu\varphi(a \wedge c) = \psi(d)$, we obtain $r^\varphi(\varphi(a \wedge c)) = r^{\varphi'}(\psi(d))$, which means that $\sigma(d) = \sigma(a \wedge c) = \beta$. Thus, $F_*(\widehat{d})/F_\beta(\widehat{d})$ has the greatest element. Since ψ transforms elements of $F_*(\mathfrak{B})$ into $\mathbf{0}$, we can take $j \in F_*(\widehat{d})$ such that $j/F_\beta(\widehat{d})$ is the greatest element of the quotient algebra $F_*(\widehat{d})/F_\beta(\widehat{d})$. Since $\tau(a) = \tau(b)$, $\tau(C(a)) = \tau(C(b))$, we can take $j_0 \leqslant b$ such that $\tau(j_0) = \langle \alpha', 0, n' \rangle$ and define a new element $d_0 = d \backslash j \vee j_0$. It is obvious that $d_0 \leqslant b$, $\tau(d_0) = \tau(a \wedge c)$, and $\tau(b \backslash d_0) = \tau(a \backslash c)$. Consequently, the pairs $\langle a \wedge c, d_0 \wedge b \rangle$ and $\langle a \backslash c, b \backslash d_0 \rangle$ belong to S. Thus, a condition of isomorphism S is fulfilled and (a) is proved.

The claim (b) follows from Proposition 1.8.11 and Theorem 1.3.1.

The proposition is proved. □

Theorem 1.8.1. *For any nonsuperatomic countable Ershov algebra $\mathfrak{A}$ there exists a normal decomposition of $\mathfrak{A}$ into the direct sum of the subalgebras $\mathfrak{A}_0$ and $\mathfrak{A}_1$; this decomposition is unique up to an isomorphism, i.e., for some other normal decomposition of the algebra $\mathfrak{A}$ into the direct sum of subalgebras $\mathfrak{B}_0$ and $\mathfrak{B}_1$ we have $\mathfrak{A}_0 \cong \mathfrak{B}_0$ and $\mathfrak{A}_1 \cong \mathfrak{B}_1$.*

PROOF. By Proposition 1.8.12, a normal decomposition exists. We prove that it is unique. Consider two cases.

CASE 1: $R_{\text{Bool}}(\mathfrak{A}) < \infty$. Then $\mathfrak{A}_0$, $\mathfrak{B}_0$ are Boolean algebras and $\mathfrak{A}_1$, $\mathfrak{B}_1$ are superatomic Ershov algebras of type (α, β, n). Since the Boolean algebras $\mathfrak{A}_0$ and $\mathfrak{B}_0$ are superatomic and the equality $\tau(\mathfrak{A}_0) = \tau(\mathfrak{B}_0)$ holds, the algebras $\mathfrak{A}_1$ and $\mathfrak{B}_1$ are isomorphic. Since $\mathfrak{A}_1$ and $\mathfrak{B}_1$ are superatomic,

1.8. Ershov Algebras and The Isomorphism Problem 87

$\mathfrak{A}_1 \subseteq F_*(\mathfrak{A})$ and $\mathfrak{B}_1 \subseteq F_*(\mathfrak{B})$. Therefore, the embeddings $i_0\colon \mathfrak{A}_0 \to \mathfrak{A}$ and $j_0\colon \mathfrak{B}_0 \to \mathfrak{A}$ induce the isomorphisms $\widehat{i_0}\colon \mathfrak{A}_0/F_*(\mathfrak{A}_0) \to \mathfrak{A}/F_*(\mathfrak{A})$ and $\widehat{j_0}\colon \mathfrak{B}_0/F_*(\mathfrak{B}_0) \to \mathfrak{B}/F_*(\mathfrak{A})$. Consequently, $\mathfrak{A}_0$ is an extension of $F_*(\mathfrak{A}) \cap \mathfrak{A}_0$ by $\mathfrak{A}/F_*(\mathfrak{B})$ and $\mathfrak{B}_0$ is an extension of $F_*(\mathfrak{B}) \cap \mathfrak{B}_0$ by $\mathfrak{A}/F_*(\mathfrak{B})$. However, $\mathfrak{A} = \mathfrak{A}_0 + \mathfrak{A}_1$, $\mathfrak{B} = \mathfrak{B}_0 + \mathfrak{B}_1$, and $\mathfrak{A}_1, \mathfrak{B}_1 \subseteq F_*(\mathfrak{A})$. Therefore, $F_*(\mathfrak{A}_0)$ and $F_*(\mathfrak{B}_0)$ are isomorphic since $\mathfrak{A}_0$ and $\mathfrak{B}_0$ are normal Ershov algebras. Then $F_*(\mathfrak{A}_0)$ and $F_*(\mathfrak{B}_0)$ are special algebras of the same superatomicity type. For these algebras we consider the functions σ_0 and σ_1 and construct the functions $r_0^0(a/F_*(\mathfrak{A}_0)) = \sigma_0(a)$ and $r_0^1(b/F_*(\mathfrak{B}_0)) = \sigma_1(b)$ for any $a \in \mathfrak{A}_0$ and $b \in \mathfrak{B}_0$. In view of the inclusion $\mathfrak{A}_1, \mathfrak{B}_1 \subseteq F_*(\mathfrak{A})$, we obtain $\sigma_0(a) = \sigma(a)$ and $\sigma_1(b) = \sigma(b)$ for any $a \in \mathfrak{A}_0$ and $b \in \mathfrak{B}_0$, where $\sigma\colon \mathfrak{A} \to \alpha + 1$ and $\alpha = \sigma(\mathfrak{A})$. Taking the isomorphism $\widehat{i_0} \circ \widehat{j_0}^{-1}$, we obtain an isomorphism between $\mathfrak{A}_0/F_*(\mathfrak{A}_0)$ and $\mathfrak{B}_0/F_*(\mathfrak{B}_0)$ such that $r_0^0(d) = r_0^1(\widehat{i} \circ \widehat{j}^{-1})(d)$ for any $d \in \mathfrak{A}_0/F_*(\mathfrak{A}_0)$. By Proposition 1.8.15, the algebras $\mathfrak{A}_0$ and $\mathfrak{B}_0$ are isomorphic.

CASE 2: $R_{\mathrm{Bool}}(\mathfrak{A}) = \infty$. Then $\mathfrak{A}/F_*(\mathfrak{A}) \cong \mathfrak{B}_\eta^\omega$ and $\mathfrak{A}_1, \mathfrak{B}_1$ are superatomic Boolean algebras such that either $\mathfrak{A}_1 = \mathfrak{B}_1 = \{0\}$ or $\mathfrak{A}_1$ and $\mathfrak{B}_1$ have the same superatomicity type. Consequently, $\mathfrak{A}_1 \cong \mathfrak{B}_1$. We show that $\mathfrak{A}_0$ and $\mathfrak{B}_0$ are also isomorphic. As in Case 1, $\mathfrak{A}_0$ is an extension of $F_*(\mathfrak{A}_0)$ by $\mathfrak{A}/F_*(\mathfrak{A})$ and $\mathfrak{B}_0$ is an extension of $F_*(\mathfrak{B}_0)$ by $\mathfrak{A}/F_*(\mathfrak{A})$. In view of the normality and the equality $\sigma(\mathfrak{A}_0) = \sigma(\mathfrak{B}_0)$, we see that $F_*(\mathfrak{A}_0)$ and $F_*(\mathfrak{B}_0)$ are isomorphic.

We now consider functions σ_0 and σ_1 of E-rank for $\mathfrak{A}_0$ and $\mathfrak{B}_0$ and the function σ of E-rank for $\mathfrak{A}$. By the equalities $\sigma_0(a) = \sigma_0(\widehat{a}) = \sigma(\widehat{a}) = \sigma(a)$ for any $a \in \mathfrak{A}_0$ and $\sigma_1(b) = \sigma_1(\widehat{b}) = \sigma(\widehat{b}) = \sigma(b)$ for any $b \in \mathfrak{B}_0$, the functions $r_0(a/F_*(\mathfrak{A}_0)) = \sigma_0(a)$ and $r_1(a/F_*(\mathfrak{B}_0)) = \sigma_1(a)$ are well defined and coincide with similar functions for $\mathfrak{A}$ under the natural isomorphisms between $\mathfrak{A}_0/F_*(\mathfrak{A}_0)$, $\mathfrak{A}/F_*(\mathfrak{A})$ and between $\mathfrak{B}_0/F_*(\mathfrak{B}_0)$, $\mathfrak{A}/F_*(\mathfrak{A})$. We show that the complementary functions also coincide. By definition, $r_1^0(a/F_*(\mathfrak{A}_0)) = \varphi_0((a)_{\mathfrak{A}}^\perp)$ and $r_1^1(a/F_*(\mathfrak{B}_0)) = \varphi_1((a)_{\mathfrak{B}_0}^\perp)$. However, $(a)_{\mathfrak{A}_0}^\perp = (a)_{\mathfrak{A}}^\perp \cap (b_0)_{\mathfrak{A}}^\perp$ and $(a)_{\mathfrak{B}_0}^\perp = (a)_{\mathfrak{A}}^\perp \cap (b_1)_{\mathfrak{A}}^\perp$, where b_0 is the greatest element of $\mathfrak{A}_1$ and b_1 is the greatest element of $\mathfrak{B}_1$. Consequently, $r_1^0(a/F_*(\mathfrak{A}_0)) = \varphi_0((a)_{\mathfrak{A}_0}^\perp) = \varphi((a)_{\mathfrak{A}}^\perp \cap (b_0)_{\mathfrak{A}}^\perp)$. We note that $\sigma((a)_{\mathfrak{A}}^\perp \cap (b_0)_{\mathfrak{A}}^\perp) = \sigma(a^\perp)$ and $\sigma((b)_{\mathfrak{A}}^\perp \cap (b_1)_{\mathfrak{A}}^\perp) = \sigma(b^\perp)$ because b_0 and b_1 are superatomic. Therefore, the complementary functions under these extensions with respect to the isomorphism $\widehat{i} \circ \widehat{j}^{-1}$ coincide. By Proposition 1.8.13, $\mathfrak{A}_0$ and $\mathfrak{B}_0$ are isomorphic. The uniqueness is proved. □

88   1. Algebraic Properties of Boolean Algebras

Corollary 1.8.3. *For any countable nonsuperatomic Boolean algebra $\mathfrak{A}$ of the superatomicity type (α, β, n) there exists a unique normal decomposition into the direct sum of subalgebras $\mathfrak{A}_0$ and $\mathfrak{A}_1$ such that $\mathfrak{A}_0, \mathfrak{A}_1$ are Boolean algebras, $\mathfrak{A}_0$ is a normal algebra, $\mathfrak{A}_1$ is a superatomic algebra, and $\tau(\mathfrak{A}_1) = \langle \alpha, 0, n \rangle$.*

The proof follows from Theorem 1.8.1 and definitions. □

Under the hypotheses of Corollary 1.8.3, $\mathfrak{A}_0$ is called the *normal part* and $\mathfrak{A}_1$ is called the *proper superatomic part* of $\mathfrak{A}$.

Corollary 1.8.4. *Two countable nonsuperatomic Boolean algebras $\mathfrak{A}$ and $\mathfrak{B}$ are isomorphic if and only if their normal parts and proper superatomic parts are isomorphic, respectively.*

Thus, we have described the isomorphism types for countable Boolean algebras. Recall the main steps of this description. We first separate the proper superatomic parts (we assume that it coincides with the whole algebra in the case of a superatomic Boolean algebra) and compare their types. If they coincide, then these parts are isomorphic and for the normal parts we may apply the characterization through additive functions on atomless Boolean algebras.

Exercises

1. Prove Proposition 1.8.1.

2. Let S be a condition of isomorphism of Ershov algebras $\mathfrak{A}$ and $\mathfrak{B}$. Prove the following assertions.

    (a) if $a < a'$, $b < b'$, and $(a'\backslash a, b'\backslash b) \in S$, then for any $c \in \mathfrak{A}$ such that $a \leqslant c \leqslant a'$ there exists d, $b \leqslant d \leqslant b'$, such that $(c\backslash a, d\backslash b) \in S$ and $(a'\backslash c, b'\backslash d) \in S$, while for any $b \leqslant d \leqslant b'$ there exists $a \leqslant c \leqslant a'$ such that $(c\backslash a, d\backslash b) \in S$ and $(a'\backslash c, b'\backslash d) \in S$,

    (b) if $\langle a, b \rangle \in S$ and $a < c$, then there exists $d > b$ such that $\langle c\backslash a, d\backslash b \rangle \in S$ and $\langle c, d \rangle \in S$.

3. Prove that for countable atomless Ershov algebras $\mathfrak{A}$ and $\mathfrak{B}$ without greatest elements the set
$$S = \{(a, b) \mid a \in \mathfrak{A},\ b \in \mathfrak{B},\ a = 0 \Leftrightarrow b = 0\}$$
is a condition of isomorphism.

1.8. Ershov Algebras and The Isomorphism Problem

4. If a is an element of an Ershov algebra $\mathfrak{A}$, then the set $a^\perp = \{x \in |\mathfrak{A}| \mid x \wedge a = 0\}$ defines a subalgebra of $\mathfrak{A}$.

5. Prove that $F_*(\mathfrak{A}) = F_\alpha(\mathfrak{A})$ for any Ershov algebra $\mathfrak{A}$ and ordinal α equal to the ordinal type of $\mathfrak{A}$ (i.e., α is the least ordinal such that $F_\alpha(\mathfrak{A}) = F_{\alpha+1}(\mathfrak{A})$).

6. Show that $\sigma(a \vee b) = \max\{\sigma(a), \sigma(b)\}$ for a function σ of E-rank of an Ershov algebra $\mathfrak{A}$ and for any elements a, b of $\mathfrak{A}$.

7. If $\mathfrak{A}$ is a special Ershov algebra, then $\sigma(\mathfrak{A}) = \sigma(a^\perp)$ for any $a \in \mathfrak{A}$.

8. If $\mathfrak{A}$ is a special Ershov algebra, then the subalgebra $a^\perp$ is a special algebra for any $a \in \mathfrak{A}$.

9. If Ershov algebras $\mathfrak{A}$ and $\mathfrak{B}$ are special, then $\mathfrak{A} \times \mathfrak{B}$ is a special Ershov algebra.

10. If $\mathfrak{A}, \mathfrak{B}$ are special Ershov algebras, $\mathfrak{A}', \mathfrak{B}'$ are arbitrary Ershov algebras, and $\sigma(\mathfrak{A}) = \sigma(\mathfrak{A}') \leqslant \sigma(\mathfrak{B}) = \sigma(\mathfrak{B}')$, $\tau(\mathfrak{A} \times \mathfrak{B}) = \tau(\mathfrak{A}' \times \mathfrak{B}')$, then $\tau(\mathfrak{B}) = \tau(\mathfrak{B}')$.

11. If $\mathfrak{A}$ is a superatomic Boolean algebra of type (α, n), then the superatomicity type of $\mathfrak{A}$ (as an Ershov algebra) is $(\alpha, 0, n)$.

12. Prove Proposition 1.8.11.

13. Prove the second case in Proposition 1.8.13.

Chapter 2

Elementary Classification of Boolean Algebras

2.1. Basic Notions and Methods of Model Theory

Before characterizing elementary theories of Boolean algebras, we consider some basic model-theoretic constructions used below (cf. [18, 36, 72]). Studying the model-theoretic properties of constructions, we consider the whole class Mod (Δ) of algebraic systems subject to given axioms. Algebraic systems on which axioms of Δ hold are called *models of the theory* Δ and the class Mod (Δ) is called *axiomatizable*. If an axiomatizable class contains an infinite algebraic system or finite systems that are as large as desired, then it also contains systems of cardinality greater than any given one. This fact follows from the Mal'tsev compactness theorem (cf. Theorem 1.1.2).

For a class of algebraic systems we introduce a certain notion that is stronger than that of an isomorphic embedding. The following reasoning justifies the necessity of this notion. The truth of formulas is invariant under isomorphisms "onto." However, this is not true in the case of isomorphic embeddings. For example, under an isomorphic embedding of a countable atomic Boolean algebra into an atomless Boolean algebra, atoms go to atomless elements.

A mapping φ from the universe A of an algebraic system $\mathfrak{A}$ into the universe B of an algebraic system $\mathfrak{B}$ is called an *elementary embedding* if for any formula $\Phi(x_1, \ldots, x_n)$ and elements $a_1, \ldots, a_n \in A$

$$\mathfrak{A} \vDash \Phi(a_1, \ldots, a_n) \Leftrightarrow \mathfrak{B} \vDash \Phi(\varphi(a_1), \ldots, \varphi(a_n))$$

An elementary embedding is denoted by $\varphi : \mathfrak{A} \preccurlyeq \mathfrak{B}$. It is easy to see that an elementary embedding is an isomorphic embedding. A subsystem $\mathfrak{A}$ of an algebraic system $\mathfrak{B}$ is called an *elementary subsystem* of $\mathfrak{B}$ and $\mathfrak{B}$ is called an *elementary extension* of $\mathfrak{A}$ if the identity embedding of A into B is an elementary embedding, i.e., $\mathfrak{A} \preccurlyeq \mathfrak{B}$.

The following theorem can be proved by induction on the complexity of formulas.

Theorem 2.1.1. *If $\mathfrak{A}_0 \preccurlyeq \mathfrak{A}_1 \preccurlyeq \ldots \preccurlyeq \mathfrak{A}_n \preccurlyeq \ldots$ is a chain of elementary embedded models, then the union $\mathfrak{A}_\omega$ of models of this chain is an elementary extension of each model $\mathfrak{A}_i$, $i = 0, 1, \ldots$.*

Let $\varphi(\mathfrak{A})$ stand for the set of all sequences of elements of $\mathfrak{A}$ on which the formula $\varphi(\overline{x})$ is true. If $\Phi(x_1, \ldots, x_n)$ is the set of formulas in free variables of the sequence $x_1, \ldots, x_n$, then we set $\Phi(\mathfrak{A}) \rightleftharpoons \bigcap\limits_{\varphi \in \Phi} \varphi(\mathfrak{A})$.

We present one more criterion for a subsystem to be elementary, which is convenient in the study of the model completeness of some algebraic theories. This criterion allows us to know whether a given subsystem is elementary without checking the truth of formulas. This criterion can be proved by induction on the complexity of formulas.

Proposition 2.1.1 [on elementariness of subsystems]. *A subsystem $\mathfrak{A}$ of an algebraic system $\mathfrak{B}$ is elementary if and only if the following Robinson condition holds:*

(R) *for any elements $a_1, \ldots, a_n$ of A and a formula of the form $(\exists y)\varphi(x_1, \ldots, x_n, y)$, if $\mathfrak{B} \vDash (\exists y)\varphi(a_1, \ldots, a_n, y)$, then there exists $b \in A$ such that $\mathfrak{B} \vDash \varphi(a_1, \ldots, a_n, b)$.*

2.1. Basic Notions and Methods of Model Theory

This proposition turns out to be a powerful tool in determining the elementariness of concrete embeddings. In particular, it is very useful in the case of "small" systems.

Let $\mathfrak{A}$ be an algebraic system of a signature σ and let $X \subseteq A$. We construct an elementary subsystem that contains X and has cardinality equal to the maximum of cardinalities of X and of the first-order language L_σ of the signature σ. To this end, we consider all formulas of the form $(\exists y)\varphi(x_1, \ldots, x_n, y)$ in the language L_σ and extend the signature σ to a signature σ^* by adding new operations F_φ^n whose numbers of places are equal to the numbers of variables occurring free in the corresponding formulas $(\exists y)\varphi(x_1, \ldots, x_n, y)$. We define the enrichment $\mathfrak{A}^*$ of $\mathfrak{A}$ by these new operations:

$$F_\varphi(a_1, \ldots, a_n) \rightleftharpoons \begin{cases} h(A) & \text{if } \mathfrak{A} \not\models (\exists y)\varphi(a_1, \ldots, a_n, y) \\ h(\{a \mid \mathfrak{A} \models \varphi(a_1, \ldots, a_n, a)\}) \\ & \text{if } \mathfrak{A} \models (\exists y)\varphi(a_1, \ldots, a_n, y) \end{cases}$$

where h is the choice function defined on the set of nonempty subsets of A. Such functions are called *Skolem functions*. We consider the subsystem $\mathrm{gr}_{\mathfrak{A}^*}(X)$ of $\mathfrak{A}^*$ generated by X. Using Proposition 2.1.1, it is easy to verify that its restriction to a system of the signature σ is an elementary subsystem of $\mathfrak{A}$. The enrichment $\mathfrak{A}^*$ of $\mathfrak{A}$ is called the *Skolem enrichment*. Iterating the above construction ω times, we find that the constructed subsystem is elementary not only in the initial language, but in the enrichment too. This fact and the Mal'tsev compactness theorem imply that, in an axiomatizable class containing infinite systems, there exist systems of any infinite cardinality, including countable and continual ones.

Two algebraic systems are called *elementarily equivalent* if the same sentences are true in them. We denote by $\mathrm{Th}(\mathfrak{A})$ the set of all sentences that are true in $\mathfrak{A}$, and the set $\mathrm{Th}(\mathfrak{A})$ is referred to as the *theory* of $\mathfrak{A}$. We note that $\mathfrak{A}$ is a model for a set of axioms Δ if and only if $\Delta \subseteq \mathrm{Th}(\mathfrak{A})$. Let T be a complete theory in a signature σ and let $p(x_1, \ldots, x_n)$ be a set of formulas having free occurrences only of variables $x_1, \ldots, x_n$. A maximal set of formulas having free occurrences of variables $x_1, \ldots, x_n$ is called a *type of the theory* T in the variables $x_1, \ldots, x_n$ if it is consistent with T. Let $S_n(T)$ denote the set of types of the theory T in n variables. If $\mathfrak{A}$ is an algebraic system of the theory T and $a_1, \ldots, a_n$ are elements of A, then the set

$$p_{\overline{a}}(x_1, \ldots, x_n) = \{\varphi(x_1, \ldots, x_n) \mid \mathfrak{A} \models \varphi(a_1, \ldots, a_n)\}$$

being a type of the theory T, is called the *type of the n-tuple* $\bar{a} = (a_1, \ldots, a_n)$. By the Gödel completeness theorem and the Mal'tsev compactness theorem (cf. Theorems 1.1.1 and 1.1.2), any type $p \in S_n(T)$ is the type of some n-tuple of elements of a suitable system $\mathfrak{A} \vDash T$. If a type $p(x_1, \ldots, x_n)$ holds on a sequence $\bar{a}$ of an algebraic system $\mathfrak{A}$, then we say that the type $p(x_1, \ldots, x_n)$ is *realized* on $\bar{a}$ in $\mathfrak{A}$.

We introduce a topology τ on the set $S_n(T)$. For a basis of open sets we take

$$\{\widehat{\varphi} \mid \varphi \text{ is a formula consistent with the theory } T \text{ in free variables of } x_1, \ldots, x_n\}$$

where $\widehat{\varphi} \rightleftharpoons \{p(x_1, \ldots, x_n) \mid \varphi \in p(x_1, \ldots, x_n) \in S_n(T)\}$. This topology is called the *Mal'tsev topology* of the set of types. By the Mal'tsev compactness theorem, $S_n(T)$ equipped with this topology is a compact topological space.

A formula $\varphi(x_1, \ldots, x_n)$ is called a *complete formula* of a theory T if it is consistent with T and, for any formula $\psi(x_1, \ldots, x_n)$, we have $T \vdash \varphi \to \psi$ or $T \vdash \varphi \to \neg\psi$. A type containing complete formulas is called *principal*. It is obvious that every principal type is realized in any system of the theory T, i.e., for any system, there is a sequence of elements on which all the formulas of a principal type hold. An algebraic system $\mathfrak{A}$ *omits a type* $p(x_1, \ldots, x_n)$ if there is no sequence $\bar{a}$ realizing the type $p(x_1, \ldots, x_n)$ in $\mathfrak{A}$.

To construct models possessing prescribed properties, the Henkin method is often used. In particular, this method is useful if we deal with those theories in enrichments of a signature by constants (called the Henkin theories) for which it is possible to construct a model of the quotient set of the set of constant symbols.

A complete theory T is called a *Henkin theory* if the following condition (called the *Henkin condition*) holds:

(H) *for any sentence of the form* $(\exists x)\varphi(x)$ *there is a constant c such that the sentence* $[\varphi(x)]_c^x$ *is deduced from T provided that the sentence* $(\exists x)\varphi(x)$ *is deduced from T.*

We consider the set C of all constant symbols of a signature σ of a Henkin theory T and the following equivalence relation: $c \sim d$ for $c, d \in C$ if $T \vdash c = d$. For the universe of the desired model we take the quotient set $M_T = C/\sim$. On M_T, the signature symbols will be interpreted as follows:

— $(c_1/\sim, \ldots, c_n/\sim) \in P$ if and only if $T \vdash P(c_1, \ldots, c_n)$ for an n-ary predicate symbol $P \in \sigma$,

2.1. Basic Notions and Methods of Model Theory 95

— for an n-ary functional symbol $F \in \sigma$ and any sequence of constant symbols $c_1, \ldots, c_n$ we set $F(c_1/\sim, \ldots, c_n/\sim) \rightleftharpoons c/\sim$, where c is a constant symbol such that $T \vdash F(c_1, \ldots, c_n) = c$ (such c exists by the Henkin condition),

— for the value of a constant symbol $c \in \sigma$ we take the congruence class $c/\sim$.

The notions just introduced are well defined. Thus, we obtain the algebraic system H_T of the signature σ, which will be referred to as the *Henkin model* of the theory T.

Proposition 2.1.2. *For any formula $\varphi(x_1 \ldots, x_n)$ of a signature σ and constant symbols $c_1, \ldots, c_n$ the following equivalence holds: $H_T \vDash \varphi(c_1/\sim, \ldots, c_n/\sim) \Leftrightarrow T \vdash \varphi(c_1, \ldots, c_n)$.*

PROOF. We can argue by induction on the complexity of the formula φ taking into account the completeness condition and the Henkin condition (H) on the theory T as well. □

Corollary 2.1.1. *All sentences of the theory T hold in the Henkin model H_T.*

The Henkin method allows us to solve the omitting-types problem. Applying the method, it is possible to prove also the Gödel completeness theorem (cf. Theorem 1.1.1) for the predicate calculus.

Theorem 2.1.2 [omitting-types theorem]. *If $p_i(\overline{x}_i)$, $i \in \mathbb{N}$, is a countable family of nonprincipal types in a countable signature σ of a consistent theory T in the signature σ, then there exists a countable model $\mathfrak{A}$ of the theory T omitting all these types.*

PROOF. We consider the enrichment σ_C of σ by new constant symbols $\{c_0, c_1, \ldots, c_n, \ldots\}$, a set Δ of sentences of the signature σ_C, and the set S of all maximal subsets of Δ that are consistent with T. On S, we introduce the structure of a topological space. For a basis of topology τ we take a set of the form $\widehat{\varphi} = \{\Delta \in S \mid \varphi \in \Delta\}$, where φ is a sentence of the signature σ_C that is consistent with T. By the Mal'tsev compactness theorem (cf. Theorem 1.1.2), the topological space (S, τ) is compact and, by the countability of the language, is a Baire space. For any sentence of the form $(\exists x)\varphi$ of the signature σ_C the set

$$U_{(\exists x)\varphi(x)} \rightleftharpoons \{\Delta \in S \mid \text{there exists } c \in C \text{ such that } \varphi(c) \in \Delta\}$$

is open and dense in (S, τ) everywhere. For any type $p_i(\overline{x}_i)$ and a sequence of constant symbols $\overline{c}$ the set

$$U_{p_i,\overline{c}} \rightleftharpoons \{\Delta \in S \mid \text{ there exists a formula } \varphi \in p_i \text{ such that } \neg \varphi(\overline{c}) \in \Delta\}$$

is open and is dense everywhere. By the Baire theorem [88, 99], there exists a set Δ of formulas such that $\Delta \in \bigcap\limits_{(\exists x)\varphi} U_{(\exists x)\varphi} \cap \bigcap\limits_{i} \bigcap\limits_{\overline{c}} U_{p_i,\overline{c}}$. Since $\Delta \in U_{(\exists x)\varphi}$ for any sentence $(\exists x)\varphi$, it follows that Δ is a Henkin theory. Since $\Delta \in \bigcap\limits_{\overline{c}} U_{p_i,\overline{c}}$, the Henkin model H_Δ omits the type p_i. Thus, the restriction of the Henkin model to a model of the signature σ yields the required algebraic system. □

We consider some particular types for countable models of complete countable theories. Let $\mathfrak{A}$ and $\mathfrak{B}$ be models of a complete theory T in a signature σ. Sequences $\overline{a}$ and $\overline{b}$ of elements of $\mathfrak{A}$ and $\mathfrak{B}$ respectively are called *elementarily equivalent* (denoted by $\overline{a} \equiv \overline{b}$) if they realize the same type. For a family X of elements of an algebraic system $\mathfrak{A}$ we denote by $(\mathfrak{A}, X)$ the system obtained from $\mathfrak{A}$ by adding new constant symbols $c_a, a \in X$, so that the value of the constant c_a is exactly a. In the case $X = \{a_1, \ldots, a_n\}$, we also use the notation $(\mathfrak{A}, a_1, \ldots, a_n)$. We note that sequences $\overline{a}$ and $\overline{b}$ are elementarily equivalent if and only if the systems $(\mathfrak{A}, \overline{a})$ and $(\mathfrak{B}, \overline{b})$ are elementarily equivalent.

An algebraic system $\mathfrak{A}$ is called ω-*homogeneous* if, for any elementarily equivalent sequences $\overline{a}$ and $\overline{b}$ of $\mathfrak{A}$ and an element a of $\mathfrak{A}$, there is an element $b \in \mathfrak{A}$ such that the extended sequences $(\overline{a}, a)$ and $(\overline{b}, b)$ are elementarily equivalent. A countable ω-homogeneous system is called *homogeneous*.

A mapping f from a subset X of an algebraic system $\mathfrak{A}$ onto a subset Y of an algebraic system $\mathfrak{B}$ is called a *partial elementary (isomorphic) embedding* of $\mathfrak{A}$ into $\mathfrak{B}$ if for any (quantifier-free) formula $\varphi(x_1, \ldots, x_n)$ and elements $x_1, \ldots, x_n$ of X

$$\mathfrak{A} \vDash \varphi(x_1, \ldots, x_n) \Leftrightarrow \mathfrak{B} \vDash \varphi(f(x_1), \ldots, f(x_n))$$

For elementarily equivalent sequences $(a_1, \ldots, a_n)$ of $\mathfrak{A}$ and $(b_1, \ldots, b_n)$ of $\mathfrak{B}$ a mapping f from $(a_1, \ldots, a_n)$ onto $(b_1, \ldots, b_n)$ such that $f(a_i) = b_i$ is a partial elementary embedding. For a system $\mathfrak{A}$ we denote by $S_\mathfrak{A}$ the set of all types realized in $\mathfrak{A}$.

Theorem 2.1.3. *Countable homogeneous algebraic systems $\mathfrak{A}$ and $\mathfrak{B}$ are isomorphic if and only if $S_\mathfrak{A} = S_\mathfrak{B}$.*

2.1. Basic Notions and Methods of Model Theory

PROOF. Applying the back and forth method, for any finite elementary embedding of $\mathfrak{A}$ into $\mathfrak{B}$ we are able to construct an elementary embedding that extends the initial one and contains a given element of $\mathfrak{A}$ or $\mathfrak{B}$ in the domain of definition or the range of values respectively. □

The following assertion is proved in a similar way.

Corollary 2.1.2. *Any finite elementary embedding of a countable homogeneous model $\mathfrak{A}$ into a countable homogeneous model $\mathfrak{B}$ such that $\mathfrak{A}$ and $\mathfrak{B}$ realize the same family of types can be extended to an isomorphism of $\mathfrak{A}$ onto $\mathfrak{B}$.*

Corollary 2.1.3. *Any finite elementary embedding of a countable homogeneous model into itself can be extended to an automorphism.*

Using the same method of constructing an extending chain of partial elementary embeddings, one can prove the following assertion.

Corollary 2.1.4. *If $\mathfrak{A}$ is a countable model, $\mathfrak{B}$ is an ω-homogeneous model of a theory T, and $S_{\mathfrak{A}} \subseteq S_{\mathfrak{B}}$, then there exists an elementary embedding of $\mathfrak{A}$ into $\mathfrak{B}$.*

The question on the existence of countable homogeneous models can be easily clarified due to the Mal'tsev compactness theorem and the diagram method. We consider the enrichment $\mathfrak{A}^* = (\mathfrak{A}, A)$ of an algebraic system $\mathfrak{A}$ of the signature $\sigma_A^* = \sigma \cup \{c_a \mid a \in A\}$ obtained by adding constants c_a, $a \in A$, to the signature σ. The theory $\text{Th}(\mathfrak{A}^*)$ is called a *complete diagram* of $\mathfrak{A}$ and is denoted by $D_\omega(\mathfrak{A})$. By the *diagram* of $\mathfrak{A}$ we mean the set

$$D_0(\mathfrak{A}) = \{\varphi \mid \mathfrak{A}^* \vDash \varphi, \varphi \text{ is an atomical formula or the negation}$$
$$\text{of an atomical formula of the extended signature } \sigma_A^*\}$$

We formulate the following obvious claims.

⟨1⟩ *An algebraic system $\mathfrak{A}$ is isomorphically embedded into a system $\mathfrak{B}$ if there exists an enrichment $\mathfrak{B}_A^*$ of $\mathfrak{B}$ of the extended signature σ_A^* such that $\mathfrak{B}_A^* \vDash D_0(\mathfrak{A})$.*

⟨2⟩ *An algebraic system $\mathfrak{A}$ is elementarily embedded into a system $\mathfrak{B}$ if there exists an enrichment $\mathfrak{B}_A^*$ of $\mathfrak{B}$ of the extended signature σ_A^* such that $\mathfrak{B}_A^* \vDash D_\omega(\mathfrak{A})$.*

Applying the diagram method, we can prove the following theorem.

Theorem 2.1.4. *Any countable algebraic system is elementarily embedded into some countable homogeneous algebraic system.*

Thus, any theory has countable homogeneous models. This is a good basis for further study of theories.

We consider homogeneous models under the additional conditions of minimality and maximality. Not all the theories are subject to these conditions, but for theories satisfying them the study is significantly simplified.

An algebraic system is called *atomic* if only principal types are realized in it. It is obvious that an atomic system is homogeneous. A model $\mathfrak{A}$ of a theory T is called *prime* if it is elementarily embedded in any model of the theory T.

From Corollary 2.1.4 and Theorem 2.1.2 on omitting nonprincipal types we obtain the following claim.

Theorem 2.1.5. *If a complete theory T has an infinite model, then a model $\mathfrak{A} \vDash T$ of T is prime if and only if it is countable and atomic.*

A countable model $\mathfrak{A}$ of a theory T is called *universal* if any countable model of the theory T is elementarily embedded into $\mathfrak{A}$.

We consider an internal maximality condition. A model $\mathfrak{A}$ is called ω-*saturated* if, for any sequence $a_1, \ldots, a_n$, any type $p(x, a_1, \ldots, a_n)$ of the theory $\mathrm{Th}\,(\mathfrak{A}, a_1, \ldots, a_n)$ is realized in $(\mathfrak{A}, a_1, \ldots, a_n)$. It is obvious that ω-saturated models are ω-homogeneous.

A countable ω-saturated model $\mathfrak{A}$ of a theory T is called a *countably saturated model* of the theory T.

From Corollary 2.1.4 we obtain the following claim.

Theorem 2.1.6. *If a complete theory T has an infinite model, then a countable model $\mathfrak{A} \vDash T$ is countably saturated if and only if it is universal and homogeneous.*

We proceed to the question on the existence of prime models and countably saturated models. Let T be a complete theory in a signature σ. We consider the Lindenbaum–Boolean algebra $\mathfrak{B}_n(T)$ of the theory T of formulas in n free variables and the set $\mathcal{F}_n(T) \rightleftharpoons \mathcal{F}(\mathfrak{B}_n(T))$ of all ultrafilters of $\mathfrak{B}_n(T)$. As above, let $S_n(T)$ stand for the set of types of the theory T in n free variables $x_1, \ldots, x_n$. The mappings

$$\mathrm{Filter}\,(p(x_1, \ldots, x_n)) \rightleftharpoons \{\varphi/{\sim_T} \mid \varphi \in p\} \text{ for } p \in S_n(T),$$
$$\mathrm{Type}\,(F) = \{\varphi \mid \varphi/{\sim_T} \in F\}, \text{ for } F \in \mathcal{F}_n(T)$$

2.1. Basic Notions and Methods of Model Theory

are reciprocal and establish a one-to-one correspondence between types and ultrafilters. Furthermore, they are homeomorphisms provided that $S_n(T)$ is equipped with the Mal'tsev topology and $\mathcal{F}_n(T)$ is equipped with the Stone topology.

A theory T is called

— *atomic* if for any n the Lindenbaum algebra $\mathfrak{B}_n(T)$ is atomic,

— *superatomic* if for any n the Lindenbaum algebra $\mathfrak{B}_n(T)$ is superatomic.

It is clear that a superatomic theory is atomic. A theory T is atomic if and only if for any n the set of principal types in n variables is dense in $S_n(T)$ everywhere. A theory T is superatomic if and only if for any n the set of types in n variables is at most countable.

Theorem 2.1.7. *A complete theory has a prime model if and only if it is atomic.*

PROOF. The assertion is valid because of Theorem 2.1.2 and the fact that prime models are atomic. □

A theory is called *countably categorical* if it has a unique up to an isomorphism countable model.

Corollary 2.1.5. *A complete theory is countably categorical if and only if for any n the Lindenbaum algebra of formulas in n free variables is finite.*

Theorem 2.1.8. *A complete theory has a countably saturated model if and only if it is superatomic.*

PROOF. The necessity follows from the fact that a countably saturated model is countable and universal. To prove the sufficiency, we note that there exists a countable model realizing all types of the theory because of Theorem 1.1.2 and the fact that the family of all types of a complete theory is countable. The homogeneous countable elementary extension of this model is a countably saturated model. □

Corollary 2.1.6. *If a theory has a countable universal model, then it has a countably saturated model.*

Corollary 2.1.7. *If a complete theory has a countably saturated model, then it has a prime model.*

Corollary 2.1.8. *A complete theory has a countably saturated model if and only if the family of all types of this theory is countable.*

Corollary 2.1.9. *If a complete theory has at most a countable number of countable models, then it has a countably saturated model and a prime model.*

We consider one more property of elementary theories: model completeness. It is of great importance for understanding structures of models and for studying their algorithmic properties as well. A theory T is called *model-complete* if for any model $\mathfrak{N}$ of the theory T the theory $T \cup D_0(\mathfrak{N})$ is complete, where $D_0(\mathfrak{N})$ is a diagram of $\mathfrak{N}$. Adding the new predicate symbol $P_\varphi(x_1, \ldots, x_n)$ to the signature of the theory T for any formula $\varphi(x_1, \ldots, x_n)$ and the new axiom

$$(\forall x_1 \ldots \forall x_n)(\varphi(x_1, \ldots, x_n) \Leftrightarrow P_\varphi(x_1, \ldots, x_n))$$

defining the above predicate in the theory T, we extend T to a theory that is semantically equivalent to T but is model-complete. We emphasize the importance of the problem of finding of the least number of new definable predicates providing the model completeness of extensions. In particular, the solution of this problem enables us to understand the structure of models of initial theories.

Let σ be a countable signature and let T be a theory in the signature σ. A formula $\varphi(x_1, \ldots, x_n)$ is called *T-equivalent* to a formula $\psi(x_1, \ldots, x_n)$ if

$$T \vdash (\forall x_1 \ldots \forall x_n)(\varphi(x_1, \ldots, x_n) \leftrightarrow \psi(x_1, \ldots, x_n)).$$

Theorem 2.1.9. *For a theory T in a signature σ the following conditions are equivalent:*

(1) *T is model-complete,*

(2) *for any at most countable model $\mathfrak{N}$ of T the theory $T \cup D_0(\mathfrak{N})$ is complete,*

(3) *if $\mathfrak{N}$ and $\mathfrak{M}$ are at most countable models of T and $\mathfrak{N}$ is a submodel of the model $\mathfrak{M}$, then $\mathfrak{N}$ is an elementary subsystem of $\mathfrak{M}$,*

(4) *if $\mathfrak{N}$ and $\mathfrak{M}$ are models of T and $\mathfrak{N}$ is a submodel of the model $\mathfrak{M}$, then $\mathfrak{N}$ is an elementary submodel of $\mathfrak{M}$,*

(5) *any $\exists$-formula is T-equivalent to a $\forall$-formula,*

2.1. Basic Notions and Methods of Model Theory

(6) *any ∀-formula is T-equivalent to an ∃-formula,*

(7) *any formula is T-equivalent to an ∃-formula,*

(8) *any formula is T-equivalent to a ∀-formula,*

(9) *any formula is T-equivalent to a ∀-formula and an ∃-formula.*

PROOF. The implication (1) ⇒ (2) follows from the definition.

The implication (2) ⇒ (3) holds since $T \cup D_0(\mathfrak{N})$ is complete and the diagram $D_0(\mathfrak{N})$ is satisfiable in the extension.

We now prove the implication (3) ⇒ (4). In the case of an uncountable model $\mathfrak{M}$, we consider an arbitrary finite sequence $a_1, \ldots, a_n$ of elements of $\mathfrak{N}$ and use the Skolem enrichment to construct an at most countable elementary subsystem $\mathfrak{N}_0 \preccurlyeq \mathfrak{N}$ containing elements $a_1, \ldots, a_n$ and an at most countable elementary subsystem $\mathfrak{M}_0 \preccurlyeq \mathfrak{M}$ containing all elements of $\mathfrak{N}_0$. Since $\mathfrak{N}_0 \subseteq \mathfrak{M}_0$ and the above models are at most countable, $\mathfrak{N}_0 \preccurlyeq \mathfrak{M}_0$. The following relations hold:

$$\mathfrak{N}_0 \vDash \varphi(a_1, \ldots, a_n) \Leftrightarrow \mathfrak{N} \vDash \varphi(a_1, \ldots, a_n)$$
$$\mathfrak{N}_0 \vDash \varphi(a_1, \ldots, a_n) \Leftrightarrow \mathfrak{M}_0 \vDash \varphi(a_1, \ldots, a_n)$$
$$\mathfrak{M}_0 \vDash \varphi(a_1, \ldots, a_n) \Leftrightarrow \mathfrak{M} \vDash \varphi(a_1, \ldots, a_n)$$

Therefore, $\mathfrak{N} \vDash \varphi(a_1, \ldots, a_n) \Leftrightarrow \mathfrak{M} \vDash \varphi(a_1, \ldots, a_n)$. Consequently, $\mathfrak{N} \preccurlyeq \mathfrak{M}$.

To prove (4) ⇒ (5) we consider an ∃-formula $\varphi(x_1, \ldots, x_n)$ and the set of formulas

$$\Phi(x_1, \ldots, x_n) \rightleftharpoons \{\psi(x_1, \ldots, x_n) \mid \psi(x_1, \ldots, x_n) \text{ is a } \forall\text{-formula and}$$
$$T \vdash (\forall x_1 \ldots \forall x_n)(\varphi(x_1, \ldots, x_n) \to \psi(x_1, \ldots, x_n))\}$$

If $T \cup \Phi(x_1, \ldots, x_n) \cup \{\neg \varphi(x_1, \ldots, x_n)\}$ is inconsistent, then there exists a finite subset $\{\psi_1, \ldots, \psi_n\} \subseteq \Phi(x_1, \ldots, x_n)$ such that $T \cup \{\psi_1, \ldots, \psi_n\} \cup \{\neg \varphi(x_1, \ldots, x_n)\}$ is inconsistent. Hence $T \cup \{\psi_1, \ldots, \psi_n\} \vdash \varphi(x_1, \ldots, x_n)$. Then $T \cup \{\&\psi_i\} \vdash \varphi(x_1, \ldots, x_n)$. Consequently, $T \vdash \&\psi_i \to \varphi(x_1, \ldots, x_n)$. Since $T \vdash \varphi(x_1, \ldots, x_n) \to \&\psi_i$, $1 \leqslant i \leqslant n$, the formula $\varphi(x_1, \ldots, x_n)$ is T-equivalent to a conjunction of ∀-formulas. It is easy to see that a conjunction of ∀-formulas is equivalent to a ∀-formula. If $T \cup \Phi(x_1, \ldots, x_n) \cup \{\neg \varphi(x_1, \ldots, x_n)\}$ are consistent, we consider a model $\mathfrak{N}$. Let $x_1, \ldots, x_n$ be elements of $\mathfrak{N}$ on which this set of formulas hold. We consider the diagram $D_0(\mathfrak{N})$ of the model $\mathfrak{N}$ and the set of formulas $D_0(\mathfrak{N}) \cup T \cup \{\varphi(x_1, \ldots, x_n)\}$. It is obvious that the set is consistent and there exists a model $\mathfrak{M}$ in which it holds. Since $D_0(\mathfrak{N})$ and T hold in $\mathfrak{M}$, we conclude that the model $\mathfrak{M}$ of the theory T extends the model $\mathfrak{N}$. Then $\mathfrak{N} \preccurlyeq \mathfrak{M}$ and $\mathfrak{M} \vDash \neg\varphi(x_1, \ldots, x_n)$. However, by construction, we have $\mathfrak{M} \vDash \varphi(x_1, \ldots, x_n)$. The implication (4) $\Rightarrow$ (5) is proved.

The implication (5) $\Rightarrow$ (6) follows from the fact that the negation of an $\exists$-formula is equivalent to a $\forall$-formula and the negation of a $\forall$-formula is equivalent to an $\exists$-formula.

The implication (6) $\Rightarrow$ (7) is proved by induction on the number of alternations of quantifiers occurring in a formula in prenex normal form.

Since the negation of an $\exists$-formula is equivalent to a $\forall$-formula, the implication (7) $\Rightarrow$ (8) is obtained from the fact that two formulas are T-equivalent if and only if their negations are T-equivalent. The implication (8) $\Rightarrow$ (9) is established in a similar way.

To complete the proof it remains to show that (9) implies (1). Let there exist a model $\mathfrak{N}$ of T such that the theory $T \cup D_0(\mathfrak{N})$ is not complete. There is a sentence $\varphi(a_1, \ldots, a_n)$ with constants a_i, $1 \leqslant i \leqslant n$, for elements of a model $\mathfrak{N}$ such that the sets of formulas

$$T \cup D_0(\mathfrak{N}) \cup \{\varphi(a_1, \ldots, a_n)\}, \quad T \cup D_0(\mathfrak{N}) \cup \{\neg\varphi(a_1, \ldots, a_n)\}$$

are consistent. Let the first set hold in $\mathfrak{A}$ and let the second one hold in $\mathfrak{B}$. Since a diagram of $\mathfrak{N}$ holds in $\mathfrak{A}$ and $\mathfrak{B}$, we can assume that $\mathfrak{N} \subseteq \mathfrak{A}$ and $\mathfrak{N} \subseteq \mathfrak{B}$. By (9), there exist $\forall$-formulas $\psi_0(x_1, \ldots, x_n)$ and $\psi_1(x_1, \ldots, x_n)$ such that

$$T \vdash (\forall x_1 \ldots \forall x_n)(\varphi(x_1, \ldots, x_n) \Leftrightarrow \psi_0(x_1, \ldots, x_n))$$
$$T \vdash (\forall x_1 \ldots \forall x_n)(\neg\varphi(x_1, \ldots, x_n) \Leftrightarrow \psi_1(x_1, \ldots, x_n))$$

Since $\mathfrak{A} \vDash T$ and $\mathfrak{B} \vDash T$, we have $\mathfrak{A} \vDash \psi_0(a_1, \ldots, a_n)$, $\mathfrak{B} \vDash \psi_1(a_1, \ldots, a_n)$. However, $\psi_0(x_1, \ldots, x_n)$ and $\psi_1(x_1, \ldots, x_n)$ are $\forall$-formulas. If they hold in a model, they also hold in its submodels. Consequently, $\mathfrak{N} \vDash \psi_0(a_1, \ldots, a_n)$ and $\mathfrak{N} \vDash \psi_1(a_1, \ldots, a_n)$. The system $\mathfrak{N}$ is a model of the theory T, hence $\mathfrak{N} \vDash \varphi(a_1, \ldots, a_n)$ and $\mathfrak{N} \vDash \neg\varphi(a_1, \ldots, a_n)$, which is impossible. The theorem is proved. □

Let T be a theory in a signature σ.

Theorem 2.1.10 [18]. *A formula $\varphi_1(x_1, \ldots, x_n)$ is T-equivalent to an $\exists$-formula of a signature σ if and only if it preserves the truth on extensions, i.e., for any models $\mathfrak{N}$ and $\mathfrak{M}$ of the theory T and elements $a_1, \ldots, a_n$ of $\mathfrak{N}$, where $\mathfrak{N}$ is a submodel of $\mathfrak{M}$ and φ is true on elements $a_1, \ldots, a_n$ in $\mathfrak{N}$, the formula φ is also true on the same elements $a_1, \ldots, a_n$ in $\mathfrak{M}$.*

2.1. Basic Notions and Methods of Model Theory

PROOF. The necessity follows from the definition of truth. To prove the sufficiency we consider the following set of formulas:

$$\Phi(x_1,\ldots,x_n) \rightleftharpoons \{\neg\psi(x_1,\ldots,x_n) \mid \psi(x_1,\ldots,x_n) \text{ is an } \exists\text{-formula},$$
$$T \vdash (\forall x_1 \ldots \forall x_n)(\psi(x_1,\ldots,x_n) \to \varphi(x_1,\ldots,x_n))\}$$

If the set $T \cup \Phi(x_1,\ldots,x_n) \cup \{\varphi(x_1,\ldots,x_n)\}$ is inconsistent, then there exists a sequence of formulas $\neg\psi_1,\ldots,\neg\psi_m$ of Φ such that the set $T \cup \{\neg\psi_1,\ldots,\neg\psi_m\} \cup \{\varphi(x_1,\ldots,x_n)\}$ is inconsistent. Therefore, $T \cup \{\underset{i=1}{\overset{m}{\&}} \neg\psi_i\} \cup \{\varphi(x_1,\ldots,x_n)\}$ is inconsistent. Consequently,

$$T \cup \{\varphi(x_1,\ldots,x_n)\} \vdash \neg(\underset{i=1}{\overset{m}{\&}} \neg\psi_i), \quad T \cup \{\varphi(x_1,\ldots,x_n)\} \vdash \bigvee_{i=1}^{m} \psi_i$$

Since $T \vdash \psi_i \to \varphi(x_1,\ldots,x_n)$, $1 \leqslant i \leqslant m$, we have

$$T \vdash (\forall x_1 \ldots \forall x_n)(\varphi(x_1,\ldots,x_n) \Leftrightarrow \bigvee_{i=1}^{m} \psi_i)$$

A disjunction of $\exists$-formulas is equivalent to an $\exists$-formula. Hence the conclusion of the theorem is true.

We assume that the set $T \cup \Phi(x_1,\ldots,x_n) \cup \{\varphi(x_1,\ldots,x_n)\}$ is satisfiable. We consider a model $\mathfrak{M}$ of the theory T and elements $a_1,\ldots,a_n$ of $\mathfrak{M}$ such that $\mathfrak{M} \vDash \Phi(a_1,\ldots,a_n)$ and $\mathfrak{M} \vDash \varphi(a_1,\ldots,a_n)$. Let $D_0(\mathfrak{M})$ be a diagram of the model $\mathfrak{M}$. Assuming that $T \cup D_0(\mathfrak{M}) \cup \{\neg\varphi(a_1,\ldots,a_n)\}$ is satisfiable, we obtain an extension $\mathfrak{N}$ of the model $\mathfrak{M}$ in which the formula $\varphi(x_1,\ldots,x_n)$ is false on the elements $a_1,\ldots,a_n$, which contradicts the hypothesis. Consequently, the set $T \cup D_0(\mathfrak{M}) \cup \{\neg\varphi(a_1,\ldots,a_n)\}$ is unsatisfiable. By the Mal'tsev compactness theorem (cf. Theorem 1.1.2), there are a finite number of formulas $\zeta_1,\ldots,\zeta_m$ of $D_0(\mathfrak{M})$ such that $T \cup \{\zeta_1,\ldots,\zeta_m\} \cup \{\neg\varphi(a_1,\ldots,a_n)\}$ is unsatisfiable. By the Gödel completeness theorem (cf. Theorem 1.1.1), it is inconsistent. Hence $T \cup \{\underset{i=1}{\overset{m}{\&}} \zeta_i\} \cup \{\neg\varphi(a_1,\ldots,a_n)\}$ is inconsistent and $T \cup \{\underset{i=1}{\overset{m}{\&}} \zeta_i\} \vdash \varphi(a_1,\ldots,a_n)$. Therefore, $T \vdash \underset{i=1}{\overset{m}{\&}} \zeta_i \to \varphi(a_1,\ldots,a_n)$.

We consider all the new constants $b_1,\ldots,b_m$ that occur in the formula $\underset{i=1}{\overset{m}{\&}} \zeta_i$ and do not belong to the sequence $a_1,\ldots,a_n$. In the proof of the formula $\&\zeta_i \to \varphi(a_1,\ldots,a_n)$ we replace all the constants $b_1,\ldots,b_m$ by the new variables $y_1,\ldots,y_m$ and the constants $a_1,\ldots,a_n$ by the new variables $x_1,\ldots,x_n$. Since formulas of T do not contain the constants $b_1,\ldots,b_m$, the proof is reduced to that of the formula

$$[\&\zeta_i \to \varphi]_{y_1,\ldots,y_m,x_1,\ldots,x_n}^{b_1,\ldots,b_m,a_1,\ldots,a_n}$$

obtained by the substitution of variables for constants. The constants $b_1,\ldots,b_m$ do not occur in φ, hence the variables $y_1,\ldots,y_m$ do not occur in it either; moreover, the last formula coincides with the formula

$$[[\&\zeta_i]_{y_1,\ldots,y_m}^{b_1,\ldots,b_m} \to \varphi]_{x_1,\ldots,x_n}^{a_1,\ldots,a_n}$$

Consequently, $T \vdash (\forall x_1 \ldots \forall x_n)(\forall y_1 \ldots \forall y_m)([\&\zeta_i]_{y_1,\ldots,y_m}^{b_1,\ldots,b_m} \to \varphi)$. Since the variables $y_1,\ldots,y_m$ do not occur in the formula φ, the latter is equivalent to the formula

$$(\forall x_1 \ldots \forall x_n)[(\exists y_1 \ldots \exists y_m)[\&\zeta_i]_{y_1,\ldots,y_m}^{b_1,\ldots,b_m} \to \varphi]_{x_1,\ldots,x_n}^{a_1,\ldots,a_n}$$

and is also deduced from T. Since the formula

$$\exists y_1 \ldots \exists y_m [[\&\zeta_i]_{y_1,\ldots,y_m,x_1,\ldots,x_n}^{b_1,\ldots,b_m,a_1,\ldots,a_n}]$$

is an $\exists$-formula, its negation belongs to Φ by the construction of Φ. The negation of this formula is true on the elements $a_1,\ldots,a_n$ in $\mathfrak{N}$. By condition, the formula $[\&\zeta_i]$ is true on the values of the constants, i.e., the formula

$$(\exists y_1 \ldots \exists y_m)\,[\&\zeta_i]_{y_1,\ldots,y_m,x_1,\ldots,x_n}^{b_1,\ldots,b_m,a_1,\ldots,a_n}$$

holds on the sequence $a_1,\ldots,a_n$. The contradiction obtained completes the proof. □

Corollary 2.1.10. *A formula $\varphi(x_1,\ldots,x_n)$ is T-equivalent to a $\forall$-formula if and only if it preserves the truth in submodels.*

A model $\mathfrak{N}$ of a theory T is called *n-complete* if for any formula $\varphi(x_1,\ldots,x_m)$ in prenex normal form with at most n alternative groups of quantifiers and a sequence of elements $a_1,\ldots,a_m$ of $\mathfrak{N}$, where $\mathfrak{N} \vDash \varphi(a_1,\ldots,a_m)$, there is an $\exists$-formula $\psi(x_1,\ldots,x_m)$ such that

$$\mathfrak{N} \vDash \psi(a_1,\ldots,a_m), \quad T \vdash (\forall x_1 \ldots \forall x_m)(\psi(x_1,\ldots,x_m) \to \varphi(x_1,\ldots,x_m))$$

A model $\mathfrak{N}$ of a theory T is called ω-*complete* if it is n-complete for any n. Any model of a model-complete theory is ω-complete. The converse assertion is also true.

Corollary 2.1.11. *A theory is model-complete if and only if any model of this theory is ω-complete.*

2.1. Basic Notions and Methods of Model Theory

A model $\mathfrak{A}$ of a theory T is said to be *almost n-complete (almost complete)* if there exists an n-complete (complete) finite enrichment $(\mathfrak{A}, a_1, \ldots, a_m)$ of $\mathfrak{A}$ by constants $a_1, \ldots, a_m$.

From the definition of n-completeness we arrive at the following claim.

Proposition 2.1.3. *Any n-complete (ω-complete) model is also n-complete (ω-complete) in any finite enrichment by constants.*

By the method of complete diagrams, it is easy to prove the following proposition.

Proposition 2.1.4. *For any elementarily equivalent models $\mathfrak{A}$ and $\mathfrak{B}$ there exists a model $\mathfrak{C}$ such that $\mathfrak{A}$ and $\mathfrak{B}$ are elementarily embedded into $\mathfrak{C}$.*

Proceeding by induction on the complexity of formulas, we can prove the followings assertion.

Proposition 2.1.5. *If $\mathfrak{A} \preccurlyeq \mathfrak{A}'$ and $\mathfrak{B} \preccurlyeq \mathfrak{B}'$, then $\mathfrak{A} \times \mathfrak{B} \preccurlyeq \mathfrak{A}' \times \mathfrak{B}'$.*

From Propositions 2.1.4 and 2.1.5 we obtain the following claim.

Corollary 2.1.12. *If $\mathfrak{A} \equiv \mathfrak{A}'$ and $\mathfrak{B} \equiv \mathfrak{B}'$, then $\mathfrak{A} \times \mathfrak{B} \equiv \mathfrak{A}' \times \mathfrak{B}'$.*

Let T and T' be theories in signatures σ and σ' respectively. The theory T' is called *definable in the theory T* if for every signature symbol $\zeta \in \sigma'$ there exists a formula Φ_ζ and a formula $\Delta(x)$ of the signature σ such that for any formula φ of the signature σ' the relation $T' \vdash \varphi$ holds if and only if $T \vdash \widehat{\varphi}$, where $\widehat{\varphi}$ is obtained from φ as follows:

For a formula φ we first consider an equivalent formula φ' in perfect disjunctive normal form $Q_1 x_1 \ldots Q_m x_m \Psi$.

Replace subformulas of Ψ of the form (1.1.1) by formulas of the signature σ. As a result, we obtain a formula Ψ'. For a constant c the formula $\Phi_c(x)$ has only one free variable. For an n-ary functional symbol $F \in \sigma'$ the formula $\Phi_F(x_1, \ldots, x_n, x_{n+1})$ has only $n+1$ free variables $x_1, \ldots, x_n, x_{n+1}$. For an n-ary predicate symbol $P \in \sigma'$ the formula $\Phi_P(x_1, \ldots, x_n)$ has only n free variables.

In the quantifier-free part Ψ of formula φ', we make the change of atomical subformulas as follows:

subformulas of the form $y = c$ are replaced by the formulas $\Phi_c(y)$,

subformulas of the form $f(x_1, \ldots, x_n) = y$ are replaced by the formulas $\Phi_f(x_1, \ldots, x_n, y)$, subformulas of the form $P(x_1,\ldots,x_n)$ are replaced by the formulas $\Phi_P(x_1,\ldots,x_n)$.

As a result, we obtain the formula $Q_1 x_1,\ldots Q_m x_m \Psi'$. The last step is to transform subsequently the prefix $Q_1 x_1,\ldots Q_m x_m \Psi'$, beginning from a formula of the form $(Q_m x)\Psi_m(x)$ on the restricted quantifiers:

subformulas of the form $(\exists x_i)\Psi_i(x_i)$ are replaced by the formulas $\Psi_{i-1} \leftrightharpoons (\exists x_i)(\Delta(x_i) \& \Psi(x_i))$,

subformulas of the form $(\forall x_i)\Psi(x_i)$ are replaced by the formulas $\Psi_{i-1} \leftrightharpoons (\forall x_i)(\Delta(x) \to \Psi_i(x))$, where $\Psi_m \leftrightharpoons \Psi'$; since the formula φ' has the form $Q_1 x_1 \ldots Q_m x_m \Psi$, where Ψ is a quantifier-free formula, we transform the "quantifier" prefix as well as the quantifier-free part of Ψ in the same way as above.

Having transformed all such formulas, we obtain the required formula $\widehat{\varphi}$ of the signature σ.

We now consider a model $\mathfrak{M}$ of a signature σ of a theory T and the subset $A = \{x \mid \mathfrak{M} \vDash \Delta(x)\}$. On the latter, using the defining formulas, we introduce an interpretation of a signature σ' and obtain a model of T' which will be denoted by $\mathfrak{M}_A$. By the Mal'tsev compactness theorem (cf. Theorem 1.1.2), from an arbitrary model $\mathfrak{N} \vDash T'$ we construct a model $\mathfrak{M} \vDash T$ such that $\mathfrak{N} \preccurlyeq \mathfrak{M}_A$. Thus, up to an elementary equivalence, we obtain all models of the theory T'. Therefore, having the classification of complete extensions of T, we can obtain such a classification for T'.

The notion of definability allows us to interpret the theory under consideration in some known theory and transfer many of its properties to the initial theory. Let a signature σ' extend a signature σ and let a theory T' be obtained from a theory T by adding the following axioms:

— $(\forall v)(v = c \Leftrightarrow \varphi_c(v))$ for a constant $c \in \sigma'_C \setminus \sigma_C$ and some formula $\varphi_c(v)$ of the signature σ,

— $(\forall v_1 \ldots \forall v_n \forall v_{n+1})(f(v_1,\ldots,v_n) = v_{n+1} \Leftrightarrow \varphi_F(v_1,\ldots,v_n,v_{n+1}))$ for an n-ary functional symbol $f \in \sigma'_F \setminus \sigma_F$ and some formula $\varphi_f(v_1,\ldots,v_n,v_{n+1})$ of the signature σ,

— $(\forall v_1 \ldots \forall v_n)(P(v_1,\ldots,v_n) \Leftrightarrow \varphi_P(v_1,\ldots,v_n))$ for an n-ary predicate symbol $P \in \sigma'_P \setminus \sigma_P$ and some formula $\varphi_P(v_1,\ldots,v_n)$ of the signature σ.

Then the theory T' is called a *definable enrichment* of the theory T to a theory of the signature σ' (or to the signature σ' for brevity). We can

2.1. Basic Notions and Methods of Model Theory

uniquely enrich any model $\mathfrak{A}$ of the theory T to a model $\mathfrak{A}'$ of the signature σ' of the theory T'. Restrictions of models of the theory T' get all models of the theory T. The axioms added to T are called *definitions* of the corresponding signature symbols.

We say that a theory T has

— a *recursive system of axioms* if there exists a system of axioms A and an algorithm that can recognize whether a given formula is an axiom,

— an *enumerable system of axioms* if there exists a system of axioms A and an algorithm enumerating all the axioms of A and only them.

A theory T is called *decidable* if there exists an algorithm that, in a finite number of steps, can recognize whether a given arbitrary sentence φ satisfies $T \vdash \varphi$.

Theorem 2.1.11. *If a complete theory T has an enumerable system of axioms, then T is decidable.*

PROOF. It is possible to clarify whether an arbitrary sequence of formulas is a proof. In addition, one must take into account that either $T \vdash \varphi$ or $T \vdash \neg\varphi$ holds for any sentence φ. We look over finite sequences of proofs as long as we obtain φ or $\neg\varphi$. If there is a proof of φ, then $T \vdash \varphi$ and if there is a proof of $\neg\varphi$, then $T \nvdash \varphi$. □

Corollary 2.1.13. *If a theory is categorical with respect to some cardinality and has no finite models, then it is complete. In addition, if it has an enumerable system of axioms, then it is decidable.*

Corollary 2.1.14. *If a theory has a countable universal model and an enumerable system of axioms, then it is complete and decidable.*

Corollary 2.1.15. *If a theory has a prime model and has an enumerable system of axioms, then it is complete and decidable.*

Proposition 2.1.6. *If a theory has an enumerable system of axioms, then it has a recursive system of axioms.*

We emphasize the following simple claim.

Proposition 2.1.7. *The following assertions hold*:

(a) *if a theory T' is undecidable and is definable in T, then T is also undecidable,*

(b) *if a theory T' is definable in a decidable theory T, then T' is also decidable.*

2.2. Definable Ershov–Tarski Ideals and Elementary Characteristics of Boolean Algebras

In this section, we consider ideals that are formally definable in Boolean algebras and examine their structure. With each Boolean algebra $\mathfrak{B}$ we associate the elementary characteristic $\operatorname{ch}(\mathfrak{B})$ which completely defines the elementary theory of $\mathfrak{B}$. Describing properties and the first-order definability of such ideals, we follow Ershov [39], who gave the elementary characterization of Boolean algebras for the first time. We also note that Tarski [194] announced a result concerning the decidability and elementary classification of Boolean algebras.

For a Boolean algebra $\mathfrak{A}$ we define a certain ideal $I(\mathfrak{A})$, called an *Ershov–Tarski ideal*. We define

$$I(\mathfrak{A}) = \{x \in A \mid \text{there exists an atomic element } a \text{ and an atomless element } b \text{ such that } x = a \vee b\}$$

Lemma 2.2.1. *The set $I(\mathfrak{A})$ is an ideal of $\mathfrak{A}$.*

PROOF. For $x \in I(\mathfrak{A})$ and $y \leqslant x$ we consider an atomic element a and an atomless element b such that $x = a \vee b$. Then $a \wedge y$ is an atomic element and $b \wedge y$ is an atomless one. Therefore, $y = (a \wedge y) \vee (b \wedge y)$ and $y \in I(\mathfrak{A})$. If $x = a_1 \vee b_1$ and $y = a_2 \vee b_2$, where a_1, a_2 are atomic elements and b_1, b_2 are atomless ones, then $x \vee y = (a_1 \vee a_2) \vee (b_1 \vee b_2)$, $a_1 \vee a_2$ is an atomic element and $b_1 \vee b_2$ is an atomless element, i.e., $x \vee y \in I(\mathfrak{A})$. □

Lemma 2.2.2. *If $a \notin I(\mathfrak{A})$, then there are infinitely many pairwise disjoint atoms under a.*

PROOF. Assume the contrary. Let the set

$$\operatorname{at}(a) = \{c \mid c \leqslant a, c \text{ is an atom}\}$$

be finite. Then $d = \bigvee_{c \in \operatorname{at}(a)} c$ is an atomic element located under a, and there is no atom under $a \setminus d$. In this case, $a = d \vee (a \setminus d)$ and an element $a \setminus d$ is atomless. Consequently, $a \in I(\mathfrak{A})$, which contradicts the assumption. □

2.2. Definable Ershov–Tarski Ideals

Lemma 2.2.3. *If $a = x \vee y = x' \vee y'$, where x, x' are atomic elements and y, y' are atomless ones, then $x = x'$ and $y = y'$.*

PROOF. We note that $x \wedge y'$ as well as $y \wedge x'$ is simultaneously an atomic element and an atomless element, which is possible only if it is zero. Therefore,

$$x = x \wedge a = x \wedge (x' \vee y') = (x \wedge x') \vee (x \wedge y') = x \wedge x'$$
$$= x' \wedge x = (x' \wedge x) \vee (x' \wedge y) = x' \wedge (x \vee y) = x' \wedge a = x'$$
$$y = y \wedge a = y \wedge (x' \vee y') = (y \wedge x') \vee (y \wedge y') = y \wedge y'$$
$$= y' \wedge y = (y' \wedge y) \vee (y' \wedge x) = y' \wedge (x \vee y) = y' \wedge a = y'$$

The lemma is proved. $\square$

We now introduce a sequence of Ershov–Tarski ideals $I_n(\mathfrak{A})$, $n \in \mathbb{N}$, of a Boolean algebra $\mathfrak{A}$ by setting

$$I_0(\mathfrak{A}) \rightleftharpoons \{0\}, \quad I_{n+1}(\mathfrak{A}) \rightleftharpoons I_n(\mathfrak{A}) \circ I\big(\mathfrak{A}/I_n(\mathfrak{A})\big)$$

By definition, $a \in I_{n+1}(\mathfrak{A})$ if and only if $a/I_n(\mathfrak{A})$ is the union of an atomic element of $\mathfrak{A}/I_n(\mathfrak{A})$ and an atomless element of $\mathfrak{A}/I_n(\mathfrak{A})$. We define the elementary characteristic $\operatorname{ch}(\mathfrak{A}) = (\operatorname{ch}_1(\mathfrak{A}), \operatorname{ch}_2(\mathfrak{A}), \operatorname{ch}_3(\mathfrak{A}))$ of a Boolean algebra $\mathfrak{A}$. The least n such that $I_{n+1}(\mathfrak{A}) = \mathfrak{A}$ is called the *first (elementary) characteristic* of $\mathfrak{A}$ and is denoted by $\operatorname{ch}_1(\mathfrak{A})$. If there is no such n, we set $\operatorname{ch}_1(\mathfrak{A}) = \infty$. For $\operatorname{ch}_1(\mathfrak{A}) = \infty$ we agree that the *second characteristic* $\operatorname{ch}_2(\mathfrak{A})$ and the *third characteristic* $\operatorname{ch}_3(\mathfrak{A})$ of $\mathfrak{A}$ are zero. Otherwise (i.e., if $\operatorname{ch}_1(\mathfrak{A}) = n$), the second characteristic $\operatorname{ch}_2(\mathfrak{A})$ is equal to the number of atoms of the Boolean algebra $\mathfrak{A}/I_n(\mathfrak{A})$ provided that this number is finite and $\operatorname{ch}_2(\mathfrak{A}) = \infty$ if there are infinitely many atoms in $\mathfrak{A}/I_n(\mathfrak{A})$. The third characteristic is 1 if there is a nonzero atomless element of $\mathfrak{A}/I_n(\mathfrak{A})$ and is 0 in the opposite case.

By analogy, for an element $a \in \mathfrak{A}$ we introduce the elementary characteristic $\operatorname{ch}(a) = (\operatorname{ch}_1(a), \operatorname{ch}_2(a), \operatorname{ch}_3(a))$, where the elementary characteristics $\operatorname{ch}_1(a)$, $\operatorname{ch}_2(a)$, and $\operatorname{ch}_3(a)$ are defined as follows:

$$\operatorname{ch}_1(a) \rightleftharpoons \begin{cases} \infty, & a \notin \bigcup_{n \geqslant 0} I_n(\mathfrak{A}) \\ n, & a \in I_{n+1}(\mathfrak{A}) \setminus I_n(\mathfrak{A}) \end{cases}$$

$$\mathrm{ch}_2(a) \rightleftharpoons \begin{cases} 0, & \mathrm{ch}_1(a) = \infty \\ n, & \text{under } a/I_m(\mathfrak{A}) \text{ there are exactly } n \text{ atoms and} \\ & m = \mathrm{ch}_1(a) < \infty, \\ \infty, & \text{under } a/I_m(\mathfrak{A}) \text{ there are infinitely many} \\ & \text{atoms and } m = \mathrm{ch}_1(a) < \infty \end{cases}$$

$$\mathrm{ch}_3(a) \rightleftharpoons \begin{cases} 0, & \mathrm{ch}_1(a) = \infty \text{ or } \mathrm{ch}_1(a) = m < \infty \\ & \text{and under } a/I_m(\mathfrak{A}) \text{ there is no nonzero} \\ & \text{atomless element} \\ 1 & \text{otherwise} \end{cases}$$

It is easy to see that $\mathrm{ch}_i(\mathfrak{A}) = \mathrm{ch}_i(\mathbf{1})$ and $\mathrm{ch}_i(\widehat{a}) = \mathrm{ch}_i(a)$, $1 \leqslant i \leqslant 3$.

Lemma 2.2.4. *For any $a, b \in \mathfrak{A}$ such that $a \wedge b = \mathbf{0}$ the following conditions hold*:

(1) $\mathrm{ch}_1(a \vee b) = \max\{\mathrm{ch}_1(a), \mathrm{ch}_1(b)\}$,

(2) *if* $\mathrm{ch}_1(a) = \mathrm{ch}_1(b)$, *then* $\mathrm{ch}_2(a \vee b) = \mathrm{ch}_2(a) + \mathrm{ch}_2(b)$ *and* $\mathrm{ch}_3(a \vee b) = \max\{\mathrm{ch}_3(a), \mathrm{ch}_3(b)\}$,

(3) *if* $\mathrm{ch}_1(a) < \mathrm{ch}_1(b)$, *then* $\mathrm{ch}_i(a \vee b) = \mathrm{ch}_i(b)$, $1 \leqslant i \leqslant 3$.

REMARK 2.2.1. We adopt the convention that $\infty + n = n + \infty = \infty + \infty = \infty$.

PROOF OF LEMMA 2.2.4. (1) If $\mathrm{ch}_1(a) = n$ and $\mathrm{ch}_1(b) = m$, then $a \in I_{n+1}(\mathfrak{A}) \setminus I_n(\mathfrak{A})$, $b \in I_{m+1}(\mathfrak{A}) \setminus I_m(\mathfrak{A})$, $a \vee b \in I_{k+1}(\mathfrak{A}) \setminus I_k(\mathfrak{A})$, $k = \max\{n, m\}$. If $\mathrm{ch}_1(a) = \infty$ or $\mathrm{ch}_1(b) = \infty$, then $a \notin \bigcup_{n \geqslant 0} I_n(\mathfrak{A})$ or $b \notin \bigcup_{n \geqslant 0} I_n(\mathfrak{A})$. Therefore, $a \vee b \notin \bigcup_{n \geqslant 0} I_n(\mathfrak{A})$.

(2) For $\mathrm{ch}_1(a) = \mathrm{ch}_1(b) = \infty$ the assertion is obvious in view of (1). For $n = \mathrm{ch}_1(a) = \mathrm{ch}_1(b) < \infty$ there are no common atoms under $a/I_n(\mathfrak{A})$ and $b/I_n(\mathfrak{A})$ since they are pairwise disjoint. Therefore, $\mathrm{ch}_2(a \vee b) = \mathrm{ch}_2(a) + \mathrm{ch}_2(b)$. For the third elementary characteristic the assertion is valid in view of the fact that for any $y < a \vee b$ the following claim holds: $y/I_n(\mathfrak{A})$ is atomless if and only if $y \wedge a/I_n(\mathfrak{A})$ and $y \wedge b/I_n(\mathfrak{A})$ are atomless elements. Therefore, a nonzero atomless element can be located under the union of disjunctive elements if and only if it is located under at least one of the members of the union.

(3) If $\mathrm{ch}_1(a) < \mathrm{ch}_1(b)$, then $a \vee b/I_n(\mathfrak{A}) = b/I_n(\mathfrak{A})$ for any $n > \mathrm{ch}_1(a)$. □

2.2. Definable Ershov–Tarski Ideals

Lemma 2.2.5. *If $a \leqslant b$, then $\mathrm{ch}_1(a) < \mathrm{ch}_1(b)$ or $\mathrm{ch}_1(a) = \mathrm{ch}_1(b)$ & $\mathrm{ch}_2(a) \leqslant \mathrm{ch}_2(b)$ & $\mathrm{ch}_3(a) \leqslant \mathrm{ch}_3(b)$.*

PROOF. We use the definition of characteristics and the fact that, with an element a, an ideal contains all elements that are less than a. □

Lemma 2.2.6. *For any element a of a Boolean algebra $\mathfrak{A}$ and a triple (m_1, m_2, m_3) there exists an element $b \leqslant a$ such that one of the following conditions holds*:

(1) *if $m_1 < \mathrm{ch}_1(a)$ & $m_2 < \infty$ & $m_3 \leqslant 1$, then $\mathrm{ch}_i(b) = m_i$, $\mathrm{ch}_i(a \backslash b) = \mathrm{ch}_i(a)$, $1 \leqslant i \leqslant 3$,*

(2) *if $m_1 = \mathrm{ch}_1(a)$ & $m_2 \leqslant \mathrm{ch}_2(a)$ & $m_3 \leqslant \mathrm{ch}_3(a)$, then $\mathrm{ch}_i(b) = m_i$ for $1 \leqslant i \leqslant 3$; moreover, if $m_2 < \mathrm{ch}_2(a)$ or $\mathrm{ch}_3(a) \neq 0$, then $\mathrm{ch}_1(a \backslash b) = \mathrm{ch}_1(a)$, $\mathrm{ch}_2(a \backslash b) = \mathrm{ch}_2(a) - m_2$, and $\mathrm{ch}_3(a \backslash b) = \mathrm{ch}_3(a)$, where $\infty - \infty = 0$.*

PROOF. (1) Since $m_1 < \mathrm{ch}_1(a)$, we have $a \notin I_{m_1+1}(\mathfrak{A})$. Therefore, the element $a/I_{m_1}(\mathfrak{A})$ is not atomic. Consequently, there is a nonzero atomless element $x_0/I_{m_1}(\mathfrak{A})$ under $a/I_{m_1}(\mathfrak{A})$. By Lemma 1.3.3, we can assume that $x_0 \leqslant a$. Since $a \notin I_{m_1+1}(\mathfrak{A})$, we have $a/I_{m_1}(\mathfrak{A}) \notin I(\mathfrak{A}/I_{m_1}(\mathfrak{A}))$. By Lemma 2.2.2, there are infinitely many atoms under $a/I_{m_1}(\mathfrak{A})$. We choose m_2 pairwise disjoint atoms that are less than $a/I_{m_1}(\mathfrak{A})$, $x_i/I_{m_1}(\mathfrak{A})$, $1 \leqslant i \leqslant m_2$. Again, by Lemma 1.3.3, we can assume that $x_i \leqslant a$ and $x_i \wedge x_j = 0$ for $1 \leqslant i \neq j \leqslant m_2$. We obtain the desired element taking $b = \bigvee_{i=1}^{m_2} x_i$ if $m_3 = 0$ and $b = \bigvee_{i=0}^{m_2} x_i$ if $m_3 = 1$.

(2) If $m_2 < \infty$, we take just m_2 pairwise disjoint atoms $a_i/I_{m_1}(\mathfrak{A})$, $1 \leqslant i \leqslant m_2$, located under $a/I_{m_1}(\mathfrak{A})$. As in the case (1), we can find elements $x_i \leqslant a$, $1 \leqslant i \leqslant m_2$, that are pairwise disjoint and $x_i/I_{m_1}(\mathfrak{A}) = a_i/I_{m_1}(\mathfrak{A})$. Then $b \rightleftharpoons \bigvee_{i=1}^{m_2} x_i$ is the desired element for $m_3 = 0$. If $m_3 = 1$, then $\mathrm{ch}_3(a) = 1$ and we can take a nonzero atomless element $x_0/I_{m_1}(\mathfrak{A})$ such that it is located under $a/I_{m_1}(\mathfrak{A})$, $x_0 \leqslant a$, $x_0 \wedge x_i = 0$, $1 \leqslant i \leqslant m_2$, and $a \backslash x_0/I_{m_1}(\mathfrak{A})$ is not atomic. In this case, $b \rightleftharpoons \bigvee_{i=0}^{m_2} x_i$ is the desired element. The proof in the cases $m_1 = \infty$ and $m_2 = \infty$ is trivial. □

Lemma 2.2.7. *If a is an element of a Boolean algebra $\mathfrak{A}$ and $\mathrm{ch}_3(a) = 1$, then there exist elements b and $c \leqslant a$ such that $b \leqslant c$,*

(1) $\mathrm{ch}_1(a) = \mathrm{ch}_1(b) = \mathrm{ch}_1(c) = \mathrm{ch}_1(a \backslash c) = \mathrm{ch}_1(a \backslash b)$,

(2) $\mathrm{ch}_2(a) = \mathrm{ch}_2(b) = \mathrm{ch}_2(c)$, $\mathrm{ch}_2(a\backslash b) = \mathrm{ch}_2(b\backslash c) = 0$,

(3) $\mathrm{ch}_3(b) = 0$, $\mathrm{ch}_3(a\backslash b) = 1$, $\mathrm{ch}_3(a\backslash c) = \mathrm{ch}_3(c) = 1$.

PROOF. We choose an element $c \leqslant a$ so that $\mathrm{ch}_1(a\backslash c) = \mathrm{ch}_1(a) = \mathrm{ch}_1(c)$, $\mathrm{ch}_2(a\backslash c) = 0$, $\mathrm{ch}_2(c) = \mathrm{ch}_2(a)$, and $\mathrm{ch}_3(c) = \mathrm{ch}_3(a\backslash c) = 1$. Such an element exists by Lemma 2.2.6. Since $a/I_m(\mathfrak{A}) \in I(\mathfrak{A}/I_m(\mathfrak{A}))$, where $m = \mathrm{ch}_1(a)$, we have $a/I_m(\mathfrak{A}) = x/I_m(\mathfrak{A}) \vee y/I_m(\mathfrak{A})$, where $x/I_m(\mathfrak{A})$ is atomic and $y/I_m(\mathfrak{A})$ is atomless. By the properties of quotient algebras, we can assume that $x \vee y = a$ and $x \wedge y = \mathbf{0}$. Taking $b \leftrightharpoons c \wedge x$, we obtain the required element. □

Arguing by induction, we introduce a sequence Atom_n, $\mathrm{Atomless}_n$, Atomic_n, I_n, $n \in \mathbb{N}$, of formulas of the signature of Boolean algebras.

BASIS OF INDUCTION:

$I_0(x) \leftrightharpoons (x = (x \wedge C(x)))$
$\mathrm{Atom}_0(x) \leftrightharpoons (\forall y)(x \wedge y = y \Rightarrow (I_0(y) \vee y = x))$ & $\neg I_0(x)$
$\mathrm{Atomless}_0(x) \leftrightharpoons \neg(\exists y)(y \wedge x = y\, \&\, \mathrm{Atom}_0(y))$
$\mathrm{Atomic}_0(x) \leftrightharpoons \neg(\exists y)(y \wedge x = y\, \&\, \neg I_0(y)\, \&\, \mathrm{Atomless}_0(y))$

INDUCTIVE STEP. Let I_n, Atom_n, Atomic_n, and $\mathrm{Atomless}_n$ have been defined. We set

$I_{n+1}(x) \leftrightharpoons (\exists y)(\exists z)(\mathrm{Atomic}_n(y)\, \&\, \mathrm{Atomless}_n(z)\, \&\, x = y \vee z)$
$\mathrm{Atom}_{n+1}(x) \leftrightharpoons \neg I_{n+1}(x)\, \&\, (\forall y)(y \wedge x = y \Rightarrow (I_{n+1}(y) \vee I_{n+1}(x\backslash y)))$
$\mathrm{Atomless}_{n+1}(x) \leftrightharpoons \neg(\exists y)(y \wedge x = y\, \&\, \mathrm{Atom}_{n+1}(y))$
$\mathrm{Atomic}_{n+1}(x) \leftrightharpoons \neg(\exists y)(y \wedge x = y\, \&\, \mathrm{Atomless}_n(y)\, \&\, \neg I_{n+1}(y))$

We use the same letter $\mathfrak{A}$ to designate the enrichment of a Boolean algebra $\mathfrak{A}$ by the above-mentioned first-order definable unary predicates A_n, B_n, C_n, $\mathcal{I}_n$ that can be defined by the formulas Atom_n, $\mathrm{Atomless}_n$, Atomic_n, I_n..

Lemma 2.2.8. *For any Boolean algebra $\mathfrak{A}$ and $n \in \mathbb{N}$ the following assertions hold:*

(1) $\mathfrak{A} \vDash I_n(a) \Leftrightarrow a/I_n(\mathfrak{A}) = \mathbf{0}$,

(2) $\mathfrak{A} \vDash \mathrm{Atom}_n(a) \Leftrightarrow a/I_n(\mathfrak{A})$ *is an atom of the quotient algebra* $\mathfrak{A}/I_n(\mathfrak{A})$,

2.2. Definable Ershov–Tarski Ideals

(3) $\mathfrak{A} \vDash \text{Atomless}_n(a) \Leftrightarrow a/I_n(\mathfrak{A})$ *is an atomless element of the quotient algebra* $\mathfrak{A}/I_n(\mathfrak{A})$,

(4) $\mathfrak{A} \vDash \text{Atomic}_n(a) \Leftrightarrow a/I_n(\mathfrak{A})$ *is an atomic element of the quotient algebra* $\mathfrak{A}/I_n(\mathfrak{A})$.

PROOF. The assertion follows from the lemma on quotient algebras and the definition of the truth of formulas in algebraic systems. □

For a formula Φ we set $[\Phi]^0 \rightleftharpoons \Phi$, $[\Phi]^1 \rightleftharpoons \neg\Phi$. For any natural numbers m_1, m_2, and $m_3 \leqslant 1$ we introduce the formula

$$E^{m_1,m_2,m_3}(x) \rightleftharpoons I_{m_1+1}(x) \,\&\, \neg I_{m_1}(x) \,\&\, (\exists x_1 \exists \ldots \exists x_{m_2})$$

$$(\underset{1 \leqslant i < j \leqslant m_2}{\&}\, x_i \wedge x_j = 0 \,\&\, \underset{i=1}{\overset{m_2}{\&}} (x_i \leqslant x \,\&\, \text{Atom}_{m_1}(x_i))$$

$$\&\, ((\exists y) \text{Atomless}_{m_1}(y) \,\&\, ((\underset{i=1}{\overset{m_2}{\&}} x_i \vee y = x) \,\&\, [I_{m_1}(y)]^{m_3}))$$

By Lemma 2.2.8, $\mathfrak{A} \vDash E^{m_1,m_2,m_3}(a) \Leftrightarrow \text{ch}(a) = (m_1, m_2, m_3)$.

We also introduce one more series of formulas $\mathbb{N}^{n,k}$ for natural n and $k \geqslant 1$ as follows:

$$\mathbb{N}^{n,k}(x) \rightleftharpoons (\exists x_1 \ldots \exists x_k)(\underset{1 \leqslant i < j \leqslant k}{\&}\, x_i \wedge x_j = 0 \,\&\, \underset{i=1}{\overset{k}{\&}} (x_i \leqslant x \,\&\, \text{Atom}_n(x_i)))$$

By Lemma 2.2.8,

$$\mathfrak{A} \vDash \mathbb{N}^{n,k}(a) \Leftrightarrow \text{ch}_1(a) > n \vee (\text{ch}_1(a) = n \,\&\, \text{ch}_2(a) \geqslant k).$$

Therefore, from Lemma 2.2.6 we obtain the following claim.

Corollary 2.2.1. *If* $\mathfrak{A} \vDash \mathbb{N}^{n,k}(a)$, *then* $\mathfrak{A} \vDash \mathbb{N}^{m,l}(a)$ *for any* m *and* l *such that* $m < n \vee (m = n \,\&\, l \leqslant k)$.

The properties just listed help us to describe elements in the cases in which one of the elementary characteristics is ∞.

Corollary 2.2.2. *For any element* a *of a Boolean algebra* $\mathfrak{A}$

$$\text{ch}(a) = (n, \infty, 0) \Leftrightarrow \mathfrak{A} \vDash \text{Atomic}_n(a) \text{ and } \mathfrak{A} \vDash \mathbb{N}^{n,k}(a) \text{ for any } k \geqslant 1$$

Corollary 2.2.3. *For any Boolean algebra* $\mathfrak{A}$ *and* $a \in \mathfrak{A}$

$$\text{ch}(a) = (\infty, 0, 0) \Leftrightarrow \mathfrak{A} \vDash \neg I_n(a) \text{ for any } n$$

Corollary 2.2.4. *If $\mathfrak{A}$ and $\mathfrak{B}$ are elementarily equivalent, then their characteristics coincide.*

Exercises

1. Prove that $\operatorname{ch}(\mathfrak{A}) = \operatorname{ch}(\mathbf{1}_\mathfrak{A})$ for an arbitrary Boolean algebra $\mathfrak{A}$.

2. Prove that $\operatorname{ch}(\widehat{a}) = \operatorname{ch}(a)$ for any element a of an arbitrary Boolean algebra $\mathfrak{A}$.

3. Complete the proof of Lemma 2.2.8.

4. Prove that $E^{m_1,m_2,m_3}(a) \Leftrightarrow \operatorname{ch}(a) = (m_1, m_2, m_3)$.

5. Prove that $\mathfrak{A} \vDash \mathbb{N}^{m,k}(a) \Leftrightarrow \operatorname{ch}_1(a) > m \vee (\operatorname{ch}_1(a) = m \,\&\, \operatorname{ch}_2(a) \geqslant k)$.

6. Prove Corollaries 2.2.2 and 2.2.3.

7. Find the elementary characteristics of the Boolean algebras $\mathfrak{B}_\omega$, $\mathfrak{B}_{\omega \times \eta}$, $\mathfrak{B}_{\omega + \eta}$, $\mathfrak{B}_{\eta + \omega}$, and $\mathfrak{B}_{n+\eta}$, $n \in \mathbb{N}$.

8. Prove that $I_n(\mathfrak{A} \times \mathfrak{B}) = I_n(\mathfrak{A}) \times I_n(\mathfrak{B})$ for any Boolean algebras $\mathfrak{A}$ and $\mathfrak{B}$.

9. Prove that $I_n(\widehat{a}) = I_n(\mathfrak{A}) \cap \widehat{a}$ for any element a of a Boolean algebra $\mathfrak{A}$.

2.3. Countably Saturated Boolean Algebras and Elementary Classification

For every complete theory of Boolean algebras, we construct a countable Boolean algebra which will be called dense. In model theory, there is the notion of a countably saturated model of a complete theory. The dense Boolean algebra that will be constructed below is exactly a saturated model of a given theory, although the conditions in its definition are weaker. However, in the case of Boolean algebras, the conditions imposed here are also sufficient for the model-theoretic saturation. Using the Vaught criterion, we can obtain a condition of isomorphism for saturated Boolean algebras, which already leads to a criterion for elementary equivalence.

2.3. Countably Saturated Boolean Algebras

We construct a countably saturated elementary extension of an arbitrary countable Boolean algebra. The proof is based on the Mal'tsev compactness theorem. A Boolean algebra $\mathfrak{A}$ extending a Boolean algebra $\mathfrak{B}$ is called a *condensation* of $\mathfrak{A}$ if for any $k \in \mathbb{N}$ and $b \in \mathfrak{B}$ there exists an element $a \leqslant b$ of $\mathfrak{A}$ such that $\operatorname{ch}(a) = (k, \infty, 0)$ if $k < \operatorname{ch}_1(b)$ and $\operatorname{ch}_1(a) = \operatorname{ch}_1(b) = \operatorname{ch}_1(b \setminus a)$ & $\operatorname{ch}_2(a) = \operatorname{ch}_2(b) = \operatorname{ch}_2(b \setminus a)$ if $(\operatorname{ch}_1(b) = \infty$ or $\operatorname{ch}_2(b) = \infty)$. A Boolean algebra $\mathfrak{A}$ is called *dense* if $\mathfrak{A}$ is a condensation of itself.

Proposition 2.3.1 [67]. *For any countable Boolean algebra $\mathfrak{A}$ there exists its countable condensation.*

PROOF. We consider the extension σ of the signature of Booleans algebras by the constant symbols $\{c_a, d_a, d_a^k \mid a \in A,\ k \in \mathbb{N}\}$ and define the set of formulas

$$\mathrm{D}_\omega(\mathfrak{A}) \rightleftharpoons \{\varphi(c_{a_1}, \ldots, c_{a_n}) \mid \mathfrak{A} \vDash \varphi(a_1, \ldots, a_n) \text{ and } \varphi \text{ is a formula}$$
$$\text{of the signature of Boolean algebras}\}$$

which is referred to as a *complete diagram* of the algebra $\mathfrak{A}$. It is easy to see that $\mathfrak{B} \vDash \mathrm{D}_\omega(\mathfrak{A})$ means the existence of the elementary embedding φ of the algebra $\mathfrak{A}$ into the algebra $\mathfrak{B}$ provided that $\varphi(a) \rightleftharpoons \operatorname{val}_\mathfrak{B}(c_a)$ is the value of the constant c_a in $\mathfrak{B}$.

We add a family of formulas to $\mathrm{D}_\omega(\mathfrak{A})$, which allows us to construct the condensation of a Boolean algebra $\mathfrak{A}$. Let

$$\Delta \rightleftharpoons \mathrm{D}_\omega(\mathfrak{A}) \cup \{d_a^n \leqslant c_a \,\&\, \mathrm{N}^{n,k}(d_a^n) \,\&\, \mathrm{N}^{n,k}(c_a \setminus d_a^n),\ \text{если } k, n \in \mathbb{N}$$
$$\text{and } (n < \operatorname{ch}_1(a) \text{ or } n = \operatorname{ch}_1(a) \,\&\, \operatorname{ch}_2(a) = \infty)\}$$
$$\cup \{d_a \leqslant c_a \,\&\, \neg I_n(d_a) \,\&\, \neg I_n(c_a \setminus d_a),\ \text{if } n \in \mathbb{N} \text{ and } \operatorname{ch}_1(a) = \infty\}$$

Let us show that Δ is locally satisfiable. If Δ_0 is a finite subset of Δ, then

$$\Delta_0 \subset \Delta^{l, a_1, \ldots, a_k, b_1, \ldots, b_k, n_1, \ldots, n_k} \rightleftharpoons \mathrm{D}_\omega(\mathfrak{A}) \cup \{d_{a_i}^{n_i} \leqslant c_{a_i} \,\&\, \mathrm{N}^{n_i, k'}(d_{a_i}^{n_i})$$
$$\&\, \mathrm{N}^{n_i, k'}(c_{a_i} \setminus d_{a_i}^{n_i}) \mid (n_i < \operatorname{ch}_1(a_i) \vee (n_i = \operatorname{ch}_1(a_i)$$
$$\&\, \operatorname{ch}_2(a_i) = \infty)) \,\&\, k' \leqslant l \text{ and } 1 \leqslant i \leqslant k\}$$
$$\cup \{d_{b_i} \leqslant c_{b_i} \,\&\, \neg I_m(d_{b_i}) \,\&\, \neg I_m(c_{b_i} \setminus d_{b_i}) \mid$$
$$\operatorname{ch}_1(b_i) = \infty,\ 1 \leqslant i \leqslant k \text{ and } 0 \leqslant m \leqslant l\} \text{ for a suitable } l$$

If we construct a model for $\Delta^{l, \overline{a}, \overline{b}, \overline{n}}$, then it is also a model for Δ_0. We consider the set of formulas $\Delta^{l, \overline{a}, \overline{b}, \overline{n}}$. By the definition of $\Delta^{l, \overline{a}, \overline{b}, \overline{n}}$, we have $\operatorname{ch}_1(b_i) = \infty$ and $n_i < \operatorname{ch}_1(a_i) \vee \operatorname{ch}_2(a_i) = \infty$, $1 \leqslant i \leqslant k$. By Lemma 2.2.6, under the elements a_i and b_i there are elements x_i and

y_i $(1 \leqslant i \leqslant k)$ such that $\operatorname{ch}(x_i) = (n_i, l, 0)$ and $\operatorname{ch}(y_i) = (l+1, 1, 0)$. By Lemma 2.2.4, $\operatorname{ch}(a_i \backslash x_i) = \operatorname{ch}(a_i)$ and $\operatorname{ch}(b_i \backslash y_i) = (\infty, 0, 0)$. For the values of the constants c_{a_i}, $d_{a_i}^{n_i}$, c_{b_i}, and d_{b_i} in $\mathfrak{A}$ we take elements a_i, x_i, b_i, and y_i. For the rest of c_a we assume that their values are equal to a and the values of d_a and d_a^k are arbitrary for those a that do not occur in $\bar{a}$, $\bar{b}$. In the enrichment $\mathfrak{A}^*$ of $\mathfrak{A}$, all the formulas of $\Delta^{l, a_1, \ldots, a_k, b_1, \ldots, b_k, n_1, \ldots, n_k}$ are satisfiable. Consequently, Δ is a locally satisfiable set. Since Δ is countable and locally satisfiable, for Δ there exists a countable model $\mathfrak{B}^*$ of the signature σ.

We now define a mapping φ from $\mathfrak{A}$ into $\mathfrak{B}^*$ by setting $\varphi(a) = \operatorname{val}_{\mathfrak{B}^*}(c_a)$. From the construction of Δ it follows that if $\Phi(x_1, \ldots, x_n)$ is a formula of the signature of Boolean algebras and $\mathfrak{A} \vDash \Phi(a_1, \ldots, a_n)$, then $\mathfrak{B}^* \vDash \Phi(c_{a_1}, \ldots, c_{a_n})$. Therefore, $\mathfrak{B}^* \vDash \Phi(\varphi(a_1), \ldots, \varphi(a_n))$. We now show that the restriction $\mathfrak{B}$ of $\mathfrak{B}^*$ to a system with the signature of Boolean algebras is the required Boolean algebra. To this end, we only need to verify that $\mathfrak{B}$ is a condensation of $\mathfrak{A}$. But this fact follows from the definition of Δ. □

Theorem 2.3.1 [67] [on existence of dense Boolean algebras]. *For any countable Boolean algebra $\mathfrak{A}$ there exists a countable elementary extension $\mathfrak{B}$ which is a dense Boolean algebra.*

PROOF. Using Proposition 2.3.1, we construct a countable sequence of countable elementary extensions $\mathfrak{A}_0 \preccurlyeq \mathfrak{A}_1 \preccurlyeq \mathfrak{A}_2 \preccurlyeq \ldots$ such that $\mathfrak{A}_0 = \mathfrak{A}$ and $\mathfrak{A}_{n+1}$ is a condensation of $\mathfrak{A}_n$ for $n \in \mathbb{N}$. In view of Corollaries 2.2.2, 2.2.3, and Lemma 2.2.8, the property of having a given characteristic can be expressed by a formula or a sequence of formulas and, consequently, this property remains valid in passing to elementary extensions. Therefore, for any $n < m$ the Boolean algebra $\mathfrak{A}_m$ is a condensation of the Boolean algebra $\mathfrak{A}_n$. By the theorem on elementary chains, $\bigcup_{n \geqslant 0} \mathfrak{A}_n$ is an elementary extension of $\mathfrak{A}_m$ for any m, hence $\bigcup_{n \geqslant 0} \mathfrak{A}_n$ is a condensation of $\mathfrak{A}_m$ for any m. Consequently, it is a condensation of $\bigcup_{n \geqslant 0} \mathfrak{A}_n$. In this case, the Boolean algebra $\mathfrak{B} \rightleftharpoons \bigcup_{n \geqslant 0} \mathfrak{A}_n$ is an elementary extension of the Boolean algebra $\mathfrak{A}$ and it is dense. □

Theorem 2.3.2 [67] [on uniqueness]. *Any two countable dense Boolean algebras of the same elementary characteristic are isomorphic.*

2.3. Countably Saturated Boolean Algebras

PROOF. Let $\mathfrak{A}$ and $\mathfrak{B}$ be countable dense Boolean algebras of the same elementary characteristic. We consider a subset of $S \subseteq A \times B$ such that

$$S \rightleftharpoons \{(a,b) \mid \text{ch}_\mathfrak{A}(a) = \text{ch}_\mathfrak{B}(b),\ \text{ch}_\mathfrak{A}(C(a)) = \text{ch}_\mathfrak{B}(C(b))\}$$

Since $\text{ch}(\mathfrak{A}) = \text{ch}(\mathfrak{B})$ and $\text{ch}_\mathfrak{A}(1) = \text{ch}_\mathfrak{B}(1)$, and the elementary characteristic of the zero element of all Boolean algebras is the same, for S the axiom Ax1 of Sec. 1.5 holds. The validity of axioms Ax2 and Ax3 for S can be proved in a similar way since $\text{ch}(a) = (0,0,0) \Leftrightarrow a = 0$. It remains to show that S satisfies Ax4 and Ax5. Since the situation is symmetric, we only prove the validity of Ax4.

Let $\text{ch}(a) = \text{ch}(b)$, $\text{ch}(C(a)) = \text{ch}(C(b))$, $0 < x < a$. Let $\text{ch}(x) = (n,m,k)$. By the symmetry of the choice of y for x and $a\backslash x$, without loss of generality we assume that $\text{ch}_1(x) \leqslant \text{ch}_1(a\backslash x)$ and $\text{ch}_2(x) \leqslant \text{ch}_2(a\backslash x)$ if $\text{ch}_1(x) = \text{ch}_1(a\backslash x)$. By Lemma 2.2.5,

$$n \leqslant \text{ch}_1(a)\ \&\ (n = \text{ch}_1(a) \Rightarrow (m \leqslant \text{ch}_2(a)\ \&\ k \leqslant \text{ch}_3(a)))$$

If $m < \infty$ and $n < \infty$, then, by Lemma 2.2.6, $\text{ch}(a) = \text{ch}(b)$ means that there exists $y < b$ such that $\text{ch}(x) = \text{ch}(y)$ and $\text{ch}(a\backslash x) = \text{ch}(b\backslash y)$. By Lemma 2.2.4, $\text{ch}(C(x)) = \text{ch}(C(y))$ since $C(y) = b\backslash y \vee C(b)$ and $\text{ch}(C(a\backslash x)) = \text{ch}(C(b\backslash y))$ since $C(b\backslash y) = (b \wedge y) \vee C(b)$. Consequently, the pairs (x,y) and $(a\backslash x, b\backslash y)$ belong to S.

If $n = \infty$, then, in view of the density of the Boolean algebra $\mathfrak{B}$, there exists an element $y < b$ such that $\text{ch}(y) = \text{ch}(b\backslash y) = \text{ch}(b)$. By Lemma 2.2.4, $\text{ch}(x) = \text{ch}(y)$, $\text{ch}(a\backslash x) = \text{ch}(b\backslash y)$ and $\text{ch}(C(x)) = \text{ch}(C(y))$, $\text{ch}(C(a\backslash x)) = \text{ch}(C(b\backslash y))$. Consequently, $(x,y) \in S$ and $(a\backslash x, b\backslash y) \in S$.

In the case $n < \infty$, $m = \infty$, it is necessary to consider two cases: $\text{ch}_1(a) = n$ and $\text{ch}_1(a) > n$.

If $\text{ch}_1(a) = n$, $m = \infty$, we consider $\text{ch}_3(a)$. If $\text{ch}_3(a) = 0$, then, as in the case $n = \infty$, by density, we can choose the required element y. If $\text{ch}_3(a) = 1$, then we take an element d such that $d < a$ and $\text{ch}(d) = (n,0,1)$, $\text{ch}(a\backslash d) = (n,\infty,0)$. We take an element e located under b and such that $\text{ch}(e) = (n,0,1)$ and $\text{ch}(b\backslash e) = (n,\infty,0)$. If $\text{ch}_3(x) = 0$, then, in view of density, we can choose an element y located under $b\backslash e$ and such that $\text{ch}(y) = (n,\infty,0)$ and $\text{ch}(b\backslash y) = (n,\infty,0)$. In accordance with Lemma 2.2.4, y is the required element. If $\text{ch}_3(x) = 1$, then, by density, we can take an element $\widetilde{y}$ located under $b\backslash e$ and such that $\text{ch}(\widetilde{y}) = \text{ch}(b\backslash(e \vee \widetilde{y})) = (n,\infty,0)$. We put $y \rightleftharpoons \widetilde{y} \vee e$ or $y \rightleftharpoons \widetilde{y} \vee e'$, where $\text{ch}(e') = \text{ch}(e) = \text{ch}(e\backslash e')$ and $e' < e$ if $\text{ch}_3(a\backslash x) = 0$ or $\text{ch}_3(a\backslash x) = 1$ respectively.

In the case $n < \text{ch}_1(a)$, in view of the density of $\mathfrak{B}$ we can choose an element $\widetilde{y} < b$ such that $\text{ch}_1(\widetilde{y}) = \text{ch}_1(x)$, $\text{ch}_2(\widetilde{y}) = \text{ch}_2(x)$ and $y =$

$\widetilde{y} \vee z_1$ or $y = \widetilde{y} \vee z_2$ provided that $\mathrm{ch}_3(x) = 1 \,\&\, \mathrm{ch}(z_1) = (n,0,1)$ or $\mathrm{ch}_3(x) = 0 \,\&\, \widetilde{y} = y \vee z_2 \,\&\, \mathrm{ch}(y) = (n,\infty,0)$ respectively. Thus, S satisfies the axioms Ax1–Ax5. By the Vaught theorem on isomorphism, $\mathfrak{A}$ and $\mathfrak{B}$ are isomorphic. □

Corollary 2.3.1. *If $\mathfrak{A}$ and $\mathfrak{B}$ are countable dense Boolean algebras of the same elementary characteristic and $\{x_i\}_{i=1}^n$, $\{y_i\}_{i=1}^n$ are the partition of unity for $\mathfrak{A}$ and $\mathfrak{B}$ such that $\mathrm{ch}(x_i) = \mathrm{ch}(y_i)$, $1 \leqslant i \leqslant n$, then there exists an isomorphism φ between $\mathfrak{A}$ and $\mathfrak{B}$ such that $\varphi(x_i) = y_i$, $1 \leqslant i \leqslant n$.*

PROOF. It suffices to show that for any element a of a dense Boolean algebra $\mathfrak{A}$ the Boolean algebra $\widehat{a}$ defined in 1.5 is dense and $\mathrm{ch}(a) = \mathrm{ch}(\widehat{a})$. This is true because in the definition of density we take only elements smaller than the given ones and all the elements that are in $\widehat{a}$ are smaller than a. Furthermore, $I_n(\widehat{a}) = I_n(\mathfrak{A}) \cap \widehat{a}$, $n \geqslant 0$. □

Corollary 2.3.2. *Any two Boolean algebras $\mathfrak{A}$ and $\mathfrak{B}$ of the same elementary characteristic are elementarily equivalent.*

PROOF. By the Löwenheim–Skolem theorem, any infinite system of at most countable signature contains a countable elementary subsystem. Let $\mathfrak{A}_0$ and $\mathfrak{B}_0$ be countable elementary subsystems of the algebras $\mathfrak{A}$ and $\mathfrak{B}$ respectively. In view of Theorem 2.3.1, there exist their elementary countable dense extensions $\mathfrak{A}_0^\sharp$ and $\mathfrak{B}_0^\sharp$. Since all the embeddings appearing here are elementary, we have $\mathrm{ch}(\mathfrak{A}_0^\sharp) = \mathrm{ch}(\mathfrak{A}_0) = \mathrm{ch}(\mathfrak{A}) = \mathrm{ch}(\mathfrak{B}) = \mathrm{ch}(\mathfrak{B}_0) = \mathrm{ch}(\mathfrak{B}_0^\sharp)$. By Theorem 2.3.2, the Boolean algebras $\mathfrak{A}_0^\sharp$ and $\mathfrak{B}_0^\sharp$ are isomorphic. Consequently, $\mathfrak{A}$ and $\mathfrak{B}$ are elementarily equivalent. □

Theorem 2.3.3. *Two Boolean algebras are elementarily equivalent if and only if their elementary characteristics coincide.*

PROOF. The assertion follows from Corollaries 2.3.2 and 2.2.4. □

Theorem 2.3.4. *A countable Boolean algebra is countably saturated if and only if it is dense.*

PROOF. If $\mathfrak{B}$ is a countably saturated Boolean algebra, then it is dense since the density property is equivalent to the fact that some locally satisfiable sets of special formulas are satisfiable. To prove the converse assertion we consider a countable dense Boolean algebra $\mathfrak{B}$. We note that $\mathrm{Th}(\mathfrak{B})$ has a countably saturated model. Indeed, let $\mathfrak{A}$ be a countable model of $\mathrm{Th}(\mathfrak{B})$. By Theorem 2.3.1, there is an elementary extension of

2.3. Countably Saturated Boolean Algebras

$\mathfrak{A}$ to a countable dense Boolean algebra $\mathfrak{B}'$. By Theorem 2.3.2, $\mathfrak{B}$ and $\mathfrak{B}'$ are isomorphic because they are elementarily equivalent and dense. Hence $\mathfrak{A}$ is elementarily embedded into $\mathfrak{B}$. Since $\mathfrak{A}$ is an arbitrary countable Boolean algebra, $\mathfrak{B}$ is a universal system. By Corollary 2.1.6, Th($\mathfrak{B}$) has a countably saturated model $\mathfrak{B}^*$. As already mentioned, a countably saturated system is dense. By Theorem 2.3.2, $\mathfrak{B}^*$ and $\mathfrak{B}$ are isomorphic. Consequently, $\mathfrak{B}$ is a countably saturated Boolean algebra. □

Corollary 2.3.3. *For any countable Boolean algebra there exists its elementary extension to a countably saturated Boolean algebra.*

Corollary 2.3.4. *An elementary theory of any Boolean algebra is decidable.*

PROOF. The property of having a prescribed elementary characteristic can be written by means of a decidable system of axioms. Therefore, a theory of any Boolean algebra is recursively axiomatizable and its completeness implies its decidability. □

Corollary 2.3.5 [on existence of elementary Boolean algebras]. *A theory of any infinite Boolean algebra has a prime model.*

PROOF. Since every complete theory of a Boolean algebra has a countably saturated model, such a theory is superatomic and, consequently, is atomic. By the theorem on existence of prime models, it has a prime model. □

We describe countable homogeneous Boolean algebras following Morozov [132].

For a homogeneous Boolean algebra $\mathfrak{B}$ we introduce the notion of an *invariant* $p(\mathfrak{B})$ as follows:

$p(\mathfrak{B}) = 0$ if there are no atomic elements with infinite number of atoms of $\mathfrak{B}$,

$p(\mathfrak{B}) = 1$ if there exists $x \in \mathfrak{B}$ such that the algebra $\hat{x}$ is isomorphic to the Boolean algebra $\mathfrak{B}_\omega$,

$p(\mathfrak{B}) = 2$ if there exists $x \in \mathfrak{B}$ such that the algebra $\hat{x}$ is isomorphic to the Boolean algebra $\mathfrak{B}_{\omega \times \eta}$.

We note that for a countable homogeneous Boolean algebra $\mathfrak{B}$ such that $\mathrm{ch}_1(\mathfrak{B}) = \infty$ one of the following cases holds:

CASE 1: for any $x \in \mathfrak{B}$ either $\mathrm{ch}_1(x) < \infty$ or $\mathrm{ch}_1(C(x)) < \infty$.

CASE 2: there exists $x \in \mathfrak{B}$ such that $\mathrm{ch}_1(x) = \mathrm{ch}_1(C(x)) = \infty$ but $\widehat{x}$ and $\widehat{C(x)}$ satisfy the conditions of Case 1.

CASE 3: for any $x \in \mathfrak{B}$ such that $\mathrm{ch}_1(x) = \infty$ there is an element $y \leqslant x$ such that $\mathrm{ch}_1(y) = \infty$ and $\mathrm{ch}_1(x \setminus y) = \infty$.

For a countable homogeneous Boolean algebra we define an *invariant* $t(\mathfrak{B})$ as follows: $t(\mathfrak{B})$ is equal to 0 if $\mathrm{ch}_1(\mathfrak{B}) < \infty$ and $t(\mathfrak{B})$ is equal to the number of the case (i), $i = 1, 2, 3$, which takes place for the algebra $\mathfrak{B}$ if $\mathrm{ch}_1(\mathfrak{B}) = \infty$.

Proposition 2.3.2. *If $\mathfrak{B}$ is a homogeneous Boolean algebra and $n + 1 < \mathrm{ch}_1(\mathfrak{B})$ or $\mathrm{ch}(\mathfrak{B}) = (n+1, \alpha, \beta)$, where $\beta = 1$ or $\alpha > 1$, then $p(\mathfrak{B}/I_n(\mathfrak{A})) = 0$ or $p(\mathfrak{B}/I_n(\mathfrak{A})) = 2$.*

PROOF. We assume that $p_n(\mathfrak{B}) = p(\mathfrak{A}/I_n(\mathfrak{A})) = 1$. Let $x \in \mathfrak{B}$ be such that $\widehat{x}/I_n \cong \mathfrak{B}_\omega$. Then $x \in I_{n+1}$. We consider two cases.

CASE 1: There exist two elements a and b of $\mathfrak{B}$ such that $a \wedge b = 0$ and a/I_{n+1}, b/I_{n+1} are atoms. Since $x \in I_{n+1}$, without loss of generality, we assume that $x \leqslant a$. Since a, b and $C(a)$, $C(b)$ respectively have the same elementary characteristics, we conclude that the types that are realized on a and b coincide since, in this case, in a countably saturated extension they can be transformed into each other by automorphisms of this countably saturated elementary extension. By the homogeneity of a Boolean algebra $\mathfrak{B}$ as a model, there exists an automorphism of $\mathfrak{B}$ transforming a into b. Then under b there exists an element $x' \leqslant b$ such that $\widehat{x'}/I_n \cong \mathfrak{B}_\omega$. It is obvious that the elementary characteristics of x and $x' \vee x$ as well as the elementary characteristics of their complements coincide. Arguing as above, we conclude that there exists an automorphism $\mathfrak{B}$ transforming $x' \vee x$ into x, but the element x cannot decompose into two elements of characteristic $(n, \infty, 0)$ while for $x' \vee x$ such a decomposition takes place. This reasoning completes the consideration of Case 1.

CASE 2: Case 1 does not take place, i.e., $\mathrm{ch}_1(\mathfrak{B}) = (n+1, \alpha, 1)$, where $\alpha \leqslant 1$. We consider elements a and b such that a/I_{n+1}, b/I_{n+1} are nonzero atomless elements and the intersection $a \wedge b$ is the zero element. Since $x \in I_{n+1}$, without loss of generality, we can assume that $x \leqslant a$. As in the case of atoms, it is easy to show that there exists an automorphism that transforms a into b since the elementary characteristics of the pairs a, b and $C(a)$, $C(b)$ coincide. Hence we obtain an element $x' \leqslant b$ such that $\mathrm{ch}(x') = \mathrm{ch}(x)$. Therefore, there exists an automorphism of the homogeneous Boolean algebra $\mathfrak{B}$ such that $x' \vee x$ goes into x, since the

2.3. Countably Saturated Boolean Algebras

elementary characteristics of x, $x' \vee x$ and $C(x)$, $C(x \vee x')$ respectively coincide. This contradicts the condition that it cannot decompose into two elements of characteristic $(n, \infty, 0)$. However, $x' \wedge x = 0$ and $\text{ch}(x) = \text{ch}(x') = \text{ch}(x \vee x') = (n, \infty, 0)$. The contradiction obtained completes the proof. □

Corollary 2.3.6. *If* $\text{ch}_1(\mathfrak{B}) = \omega$ *for a countable homogeneous Boolean algebra* $\mathfrak{B}$, *then* $p(\mathfrak{B}/I_{n+1}) \in \{0, 2\}$ *for any* n.

Corollary 2.3.7. *If* $\text{ch}_1(\mathfrak{B}) = (n+1, \alpha, \beta)$, *where* $\alpha > 2$ *or* $\beta = 1$, *then* $p_k(\mathfrak{B}) = p(\mathfrak{A}/I_k) \in \{0, 2\}$ *for a homogeneous Boolean algebra* $\mathfrak{B}$ *and every* $k \leqslant n$.

Thus, we have shown that for a homogeneous Boolean algebra $\mathfrak{B}$ the conditions $p_n(\mathfrak{B}) = 1$ and $\text{ch}_1(\mathfrak{B}) = n+1$ imply $\text{ch}(\mathfrak{B}) = (n+1, 1, 0)$.
We now describe homogeneous atomic Boolean algebras.

Proposition 2.3.3. *The Boolean algebras* $\mathfrak{B}_\omega$, $\mathfrak{B}_{\omega+\omega}$, *and* $\mathfrak{B}_{\omega \times \eta}$ *are countable, homogeneous, and atomic. Any countable homogeneous atomic Boolean algebra is isomorphic to one of the algebras just listed.*

PROOF. A Boolean algebra $\mathfrak{B}_{\omega \times \eta}$ is a dense Boolean algebra and, consequently, countably saturated and homogeneous. By the existence of a countably saturated algebra, there is a prime model in the theory of infinite atomic models. However, any elementary subsystem of the Boolean algebra $\mathfrak{B}_\omega$ is isomorphic to this algebra. Indeed, since for any element $a \in \mathfrak{B}_\omega$ either a or $C(a)$ is a union of a finite number of atoms, the Boolean algebra $\mathfrak{B}_\omega$ is superatomic and has type $(1,1)$. Hence $\mathfrak{B}_\omega$ is a prime model. Consequently, it is a homogeneous model. The homogeneity of finite models is obvious. To prove that $\mathfrak{B}_{\omega+\omega}$ is homogeneous we consider two decompositions $1 = \bigvee_{i=1}^{n} a_i = \bigvee_{i=1}^{n} b_i$ and $a_i \wedge a_j = b_i \wedge b_j = 0$ for any $i \neq j$. If the type $p(x_1, \ldots, x_n)$ that is realized on the elements $a_1, \ldots, a_n$ coincides with the type $q(x_1, \ldots, x_n)$ that is realized on the elements $b_1, \ldots, b_n$, then $\text{ch}_2(a_i) = \text{ch}_2(b_i)$. It is obvious that there exists an automorphism φ of the Boolean algebra $\mathfrak{B}_{\omega+\omega}$ transforming $a_1, \ldots, a_n$ into $b_1, \ldots, b_n$. If $a_1, \ldots, a_n$ and $b_1, \ldots, b_n$ are arbitrary sequences of $\mathfrak{B}_{\omega+\omega}$ realizing the same type, then, taking the subalgebra generated by these sequences and atoms that are defined by the same terms of $a_1, \ldots, a_n$ and $b_1, \ldots, b_n$, we again obtain elementarily equivalent sequences but they are subject to the condition that their elements do not intersect. Since the atom $t(a_1, \ldots, a_n)$ corresponds to the atom $t(b_1, \ldots, b_n)$ that is defined by the same term, the automorphism defined for atoms transforms a_1 into b_1, a_2 into b_2, ..., a_n into b_n.

It remains to note that there are no other homogeneous models. Let an infinite atomic Boolean algebra $\mathfrak{B}$ be countable and homogeneous. We consider the quotient algebra $\mathfrak{B}/F(\mathfrak{B})$ of $\mathfrak{B}$ by the Frechét ideal $F(\mathfrak{B})$. If $\mathfrak{B}/F(\mathfrak{B})$ is atomic and consists of just two elements, then $\mathfrak{B} \cong \mathfrak{B}_\omega$, which follows from the characterization of countable superatomic Boolean algebras. Arguing as above, we find that $\mathfrak{B} \cong \mathfrak{B}_{\omega+\omega}$ if $\mathfrak{B}/F(\mathfrak{B})$ consists of just two atoms. If these conditions fail, then $\mathfrak{B}$ satisfies one of the following conditions:

(1) $\mathfrak{B}/F(\mathfrak{B})$ is an atomless Boolean algebra,

(2) $\mathfrak{B}/F(\mathfrak{B})$ contains an atom and a nonzero atomless element,

(3) $\mathfrak{B}/F(\mathfrak{B})$ is atomic and contains at least three atoms.

In case (1), by Proposition 1.5.2, we obtain $\mathfrak{B} \cong \mathfrak{B}_{\omega \times \eta}$. We show that cases (2) and (3) cannot be realized for homogeneous countable atomic Boolean algebras. We assume that (2) holds and consider elements a and b of $\mathfrak{B}$ such that $a \wedge b = 0$ and $a/F(\mathfrak{B})$ is an atom, $b/F(\mathfrak{B})$ is a nonzero atomless element. Then $\mathrm{ch}\,(a) = \mathrm{ch}\,(b)$ and $\mathrm{ch}\,(C(a)) = \mathrm{ch}\,(C(b))$. In this case, their types coincide. By homogeneity, there exists an automorphism f of the Boolean algebra $\mathfrak{B}$ such that $f(a) = b$. However, under the isomorphism f the elements of $F(\mathfrak{B})$ go to the elements of $F(\mathfrak{B})$. Consequently, f induces an automorphism $f/F(\mathfrak{B})$ of $\mathfrak{B}/F(\mathfrak{B})$. But an atom does not go to an atomless element. Consequently, case (2) is impossible.

In case (3), we can find a_1, a_2, a_3 of $\mathfrak{B}$ such that $a_i \wedge a_j = 0$, $i \neq j$, and $a_i/F(\mathfrak{B})$ is an atom. It is obvious that $\mathrm{ch}_2(a_1) = \mathrm{ch}_2(a_2 \vee a_3) = \infty$ and $\mathrm{ch}_2(C(a_1)) = \mathrm{ch}_2(C(a_2 \vee a_3)) = \infty$. Therefore, there exists an automorphism $\mathfrak{B}$ transforming a_1 into $a_2 \vee a_3$. In this case, the automorphism induced by f on the quotient algebra $f/F(\mathfrak{B})$ by the Frechét ideal transforms an atom into an element which is not an atom. Hence we arrive at a contradiction. □

For any countable homogeneous Boolean algebra $\mathfrak{B}$ we define the *homogeneity type* $t_h(\mathfrak{B})$ as a pair $\langle \alpha, \overline{p} \rangle$, where $\alpha \in \{0, 1, 2, 3\}$, and $\overline{p}$ is a finite or infinite sequence. We set

$\alpha = 0$ if $\mathrm{ch}_1(\mathfrak{B}) < \infty$ and $\alpha = t(\mathfrak{B})$ if $\mathrm{ch}_1(\mathfrak{B}) = \infty$;

if $\mathrm{ch}_1(\mathfrak{B}) = \infty$, then $\overline{p} = (p_0, p_1, \ldots, p_n, \ldots)$ is an infinite sequence of $p_k \in \{0, 2\}$, where $p_k = p(\mathfrak{B}/I_k(\mathfrak{B}))$;

2.3. Countably Saturated Boolean Algebras

if $n = \mathrm{ch}_1(\mathfrak{B}) < \infty$, $\mathrm{ch}\,(\mathfrak{B}) = (n,1,0)$, then $\bar{p} = (p_0, \ldots, p_{n-1}, p_n)$, $p_k = p(\mathfrak{B}/I_k(\mathfrak{B}))$ for any $k \leqslant n-1$ and $p_n = 1$; in this case, $p_i \in \{0,2\}$, $i < n-1$, and $p_{n-1} \in \{0,1,2\}$;

if $n = \mathrm{ch}_1(\mathfrak{B}) < \infty$ and $\mathrm{ch}\,(\mathfrak{B}) \neq (n,1,0)$, then the length of $\bar{p}$ is $n+1$ and $\bar{p} = (p_0, p_1, \ldots, p_{n-1}, p_n)$, where $p_i = p(\mathfrak{B}/I_i(\mathfrak{B}))$, $i < n$ (in this case, $p_i \in \{0,2\}$) and

$$p_n = \begin{cases} k & \text{if } \mathfrak{B}/I_n(\mathfrak{B}) \text{ contains exactly} \\ & k \text{ atoms, where } k < \infty \\ \omega & \text{if } \mathfrak{B}/I_n(\mathfrak{B}) \text{ is isomorphic to } \mathfrak{B}_\omega \text{ or } \mathfrak{B}_\omega \times \mathfrak{B}_\eta \\ \omega + \omega & \text{if } \mathfrak{B}/I_n(\mathfrak{B}) \text{ is isomorphic to } \mathfrak{B}_{\omega+\omega} \text{ or } \mathfrak{B}_{\omega+\omega} \times \mathfrak{B}_\eta \\ \omega \times \eta & \text{if } \mathfrak{B}/I_n(\mathfrak{B}) \text{ is isomorphic to } \mathfrak{B}_{\omega\times\eta} \text{ or } \mathfrak{B}_{\omega\times\eta} \times \mathfrak{B}_\eta \end{cases}$$

It is obvious that we have listed all possible cases. Before characterizing the isomorphism types for countable homogeneous models, we introduce a definition and mention a simple but useful property. Consider the family of types consistent with the theory T of Boolean algebras. The type $p(x_1, \ldots, x_n)$ of this family is called a *decomposition type* if $x_i \wedge x_j = 0 \in p(x_1, \ldots, x_n)$, $i \neq j$, and $x_1 \vee x_2 \vee \ldots \vee x_n = 1 \in p(x_1, \ldots, x_n)$.

Proposition 2.3.4. *Let $S_\mathfrak{A}$ and $S_\mathfrak{B}$ be two families of types realized in $\mathfrak{A}$ and $\mathfrak{B}$. Then $S_\mathfrak{A}$ and $S_\mathfrak{B}$ coincide if and only if all the decomposition types realized in $\mathfrak{A}$ and $\mathfrak{B}$ coincide.*

PROOF. If $S_\mathfrak{A} = S_\mathfrak{B}$, then it is obvious that the decomposition types coincide. Let us prove the converse assertion. It suffices to prove that any type realized in $\mathfrak{A}$ is also realized in $\mathfrak{B}$. Let a type $p(x_1, \ldots, x_n)$ be realized on elements $a_1, \ldots, a_n$ in $\mathfrak{A}$ and let $\mathrm{gr}_\mathfrak{B}(a_1, \ldots, a_n)$ denote the subalgebra of $\mathfrak{A}$ generated by $a_1, \ldots, a_n$, and $d_1, \ldots, d_r$ denote all its atoms. We consider terms $t_{\bar{\varepsilon_1}}, \ldots, t_{\bar{\varepsilon_r}}$ such that $d_i = t_{\bar{\varepsilon_i}}(a_1, \ldots, a_n)$. Since $d_1, \ldots, d_r$ are atoms, for every $1 \leqslant i \leqslant n$ there exists K_i such that $a_i = \bigvee_{j \in K_i} d_{\bar{\varepsilon_j}}$. Elements $d_1, \ldots, d_r$ determine some type $q(y_1, \ldots, y_r)$. Since the type q is realized in $\mathfrak{A}$, it is a decomposition type in view of the choice of $d_1, \ldots, d_r$. Consequently, the type $q(y_1, \ldots, y_r)$ is also realized in $\mathfrak{B}$ on some sequence $d'_1, \ldots, d'_r$ which is the partition of unity. This fact follows from the imposed condition. We define elements $a'_i = \bigvee_{j \in K_i} d'_j$.

Since

$$\mathfrak{B} \vDash \varphi(a'_1, \ldots, a'_n) \Leftrightarrow \mathfrak{B} \vDash \varphi\left(\bigvee_{j \in K_1} d'_j, \ldots, \bigvee_{j \in K_n} d'_j\right)$$

$$\Leftrightarrow \varphi\left(\bigvee_{j\in K_1} y_j, \ldots, \bigvee_{j\in K_n} y_j\right) \in q(y_1, \ldots, y_n)$$

$$\Leftrightarrow \mathfrak{A} \vDash \varphi\left(\bigvee_{j\in K_1} d_j, \ldots, \bigvee_{j\in K_n} d_j\right) \Leftrightarrow \mathfrak{B} \vDash \varphi(a_1, \ldots, a_n)$$

for any formula $\varphi(x_1, \ldots, x_n)$ the sequence $a'_1, \ldots, a'_n$ realizes the same type in $\mathfrak{B}$ as the sequence $a_1, \ldots, a_n$ in $\mathfrak{A}$. □

If we know all the decomposition types of $S^1_{\mathfrak{A}}$ that are realized in $\mathfrak{A}$, we easily construct all the types of $S_{\mathfrak{A}}$ that are realized in $\mathfrak{A}$. Namely, if $d(y_1, \ldots, y_n)$ is a decomposition type realized in $\mathfrak{A}$, then, for an arbitrary system of terms $t_i(y_1, \ldots, y_n)$, $i \leqslant m$, we introduce the type

$$p_{d,t}(x_1, \ldots, x_n) = \{\varphi(x_1, \ldots, x_n) \mid \varphi(t_1(y_1, \ldots, y_n), \ldots,$$
$$t_n(y_1, \ldots, y_n)) \in d(y_1, \ldots, y_n)\}$$

It is obvious that $p_{d,t}$ is a type and any type realized in $\mathfrak{A}$ can be obtained from a decomposition type in such a way.

Theorem 2.3.5. *If $\mathfrak{A}$ and $\mathfrak{B}$ are two countable homogeneous Boolean algebras of the same elementary characteristics, then $\mathfrak{A}$ and $\mathfrak{B}$ are isomorphic if and only if their homogeneity types coincide, i.e., $t_h(\mathfrak{A}) = t_h(\mathfrak{B})$.*

PROOF. If $\mathfrak{A}$ and $\mathfrak{B}$ are isomorphic, the homogeneity types coincide because they are characterized by the families $S_{\mathfrak{A}}$ and $S_{\mathfrak{B}}$ of elementary types realized in these systems. Conversely, if the homogeneity types of homogeneous models coincide, then the same decomposition types are realized in these algebras since decomposition types can be characterized by the elementary characteristics of their components in view of the theorem on an isomorphism of countable dense models. The principal difficulty is to choose elements of infinite characteristic. However, the equality $t_h(\mathfrak{A}) = t_h(\mathfrak{B})$ guarantees that in $\mathfrak{A}$ and $\mathfrak{B}$ there exist decompositions whose components have the same characteristics. □

Corollary 2.3.8. *A homogeneous Boolean algebra $\mathfrak{B}$ is countably saturated if and only if one of the following conditions holds:*

(1) $\mathrm{ch}_1(\mathfrak{B}) = \infty$ and $t_h(\mathfrak{B}) = \langle 3, \{p_n \mid n \in \mathbb{N}\}\rangle$, $p_n = 2$ for all n,

(2) $\mathrm{ch}(\mathfrak{B}) = (n, \infty, \varepsilon)$ and $t_h(\mathfrak{B}) = \langle 0, \langle 2, \ldots, 2, \omega \times \eta\rangle\rangle$,

(3) $\mathrm{ch}(\mathfrak{B}) = (n, k, \varepsilon)$, $k \in \mathbb{N}$, and $t_h(\mathfrak{B}) = \langle 0, \langle 2, \ldots, 2, k\rangle\rangle$.

2.3. Countably Saturated Boolean Algebras

Corollary 2.3.9. *A homogeneous Boolean algebra $\mathfrak{B}$ is elementary if and only if one of the following conditions holds:*

(1) $\mathrm{ch}_1(\mathfrak{B}) = \infty$ *and* $t_h(\mathfrak{B}) = \langle 1, \{p_n \mid n \in \mathbb{N}\}\rangle$, $p_n = 0$ *for all* n,

(2) $\mathrm{ch}(\mathfrak{B}) = (n, \infty, \varepsilon)$ *and* $t_h(\mathfrak{B}) = \langle 0, \langle 0, \ldots, 0, \omega\rangle\rangle$,

(3) $\mathrm{ch}(\mathfrak{B}) = (n, k, \varepsilon)$, $k \in \mathbb{N}$, *and* $t_h(\mathfrak{B}) = \langle 0, \langle 0, \ldots, 0, k\rangle\rangle$.

The proof of the assertion that for a given admissible pair $\langle \alpha, \overline{p}\rangle$ there exists a homogeneous countable Boolean algebra of the homogeneity type $\langle \alpha, \overline{p}\rangle$ is left to the reader as an exercise. (Hint: Proceeding by induction on n and using linear orders, construct the required Boolean algebra in the case of finite elementary characteristics; further, examine the case of infinite characteristic $\mathrm{ch}_1(\mathfrak{B}) = \infty$.)

In the conclusion of the section, we describe one more criterion for a countable Boolean algebra to be homogeneous, which is simpler to prove.

A countable Boolean algebra $\mathfrak{A}$ is called *relatively dense* if the following conditions hold:

(1) for any n, if there exists an element $a \in \mathfrak{A}$ such that $\mathrm{ch}\,(a) = (n, \infty, 0)$, then

  (1.1) for any $b \in \mathfrak{A}$ such that $\mathrm{ch}_1(b) > n$ there is $a' \leqslant b$ such that $\mathrm{ch}\,(a') = (n, \infty, 0)$,

  (1.2) if either $n < \mathrm{ch}_1(\mathfrak{A})$ or $n = \mathrm{ch}_1(\mathfrak{A})$ and there exist three elements a_1, a_2, a_3 such that $a_i \wedge a_j = 0$, $i \neq j$, and $\mathrm{ch}\,(a_i) = (n, \infty, 0)$, then for any $a' \in \mathfrak{A}$ such that $\mathrm{ch}\,(a') = (n, \infty, 0)$, there is $a'' \leqslant a'$ such that $\mathrm{ch}\,(a'') = (n, \infty, 0)$ and $\mathrm{ch}\,(a' \setminus a'') = (n, \infty, 0)$,

(2) if the first characteristic $\mathrm{ch}_1(\mathfrak{A})$ of $\mathfrak{A}$ is finite and in $\mathfrak{B}$ there is the partition of unity into the elements a_1, a_2, a_3 such that $\vee a_i = 1$, $a_i \wedge a_j = 0$, $i \neq j$, and $\mathrm{ch}\,(a_i) = (\infty, 0, 0)$, for any $a \in \mathfrak{B}$ such that $\mathrm{ch}\,(a) = (\infty, 0, 0)$, then there is $a' \leqslant a$ such that $\mathrm{ch}\,(a) = \mathrm{ch}\,(a') = \mathrm{ch}\,(a \setminus a')$.

Proposition 2.3.5. *For any two partitions of unity $a_1, \ldots, a_n$ and $b_1, \ldots, b_n$ in a relatively dense Boolean algebra $\mathfrak{A}$ such that $\mathrm{ch}\,(a_i) = \mathrm{ch}\,(b_i)$, $1 \leqslant i \leqslant n$, there exists an automorphism φ of $\mathfrak{A}$ such that $\varphi(a_i) = b_i$, $1 \leqslant i \leqslant n$.*

PROOF. We define the condition of the Vaught isomorphism as follows:

$$S \doteq \{(a, b) \mid \mathrm{ch}\,(a) = \mathrm{ch}\,(b),\ \mathrm{ch}\,(C(a)) = \mathrm{ch}\,(C(b))\}$$

For S the validity of the requirement of the Vaught criterion follows immediately from the relative density. □

From Proposition 2.3.5 and the possibility of reduction of any type to a decomposition type, we obtain the following assertion.

Theorem 2.3.6. *A countable Boolean algebra is homogeneous if and only if it is relatively dense.*

PROOF. The relative density of a countable homogeneous Boolean algebra follows from the fact that a decomposition type is completely determined by the elementary characteristics of its components, which enables us to transform $a \vee b$ into b provided that $\mathrm{ch}_1(b) > \mathrm{ch}_1(a)$ or $\mathrm{ch}_1(b) = \mathrm{ch}_1(a) = \infty \,\&\, \mathrm{ch}\,(C(a \vee b)) = \mathrm{ch}\,(C(b))$. The converse assertion follows from Proposition 2.3.5 and the possibility of reducing an arbitrary type to a decomposition type. □

Exercises

1. Prove that $\widehat{a}$ is a dense Boolean algebra provided that $\mathfrak{A}$ is a dense Boolean algebra and a is a nonzero element of $\mathfrak{A}$.

2. Prove that $\mathfrak{A}$ is a dense Boolean algebra if and only if $\widehat{a}$ is a dense Boolean algebra for any $a \in \mathfrak{A}$.

3. Prove that $\mathfrak{B}_{\omega \times \eta}$ is a dense Boolean algebra.

4. Prove that $\mathfrak{B}_{\omega}$ is not a dense Boolean algebra.

5. Let L_1 and L_2 be two linear orders. Prove that $\mathfrak{B}_{L_1+1+L_2}$ is a dense Boolean algebra provided that $\mathfrak{B}_{L_1}$ and $\mathfrak{B}_{L_2}$ are dense Boolean algebras.

6. Prove that if $\mathfrak{A}$ and $\mathfrak{B}$ are dense Boolean algebras, then the Boolean algebra $\mathfrak{A} \times \mathfrak{B}$ is dense.

7. Prove that if $\mathrm{ch}_1(\mathfrak{A}) \geqslant n$ and $\mathrm{ch}_1(\mathfrak{B}) \geqslant n$, then $\mathfrak{A} \equiv_{4n} \mathfrak{B}$, i.e., in $\mathfrak{A}$ and $\mathfrak{B}$, the same formulas with less than $4n$-alternations of quantifiers in prenex normal form are true.

8. Construct elementary embeddings of $\mathfrak{B}_{\omega}$ and $\mathfrak{B}_{\omega^2}$ into $\mathfrak{B}_{\omega \times \eta}$, $\mathfrak{B}_{\omega}$ into $\mathfrak{B}_{\omega^2}$, $\mathfrak{B}_{\omega+\eta}$ into $\mathfrak{B}_{\omega \times \eta+\eta}$, $\mathfrak{B}_{\omega^\alpha}$ into $\mathfrak{B}_{\omega^{\alpha+\beta}}$.

2.4. Model-Complete Theories of Boolean Algebras

The model completeness is one of the important properties of elementary theories. It clarifies the structure of models of these systems of axioms and allows us to solve various algorithmic problems in such systems as well.

Theorem 2.1.9 asserts that a theory T is model-complete if and only if for any formula ψ there exists an $\exists$-formula φ such that $T \vdash \psi \Leftrightarrow \varphi$. Due to this fact, the considered class of theories is important.

It is clear that the theory of Boolean algebras of the ordinary signature $\langle \wedge^2, \vee^2, C^1 \rangle$ is not model-complete since finite Boolean subalgebras exist in any infinite Boolean algebra but they are not elementary subsystems. However, this theory becomes model-complete in the first-order enrichment of the theory of Boolean algebras.

We consider the extension of the signature by unary predicate symbols A_n^1, B_n^1, C_n^1, $\mathcal{I}_n^1$, $n \in \mathbb{N}$. We define these predicate symbols with the help of formulas introduced in Sec. 2.2. In this extended signature, for axioms of Boolean algebras we take ordinary axioms of Boolean algebras as well as the following axioms defining the above predicates:

(a) $A_n(x) \Leftrightarrow \mathrm{Atom}_n(x)$,

(b) $B_n(x) \Leftrightarrow \mathrm{Atomless}_n(x)$,

(c) $C_n(x) \Leftrightarrow \mathrm{Atomic}_n(x)$,

(d) $\mathcal{I}_n(x) \Leftrightarrow I_n(x)$

for all $n \in \mathbb{N}$. Thus, we obtain the definable extension BA* of the theory of Boolean algebras. The above-mentioned predicates give no new information concerning the structure of the theory. They were introduced in order to obtain the model completeness property. Under the condition of model completeness it is required that a submodel be an elementary subsystem provided that it has the same characteristic as the model under consideration. In this case, all elements of the submodel must have the same type in it as they have in the model. To satisfy this requirement, we add new predicates to the signature. As we show below, this condition is sufficient for the elementariness of a given submodel.

We now use Proposition 2.1.1 to study the elementariness of a subalgebra in the case of Boolean algebras.

Theorem 2.4.1. *If $\mathfrak{A}$ is a subalgebra of a countable Boolean algebra $\mathfrak{B}$ and for any $a \in A$ and $n \in \mathbb{N}$ the conditions below hold*

(1) $\mathfrak{A} \vDash \mathrm{Atom}_n(a) \Leftrightarrow \mathfrak{B} \vDash \mathrm{Atom}_n(a)$,

(2) $\mathfrak{A} \vDash \mathrm{Atomless}_n(a) \Leftrightarrow \mathfrak{B} \vDash \mathrm{Atomless}_n(a)$,

(3) $\mathfrak{A} \vDash \mathrm{Atomic}_n(a) \Leftrightarrow \mathfrak{B} \vDash \mathrm{Atomic}_n(a)$,

(4) $\mathfrak{A} \vDash I_n(a) \Leftrightarrow \mathfrak{B} \vDash I_n(a)$,

then $\mathfrak{A}$ is an elementary subsystem of $\mathfrak{B}$.

PROOF. We divide the proof into several steps. First of all, we note that the elementary characteristics of $\mathfrak{A}$ and $\mathfrak{B}$ coincide. Indeed, if $\mathrm{ch}(\mathfrak{A}) = (\infty, 0, 0)$, then $\mathfrak{A} \vDash \neg I_n(\mathbf{1})$ for all n. By (4), $\mathfrak{B} \vDash \neg I_n(\mathbf{1})$ for all n. Consequently, $\mathrm{ch}(\mathfrak{B}) = \mathrm{ch}_\mathfrak{B}(\mathbf{1}) = (\infty, 0, 0)$. If $\mathrm{ch}(\mathfrak{A}) = (n, \alpha, i)$, then $\mathfrak{A} \vDash I_{n+1}(\mathbf{1}) \wedge \neg I_n(\mathbf{1})$. By (4), we have $\mathfrak{B} \vDash I_{n+1}(\mathbf{1}) \wedge \neg I_n(\mathbf{1})$, hence $\mathrm{ch}_1(\mathfrak{B}) = \mathrm{ch}_1(\mathbf{1}) = n$. If $\alpha \in \mathbb{N}$, we consider $\alpha + 1$ elements $b_0, b_1, \ldots, b_\alpha \in A$ such that

$\mathfrak{A} \vDash \mathrm{Atomless}_n(b_0)$ and $b_0/I_n(\mathfrak{A}) \neq \mathbf{0}$ for $i = 1$ and $b_0 = \mathbf{0}$ for $i = 0$,
$\mathfrak{A} \vDash \mathrm{Atom}_n(b_j)$ for $1 \leqslant j \leqslant \alpha$,
$\bigvee_{j=0}^{\alpha} b_j = \mathbf{1}$, but $b_k \wedge b_j = \mathbf{0}$ for all $k \neq j$.

Then $\mathfrak{B} \vDash \mathrm{Atom}_n(b_j)$ for all $1 \leqslant j \leqslant \alpha$ by (1) and $\mathfrak{B} \vDash \mathrm{Atomless}_n(b_0)$, $\mathfrak{B} \vDash I_n(b_0) \Leftrightarrow \mathfrak{A} \vDash I_n(b_0)$ in view of (2) and (4). Therefore, $\mathrm{ch}(\mathfrak{B}) = (n, \alpha, i)$. If α is the infinity, we consider elements a and b such that $a \vee b = \mathbf{1}$, $\mathfrak{A} \vDash \mathrm{Atomic}_n(a)$, and $\mathfrak{A} \vDash \mathrm{Atomless}_n(b)$; moreover, $b = \mathbf{0}$ if $i = 0$ and $b/I_n(\mathfrak{A}) \neq \mathbf{0}$ if $i = 1$. By (2) and (3), we have $\mathfrak{B} \vDash \mathrm{Atomic}_n(a)$ and $\mathfrak{B} \vDash \mathrm{Atomless}_n(b) \& [I_n(b)]^i$. Furthermore, for any $k \geqslant 1$ it is possible to choose k elements $a_1, \ldots, a_k$ located under $a/I_n(\mathfrak{A}) \in \mathfrak{A}/I_n(\mathfrak{A})$ such that $a_i \wedge a_j = \mathbf{0}$, $i \neq j$, $a_i \leqslant a$, and $\mathfrak{A} \vDash \mathrm{Atom}_n(a_i)$, $1 \leqslant i \leqslant k$. The types of these elements in $\mathfrak{B}$ and in $\mathfrak{A}$ are the same. Therefore, $\mathrm{ch}(\mathfrak{B}) = (n, \alpha, i)$.

We consider a model $\mathfrak{A}$ and assume that its elementary extension $\mathfrak{A}^*$ is a dense countable Boolean algebra. Such a system $\mathfrak{A}$ exists by Theorem 2.3.1. Without loss of generality, we can assume that $A^* \cap B = A$. We consider the complete diagram $\mathrm{D}_\omega(\mathfrak{B})$ of the Boolean algebra $\mathfrak{B}$ and the set of formulas

$\Delta \rightleftharpoons \{\mathrm{Atom}_n^\varepsilon(a) \mid \mathfrak{A}^* \vDash \mathrm{Atom}_n^\varepsilon(a),\ \varepsilon \in \{0,1\},\ n \in \mathbb{N},\ a \in A^*\}$
$\cup \{\mathrm{Atomless}_n^\varepsilon(a) \mid \mathfrak{A}^* \vDash \mathrm{Atomless}_n^\varepsilon(a),\ n \in \mathbb{N},\ a \in A^*,\ \varepsilon \in \{0,1\}\}$
$\cup \{\mathrm{Atomic}_n^\varepsilon(a) \mid \mathfrak{A}^* \vDash \mathrm{Atomic}_n^\varepsilon(a),\ n \in \mathbb{N},\ a \in A^*,\ \varepsilon \in \{0,1\}\}$
$\cup \{I_n^\varepsilon(a) \mid \mathfrak{A}^* \vDash I_n^\varepsilon(a),\ n \in \mathbb{N},\ a \in A^*,\ \varepsilon \in \{0,1\}\} \cup D_0(\mathfrak{A}^*)$

2.4. Model-Complete Theories

where

$$D_0(\mathfrak{A}^*) \rightleftharpoons \{\varphi(a_1,\ldots,a_n) \in D_\omega(\mathfrak{A}^*) \mid \varphi \text{ is a quantifier-free formula}$$

Let us show that the set of formulas $\Delta \cup D_\omega(\mathfrak{B})$ is satisfiable. In view of the Mal'tsev compactness theorem (cf. Theorem 1.1.2), it suffices to prove that it is locally satisfiable. Let $\Delta_0 \subseteq \Delta \cup D_\omega(\mathfrak{B})$. The set of formulas Δ_0 is finite and

$$\Delta_0 = \{\Phi_0(\overline{b}_0), \ldots, \Phi_n(\overline{b}_n)\} \cup \{\text{Atom}_{n_i}^{\varepsilon_i}(a_i^1) \mid i \leqslant k_1\}$$
$$\cup \{\text{Atomless}_{m_i}^{\delta_i}(a_i^2) \mid i \leqslant k_2\} \cup \{\text{Atomic}_{s_i}^{\alpha_i}(a_i^3) \mid i \leqslant k_3\}$$
$$\cup \{I_{p_i}^{\beta_i}(a_i^4) \mid i \leqslant k_4\} \cup \{\varphi_0(\overline{a}_0), \ldots, \varphi_m(\overline{a}_m)\}$$

where $\overline{b}_i = (b_1^i, \ldots, b_{l_i}^i)$, $\overline{a}_i = (\widetilde{a}_1^i, \ldots, \widetilde{a}_{t_i}^i)$, $t_i^j \in B$, $\widetilde{a}_i^j \in A^*$, $a_i^j \in A^*$, and $\varepsilon_i, \delta_i, \alpha_i, \beta_i \in \{0,1\}$, $\Phi_i(\overline{b}_i) \in D_\omega(\mathfrak{B})$; moreover, $\varphi_i(\overline{a}_i)$ are equivalent quantifier-free formulas in $D_0(\mathfrak{A}^*)$. We consider the conjunction Ψ of all the formulas from

$$\Delta_0 \setminus \{\Phi_0(\overline{b}_0), \ldots, \Phi_n(\overline{b}_n)\} \cup \{\text{the diagram of the subalgebra } \mathfrak{A}_0$$

of the algebra $\mathfrak{A}$ obtained by intersection of $\mathfrak{A}$ with the subalgebra generated by elements of $\overline{b}_i$, $\overline{a}_i$, $\{a_i^j \mid 1 \leqslant j \leqslant 4 \,\&\, 1 \leqslant i \leqslant k_j\}$ in $\mathfrak{B}\}$.

Acting in such a way, we add only a finite number of formulas. Thus, we obtain the formula $\Psi(\overline{a}, \overline{a}^*)$, where the values of $\overline{a}$ are in A and those of $\overline{a}^*$ are in $A^* \setminus A$. However, $\mathfrak{A}$ is an elementary subsystem of $\mathfrak{A}^*$. Consequently, there are elements $\overline{a}_1$ of $\mathfrak{A}$ such that $\mathfrak{A} \vDash \Psi(\overline{a}, \overline{a}_1)$; moreover, all the elements of $\overline{a}$ differ from the elements of $\overline{a}_1$.

For formulas of Δ_0 we consider the permutation that keeps elements of B unchanged and replaces the elements of $A^* \setminus A$ by the corresponding elements of $\overline{a}_1$. By construction, all the formulas of $\Delta_0 \setminus \{\Phi_0(\overline{b}_0), \ldots, \Phi_n(\overline{b}_n)\}$ hold in $\mathfrak{A}$ since these formulas are either quantifier-free ones or have the form $\text{Atom}_n^\varepsilon$, $\text{Atomless}_n^\varepsilon$, $\text{Atomic}_n^\varepsilon$, I_n^ε, where $n \in \mathbb{N}$ and $\varepsilon \in \{0,1\}$. Furthermore, they are true in $\mathfrak{B}$. The formulas $\{\Phi_0(\overline{b}_0), \ldots, \Phi_n(\overline{b}_n)\}$ are true in $\mathfrak{B}$ by construction. Hence Δ_0 and, consequently, Δ is satisfiable.

We now consider a countable model $\mathfrak{B}'$ in which all the formulas of Δ hold. Then $\mathfrak{B}$ is a prime submodel of $\mathfrak{B}'$ and $\mathfrak{A}^*$ is a subalgebra of $\mathfrak{B}'$; moreover, for any $a \in A^*$ the conditions (1)–(4) hold.

For the Boolean algebra $\mathfrak{B}''$ obtained from $\mathfrak{B}'$ by the restriction to the signature of Boolean algebras, we consider a countable dense Boolean algebra $\mathfrak{B}^*$ which elementariness extends $\mathfrak{B}'$. Then $\mathfrak{A}^*$ is a subalgebra of $\mathfrak{B}^*$ because it is a subalgebra $\mathfrak{B}''$ of the subalgebra $\mathfrak{B}^*$. We have $\mathfrak{A}^* \vDash \Phi(a) \Leftrightarrow \mathfrak{B}' \vDash \Phi(a) \Leftrightarrow \mathfrak{B}^* \vDash \Phi(a)$ for $a \in A^*$ and any formula Φ

of the form $\mathrm{Atom}_n^\varepsilon(x)$, $\mathrm{Atomless}_n^\varepsilon(x)$, $\mathrm{Atomic}_n^\varepsilon(x)$, $I_n^\varepsilon(x)$, where $n \in \mathbb{N}$ and $\varepsilon \in \{0,1\}$. The first equivalence is valid by the construction of $\mathfrak{B}'$ and the second one holds since the extension is elementary. Thus, we have constructed the countable Boolean algebras $\mathfrak{A}^*$ and $\mathfrak{B}^*$ such that $\mathfrak{A} \preccurlyeq \mathfrak{A}^*$, $\mathfrak{B} \preccurlyeq \mathfrak{B}^*$, $\mathfrak{A} \subseteq \mathfrak{B}$, and $\mathfrak{A}^* \subseteq \mathfrak{B}^*$; moreover, for any elements $a \in A^*$ and formula Φ of the form $\mathrm{Atom}_n^\varepsilon(x)$, $\mathrm{Atomless}_n^\varepsilon(x)$, $\mathrm{Atomic}_n^\varepsilon(x)$, $I_n^\varepsilon(x)$, we have $\mathfrak{A}^* \vDash \Phi(a) \Leftrightarrow \mathfrak{B}^* \vDash \Phi(a)$. It suffices to show that $\mathfrak{B}^*$ is an elementary extension of the Boolean algebra $\mathfrak{A}^*$. Indeed, in this case, for any elements $a_1, \ldots, a_n$ of A and formula Φ of the signature of Boolean algebras the following equivalences hold:

$$\mathfrak{A} \vDash \Phi(a_1, \ldots, a_n) \Leftrightarrow \mathfrak{A}^* \vDash \Phi(a_1, \ldots, a_n)$$

(in view of the elementariness of the embedding of $\mathfrak{A}$ into $\mathfrak{A}^*$),

$$\mathfrak{A}^* \vDash \Phi(a_1, \ldots, a_n) \Leftrightarrow \mathfrak{B}^* \vDash \Phi(a_1, \ldots, a_n)$$

(in view of the elementariness of submodels of $\mathfrak{A}^*$ in $\mathfrak{B}^*$),

$$\mathfrak{B}^* \vDash \Phi(a_1, \ldots, a_n) \Leftrightarrow \mathfrak{B} \vDash \Phi(a_1, \ldots, a_n)$$

(in view of the elementarity of submodels of $\mathfrak{B}$ in $\mathfrak{B}^*$).

This shows that $\mathfrak{A}$ is an elementary subalgebra of the countable algebra $\mathfrak{B}$.

It remains to prove that $\mathfrak{A}^*$ is a prime submodel of $\mathfrak{B}^*$. By Proposition 2.1.1, for any formula $\Phi(x_1, \ldots, x_n, y)$ and the sequence of elements $a_1, \ldots, a_n \in A^*$ such that $\mathfrak{B}^* \vDash (\exists y)\Phi(a_1, \ldots, a_n, y)$ it is required to find an element $b \in A^*$ such that $\mathfrak{B}^* \vDash \Phi(a_1, \ldots, a_n, b)$. Let Φ be a formula and elements $a_1, \ldots, a_n$ of A^* and b of B^* be such that $\mathfrak{B}^* \vDash \Phi(a_1, \ldots, a_n, b)$. The subalgebra generated by $\{a_1, \ldots, a_n\}$ is finite and atomic. Let $e_1, \ldots, e_m$ be all its atoms. It is clear that $e_1, \ldots, e_m$ are elements of A^*. For $1 \leqslant i \leqslant m$ we consider an element $z_i \in A^*$ such that $\mathrm{ch}\,(e_i \setminus z_i) = \mathrm{ch}\,(e_i \setminus b)$ and $\mathrm{ch}\,(e_i \wedge z_i) = \mathrm{ch}\,(e_i \wedge b)$. Since $\mathfrak{A}^*$ and $\mathfrak{B}^*$ have the same elementary characteristic, in view of Lemmas 2.2.5, 2.2.6, and the completeness of $\mathfrak{A}^*$ such elements z_i exist. We consider the set of pairs

$$\{\langle z_i \wedge e_i,\ e_i \wedge b\rangle \mid 1 \leqslant i \leqslant m\} \cup \{\langle e_i \setminus z_i,\ e_i \setminus b\rangle \mid 1 \leqslant i \leqslant m\}$$

which is a subset of $S \rightleftharpoons \{(a,b) \mid \mathrm{ch}\,(a) = \mathrm{ch}\,(b)\}$ by Lemma 2.2.4 and $\mathrm{ch}\,(C(a)) = \mathrm{ch}\,(C(b))\} \subseteq B^* \times B^*$. In Theorem 2.3.2 we have proved that S is the condition of isomorphism for countable dense Boolean algebras of the same elementary characteristic. By the Vaught theorem, there exists an isomorphism φ of the Boolean algebra $\mathfrak{B}^*$ onto the Boolean algebra

2.4. Model-Complete Theories

$\mathfrak{B}^*$ such that $\varphi(z_i \wedge e_i) = b \wedge e_i$, $\varphi(e_i \setminus z_i) = e_i \setminus b$ for any $1 \leqslant i \leqslant m$. Therefore, $\varphi(e_i) = e_i$. Consequently, $\varphi(a_i) = a_i$ for $i \leqslant n$. We consider an element $a \in A^*$ such that $\varphi(a) = b$. We have

$$\mathfrak{B}^* \vDash \Phi(a_1, \ldots, a_n, a) \Leftrightarrow \mathfrak{B}^* \vDash \Phi(\varphi(a_1), \ldots, \varphi(a_n), \varphi(a))$$

However, $\mathfrak{B}^* \vDash \Phi(a_1, \ldots, a_n, b)$. Consequently, $\mathfrak{B}^* \vDash \Phi(a_1, \ldots, a_n, a)$, and the condition of Proposition 2.1.1 holds. □

Corollary 2.4.1. *If $\mathfrak{A}$ is a countable subalgebra of an algebra $\mathfrak{B}$ and $\text{ch}_{\mathfrak{A}}(a) = \text{ch}_{\mathfrak{B}}(a)$ for any $a \in A$, then $\mathfrak{A}$ is an elementary subalgebra of the algebra $\mathfrak{B}$.*

Here, we use the notation $\text{ch}_{\mathfrak{A}}(a)$ for the characteristic of an element a of the Boolean algebra $\mathfrak{A}$. We now consider the system $\mathfrak{B}^*$ and its subsystem $\mathfrak{A}^*$ in the extended language of Boolean algebras.

Corollary 2.4.2. *If $\mathfrak{A}^*$ is a subsystem of the countable system $\mathfrak{B}^*$ of the theory BA^*, then $\mathfrak{A}^*$ is an elementary subsystem of the system $\mathfrak{B}^*$.*

PROOF. In the extended language, there are predicate symbols that are equivalent to the formulas Atom_n, Atomless_n, Atomic_n, I_n, $n \geqslant 0$. Therefore, the restrictions of $\mathfrak{A}^*$ and $\mathfrak{B}^*$ to the signature of Boolean algebras satisfy the hypotheses of Theorem 2.4.1. Consequently, for the restricted signature the embedding is elementary. The added predicates are equivalent to the formulas of the signature of Boolean algebras. Therefore, in the extended signature, this embedding is also elementary. □

Theorem 2.4.2. *The theory BA^* is model-complete.*

PROOF. We consider two models $\mathfrak{A}^*$ and $\mathfrak{B}^*$ of the theory BA^* such that $\mathfrak{A}^*$ is a subsystem of the $\mathfrak{B}^*$. We show that $\varphi \cdot \mathfrak{A}^* \vDash \varphi(a_1, \ldots, a_n) \Leftrightarrow \mathfrak{B}^* \vDash \varphi(a_1, \ldots, a_n)$ for any $a_1, \ldots, a_n \in A$. Let $a_1, \ldots, a_n$ be elements of A and let φ be a formula of the extended signature. In view of the Löwenheim–Skolem theorem, there exists a countable elementary subsystem $\mathfrak{A}_0^* \preccurlyeq \mathfrak{A}^*$ containing elements $a_1, \ldots, a_n$. The universe of the system $\mathfrak{A}_0^*$ is a countable subset of the system $\mathfrak{B}^*$. In view of the Löwenheim–Skolem theorem, there exists a countable elementary subsystem $\mathfrak{B}_0^* \preccurlyeq \mathfrak{B}^*$ containing $\mathfrak{A}_0^*$ as its subsystem. However, we can apply Corollary 2.4.2 to $\mathfrak{A}_0^*$ and $\mathfrak{B}_0^*$ and conclude that

$$\mathfrak{A}_0^* \vDash \varphi(a_1, \ldots, a_n) \Leftrightarrow \mathfrak{B}_0^* \vDash \varphi(a_1, \ldots, a_n)$$

In accordance with the choice of elementary subsystems, we have
$$\mathfrak{A}^* \vDash \varphi(a_1, \ldots, a_n) \Leftrightarrow \mathfrak{A}_0^* \vDash \varphi(a_1, \ldots, a_n)$$
$$\mathfrak{B}^* \vDash \varphi(a_1, \ldots, a_n) \Leftrightarrow \mathfrak{B}_0^* \vDash \varphi(a_1, \ldots, a_n)$$
The equivalence $\mathfrak{A}^* \vDash \varphi(a_1, \ldots, a_n) \Leftrightarrow \mathfrak{B}^* \vDash \varphi(a_1, \ldots, a_n)$ is proved. □

Exercises

1. Prove that the theory of atomic Boolean algebras with predicate distinguishing atoms is model-complete.

2. Prove that the theory of atomless Boolean algebras is model-complete.

3. Prove that the Boolean algebras $\mathfrak{B}_\omega$ and $\mathfrak{B}_{\omega \times \eta}$ are elementarily equivalent.

4. Construct elementary embeddings of $\mathfrak{B}_\omega$ into $\mathfrak{B}_{\omega^2}$ and of $\mathfrak{B}_{\omega^n}$ into $\mathfrak{B}_{\omega^{n+1}}$.

5. Prove that $\mathfrak{B}_{\omega+\eta}$ and $\mathfrak{B}_{\omega+\omega \times \eta+\eta}$ are elementarily equivalent.

6. Prove that there exists an elementary embedding of $\mathfrak{B}_\omega$ into an infinite atomic Boolean algebra $\mathfrak{B}$.

7. Prove that any countable infinite atomic Boolean algebra is elementarily embedded into the Boolean algebra $\mathfrak{B}_{\omega \times \eta}$.

8. Prove that the Boolean algebra $\mathfrak{B}_{\omega+\eta}$ is an elementary subsystem in the theory of Boolean algebras of characteristic $(1, 1, 0)$.

9. Prove that an arbitrary Boolean algebra contains a subsystem which is elementary in the theory of this Boolean algebra.

10. Prove that the subalgebra of an atomic Boolean algebra generated by its atoms is an elementary subalgebra.

2.5. Consistent Complete Theories of Boolean Algebras

As was shown in Sec. 2.3, any two Boolean algebras of the same elementary characteristic are elementarily equivalent. In this section, we

2.5. Consistent Complete Theories

study the question of what sequences (n, m, k) can be realized in the form of elementary characteristics. In accordance with the definition of an elementary characteristic, if $n \in \mathbb{N} \cup \{\infty\}$, $m \in \mathbb{N} \cup \{\infty\}$, $k \in \{0, 1\}$, and (n, m, k) is an elementary characteristic of some Boolean algebra, then for $n = \infty$ the values of m and k are zero. The triples satisfying the above-mentioned conditions will be called *admissible*.

Let us show that all admissible triples are realized as elementary characteristics of Boolean algebras. To prove Proposition 2.5.1 (below) we need the following lemma.

Lemma 2.5.1. *If $I_n(\widehat{a}) = I_n(\mathfrak{B}) \cap \widehat{a}$, $x \leqslant a$, then*

(a) $x/I_n(\widehat{a})$ *is an atom* $\Leftrightarrow$ $x/I_n(\mathfrak{B})$ *is an atom,*

(b) $x/I_n(\widehat{a})$ *is atomless* $\Leftrightarrow$ $x/I_n(\mathfrak{B})$ *is atomless,*

(c) $x/I_n(\widehat{a})$ *is atomic* $\Leftrightarrow$ $x/I_n(\mathfrak{B})$ *is atomic.*

PROOF. (a) Let $x/I_n(\widehat{a})$ be an atom. Then $x \notin I_n(\widehat{a})$. Consequently, $x \notin I_n(\mathfrak{B})$ since $x \leqslant a$. Therefore, $x/I_n \neq 0$. Assume that $x/I_n(\mathfrak{B})$ is not an atom. Then there exists an element y/I_n such that $y/I_n(\mathfrak{B}) \leqslant x/I_n(\mathfrak{B})$, $y/I_n(\mathfrak{B}) \neq 0$, and $x \backslash y/I_n(\mathfrak{B}) \neq 0$. In this case, $x \wedge y/I_n(\mathfrak{B}) = y/I_n(\mathfrak{B})$ and $x \wedge y \notin I_n(\widehat{a})$, $x \backslash y \in \widehat{a}$ and $x \backslash y \notin I_n(\widehat{a})$. We arrive at the contradiction because $x \backslash y/I_n(\widehat{a}) \bigwedge x \wedge y/I_n(\widehat{a}) = 0$, but $x \backslash y/I_n(\widehat{a}) \leqslant x/I_n(\widehat{a})$, $x \wedge y/I_n(\widehat{a}) \leqslant x/I_n(\widehat{a})$. Thus, $x/I_n(\widehat{a})$ is an atom.

Let $x/I_n(\mathfrak{B})$ be an atom. Then $x \leqslant a$ and $x \notin I_n(\mathfrak{B})$. But in this case, $x \notin I_n(\widehat{a})$ and $x/I_n(\widehat{a}) \neq 0$. If $x/I_n(\widehat{a})$ is not an atom, then there is an element y such that $y/I_n(\widehat{a}) < x/I_n(\widehat{a})$ and $y/I_n(\widehat{a}) \neq 0$. We can assume that $y \leqslant a$ and $x \backslash y \leqslant a$, hence $y \notin I_n(\mathfrak{B})$, $x \backslash y \notin I_n(\mathfrak{B})$, $x \backslash y \wedge y = 0$, and $x \backslash y \vee y = x$. Consequently, $x/I_n(\widehat{a})$ is not an atom.

(b) Let $x/I_n(\mathfrak{B})$ be an atomless element. If $x/I_n(\mathfrak{B}) = 0$, then $x \in I_n(\mathfrak{B})$ and $x \in I_n(\widehat{a}) = I_n(\mathfrak{B}) \cap \widehat{a}$. In the case $x/I_n(\mathfrak{B}) \neq 0$, we have $x \notin I_n(\widehat{a})$. Let $x/I_n(\widehat{a})$ be not atomless. Then there exists an atom $y/I_n(\widehat{a}) \leqslant x/I_n(\widehat{a})$. By (a), $y/I_n(\mathfrak{B})$ is an atom, but $y/I_n(\mathfrak{B}) = x \wedge y/I_n(\mathfrak{B}) \leqslant x/I_n(\mathfrak{B})$. We arrive at a contradiction with the fact that x/I_n is atomless. Let $x/I_n(\widehat{a})$ be atomless, but $x/I_n(\mathfrak{B})$ be not atomless. We consider an element $y/I_n(\mathfrak{B})$ that is an atom of $\mathfrak{B}/I_n(\mathfrak{B})$ less than $x/I_n(\mathfrak{B})$. Then $y \wedge x/I_n(\mathfrak{B}) = x/I_n(\mathfrak{B})$. We can assume that $y \leqslant x$. Then $y \in \widehat{a}$ and $y/I_n(\widehat{a}) \neq 0$ is an atom in view of (a), which contradicts the fact that $x/I_n(\widehat{a})$ is atomless.

(c) Let $x/I_n(\widehat{a})$ be atomic and $x/I_n(\mathfrak{B})$ be not atomic. Then there exists an atomless element $y/I_n(\mathfrak{B})$ which is different from zero. Without loss of generality, we can assume that $y \leqslant x$. By (b), $y/I_n(\widehat{a})$ is also atomless and $y/I_n(\widehat{a}) \neq 0$, which contradicts our assumption. Let $x/I_n(\widehat{a})$ be not atomic and $x/I_n(\mathfrak{B})$ be atomic. Consider an element $y/I_n(\widehat{a}) \leqslant x/I_n(\widehat{a})$ such that $y/I_n(\widehat{a}) \neq 0$ and $y/I_n(\widehat{a})$ are atomless. In view of (b), $y/I_n(\mathfrak{B}) \neq 0$ and $y/I_n(\mathfrak{B}) \leqslant x/I_n(\mathfrak{B})$ are atomless, which contradicts our assumption. □

Proposition 2.5.1. *If $\mathfrak{B}$ is a Boolean algebra, $I_n \rightleftharpoons I_n(\mathfrak{B})$ is the nth Ershov–Tarski ideal, and a is an element of $\mathfrak{B}$, then $I_n(\widehat{a}) = I_n \cap \widehat{a}$.*

PROOF. We apply induction on n. For $n = 0$ the assertion is obvious since $I_0(\widehat{a}) = I_0(\mathfrak{B}) = \{0\}$. We assume that the assertion is valid for n and prove it for $n+1$. Let $x \in I_{n+1} \cap \widehat{a}$. Then $x \leqslant a$ and $x = b \vee c$, where b and c are such that b/I_n is atomic and c/I_n is atomless. By Lemma 2.5.1 and the induction hypothesis, $b/I_n(\widehat{a})$ is atomic and $c/I_n(\widehat{a})$ is atomless. Consequently, $x \in I_{n+1}(\widehat{a})$.

We now establish the inverse inclusion. Let $x \in I_{n+1}(\widehat{a})$. Then $x \leqslant a$ and $x = b \vee c$, where b and c are such that $b/I_n(\widehat{a})$ is atomic and $c/I_n(\widehat{a})$ is atomless. In view of Lemma 2.5.1 and the induction hypothesis, b/I_n is atomic and c/I_n is atomless. Consequently, $x \in I_n \cap \widehat{a}$. □

Let $\boldsymbol{L}$ be a linear order with the least element such that $\mathfrak{B}_{\boldsymbol{L}}$ has the elementary characteristic $(n, 1, 0)$. We define the linear order $\boldsymbol{L}_m \rightleftharpoons \underbrace{\boldsymbol{L} + \ldots + \boldsymbol{L}}_{m \text{ times}}$, $m > 0$, and the linear order $\boldsymbol{L}^\omega \rightleftharpoons \underbrace{\boldsymbol{L} + \boldsymbol{L} + \ldots}_{\omega \text{ times}}$, which can be obtained if we take ω copies of $\boldsymbol{L}$, i.e., $\boldsymbol{L}^\omega \rightleftharpoons \langle \{\langle l, i \rangle \mid l \in \boldsymbol{L},\ i \in \mathbb{N}\}; \leqslant \rangle$, where $\langle l, i \rangle \leqslant \langle l', i' \rangle \rightleftharpoons i < i' \vee (i = i' \wedge l \leqslant l')$.

Proposition 2.5.2. *A Boolean algebra $\mathfrak{B}_{\boldsymbol{L}_m}$ has the elementary characteristic $(n, m, 0)$ provided that $\operatorname{ch}(\mathfrak{B}_{\boldsymbol{L}}) = (n, 1, 0)$.*

PROOF. Since each $\boldsymbol{L}$ is an order with the least element, it is possible to choose m elements $a_1, a_2, \ldots, a_m$ such that a_1 is the least element of $\boldsymbol{L}_m$, a_2 is the least element of the second copy of $\boldsymbol{L}, \ldots, a_m$ is the least element of the mth copy of $\boldsymbol{L}$. Then the segments

$$\alpha_1 \rightleftharpoons [a_1, a_2[,\ \alpha_2 \rightleftharpoons [a_2, a_3[, \ldots, \alpha_{m-1} \rightleftharpoons [a_{m-1}, a_m[,\ \alpha_m \rightleftharpoons [a_m, \infty[$$

are isomorphic to $\boldsymbol{L}$, but $I_n(\mathfrak{B}_{\boldsymbol{L}_m}) \cap \widehat{\alpha}_i = I_n(\widehat{\alpha}_i)$ in view of Proposition 2.5.1. Since $\widehat{\alpha}_i \cong \mathfrak{B}_{\boldsymbol{L}}$ and $\operatorname{ch}(\mathfrak{B}_{\boldsymbol{L}}) = (n, 1, 0)$ for any $x \leqslant \alpha_i$, we have

2.5. Consistent Complete Theories 135

$x \in I_n(\widehat{\alpha}_i)$ or $C(x) \wedge \alpha_i \in I_n(\widehat{\alpha}_i)$; moreover, $\alpha_i \notin I_n(\widehat{\alpha}_i)$. By Lemma 2.5.1, we conclude that $\alpha_i / I_n(\mathfrak{B})$ is an atom, i.e., the Boolean algebra $\mathfrak{B}_{L_m}$ has the elementary characteristic $(n, m, 0)$. □

Proposition 2.5.3. *The Boolean algebra* $\mathfrak{B}_{L^\omega}$ *has the elementary characteristic* $(n, \infty, 0)$ *provided that* $\operatorname{ch}(\mathfrak{B}_L) = (n, 1, 0)$.

PROOF. Let a be the least element of L. We consider $x \in \mathfrak{B}_{L^\omega}$. By the definition of a Boolean algebra by a linear order, $x = \bigvee_{i=1}^{m} [\gamma_i, \delta_i[$, where $\gamma_i < \delta_i$, $1 \leqslant i \leqslant m$, and $\delta_i < \gamma_{i+1}$, $1 \leqslant i \leqslant m - 1$. If $\delta_m < \infty$, then there exists s such that $\delta_m < (a, s)$. Let $\alpha_i \rightleftharpoons [(a, i), (a, i+1)[$. For $\beta_i \rightleftharpoons x \wedge \alpha_i$, $i \leqslant s$, we have $\bigvee_{i=0}^{s} \beta_i = x$. However, $\beta_i / I_n(\mathfrak{B}_L)$ is zero or is an atom, hence x/I_n cannot be atomless. If $\delta_m = \infty$, then there is s such that $\gamma_m < (a, s)$. Then $\alpha_s \subseteq [\gamma_m, \delta_m[\subseteq x$. Consequently, $\alpha_s/I_n \leqslant x/I_n$, but α_s/I_n is an atom. Thus, $\mathfrak{B}_{L^\omega}/I_n$ is an atomic Boolean algebra. Since $\alpha_i/I_n(\mathfrak{B}_{L^\omega})$ are pairwise disjoint atoms, we conclude that $\operatorname{ch}(\mathfrak{B}_{L^\omega}) = (n, \infty, 0)$. □

Let L be a linear order with the least element and let S be the linear order such that the first characteristic of $\mathfrak{B}_S$ is n. We define two linear orders $L^\omega + S \times \eta$ and $\Delta = 1 + S \times \eta$. Let

$$L^\omega = \langle \{\langle l, i \rangle \mid l \in L,\ i \in \mathbb{N}\}; \leqslant \rangle$$
$$\langle l, i \rangle \leqslant \langle l', i' \rangle \rightleftharpoons i < i' \vee (i = i' \,\&\, l \leqslant l')$$
$$S \times \eta \rightleftharpoons \langle \{\langle s, r \rangle \mid s \in S,\ r \in \eta\},\ \leqslant \rangle$$
$$\langle s, r \rangle \leqslant \langle s', r' \rangle \rightleftharpoons r < r' \vee (r = r' \,\&\, s \leqslant s')$$

Proposition 2.5.4. *If* $\operatorname{ch}(\mathfrak{B}_L) = (n, 1, 0)$, *then*

(1) $\operatorname{ch}(\mathfrak{B}_{L^\omega + S \times \eta}) = (n + 1, 1, 0)$ *if* $\operatorname{ch}(\mathfrak{B}_S) = (n, 0, 1)$,

(2) $\operatorname{ch}(\mathfrak{B}_{1 + S \times \eta}) = (n + 1, 0, 1)$ *if* $\operatorname{ch}(\mathfrak{B}_S) = (n, k + 2, 1)$ *and* $k \in \mathbb{N}$.

PROOF. (1) For the linear order $L^\omega + S \times \eta$ we consider

$$\langle \{\langle l, i \rangle \mid l \in L \,\&\, i \in \mathbb{N}\} \cup \{\langle s, r \rangle \mid s \in S \,\&\, r \in \mathbb{Q}\}; \leqslant \rangle$$

under the assumption that $L \cap S = \varnothing$, where $\mathbb{N}$ denotes the set of natural numbers and $\mathbb{Q}$ stands for the set of rational numbers with natural orders

2. Elementary Classification of Boolean Algebras

$<_\mathbb{R}$ and $<_\mathbb{Q}$. The order on this set of pairs is defined as follows:

$$(x,y) \leqslant (a,b) \rightleftharpoons (x \in L \;\&\; a \in S) \vee (x \in L \;\&\; a \in L \;\&\; (y <_\mathbb{N} b$$
$$\vee (y = b \;\&\; x \leqslant_L a))) \vee (x \in S \;\&\; a \in S \;\&\; ((y <_\mathbb{Q} b)$$
$$\vee (y = b \;\&\; x \leqslant_S a)))$$

We show that the elements $[(s',r'), (s'',r'')[, r' < r''$ and $[(s,r), \infty[$ have the characteristic $(n,0,1)$, where $r, r', r'' \in \mathbb{Q}$ and $s, s', s'' \in S$. Indeed, for any $s_1, s_2 \in S$ and $r \in \mathbb{Q}$ the order on the segment $[(s_1,r), (s_2,r)[$ is isomorphic to the order on the interval $[s_1, s_2[$. Consequently, the Boolean algebra $\widehat{[(s_1,r), (s_2,r)[}$ is isomorphic to the Boolean algebra $\widehat{[s_1, s_2[}$. Since ch $(\mathfrak{B}_S) = (n,0,1)$, there exist elements s_1 and s_2 such that ch $([s_1, s_2[) = (n,0,1)$. Let I_n be the nth Ershov–Tarski ideal of the Boolean algebra $\mathfrak{B}_{L^\omega + S \times \eta}$. By Lemma 2.5.1, $[(s_1,r), (s_2,r)[/I_n$ is an atomless element different from zero. For any $(s',r'), (s'',r'')$ such that $r' < r''$ we have $r' < r < r''$ and $[(s_1,r), (s_2,r)[\leqslant [(s',r'), (s'',r'')[$. Consequently, the element $[(s',r'), (s'',r'')[/I_n$ is different from zero and is atomless. In the case $a/I_n \leqslant [(s',r'), (s'',r'')[/I_n$ and $a/I_n \neq \mathbf{0}$ we may assume that $a \leqslant [(s',r'), (s'',r'')[$. Then $a = \bigcup_{i=1}^k [(s'_i, r'_i), (s''_i, r''_i)[$. Since $a/I_n \neq \mathbf{0}$, there exists i such that $r'_i < r''_i$ or $r'_i = r''_i$ for all i, but $[(s'_{i_0}, r'_{i_0}), (s''_{i_0}, r''_{i_0})[/I_n \neq \mathbf{0}$ for some i_0. In the first case, there are two rational numbers $r_i < p_1 < p_2 < r''_i$ such that the elements $[(s_1, p_1), (s_2, p_1)[/I_n$ and $[(s_1, p_2), (s_2, p_2)[/I_n$ are different from zero, their intersection is zero, and they are located under a/I_n. Hence a/I_n is not an atom. In the second case, we note that $\langle [(s'_i, r'_i), (s''_i, r'_i)[, \leqslant \rangle$ is isomorphic to $\langle [s'_i, s''_i[, \leqslant \rangle$ and ch $(\mathfrak{B}_S) = (n,0,1)$. Hence $[(s'_{i_0}, r'_{i_0}), (s''_{i_0}, r'_{i_0})[/I_n$ is an atomless element different from zero, i.e., a/I_n is not an atom.

It is clear that the order on the interval $[(l,i), (l',i)[$ is isomorphic to the order on the interval $[l,l'[$. Therefore, the Boolean algebras $\widehat{[l,l'[}$ and $\widehat{[(l,i), (l',i)[}$ constructed in $\mathfrak{B}_L$ and $\mathfrak{B}_{L^\omega + S + \eta}$ are isomorphic. Similarly, if l_0 is the least element of L and $l \in \omega$, then the orders on $[l_0, \infty[$ and $[(l_0, i), (l_0, i+1)[$ as well as the corresponding Boolean algebras $\widehat{[l_0, \infty[}$ and $\widehat{[(l_0, i), (l_0, i+1)[}$ are isomorphic. From the aforesaid we see that the element $[(l_0, i), (l_0, i+1)[/I_n$ is an atom and the elements $[(s,r), (s',r')[/I_n$ and $[(s,r), \infty[/I_n$ are atomless in $\mathfrak{B}_{L^\omega + S \times \eta}/I_n(\mathfrak{A})$.

We consider the element $\mathbf{1}$ of the Boolean algebra $\mathfrak{B}_{L^\omega + S \times \eta}$ and show that $\mathbf{1} \notin I_{n+1}$. Assume the contrary, i.e., $\mathbf{1} = a \vee b$, where a/I_n is atomic,

2.5. Consistent Complete Theories

but b/I_n is atomless. Since

$$a = \bigcup_{i=1}^{k}[a'_i, a''_i[, \quad b = \bigcup_{i=k+1}^{l}[a'_i, a''_i[$$

we consider the greatest m for which there exists i such that for all $s \geqslant m$ the element $[a'_i, a''_i[$ contains all (l_0, s). Then $a''_i = (s, r)$ and $a'_i = (l, p)$, where $p \leqslant m$, $r \in \mathbb{Q}$, $s \in S$, $l \in L$. In this case, $[a'_i, a'''_i[/I_n$ contains an atom (namely, the element $[(l_0, m), (l_0, m+1)[/I_n)$ and the atomless element $[(s, r'), (s, r'')[/I_n$, where $r' < r''$, $r', r'' \in \mathbb{Q}$, $s \in S$, which contradicts the choice of a and b. If $x \leqslant 1$, then $x \in I_{n+1}$ or $C(x) \in I_{n+1}$. From the elements

$$x = \bigcup_{i=1}^{n}[x'_i, x''_i[, \quad C(x) = \bigcup_{i=1}^{m}[y'_i, y''_i[$$

we choose an element for which there is an interval containing all the elements of (l_0, k), beginning from some k_0. Then another element can decompose into the union of a finite number of atoms and an atomless element.

(2) We consider the Boolean algebra $\mathfrak{B}_\Delta$, where $\Delta = 1 + S \times \eta$ and $\operatorname{ch}(\mathfrak{B}_S) = (n, k+2, 1)$. The interval $[(l, r), (l', r)[$ in Δ with the induced order is isomorphic to the interval $[l, l'[$ in S. Therefore, the Boolean algebra $\widehat{[(l, r), (l', r)[}$ is isomorphic to the Boolean algebra $\widehat{[l, l'[}$ in $\mathfrak{B}_S$. Consequently, $[(\mathfrak{l}, \mathfrak{r}), (\mathfrak{l}', \mathfrak{r})[I_{n+1}(\mathfrak{B}_\delta) = 0$. However, $[(\mathfrak{l}, \mathfrak{r}), (\mathfrak{l}', \mathfrak{r}')[I_{n+1}(\mathfrak{B}_\delta) \neq 0$ for any l, l' and $r < r'$. Therefore, $\mathfrak{B}_\Delta / I_{n+1}(\mathfrak{B}_\Delta)$ is an atomless nontrivial Boolean algebra and $\operatorname{ch}(\mathfrak{B}_{\Delta_A}) = (n+1, 0, 1)$. $\square$

Proposition 2.5.5. *If L, S are linear orders such that $\operatorname{ch}(\mathfrak{B}_L) = (n, s, 0)$, where $s > 0$, $\operatorname{ch}(\mathfrak{B}_S) = (n, 0, 1)$ and S does not have the least element, then the Boolean algebra $\mathfrak{B}_{L+1+S}$ has the elementary characteristic $(n, s, 1)$.*

PROOF. The assertion can be proved with the help of the partition of unity into two intervals $]-\infty, a[$ and $[a, \infty]$ which are isomorphic to L and S. $\square$

Proposition 2.5.6. *If for every n the linear orders L_n with the least elements are such that $\operatorname{ch}(\mathfrak{B}_{L_n}) = (n, 1, 0)$, the Boolean algebra $\mathfrak{B}_L$ constructed by the linear order $L = \sum_{i \geqslant 0} L_i$ has the characteristic $(\infty, 0, 0)$.*

PROOF. For any n we take the least element l_n of $\boldsymbol{L}_n$. It is obvious that the interval $[l_n, l_{n+1}[$ in $\boldsymbol{L}$ is isomorphic to the linear order $\boldsymbol{L}_n$. In this case, the Boolean algebras $\widehat{[l_n, l_{n+1}[}$ for $\mathfrak{B}_{\boldsymbol{L}}$ and $\mathfrak{B}_{\boldsymbol{L}_n}$ are isomorphic. Then $\widehat{[l_n, l_{n+1}[} \notin I_n(\mathfrak{B}_{\boldsymbol{L}})$. Consequently, $\mathbf{1} \notin I_n(\mathfrak{B}_{\boldsymbol{L}})$ for any n, hence ch $(\mathfrak{B}_{\boldsymbol{L}}) = (\infty, 0, 0)$. □

Theorem 2.5.1. *For any admissible triple (n, m, k) there exists a Boolean algebra $\mathfrak{B}$ such that* ch $(\mathfrak{B}) = (n, m, k)$.

PROOF. We proceed by induction on n. For an arbitrary admissible triple (n, m, k) we construct a linear order $\boldsymbol{L}_{n,m,k}$ such that ch $(\mathfrak{B}_{\boldsymbol{L}_{n,m,k}}) = (n, m, k)$. Let $n = 0$. Then $\boldsymbol{L}_{0,1,0}$ is a single-element linear order and $\boldsymbol{L}_{0,1,0}$ is the natural linear order on the set of rational numbers. From Proposition 2.5.2 we obtain a linear order for $(0, k, 0)$ and Propositions 2.5.3 and 2.5.5 yield the order for $(0, \infty, 0)$ and $(0, k, 1)$ for all $k \in \mathbb{N} \cup \{\infty\}$. We assume that all admissible characteristics (s, m, k), where $s \leqslant n$, have been realized. We show that the admissible characteristics $(n+1, m, k)$ are also realized. By Proposition 2.5.4 and the inductive hypothesis, we can obtain the linear orders $\boldsymbol{L}_{n+1,1,0}$ and $\boldsymbol{L}_{n+1,0,1}$. In view of Propositions 2.5.2, 2.5.3, and 2.5.5, we also obtain linear orders for other admissible characteristics $(n + 1, m, k)$. It remains to construct a Boolean algebra of characteristic $(\infty, 0, 0)$. But if we have Boolean algebras of characteristic $(n, 1, 0)$, then we can complete the proof by applying Proposition 2.5.6. □

Theorem 2.5.1 [39]. *For any complete elementary theory T of Boolean algebras there exists an ideal I_T of the Boolean algebra of subsets $\mathcal{P}(N) = \langle \mathcal{P}(N), \cup, \cap, C \rangle$ such that the quotient algebra $\mathcal{P}(N)/I_T$ is a model of the theory T.*

Exercises

1. Prove that the system $\hat{a}$ is a Boolean algebra and ch $(a) =$ ch $(\hat{a})$.

2. Prove that the Boolean algebra $\mathfrak{B}$ is isomorphic to $\hat{a} \times \widehat{C(a)}$ for any element $a \in \mathfrak{B}$ and $\mathfrak{B}$ is a prime (countably saturated) model if $\hat{a}$ and $C(\hat{a})$ are prime (countably saturated) models.

3. Prove that $\mathfrak{B}_\omega$ is a prime model and $\mathfrak{B}_{\omega \times \eta}$ are countably saturated models of the class of atomic infinite Boolean algebras.

4. Construct prime and countably saturated Boolean algebras of all admissible elementary characteristics in terms of linear orders.

2.6. Restricted Theories

5. Prove that the theory of atomic Boolean algebras with predicate distinguishing atoms is model-complete.

6. Prove that two infinite atomic Boolean algebras with dense ideals $(\mathfrak{B}, I)$ and $(\mathfrak{A}, I)$ are elementarily equivalent if and only if the quotient algebras $\mathfrak{B}/I$ and $\mathfrak{B}/J$ are elementarily equivalent. An ideal I of the Boolean algebra $\mathfrak{B}$ is called *dense* if for any $a \neq 0$ there exists $b \neq 0$ such that $b < a$ and $b \in I$.

7. Prove that for any atomic Boolean algebra $\mathfrak{A}$ with dense ideal I and a countable Boolean algebra $\mathfrak{B}_0$ such that $\mathfrak{A}/I$ and $\mathfrak{B}_0$ are elementarily equivalent, there exists a Boolean algebra $\mathfrak{B}$ with dense ideal J such that $(\mathfrak{B}, J)$ and $(\mathfrak{A}, I)$ are elementarily equivalent and the Boolean algebra $\mathfrak{B}/J$ is isomorphic to the Boolean algebra $\mathfrak{B}_0$.

8. Prove that a countable atomic Boolean algebra is countably saturated if and only if any of its element containing infinitely many atoms can be divided into two elements with the same property.

9. Prove that $\operatorname{ch} \langle \mathcal{P}(N), \cup, \cap, V \rangle = (0, \infty, 0)$.

10. Prove that $\bigl(\operatorname{ch} \langle \mathcal{P}(N), \cup, \cap, V\rangle / F(\mathcal{P}(N))\bigr) = (0, 0, 1)$, where $F(\mathcal{P}(N))$ is the Frechét ideal.

11. Prove that $F_n \rightleftharpoons \{X \subseteq N \mid \{0, 1, \ldots, n\} \subseteq X\}$ implies
$$\operatorname{ch}\bigl(\langle \mathcal{P}(N), \cup, \cap, C\rangle / F_n\bigr) = (0, n+1, 0).$$

12. Let I and J be ideals of $\langle \mathcal{P}(N), \cup, \cap, C \rangle$ such that
$$\operatorname{ch}\bigl(\langle \mathcal{P}(N), \cup, \cap, C\rangle / I\bigr) = (n, \infty, 0),\ \operatorname{ch}\bigl(\langle \mathcal{P}(N), \cup, \cap, C\rangle / J\bigr) = (n, 0, 1)$$
Define an ideal $K_m^\varepsilon(I, J)$ such that $\operatorname{ch}\bigl(\langle \mathcal{P}(N), \cup, \cap, C\rangle / K_m^\varepsilon\bigr) = (n+1, m, \varepsilon)$, where $m \leqslant \infty$ and $\varepsilon \in \{0, 1\}$.

2.6. Restricted Theories of Boolean Algebras

In studying Boolean algebras of small characteristic as well as formulas with restricted number of alternations of quantifiers in formulas in prenex normal form, it is often useful to consider only a part of the predicates that are necessary for the model completeness. As we show below, definable predicates have more complicated quantifier prefixes for increasing n and they characterize the corresponding restricted layer of an elementary theory.

We introduce the extending chain of signatures as follows:

$$\sigma \subset \sigma_0 \subset \sigma_1 \subset \ldots \subset \sigma_n \subset$$

where

σ is the signature of Boolean algebras $\langle \wedge, \vee, C \rangle$

$\sigma_0 = \langle \wedge, \vee, C, \mathbf{0}, \mathbf{1} \rangle$

$\sigma_{4n+1} = \sigma_{4n} \cup \{A_n\}$

$\sigma_{4n+2} = \sigma_{4n+1} \cup \{B_n\}$

$\sigma_{4n+3} = \sigma_{4n+2} \cup \{C_n\}$

$\sigma_{4n+4} = \sigma_{4n+3} \cup \{\mathcal{I}_{n+1}\}$

$\sigma^* = \sigma_n$

For any n we consider the extensions BA_n^* of the theory of Boolean algebras BA which are obtained by adding a definition of every symbol. It is clear that $\cup \mathrm{BA}_n^*$ is a model-complete definable extension BA^* of the theory of Boolean algebras BA.

Proposition 2.6.1. *The following assertions hold:*

(a) *for any Boolean algebra $\mathfrak{A}$ of characteristic $(n, 0, 1)$ the theory $\mathrm{Th}\,(\mathfrak{A}) \cup \mathrm{BA}_{4n}^*$ is model-complete,*

(b) *for any Boolean algebra $\mathfrak{A}$ of characteristic $(n, \infty, 0)$ or $(n, k, 0)$ the theory $\mathrm{Th}\,(\mathfrak{A}) \cup \mathrm{BA}_{4n+1}^*$ is model-complete,*

(c) *for any Boolean algebra $\mathfrak{A}$ of characteristic $(n, k, 1)$, $k \in \mathbb{N}$, the theory $\mathrm{Th}\,(\mathfrak{A}) \cup \mathrm{BA}_{4n+2}^*$ is model-complete,*

(d) *for any Boolean algebra $\mathfrak{A}$ of characteristic $(n, \infty, 1)$ the theory $\mathrm{Th}\,(\mathfrak{A}) \cup \mathrm{BA}_{4n+3}^*$ is model-complete.*

PROOF. Since the theory BA^* is model-complete, for any formula $\varphi(x_1, \ldots, x_n)$ of a Boolean algebra we can construct an equivalent $\exists$-formula $\Phi(x_1, \ldots, x_n)$ of the extended signature. On Boolean algebras of characteristic just indicated, all the remaining predicates are either false everywhere, or true everywhere and they can be replaced by equivalent formulas in these $\exists$-formulas but in the signature of Boolean algebras. Thus, it is necessary to add only a finite number of predicates of the signature that corresponds to this characteristic. □

2.6. Restricted Theories

We indicate connections between restricted theories of Boolean algebras and enrichments BA_n^* of the theory of Boolean algebras. We prove several simple lemmas. Let $\mathfrak{A}$ be a subalgebra of the algebra $\mathfrak{B}$ in the natural enrichment to a model of the signature σ_n, i.e., an algebra of the theory BA_n^*.

Lemma 2.6.1. *If for a formula of the signature σ_n of the form*

$$(\exists y_1 \ldots \exists y_m) D(x_1, \ldots, x_n, y_1, \ldots, y_m)$$

where $D(x_1, \ldots, x_n, y_1, \ldots, y_m)$ is a perfect conjunction, its truth in $\mathfrak{B}$ on elements $a_1, \ldots, a_n$ of $\mathfrak{A}$ means its truth in $\mathfrak{A}$ on the elements $a_1, \ldots, a_n$, then the same assertion remains valid for any $\exists$-formulas.

PROOF. Let $\varphi(x_1, \ldots, x_k)$ be an $\exists$-formula of the signature σ_n. It is equivalent to $(\exists y_1 \ldots y_m) \Psi(x_1, \ldots, x_k, y_1, \ldots, y_m)$, where the formula Ψ is a quantifier-free one. We consider all terms $t(x_1, \ldots, x_k)$ defining elements of the subalgebra generated by $x_1, \ldots, x_k$. To the formula Ψ we add new conjunctive members of the form $z_t = t(x_1, \ldots, x_k)$ for every term t and new variable z_t. Closing the formula obtained with respect to variables of types y and z by the existential quantifier, we obtain a formula that is equivalent to a given one. Therefore, we assume that Ψ contains, as conjunctive members, the equalities for all terms defining elements of the subalgebra generated by $x_1, \ldots, x_k$. For an inconsistent formula $\varphi(x_1, \ldots, x_k)$ the conclusion of the lemma is obvious. If the formula is consistent, the formula $\Psi(x_1, \ldots, x_k, y_1, \ldots, y_m)$ is also consistent. The signature σ_n is finite. Hence the formula $\Psi(x_1, \ldots, x_k, y_1, \ldots, y_m)$ is equivalent to the perfect disjunctive normal form $\bigvee_{i=1}^{s} D_s$, where D_s are perfect conjunctions. The formula $\varphi(x_1, \ldots, x_k)$ is equivalent to the formula

$$\bigvee_{i=1}^{s} (\exists y_1 \ldots \exists y_m) D_s(x_1, \ldots, x_k, y_1, \ldots, y_m)$$

If $\exists y_1 \ldots \exists y_m \Psi(x_1, \ldots, x_k, y_1, \ldots, y_m)$ is true in $\mathfrak{B}$ on $a_1, \ldots, a_k \in \mathfrak{A}$, then there exists s such that in $\mathfrak{B}$ the formula

$$(\exists y_1 \ldots y_m) D_s(x_1, \ldots, x_k, y_1, \ldots, y_m)$$

is true on $a_1, \ldots, a_n$. It is also true on these elements in $\mathfrak{A}$. But in this case, all the disjunction is true on these elements in $\mathfrak{A}$. □

We consider a model $\mathfrak{B}$ of the signature σ_n of the theory BA_n^* and a submodel $\mathfrak{A}$ of $\mathfrak{B}$ of the same theory. Let $a_1, \ldots, a_k$ be elements of $\mathfrak{A}$ and

$$\mathfrak{B} \vDash (\exists y_1 \ldots \exists y_m) D(a_1, \ldots, a_k, y_1, \ldots, y_m)$$

where $D(x_1, \ldots, x_k, y_1, \ldots, y_m)$ is a perfect conjunction. We consider the subsystem $\mathrm{gr}_{\mathfrak{B}}(a_1, \ldots, a_k)$ of $\mathfrak{B}$ generated by $a_1, \ldots, a_k$. Since $\mathfrak{B}$ is a model of the theory $\mathrm{BA} \subseteq \mathrm{BA}_n^*$, it is a subalgebra of $\mathfrak{A}$. Let $b_1, \ldots, b_s$ be all the atoms of this Boolean algebra. Then

$$\mathfrak{B} \cong \widehat{b_1} \times \widehat{b_2} \times \ldots \times \widehat{b_s}, \quad \mathfrak{A} \cong \widehat{(b_1)}_{\mathfrak{A}} \times \widehat{(b_2)}_{\mathfrak{A}} \times \ldots \times \widehat{(b_s)}_{\mathfrak{A}}$$

where $\widehat{(b_i)}_{\mathfrak{A}} \subseteq \widehat{b_i}$, $1 \leqslant i \leqslant s$ (decompositions are considered in the signature σ).

For a formula of the form $(\exists y_1 \ldots \exists y_m) D(x_1, \ldots, x_k, y_1, \ldots, y_m)$, we take elements $y_1, \ldots, y_m$ for which the formula is true in $\mathfrak{A}$. We consider $\mathrm{gr}_{\mathfrak{B}}(a_1, \ldots, a_n, y_1, \ldots, y_m)$ as a subalgebra of $\mathfrak{B}$ of the signature of Boolean algebras and all atoms $e_0, \ldots, e_s$ of this decomposition. Then each atom is located under only one element of b_i, $1 \leqslant i \leqslant n$. We obtain s subsets $K_1, \ldots, K_s \subseteq \{0, \ldots, m\}$ such that $K_i \cap K_j = \varnothing$ for $i \neq j$, $\bigvee_{i=1}^{s} K_i = \{0, \ldots, m\}$ and $\bigvee_{j \in K_i} e_j = b_i$. Since $a_1, \ldots, a_k$ belong to $\mathfrak{A}$, the atoms $b_1, \ldots, b_s$ of $\mathrm{gr}_{\mathfrak{B}}(a_1, \ldots, a_k)$ belong to $\mathfrak{A}$. For every $1 \leqslant i \leqslant s$ we find $|K_i|$ elements e'_j, $j \in K_i$, of the subalgebra $\mathfrak{A}$ such that they form the partition of b_i (i.e., $\bigvee_{j \in K_i} e'_j = b_i$ and for any $i \neq j$ from K_i $e'_i \wedge e'_j = 0$) and predicates of the signature σ_n satisfy the following condition: those predicates of σ_n are satisfiable on e'_i that are satisfiable on e_i. We set $y'_i \rightleftharpoons \bigvee_{e_j \leqslant y_i} e'_j$. On y'_i and y_i the same predicates of σ_n are satisfiable since the truth of any predicate of σ_n is uniquely defined from its definition on summands. Indeed if $a = \bigvee_{i=1}^{k} a_i$ and $a_i \wedge a_j = 0$, $i \neq j$, then

$\mathrm{A}_n(a) \Leftrightarrow$ there exists a unique i such that $\mathrm{A}_n(a_i)$
and $\mathcal{I}_n(a_j)$ for any $j \leqslant k$ and $j \neq i$

$\mathrm{B}_n(a) \Leftrightarrow \mathrm{B}_n(a_i)$ for any i

$C_n(a) \Leftrightarrow C_n(a_i)$ for any i

$\mathcal{I}_n(a) \Leftrightarrow \mathcal{I}_n(a_i)$ for any i

The relations between elements are preserved since the corresponding elements are obtained from the same e'_i. By Lemma 2.6.1, we arrive at the following assertion.

2.6. Restricted Theories

Lemma 2.6.2. *Let for any $a \in \mathfrak{A}$ and its partition $e_1, \ldots, e_m$ in $\mathfrak{B}$ there exist a partition e'_i, $1 \leqslant i \leqslant m$, of the element a in $\mathfrak{A}$ such that e'_i, $1 \leqslant i \leqslant m$, belong to $\mathfrak{A}$ and for any i, $1 \leqslant i \leqslant m$, the same predicates of σ_n are true on e_i and on e'_i. Then the truth of any $\exists$-formula of the signature σ_n on $\mathfrak{B}$ means its truth on a subalgebra of $\mathfrak{A}$ of the signature σ_n.*

Proposition 2.6.2. *The following assertions hold:*

(a) *for any $\exists$-formula $\varphi(x_1, \ldots, x_m)$ of the signature σ_n there exists a $\forall$-formula $\psi(x_1, \ldots, x_m)$ of the signature σ_{n+1} such that*

$$\mathrm{BA}^*_{n+1} \vdash (\forall x_1 \ldots \forall x_m)(\varphi(x_1, \ldots, x_m) \Leftrightarrow \psi(x_1, \ldots, x_m))$$

(b) *for any $\forall$-formula $\varphi(x_1, \ldots, x_m)$ of the signature σ_n there exists an $\exists$-formula $\psi(x_1, \ldots, x_m)$ of the signature σ_{n+1} such that*

$$\mathrm{BA}^*_{n+1} \vdash (\forall x_1 \ldots \forall x_m)(\varphi(x_1, \ldots, x_m) \Leftrightarrow \psi(x_1, \ldots, x_m))$$

PROOF. Since the negation of an $\exists$-formula is equivalent to a $\forall$-formula and the negation of a $\forall$-formula is an $\exists$-formula, the conditions (a) and (b) are equivalent. Therefore, it suffices to prove only one of them. In view of Corollary 2.1.10, it suffices to prove that any $\exists$-formula of the signature σ_n preserves the truth under taking submodels of the signature σ_{n+1} of the theory BA^*_{n+1}. By Lemma 2.6.2, for any $b \in \mathfrak{A}$ and the partition $d_1, \ldots, d_r$ of b into elements $d_1, \ldots, d_r \in \mathfrak{B}$ it suffices to find elements $d'_1, \ldots, d'_r \in \mathfrak{A}$ such that they form the partition of b (i.e., $\vee d'_i = b$ and $d'_i \wedge d'_j = 0$ for $i \neq j$) and for any i the same predicates of the signature σ_n are satisfiable on d_i and d'_i. By Lemma 2.6.2, for any submodel $\mathfrak{A} \subseteq \mathfrak{B}$ of the model of the theory BA^*_{n+1} the truth of an $\exists$-formula of the signature σ_n in $\mathfrak{B}$ implies its truth on submodels of $\mathfrak{A}$. This means the validity of the required claim. It remains to consider partitions. Let $b \in \mathfrak{A}$ and $d_1, \ldots, d_r$ be a partition of an element b in a model $\mathfrak{B}$, where $\mathfrak{A} \subseteq \mathfrak{B}$ and $\mathfrak{A}, \mathfrak{B} \vDash \mathrm{BA}^*_{n+1}$, i.e., all the predicates of the signature σ_{n+1} are equivalent to their definitions in $\mathfrak{A}$ and in $\mathfrak{B}$ as well. We consider different variants of the choice of an element b. By $\widehat{b}$ we denote the algebra obtained from elements of $\mathfrak{B}$, i.e., $\widehat{b} = \{x \mid x \in |\mathfrak{A}| \text{ and } x \leqslant b\}$. Let $\widehat{(b)}_{\mathfrak{A}}$ be a similar algebra but in $\mathfrak{A}$, i.e., $\widehat{(b)}_{\mathfrak{A}} = \{x \mid x \in |\mathfrak{B}| \text{ and } x \leqslant b\}$. Predicates of the signature σ_{n+1} are defined on $\widehat{(b)}_{\mathfrak{A}}$ and $\mathfrak{A}$ ($\widehat{(b)}_{\mathfrak{B}}$ and $\mathfrak{B}$) as above since their definitions depend only on the properties of those elements that are less than a given one. Denote by I_n the nth Ershov–Tarski ideal of the Boolean algebra $\mathfrak{B}$.

2. Elementary Classification of Boolean Algebras

CASE 1: $n = 4k$ and ch $(b) = (m, \alpha, \beta)$, but $m < k$ or $m = k \& \alpha = 0 \& \beta = 1$. If $\text{ch}_1(\widehat{b}) < k$, then $b \in I_{i+1} \setminus I_i$ and $i + 1 \leqslant k$. Hence $\text{ch}_1(\widehat{(b)}_\mathfrak{A}) = \text{ch}_1(\widehat{b})$. If $\text{ch}_1(\widehat{b}) = k$, then $b \notin I_k$. Consequently, $b \notin I_k(\mathfrak{A})$. Therefore, $\text{ch}_1(\widehat{(b)}_\mathfrak{A}) \geqslant k$. Under $b/I_k(\mathfrak{A})$ there are no atoms in $\mathfrak{B}/I_k(\mathfrak{A})$, $\mathfrak{A}$ is a subalgebra of $\mathfrak{B}$, and the predicate A_k distinguishing atoms belongs to the signature σ_{n+1}. Then under b/I_k there are no atoms in $\mathfrak{A}/I_k$. Consequently, the characteristic $\text{ch}(\widehat{(b)}_\mathfrak{A})$ is equal to $\text{ch}(\widehat{b})$. By Proposition 2.6.1, the theory $\text{Th}\,(\widehat{b}) \cup \text{BA}_{n+1}^*$ is model-complete, $\widehat{b}$ and $\widehat{(b)}_\mathfrak{A}$ are models of this theory. Hence $\widehat{(b)}_\mathfrak{A} \preccurlyeq \widehat{b}$ and for $b \in \mathfrak{A}$ there is the partition with the same distribution of predicate.

CASE 2: $n = 4k$ and ch $(b) = (k, l, 0)$. The algebra $\widehat{b}/I_k(\widehat{b})$ is finite and the predicate I_k is in the signature. Consequently, the quotient algebra $\widehat{(b)}_\mathfrak{A}/I_k(\mathfrak{A})$ is finite; otherwise, under b in $\mathfrak{A}$ the number of disjoint elements that do not belong to $I_k(\mathfrak{A})$ and possess the same properties in $\mathfrak{B}/I_k$ is larger than l, which is impossible. In $\mathfrak{A}/I_k(\mathfrak{A})$ such elements cannot be less than l. Otherwise, the element $b/I_k(\mathfrak{A})$ in $\mathfrak{A}/I_k(\mathfrak{A})$ must be equal to the union of a finite number of disjoint elements that give atoms in the quotient $\mathfrak{A}/I_k(\mathfrak{A})$. Furthermore, the predicate A_k must be true on them and the same partition must take place in $\mathfrak{B}$. But we arrive at a contradiction with the fact that in $\mathfrak{B}$ there is a partition of b into l atoms in $\mathfrak{B}/I_k(\mathfrak{B})$. Thus, $\text{ch}(\widehat{(b)}_\mathfrak{A}) = (k, l, 0)$ and, as in Case 1, $\widehat{(b)}_\mathfrak{A} \preccurlyeq \widehat{b}$, which guarantees the satisfiability of $\exists$-formulas and the existence of the required partition.

CASE 3: $n = 4k$ and the following condition holds:

$$\text{ch}_1(\widehat{b}) > k \text{ or ch }(\widehat{b}) = (k, \alpha, \beta), \text{ where } \alpha = \infty \text{ or } \beta = 1 \qquad (2.6.1)$$

Since $I_k \in \sigma_{n+1}$, we have $\text{ch}_1(\widehat{(b)}_\mathfrak{A}) \geqslant k$. Let us show that for $\text{ch}_1(\widehat{(b)}_\mathfrak{A}) = k$ the inequality ch $(\widehat{(b)}_\mathfrak{A}) \neq (k, l, 0)$ holds for a finite l. As in Case 2, if ch $(\widehat{(b)}_\mathfrak{A}) = (k, l, 0)$, the unity in $\widehat{(b)}_\mathfrak{A}$ must be partitioned into a union of l disjoint elements on which the predicate A_k is satisfied. They must also have the same properties in $\mathfrak{B}$, and ch $(b) = (k, l, 0)$, which contradicts the assumption. Consequently, for $\widehat{(b)}_\mathfrak{A}$ (2.6.1) holds. We consider the partition $d_1, \ldots, d_r$. If $d_i \notin I_k$ for all $i \leqslant r$, then all the predicates of $\sigma_n \setminus \sigma$ are false on all d_i and b is represented as a union of any finite number of elements with such a property in the Boolean algebra $\mathfrak{A}$. We assume that the condition (2.6.1) holds and, under $\widehat{b}$, there is at least one $d_i \in I_k$.

2.6. Restricted Theories

If $d_i \in I_k$, then under any element $x \leqslant b$ of $\widehat{(b)}_{\mathfrak{A}}$, such that $x \notin I_k$, there exists $y \leqslant x$ such that $y \in I_k$, and all the predicates of $\sigma_n \setminus \sigma$ are satisfiable on y if and only if they are satisfiable on d_i. Thus, there exist disjoint elements $d'_1, \ldots, d'_r$ of $\widehat{(b)}_{\mathfrak{A}}$ the union of which yields b_i and all the predicates of $\sigma_{n+1} \setminus \sigma$ are defined on d'_i and d_i in the same way.

CASE 4: $n = 4k+1$ and ch $\widehat{(b)} = (m, \alpha, \beta)$, where $m < k$ or $m = k$ and α is finite. If $\text{ch}_1\widehat{(b)} < k$, then $\text{ch}_1(\widehat{(b)}_{\mathfrak{A}}) = \text{ch}_1\widehat{(b)}$. We assume that $\text{ch}_1\widehat{(b)} = k$. The predicate distinguishing atoms in the kth quotient algebra by the ideal I_k is contained in the signature σ_{n+1}. Therefore, new atoms in the kth quotient algebra of the subalgebra $\widehat{(b)}_{\mathfrak{A}}$ do not appear. Consequently, $\text{ch}_1(\widehat{(b)}_{\mathfrak{A}}) = k$ and $\text{ch}_2(\widehat{(b)}_{\mathfrak{A}}) \leqslant \alpha$. But it is impossible that they are less than α. Otherwise, there exists a partition into a given number of elements that yield atoms in the kth quotient algebra and the zero or a nonzero atomless element of the kth quotient algebra of the Boolean algebra $\widehat{(b)}_{\mathfrak{A}}$. The predicate B_k distinguishing atomless elements in the kth quotient is contained in the signature. Therefore, the same partition must be satisfiable in the algebra $\mathfrak{B}$, which contradicts the condition on the characteristic of the element b. Consequently, $\text{ch}_2(\widehat{(b)}_{\mathfrak{A}}) = \text{ch}_2\widehat{(b)}$. Similarly, $\text{ch}_3\widehat{(b)} = \text{ch}_3(\widehat{(b)}_{\mathfrak{A}})$. Therefore, ch $(\widehat{(b)}_{\mathfrak{A}}) = $ ch $\widehat{(b)}$. The theory Th $\widehat{(b)} \cup \text{BA}^*_{n+1}$ is model-complete for a given characteristic. Hence $\widehat{(b)}_{\mathfrak{A}} \preccurlyeq \widehat{b}$, i.e., the same $\exists$-formulas are satisfiable on this summand $\widehat{(b)}_{\mathfrak{A}}$ and the required partition exists.

CASE 5: $n = 4k+1$ and ch $\widehat{(b)} = (k, \infty, 0)$. In this case, $\text{ch}_1(\widehat{(b)}_{\mathfrak{A}}) \geqslant k$ because $b \notin I_k(\mathfrak{A})$. Consequently, $b \notin I_k(\mathfrak{A})$. But under b in $\mathfrak{A}$ there are no nonzero atomless elements in the kth quotient algebra $\mathfrak{A}/I_k(\mathfrak{A})$. Indeed, the predicate B_k is contained in the signature, under b and in $\mathfrak{B}$ there are not nonzero atomless elements in the kth quotient algebra $\mathfrak{B}/I_k$. Consequently, $\text{ch}_1(\widehat{(b)}_{\mathfrak{A}}) = k$ and $\text{ch}_3(\widehat{(b)}_{\mathfrak{A}}) = 0$. Since the predicate A_k is contained in the signature, b can be partitioned in $\mathfrak{A}$ into a finite number of elements that are atoms in the kth quotient algebra. Otherwise, such a partition is also a partition in $\mathfrak{B}$, which contradicts the equality ch $\widehat{(b)} = (k, \infty, 0)$. Consequently, $\text{ch}_2\widehat{(b)} = \text{ch}_2(\widehat{(b)}_{\mathfrak{A}})$. In view of the model completeness of the theory of Boolean algebras in the enrichment definable by σ_{n+1}, we obtain $\widehat{(b)}_{\mathfrak{A}} \preccurlyeq \widehat{b}$. Hence the same $\exists$-formulas are true on this summand in $\widehat{(b)}_{\mathfrak{A}}$ and the desired partition exists.

CASE 6: $n = 4k + 1$ and $\text{ch}_1\widehat{(b)} \geqslant k$ or ch $\widehat{(b)} = (k, \infty, 1)$. We note that for $\widehat{(b)}_{\mathfrak{A}}$ the following condition holds: either $\text{ch}_1(\widehat{(b)}_{\mathfrak{A}}) > k$ or

$\mathrm{ch}_1(\widehat{(b)}_\mathfrak{A}) = k$ and $\mathrm{ch}_2(\widehat{(b)}_\mathfrak{A}) = \infty$. It is obvious that $\mathrm{ch}_1(\widehat{(b)}_\mathfrak{A}) \geqslant k$. Let $\mathrm{ch}_1(\widehat{(b)}_\mathfrak{A}) = k$. We note that $\mathrm{ch}_2(\widehat{(b)}_\mathfrak{A}) = \infty$. Indeed, in the opposite case, b in $\mathfrak{A}$ is partitioned into elements $x_1, \ldots, x_l, y$ which are pairwise disjoint, and their union yields b; moreover, $x_i/I_k(\mathfrak{A})$ are atoms, and $y/I_k(\mathfrak{A})$ is atomless in the kth quotient algebra $\mathfrak{A}/I_k(\mathfrak{A})$. Since in the signature σ_{n+1} there are predicates A_k and B_k, the same is satisfiable on these elements in $\mathfrak{B}$. Therefore, $\mathrm{ch}_2(\widehat{b})$ is finite, which contradicts the above assumption.

We consider elements of the partition $d_1, \ldots, d_r$ of b in $\mathfrak{B}$. There exists i such that $d_i \notin I_k$ is not an atom in the kth quotient algebra. For any number d_i of atoms in the kth quotient and elements d_i, where $\mathrm{ch}_1(d_i) < k$, we choose disjoint elements $d'_i \leqslant b \in \mathfrak{A}$ such that the same predicates of the signature σ are true on d'_i and on d_i. Let K be the set of subscripts of all such d_i. We have elements that are not atoms in the kth quotient and, as was mentioned, $\mathrm{ch}_1(d_i) = k$. We consider all such d_i. Let $d_{i_1}, \ldots, d_{i_s}$ be these elements.

From the condition on $\mathrm{ch}\,(\widehat{(b)}_\mathfrak{A})$ we have $\mathrm{ch}\,(\widehat{(b\setminus \bigvee_{i\in K} d'_i)})_\mathfrak{A} = \mathrm{ch}\,(\widehat{(b)}_\mathfrak{A})$. Consequently, for $b\setminus \bigvee_{i\in K} d'_i \in \mathfrak{A}$ there is a partition into s elements that do not belong to the ideal I_k and are not atoms in the kth quotient. The desired partition of b in $\mathfrak{A}$ has been constructed.

CASE 7: $n = 4k + 2$ and $\mathrm{ch}\,(\widehat{b}) = (m, \alpha, \beta)$, where $m < k$ or $m = k$ and α is finite or $\beta = 0$. For $m < k$ we have, as above, $\widehat{(b)}_\mathfrak{A} \preccurlyeq \widehat{b}$. The condition on the truth of $\exists$-formulas in a subalgebra is satisfiable and the required partition exists in $\mathfrak{A}$. If $m = k$ and $\beta = 0$, then b is an atomic element of the kth quotient algebra and the predicate C_k is satisfiable on it. Therefore, C_k is satisfiable on b and in $\mathfrak{A}$. Consequently, $\mathrm{ch}_1(\widehat{(b)}_\mathfrak{A}) = k$ and $\mathrm{ch}_3(\widehat{(b)}_\mathfrak{A}) = 0$. Since the predicates distinguishing atoms and atomless elements in the kth quotient algebra are contained in the signature, we have $\mathrm{ch}(\widehat{b}) = \mathrm{ch}(\widehat{(b)}_\mathfrak{A})$. Therefore, $\widehat{(b)}_\mathfrak{A} \preccurlyeq \widehat{b}$. The condition on the truth of $\exists$-formulas in a subalgebra is satisfiable and the required partition exists in $\mathfrak{A}$.

CASE 8: $n = 4k + 2$ and $\mathrm{ch}\,(\widehat{b}) = (k, \infty, 1)$. As in Case 6, we obtain $\mathrm{ch}_1(\widehat{(b)}_\mathfrak{A}) \geqslant k$. If $\mathrm{ch}_1(\widehat{(b)}_\mathfrak{A}) = k$, then b in $\mathfrak{A}$ can be partitioned into x and y so that $x/I_k(\mathfrak{A})$ is atomic and $y/I_k(\mathfrak{A})$ is atomless. They define atomic and atomless elements in $\mathfrak{B}/I_k$. Since under B/I_k there are infinitely many atoms in $\mathfrak{A}/I_k$, it is impossible that a finite number of atoms are located under $b/I_k(\mathfrak{A})$ in $\mathfrak{A}/I_k(\mathfrak{A})$. Otherwise, the number of such atoms of $\mathfrak{B}$ that are located under b/I_k is finite since A_k and B_k are in the

2.6. Restricted Theories   147

signature. Consequently, $\text{ch}_2(\mathfrak{A}) = \infty$. The predicate B_k is contained in the signature. Since b/I_k is not atomic in $\mathfrak{B}/I_k$, we conclude that $b/I_k(\mathfrak{A})$ is not atomic in $\mathfrak{A}/I_k(\mathfrak{A})$. Therefore, $\text{ch}\,((\widehat{b})_{\mathfrak{A}}) = (k, \infty, 1)$ and $(\widehat{b})_{\mathfrak{A}} \preccurlyeq \widehat{b}_i$. The same $\exists$-formulas hold in this subalgebra. The required partition has been found.

Let $\text{ch}_1((b)_{\mathfrak{A}}) > k$. We consider the partition $d_1, \ldots, d_r$ of an element b in $\mathfrak{A}$. At least one of these elements does not define an atom or a nonzero atomless element in the kth quotient algebra $\mathfrak{B}/I_k$ and does not belong to the kth ideal I_k. We consider all the elements $d_{i_1}, \ldots, d_{i_s}$ with such a property. For the remaining i either $\text{ch}_1(d_i) < k$ or d_i/I_k is an atom of $\mathfrak{B}/I_k$; otherwise, d_i/I_k is an atomless nonzero element in $\mathfrak{B}/I_k$. For all such i, $i \in \{1, \ldots, r\} \setminus \{i_1, \ldots, i_s\}$, we can find the pairwise disjoint elements $d_i' \leqslant b$ of $\mathfrak{A}$ such that on these elements the same predicate of σ_n is true as that on d_i. The difference $b \setminus \vee \{d_i' : i \in \{1, \ldots, r\} \setminus \{i_1, \ldots, i_s\}\}$ gives an element that does not belong to $I_k(\mathfrak{A})$. Consequently, it can be partitioned into s disjoint elements $\{v_i : i \in \{i_1, \ldots, i_s\}\}$ that define nonzero elements in the quotient algebra $\mathfrak{A}/I_k(\mathfrak{A})$ and are neither atoms nor atomless elements in this quotient algebra. We take elements of this partition as values of d_i', $i \in \{i_1, \ldots, i_s\}$. It is obvious that this partition is the required one for b in $\mathfrak{A}$.

CASE 9: $n = 4k + 2$ and $\text{ch}_1(\widehat{b}) > k$. Since in the signature σ_{n+1} there is the predicate $\mathcal{I}_k$, we have $\text{ch}_1((b)_{\mathfrak{A}}) \geqslant k$. If $\text{ch}_1((b)_{\mathfrak{A}}) = k$, then b decomposes into x and y in $\mathfrak{A}$ so that x is in the kth quotient algebra, $\mathfrak{A}/I_k$ yields an atomic element, and y gives an atomless element. The predicates B_k and C_k are contained in the signature σ_{n+1}. The partition has the same properties in $\mathfrak{B}$. Therefore, $\text{ch}_1(\widehat{b}) = k$, which contradicts the above assumption. Consequently, $\text{ch}_1((b)_{\mathfrak{A}}) > k$. Just we handled $\text{ch}_1((\widehat{b})_{\mathfrak{A}}) > k$ in Case 8, we show that the corresponding partition also exists in $\mathfrak{A}$ for b.

CASE 10: $n = 4k + 3$ and $\text{ch}\,(\widehat{b}) = (m, \alpha, \beta)$, where $m \leqslant k$. Since all the necessary predicates are in σ_{n+1}, we have $\text{ch}\,((\widehat{b})_{\mathfrak{A}}) = \text{ch}\,(\widehat{b})$. The theory $\text{Th}\,(\widehat{b}) \cup \text{BA}^*_{n+1}$ is model-complete and $(\widehat{b})_{\mathfrak{A}} \preccurlyeq \widehat{b}$, which yields the required partition of b in $\mathfrak{A}$.

CASE 11: $n = 4k + 3$ and $\text{ch}_1(\widehat{b}) > k$. In this case, $b \notin I_{k+1}$ and $b \notin I_{k+1}(\mathfrak{A})$. We consider an arbitrary partition of b into elements $d_1, \ldots, d_r$ in $\mathfrak{B}$. By the definition of characteristics, there exists d_i such that $d_i \notin I_{k+1}$. Let $d_{i_1}, \ldots, d_{i_s}$ be all d_i such that either $\text{ch}_1(d_i) > k$ or $\text{ch}\,(d_i) = k$ and d_i/I_k is neither an atomic element nor an atomless one in $\mathfrak{B}/I_k$. For any $i \in \{1, \ldots, r\} \setminus \{i_1, \ldots, i_s\}$ one of the following conditions holds:

(1) $\mathrm{ch}_1(d_i) < k$,

(2) $\mathrm{ch}_1(d_i) = k$ and d_i/I_k is an atom,

(3) $\mathrm{ch}_1(d_i) = k$ and d_i/I_k is an atomic element, but not an atom,

(4) $\mathrm{ch}_1(d_i) = k$ and d_i/I_k is atomless.

Since $\mathrm{ch}_1(\widehat{(b)}_{\mathfrak{A}}) > k$, we choose a set of elements $d'_i \leqslant b$ of $\mathfrak{A}$, $i \in \{1,\ldots,r\} \setminus \{i_1,\ldots,i_s\}$, such that they are pairwise disjoint and the same predicates of σ_{n+1} are true on d'_i and on d_i as well. Since $b \notin I_k(\mathfrak{A})$, we have $b \setminus \vee \{d'_i \mid i \in \{1,\ldots,r\} \setminus \{i_1,\ldots,i_s\}\} \notin I_k(\mathfrak{A})$. Under $b/I_k(\mathfrak{A})$ there are many atoms of $\mathfrak{A}/I_k$, and in this quotient algebra there is a nonzero atomless element which, in turn, can be partitioned into any number of nonzero atomless elements in this quotient. Thus, $b \setminus \vee \{d'_i \mid i \in \{1,\ldots,r\} \setminus \{i_1,\ldots,i_s\}\}$ is partitioned into s elements $d'_{i_1},\ldots,d'_{i_s}$ that are neither atomic elements nor atomless ones in the kth quotient algebra $\mathfrak{A}/I_k(\mathfrak{A})$. Thus, we have found the required partition of b in $\mathfrak{A}$. □

Corollary 2.6.1. *The following assertions hold*:

(a) *for any Σ_{n+1}-formulas $\varphi(\overline{x})$ of the signature of Boolean algebras σ there exists a $\forall$-formula $\psi(\overline{x})$ of the signature σ_{n+1} such that*

$$\mathrm{BA}^*_{n+1} \vdash (\forall x_1 \ldots x_n)(\varphi(\overline{x}) \Leftrightarrow \psi(\overline{x}))$$

(b) *for any Π_{n+1}-formulas $\varphi(\overline{x})$ of the signature of Boolean algebras σ there exists an $\exists$-formula $\psi(\overline{x})$ of the signature σ_{n+1} such that*

$$\mathrm{BA}^*_{n+1} \vdash (\forall x_1 \ldots x_n)(\varphi(\overline{x}) \Leftrightarrow \psi(\overline{x}))$$

PROOF. The assertion can be derived from Proposition 2.6.2 by induction on n since the negation of a Σ_n-formula is equivalent to a Π_n-formula and the converse assertion is also true. □

Chapter 3

Constructive Boolean Algebras

In this chapter, we consider the algorithmic complexity of relations defined on Boolean algebras. One way to investigate the algorithmic properties of a system is to study its numberings and consider the algorithmic complexity of the set of numbers of elements appearing in the relation under consideration.

First of all, we discuss the decidability of basic relations and elementary properties of Boolean algebras. Algebras equipped with numberings in which these relations (and properties) are decidable are called constructive (strongly constructive), while the algebras themselves are referred to as constructivizable (strongly constructivizable). Then we indicate different methods of numbering systems, which provides different algorithmic properties of the systems.

3.1. Basic Notions of the Theory of Algorithms and Constructive Models

We use the terminology of the theory of constructive models and the algorithm theory following [109, 178]. In this section, we review the main notions and results that are necessary in the subsequent discussion.

We begin with the notion of the relative computability of functions defined on subsets of natural numbers. Any function $f: X \to \mathbb{N}$, $X \subseteq \mathbb{N}^k$, is called *partial*. A partial function f is called *partially recursive* (with respect to the class of partial functions F) if it belongs to the least class containing the functions $0(x) \equiv 0$, $s(x) \leftrightharpoons x+1$, $I_m^n(x_1, \ldots, x_n) \leftrightharpoons x_m$ (and the class F as well) and is closed under composition, primitive recursion, and minimization. Any partially recursive (with respect to F) function is called simply *recursive* (with respect to F) if it is defined everywhere. A function f is partially recursive with respect to a family X of subsets of $\mathbb{N}$ if it is partially recursive with respect to the class $\{\chi_A \mid A \in X\}$, where χ_A stands for the characteristic function of the set A, i.e., $\chi_A(x) \leftrightharpoons 1$ if $x \in A$ and $\chi_A(x) \leftrightharpoons 0$ otherwise. We often write "(partially) F-recursive" instead of "partially recursive with respect to F" for brevity.

Two partial functions φ and ψ are *equal* (in symbols, $\varphi(x) \equiv \psi(x)$) if their values coincide or both functions are undefined. By means of a Turing machine, we can introduce an equivalent definition of functions that are partially recursive with respect to the sets $A_1 \ldots, A_n$. To ordinary commands we add conditional ones in which the following question appears as a condition: Does the number of unity elements located on the tape and reading out at the moment belong to the set A? Depending on the answer, we carry out one of the ordinary elementary actions.

A subset $A \subseteq \mathbb{N}^k$ is called *recursive* (with respect to F) if its characteristic function χ_A is recursive (with respect to F).

Proposition 3.1.1. *If sets A, $B \subseteq \mathbb{N}^k$, and $C \subseteq \mathbb{N}^m$ are recursive (with respect to F), then the sets $A \cap B$, $A \cup B$, $\mathbb{N}^k \backslash A$, and $A \times C$ are also recursive (with respect to F).*

PROOF. It suffices to express the characteristic functions of the sets $A \cap B$, $A \cup B$, $\mathbb{N}^k \backslash A$, and $A \times C$ in terms of the characteristic functions of the sets A, B, and C. □

3.1. Basic Notions

A subset $A \subseteq \mathbb{N}^k$ is called a $\Sigma_n^{(B)}$-set (Π_n^B-set) if there exists a relation $P(x_1, \ldots, x_k, y_1, \ldots, y_n)$ that is recursive with respect to B and

$$(x_1, \ldots, x_k) \in A \Leftrightarrow \exists y_1 \forall y_2 \ldots P(x_1, \ldots, x_k, y_1, \ldots, y_n)$$
$$((x_1, \ldots, x_k) \in A \Leftrightarrow \forall y_1 \exists y_2 \ldots P(x_1, \ldots, x_k, y_1, \ldots, y_n))$$

A $\Sigma_1^{(B)}$-subset is also called *recursively enumerable (with respect to B)*. There are subsets that are recursively enumerable but not recursive with respect to B. Thus, the classes of $\Sigma_n^{(B)}$-sets and $\Pi_n^{(B)}$-sets are arranged as in the following diagram:

$$\Delta_0^B \rightleftharpoons \Pi_0^B \cap \Sigma_0^B \quad \begin{matrix} \Sigma_0^B \\ \Pi_0^B \end{matrix} \quad \Delta_1^B \rightleftharpoons \Pi_1^B \cap \Sigma_1^B \quad \begin{matrix} \Sigma_1^B \\ \Pi_1^B \end{matrix} \quad \Delta_2^B \rightleftharpoons \Pi_2^B \cap \Sigma_2^B \ldots$$

We can define recursive functions $c \colon \mathbb{N}^2 \xrightarrow[\text{onto}]{1-1} N$ and $l, r \colon \mathbb{N} \to \mathbb{N}$ such that $c(l(n), r(n)) = n$, $lc(x, y) = x$, $rc(x, y) = y$, i.e., c, l, and r define an effective numbering of all pairs of natural numbers. With the help of the numbering c of pairs of natural numbers, it is easy to construct numberings of n-tuples of natural numbers by setting

$$c^1(x_1) \rightleftharpoons x_1, \quad c^{n+1}(x_1, \ldots, x_{n+1}) \rightleftharpoons c(c^n(x_1, \ldots, x_n), x_{n+1})$$

Using this computable sequence of functions, with every ordered sequence of natural numbers $(x_1, \ldots, x_n)$ we can associate a number $\langle x_1, \ldots, x_n \rangle$ such that, starting from it, we are able to restore effectively all the elements and the length of a sequence (cf. [109, 178]). Hereafter, we apply such numberings.

Let $D_0, D_1, \ldots, D_n, \ldots$ be a canonical numbering of all finite subsets of $\mathbb{N}$ such that, starting from n, it is possible to find the number (denoted by $|D_n|$) of elements of D_n and, for a given m, it is possible to define uniformly on n whether m belongs to D_n, i.e., the function $n \to |D_n|$ and the set $\{(n, m) \mid m \in D_n\}$ are recursive.

For a numbering $n \to D_n$ we can take $D_n \rightleftharpoons \{x_1, x_2, \ldots, x_n\}$ if $n = 2^{x_1} + 2^{x_2} + \ldots + 2^{x_k}$, where $x_1 > x_2 > \ldots > x_k$.

Theorem 3.1.1 [109, 178] [on universal functions]. *There exists a partially recursive (with respect to B) function $K(x, \overline{y})$ such that for any partially recursive (with respect to B) function $f(\overline{y})$ there is x such that $K(x, \overline{y}) \equiv f(\overline{y})$ for any $\overline{y}$.*

PROOF. The reasoning is based on constructing numberings of all Turing machines (equipped with the oracle B) that computes partially recursive functions (with respect to B) [109]. □

For a fixed K we write $\{x\}^B(\overline{y})$ instead of $K(x, \overline{y})$.

From the proof of Theorem 3.1.1 [109] we obtain the following theorem.

Theorem 3.1.2 [109] [on the Kleene normal form]. *For any partially recursive (with respect to B) function $\varphi(\overline{x})$ there is a recursive (with respect to B) relation $P \subseteq \mathbb{N}^{k+1}$ such that*

$$\varphi(x) = l\mu y(\langle x_1, \ldots, x_k, y \rangle \in P)$$

By Theorem 3.1.2, for a universal function K we can define a recursive (with respect to B) relation P such that $(\exists t)(\langle x, \overline{y}, c(z,t)\rangle \in P) \Leftrightarrow K(x, \overline{y}) = z$. Therefore, we can introduce a function $K^t(x, \overline{y})$ such that $K^t(x, \overline{y}) = z$ if there exists $t_0 \leqslant t$ for which $P(x, \overline{y}, c(z, t_0))$. In this case, we say that $K(x, \overline{y})$ *is computed in t steps and is equal to z.* In the opposite case, we say that $K^t(x, \overline{y})$ *is undefined and $K(x, \overline{y})$ is not computed in t steps.*

We consider a Kleene universal function $K^{nB}(x, y_1, \ldots, y_n)$ for n-ary partially recursive functions with oracle B constructed by Turing machines.

Theorem 3.1.3. [109] [s-m-n-theorem]. *There exists a primitive recursive function $s(k, x_1, \ldots, x_m)$ such that*

$$K^{(n)B}(k, x_1, \ldots, x_m, x_{m+1}, \ldots, x_{m+n})$$
$$\equiv K^{(n-m)B}(s(k, x_1, \ldots, x_m), x_{m+1}, \ldots, x_{m+n})$$

for any $k, x_1, \ldots, x_m, x_{m+1}, \ldots, x_n \in \mathbb{N}$.

Given a universal function $K^B(x, y)$ of all unary partially B-recursive functions, we can define a computable numbering $\varkappa$ of all partially recursive functions by setting $\lambda y \varkappa_n^B(y) \rightleftharpoons \lambda y K^B(n, y)$.

By a B-computable numbering $\nu: \mathbb{N} \to S$, where S is some class of partially B-recursive functions, we mean a numbering from $\mathbb{N}$ onto S such that there is a partially B-recursive function $\varphi(x, y)$ such that $[\nu(n)](y) \rightleftharpoons \varphi(n, y)$ for any n, y. From universality and Theorem 3.1.3 it is obvious that for any B-computable numbering ν there exists a recursive function $g(x)$ such that $\nu(n) = \varkappa_{g(n)}^B$. This property will be called the *universality property*.

3.1. Basic Notions

In the case $B = \emptyset$, we often omit B in the notation and simply write $\varkappa_n$ for short. $\emptyset$-Computable numberings are referred to as *computable numberings*.

For a universal Kleene numbering $\varkappa_n^B$ we can define a numbering of recursively B-enumerable subsets of $\mathbb{N}$ as follows: $W_n^B \rightleftharpoons \operatorname{dom} \varkappa_n^B$ for any n.

Let RE^B denote the class of all B-recursively enumerable subsets of $\mathbb{N}$. If $S \subseteq \mathrm{RE}^B$, then the numbering $\nu \colon \mathbb{N} \xrightarrow[\text{onto}]{} S$, $n \in \mathbb{N}$, is called B-*computable* provided that the set of pairs $\{(n, m) \mid m \in \nu n,\ n, m \in \mathbb{N}\}$ is recursively B-enumerable. It is obvious that the numbering W_n^B is B-computable. Furthermore, any B-computable numbering ν of B-recursively enumerable sets $S \subseteq \mathrm{RE}^B$ is reduced to W_n^B, i.e., there exists a recursive function g such that $\nu(n) = \operatorname{dom} \varkappa_{g(n)}^B$ for any $n \in \mathbb{N}$. Numberings of recursively B-enumerable sets subject to this property are called *principal*. We approximate W_n by finite sets $W_n^t \rightleftharpoons \{m \mid m \leqslant t,\ K^t(n, m) \text{ is defined}\}$. It is obvious that

$$W_n^0 \subseteq W_n^1 \subseteq \cdots \subseteq W_n^k \subseteq \cdots, \quad \bigcup_t W_n^t = W_n$$

It is also obvious that the double sequence $\{W_n^t \mid n, t \in \mathbb{N}\}$ has the property that from n, t it is possible to compute effectively how many elements are contained in the sets W_n^t and indicate all these elements as well.

We now introduce one more classification of subsets of $\mathbb{N}$ by means of the algorithmic complexity of hyperarithmetic sets. For a subset $X \subseteq \mathbb{N}$ we use the notation $X' \rightleftharpoons \{n \mid \{n\}^X(n) \text{ is defined}\}$. Define inductively the following sequence:

$$X^0 \rightleftharpoons X, \quad X^{(n+1)} \rightleftharpoons (X^{(n)})', \quad X^\omega \rightleftharpoons \{\langle n, m\rangle \mid m \in X^{(n)}\}$$

Feiner [45] put forward a hierarchy of $\emptyset^\omega$-recursive functions and sets which is useful in studying algorithmic properties of constructive and positive models since it takes account of delicate algorithmic properties of such models.

Feiner hierarchy

Let $e \in \mathbb{N}$ and $(a, b) \in \mathbb{N}^2$. We say that a *function* φ *with* $\emptyset^\omega$-*number* e *has type* (a, b) (denoted by $\varphi_e^{\emptyset^\omega} \in \Phi(a, b)$) if the function $\{e\}^{\emptyset^\omega}$ is defined everywhere and for any $n, m,\ m > a + nb$, and q we do not put the question "$\langle m, q\rangle \in \emptyset^\omega$?" to the oracle $\emptyset^\omega$ in computing $\{e\}^{\emptyset^\omega}$ at the point n. If $X \subseteq \mathbb{N}$ and e is such that $\chi_X = \{e\}^{\emptyset^\omega}$ and $\varphi_e^{\emptyset^\omega} \in \Phi(a, b)$, then the *set* X *has type* (a, b) (denoted by $X \in \Phi(a, b)$ and $X \notin \Phi(a, b)$ otherwise).

We now indicate the connection between the arithmetic hierarchy and the Feiner hierarchy.

Lemma 3.1.1. *There exists a recursive function* $\varphi\colon \mathbb{N}^2 \to \mathbb{N}$ *such that if* $\{m\}^\varnothing$ *is defined everywhere, then for any* m, n

$$(\exists z_1)(\forall z_2)\ldots\{m\}^\varnothing(\langle z_1,\ldots,z_n\rangle) = 1 \Leftrightarrow \{\varphi(m,n)\}^{\varnothing^n}(n) = 1$$

PROOF. Let $\{m\}^\varnothing$ be defined everywhere and

$$B_n \rightleftharpoons \{z \mid z = z \,\&\, (\exists z_1)(\forall z_2)\ldots\{m\}^\varnothing(\langle z_1,\ldots,z_n\rangle) = 1\}$$

It is clear that $B_n = \mathbb{N}$ or $B_n = \varnothing$. Starting from $\langle m,n \rangle$, we find an index i of B_n (regarded as a set recursively enumerable in $\varnothing^{(n-1)}$ [178]) and, for a given i, we determine a number (taken as the value of φ at $\langle m,n \rangle$) such that $\{\varphi(m,n)\}^{\varnothing^{(n)}}$ is the characteristic function of B_n. Therefore, $\{\varphi(m,n)\}^{\varnothing^{(n)}}(n) = 1 \Leftrightarrow B_n \neq \varnothing$. □

Proposition 3.1.2. *If* $f\colon \mathbb{N} \to \mathbb{N}$ *is a recursive function and* $X \rightleftharpoons \{n \mid (\exists z_1)(\forall z_2)\ldots f(\langle z_1,\ldots,z_{bn+a}\rangle) = 1\}$, *then* $X \in \Phi(a,b)$.

PROOF. Let $f = \{e\}^\varnothing$. By Lemma 3.1.1, we have the equality $X = \{n \mid \{\varphi(e, a+nb)\}^{\varnothing^{(a+nb)}}(a+nb) = 1\}$, which makes it clear how to construct a Turing machine with number q so that $\varphi_q^{\varnothing^\omega} \in \Phi(a,b)$. Moreover, $\{q\}^{\varnothing^\omega}$ is defined everywhere, takes the values 0, 1, and $n \in X \Leftrightarrow \{q\}^{\varnothing^\omega}(n) = 1$. □

Let

$$A_n \rightleftharpoons \{\langle x,m \rangle \mid x \in \varnothing^{(m)} \,\&\, m \leqslant n\}$$
$$X(a,b) \rightleftharpoons \{n \mid n \text{ is even and } \{n/2\}^{A_{a+nb}}(n) = 0\}$$

Proposition 3.1.3. *If* $Y \subseteq \mathbb{N}$ *and only odd numbers occur in the symmetric difference* $X(a,b)\Delta Y$, *then* $Y \notin \Phi(a.b)$.

PROOF. We assume that there exists a set $Y \subseteq \mathbb{N}$ such that the difference $X(a,b)\Delta Y$ contains only odd numbers, but $Y \in \Phi(a.b)$, i.e., there exists e such that $\chi_Y = \{e\}^{\varnothing^\omega}$ and $\varphi_e^{\varnothing^\omega} \in \Phi(a,b)$. By condition, for an even number m we have $m \in Y \Leftrightarrow m \in X(a,b)$. We now check the condition "$2e \in X(a,b)$." If $2e \in X(a,b)$, then $2e \in Y$ and $1 = \chi_Y(2e) = \{e\}^{\varnothing^\omega}(2e)$. However, by definition, $\{e\}^{\varnothing^\omega}(2e) = \{e\}^{A_{a+2eb}}(2e)$. By the construction of $X(a,b)$, we have $2e \notin X(a,b)$, which yields a contradiction. If $2e \notin X(a,b)$, then $2e \notin Y$. As above, we show that $0 = \chi_Y(2e) =$

3.1. Basic Notions

$\{e\}^{\emptyset^\omega}(2e) = \{e\}^{A_a + 2eb}(2e)$. By the construction of $X(a,b)$, the number $2e$ belongs to $X(a,b)$. The contradiction obtained shows that such e does not exist. □

Proposition 3.1.4. *There exists a recursive function r_k such that*

$$2m \in X(2k+1, 2) \Leftrightarrow (\exists i)(\forall j)^\neg (\exists^\omega z_1) \ldots$$
$$(\exists^\omega z_{2m+k}) r_k(\langle i, j, z_1, \ldots, z_{2m+k}\rangle) = 1$$

where the expression $\exists^\omega x P(x)$ means that infinitely many x have the property P.

PROOF. Using Theorem VIII of [178], we construct a recursive function ρ such that

$$(\forall m)(\forall n)(m \in \emptyset^{(n)}) \Leftrightarrow (\exists z_1)(\forall z_2)\ldots (\rho(\langle z_1, \ldots, z_n, m\rangle) = 1)$$

Using the Tarski–Kuratowski algorithm and inserting ρ into the definition of $X(2k+1, 2)$, it is easy to construct a recursive function $\rho_{(2k+1,2)}$ such that

$$(\forall n)(n \in X(2k+1, 2)) \Leftrightarrow (\exists z_1)(\forall z_2)\ldots (\rho_{(2k+1,2)}(\langle z_1 \ldots, z_{2k+2+2n}\rangle) = 1)$$

Applying the Kreisel–Shoenfield–Wang algorithm of reducing Π_{2n}-forms to U_n-forms [178], we obtain the form

$$(\exists z_1)(\forall z_2)^\neg (\exists^\omega u_1)\ldots (\exists^\omega u_{m+k})(r_k(\langle z_1, z_2, u_1, \ldots, u_{m+k}\rangle) = 1)$$

where r_k is the required recursive function. □

We now introduce the notion of constructibility of an arbitrary algebraic system. The notion is considered with respect to X, where X is some subset of $\mathbb{N}$, or a class of sets, or a class of functions. This means that, in computations below, we can use information about the membership of natural numbers appearing in algorithms to the set X (or a set of the class X) which may have any algorithmic complexity. As we show below, such a generalization allows us to obtain interesting results in the case of ordinary constructivization.

There are two equivalent approaches to the study of algorithmic properties of systems. The first approach deals with abstract systems of arbitrary elements. Various numberings of universes by natural numbers are chosen and their properties are investigated. The second approach admits only those systems whose universes consist of natural numbers. The first approach leads to constructive and strongly constructive systems and the second one leads to (equivalent) recursive and decidable systems.

Let $\mathfrak{A} = \langle A; P_0, \ldots, P_n; F_0, \ldots, F_k; c_0, \ldots, c_s \rangle$ be an algebraic system of a signature $\sigma \rightleftharpoons \langle P_0^{m_0}, \ldots, P_n^{m_n}; \; F_0^{n_0}, \ldots, F_k^{n_k}; \; c_0, \ldots, c_s \rangle$, where P_i is the symbol of an m_i-ary relation, F_i is the symbol of an n_i-ary operation, and c_i denotes the constant symbol. If the signature is infinite, it is required that the functions $i \to m_i$ and $i \to n_i$ be recursive. A pair $\langle \mathfrak{A}, \nu \rangle$, where ν is a mapping of $\mathbb{N}$ (or of an initial segment of $\mathbb{N}$) onto the universe A of the system $\mathfrak{A}$ is called an *enumerated system* and ν is called a *numbering* of $\mathfrak{A}$.

Let B be a set or a family of sets and let $\mathfrak{A}$ be a system of a signature σ. An enumerated system $\langle \mathfrak{A}, \nu \rangle$ is called B-*positive* if the sets

$$\eta_\nu \rightleftharpoons \{(n,m) \mid \nu n = \nu m\}$$
$$\nu^{-1}(P_i) = \{\langle l_1, \ldots, l_{m_i} \rangle \mid \langle \nu l_1, \ldots, \nu l_{m_i} \rangle \in P_i\} \text{ for } i \leqslant n$$

are recursively enumerable with respect to sets of B or the set B and there exist B-recursive functions f_i, $i \leqslant k$, such that $\nu f_i(l_1, \ldots, l_{n_i}) = F_i(\nu l_1, \ldots, \nu l_{n_i})$ for any $l_1, \ldots l_{n_i} \in \mathbb{N}$.

A B-positive system $\langle \mathfrak{A}, \nu \rangle$ is called B-*constructive* if the sets η_ν and $\nu^{-1}(P_i)$ just introduced are B-recursive. If σ is infinite, then it is necessary to impose the conditions of the uniform recursive enumeration and the recursivity in the above-mentioned definitions respectively.

We define a numbering γ of all formulas of the signature σ in variables of the set $\{v_0, v_1, \ldots\}$ so that all questions concerning the structure of formulas can be decidable by means of the numbers of these formulas. Such a numbering is called the *Gödel numbering*.

Let σ be a signature of the form

$$\langle P_0^{n_0}, \ldots, P_k^{n_k}, \ldots; F_0^{m_0}, \ldots, F_s^{m_s}, \ldots; c_0, \ldots, c_n, \ldots \rangle$$

such that there exist partially recursive functions $[n]$ and $[m]$ that are defined as follows:

$[n](i) = n_i$, where n_i is the number of places of the predicate symbol P_i,

$[m](i) = m_i$, where m_i is the number of places of the functional symbol F_i.

If there is only a finite number of predicate symbols and functional ones, such functions $[n]$ and $[m]$ exist. If a countable number of symbols is used in our language, then it is required, for any symbol, to determine effectively the number of places from the index in the signature. This is possible due to the requirement of the recursivity imposed on the functions $[n]$ and $[m]$. We consider a set V of variables $v_0, v_1, \ldots, v_n, \ldots$. Introduce the set $\text{Term}_\sigma(V)$ of terms of the signature σ in variables of V and the

3.1. Basic Notions

set of formulas $\text{Form}_\sigma(V)$ of the signature σ in variables of V. The Gödel numbering γ is introduced as a mapping $\gamma_\sigma \colon \text{Term}_\sigma(V) \cup \text{Form}_\sigma(V) \overset{1-1}{\to} \mathbb{N}$ such that we can effectively recognize if a given number is the number of a formula or that of a term; moreover, we can obtain some information about the structure of the corresponding formula or term.

We construct γ by induction on the complexity of formulas. We first define γ on $\text{Term}_\sigma(V)$ as follows:

(1) $\gamma(v_i) = c(0, c(0, i))$,

(2) $\gamma(c_i) = c(0, c(1, i))$ for i such that $c_i \in \sigma$,

(3) if a term t has the form $F_i(t_1, \ldots, t_{m_i})$, where F_i is an m_i-ary predicate symbol and $\gamma(t_1) = l_1, \ldots, \gamma(t_{m_i}) = l_{m_i}$ are the Gödel numbers of $t_1, \ldots, t_{m_i}$, then $\gamma(t) = c(0, c((i+2), c^{m_i}(l_1, \ldots, l_{m_i})))$.

It is obvious that the set of numbers of terms is recursive and, starting from the number of a term, it is possible to recognize variables and their indices as well as constants and their indices. Furthermore, it is possible to find the index of the operation and the numbers of those subterms from which the term is constructed with the help of the symbol of this operation.

We define γ on a set of formulas as follows:

(1) if t and q are terms such that $\gamma(t) = n$ and $\gamma(q) = m$, then the Gödel number of the formula $t = q$ is defined as follows: $\gamma(t = q) \rightleftharpoons c(1, c(0, c(n, m)))$,

(2) if P_i is an n_i-ary predicate symbol and $t_1, \ldots, t_{n_i}$ are terms with the Gödel numbers $\gamma(t_1) = l_1, \ldots, \gamma(t_{n_i}) = l_{n_i}$, then we define the Gödel number of the formula $P_i(t_1, \ldots, t_{n_i})$ as follows:

$$\gamma(P_i(t_1, \ldots, t_{n_i})) \rightleftharpoons c(1, c(1, c(i, c^{n_i}(l_1, \ldots, l_{n_i}))))$$

(3) for formulas φ and ψ with the Gödel numbers $\gamma(\varphi) = n$ and $\gamma(\psi) = m$, we set

$$\gamma((\varphi \& \psi)) \rightleftharpoons c(1, c(2, c(n, m)))$$
$$\gamma((\varphi \vee \psi)) \rightleftharpoons c(1, c(3, c(n, m)))$$
$$\gamma((\varphi \to \psi)) \rightleftharpoons c(1, c(4, c(n, m)))$$
$$\gamma(\neg \varphi) \rightleftharpoons c(1, c(5, n))$$
$$\gamma((\exists v_i)\varphi) \rightleftharpoons c(1, c(6, c(i, n)))$$
$$\gamma((\forall v_i)\varphi) \rightleftharpoons c(1, c(7, c(i, n)))$$

By induction on the complexity of formulas, it is easy to show that every formula of Form$_\sigma(V)$ asquires its Gödel number. Furthermore, we can recognize whether a given number is the Gödel number of a formula and obtain further information about the structure of the formula with this number: the presence of free variables (indicating all such variables), constants, the form of the formula, the presence of quantifiers, the complexity of the prefix formed by quantifiers, and the numbers of those formulas that are obtained as a result of substitutions.

For any standard equivalence

$$\varphi \to \psi \equiv \neg \varphi \vee \psi$$
$$\neg(\varphi \vee \psi) \equiv \neg\varphi \& \neg\psi$$
$$\neg(\varphi \& \psi) \equiv \neg\varphi \vee \neg\psi$$
$$\neg\neg\varphi \equiv \varphi$$
$$(\exists v)(\varphi \& \psi) \equiv (\exists v)\varphi \& \psi \text{ if } v \text{ does not occur free in } \psi$$
$$(\exists v)(\varphi \vee \psi) \equiv (\exists v)\varphi \vee (\exists v)\psi$$
$$(\forall v)(\varphi \vee \psi) \equiv (\forall v)\varphi \vee \psi \text{ if } v \text{ does not occur free in } \psi$$
$$(\forall v)(\varphi \& \psi) \equiv (\forall v)\varphi \& (\forall v)\psi$$

and other similar equivalences [36, 128], from the number of a formula we can find the numbers of formulas obtained as a result of the replacement of its subformulas by equivalent formulas. Consequently, starting from the number of any formula, it is possible to find the number of an equivalent formula in prenex normal form.

Having the Gödel numbering, we can discuss the notion of decidability from the mathematical point of view. A subset $X \subseteq$ Form$_\sigma(V) \cup$ Term$_\sigma(V)$ is called *decidable* if the set of Gödel numbers $\gamma(X) = \{\gamma(q) \mid q \in X\}$ is recursive, and it is called *enumerable* if $\gamma(X)$ is recursively enumerable. With some hierarchy of the complexity of subsets of $\mathbb{N}$ (the arithmetic hierarchy, the analytic hierarchy, the Ershov hierarchy, etc.) in mind, we say that X *belongs to some complexity class* Δ if the set $\gamma(X)$ belongs to Δ.

Starting from the Gödel number of a formula, we can recognize whether the formula is an axiom of the predicate calculus PC$^\sigma$. Starting from a sequence of numbers, we also can recognize whether a formula is obtained from formulas of a given finite set by some rules of PC$^\sigma$. Hence we can recognize if a sequence of formulas with given Gödel numbers is a proof in PC$^\sigma$. Therefore, the following proposition holds.

3.1. Basic Notions

Proposition 3.1.5. *A set of formulas that are provable in* PC^σ *is enumerable.*

Proposition 3.1.5 leads to the following claim.

Proposition 3.1.6. *If a set of axioms A is enumerable, then the theory $T_A \rightleftharpoons \{\varphi \mid A \vdash \varphi\}$ is also enumerable.*

Let σ be a signature and let σ' be an extension of σ. If, from the indices of predicate symbols, functional symbols, and constant ones of the signature σ, we can compute the indices of these symbols regarded as symbols of the signature σ', then, starting from the Gödel numbers of formulas and terms of the signature σ, we can compute the Gödel numbers of the corresponding formulas and terms with respect to the signature σ'. Moreover, if, starting from indices of predicate symbols, functional symbols, and constant ones of the signature σ', we can recognize whether they belong to σ, then, starting from any formula of the signature σ', we can recognize whether this formula is of the signature σ and find its number with respect to σ. Considering extensions of signatures, we always mean that the properties just listed are valid. In this case, we say that the *Gödel numbering of formulas and terms of the signature σ' extends the Gödel numbering of formulas and terms of the signature σ.*

For a decidable theory T we introduce the principal computable numbering $p_0(\overline{x}_0), \ldots, p_n(\overline{x}_n), \ldots$ of the set of all partial enumerable types consistent with the theory T.

By a *partial type* $p(\overline{x})$ of a theory T we mean the set of formulas in variables of $\overline{x}$ such that the set $p(\overline{x}) \cup T$ is consistent. The numbering $d_0(\overline{x}_0), \ldots, d_n(\overline{x}_n), \ldots$ of partial types of the theory T is called *computable* if $\overline{d}_0, \overline{d}_1, \ldots, \overline{d}_n, \ldots$ is a computable numbering of recursively enumerable sets, where $\overline{d}_i = \{n \mid n$ is the Gödel number of a formula in $d_i\}$ and there exists a recursive function v such that for any n the value $v(n)$ is equal to the number of the sequence $\langle i_1, \ldots, i_{m_n} \rangle$ of indices such that $\overline{x}_n = (v_{i_1}, \ldots, v_{i_{m_n}})$. The numbering $p_0(\overline{x}_0), \ldots, p_n(\overline{x}_n), \ldots$ of partial types of the theory T is called *principal* if for any computable numbering $d_0(\overline{x}'_0), \ldots, d_n(\overline{x}'_n), \ldots$ of partial types of the theory T there is a recursive function $f(n)$ such that $d_n(\overline{x}'_n) = p_{f(n)}(\overline{x}_{f(n)})$ for any n.

To define $p_n(\overline{x}_n)$ we construct an extending sequence of finite sets

$$\varnothing = p_n^0(\overline{x}_n) \subseteq p_n^1(\overline{x}_n) \subseteq \ldots \subseteq p_n^t(\overline{x}_n) \subseteq \ldots$$

such that $p_n(\overline{x}_n) \rightleftharpoons \bigcup_t p_n^t(\overline{x}_n)$. For any n we take i and k such that $c(i, k) = n$. We use i as the number of the ith recursively enumerable set

W_i and k as the number of the sequence $(i_1, \ldots, i_s)$ with respect to the numbering of all sequences of finite length. For any $i, k, t \in \mathbb{N}$ we set

$$W_i^t \upharpoonright k \rightleftharpoons \{m \in W_i^t \mid m \text{ is the Gödel number of a formula in free}$$
$$\text{variables with indices in } \langle i_1, \ldots, i_s \rangle \text{ and with the number } k\}$$

$$p_n^t(\overline{x}_n) \rightleftharpoons \{\varphi \mid \varphi \text{ has the Gödel number in } W_{l(n)}^m \upharpoonright r(n)\}$$

where m is the maximal number among the numbers less than $t+1$ and such that the set

$$T \cup \{\varphi \mid \varphi \text{ has the Gödel number in } W_{l(n)}^m \upharpoonright r(n)\}$$

is consistent. By the decidability of T, the consistency condition is decidable. Consequently, for n and t we can recognize whether a formula belongs to $p_n^t(\overline{x}_n)$ and indicate the number of such formulas, i.e., write out the list of all formulas of $p_n^t(\overline{x}_n)$. It is obvious that the Gödel number of such a formula is less than $t+1$ in view of the condition imposed on W_n^t. The definition of p_n in terms of $W_{l(n)}$, the possibility of computing exactly a sequence of free variables with respect to a computable numbering of a family of finite types, and the fact that $\{W_n\}_{n \in \mathbb{N}}$ is principal imply that p_n is principal.

Proposition 3.1.7. *A family S of types of a theory T is computable, i.e., there exists a computable numbering*

$$d_0(\overline{x}_0'), \ldots, d_n(\overline{x}_n'), \ldots$$

such that $S = \{d_0(\overline{x}_0'), \ldots, d_n(\overline{x}_n'), \ldots\}$ if and only if there exists a recursively enumerable set W such that $S = \{p_n(\overline{x}_n) \mid n \in W\}$.

PROOF. The claim follows from the definition since the numbering $\{p_n\}_{n \in \mathbb{N}}$ is principal. □

To study algorithmic properties of models it is necessary to solve the problem on the existence of an algorithm that verifies the truth of formulas. The above approach leads to a notion that is stronger than that of constructibility. We introduce into consideration the notions of relative constructibility and strong constructibility. Let $\mathcal{B}$ be a class of subsets of $\mathbb{N}$. We adopt the convention that a *quantifier-free formula has 0 alternating quantifiers*. We say that a formula Φ *has n alternating quantifiers* if n groups of alternating quantifiers are contained in prenex normal form of Φ. Let $\mathfrak{F}_n$ stand for the set of formulas with n alternating quantifiers and let $\mathfrak{F}_\omega$ denote the set of all formulas called *fragments* (of

3.1. Basic Notions

a language). The above sets $\mathfrak{F}_n$ are said to be *restricted fragments* (of a language). Let $\mathfrak{F}$ be a set of formulas of a signature σ. An enumerated system $\langle \mathfrak{A}, \nu \rangle$ is called *$\mathcal{B}$-$\mathfrak{F}$-constructive* or *$\mathfrak{F}$-constructive with respect to $\mathcal{B}$* if the set

$$\{\langle s, l_1, \ldots, l_k \rangle \mid s \text{ is the number of the formula } \Phi(x_1, \ldots, x_k)$$
$$\text{of } \mathfrak{F} \text{ with } k \text{ free variables and } \mathfrak{A} \models \Phi(\nu l_1, \ldots, \nu l_k)\}$$

belongs to $\mathcal{B}$. It is easy to see that systems $\mathfrak{F}_0$-constructive with respect to $\mathcal{B}$ are exactly $\mathcal{B}$-constructive systems. We say simply $\mathfrak{F}$-constructibility if the latter is taken with respect to the class of recursive relations, and $\mathcal{B}$-constructibility if $\mathfrak{F} = \mathfrak{F}_0$. If we do not specify $\mathfrak{F}$ or $\mathcal{B}$ we mean that $\mathfrak{F} = \mathfrak{F}_0$ or $\mathcal{B}$ is the class of recursive relations respectively.

$\mathcal{B}$-$\mathfrak{F}_\omega$-constructive systems are referred to as *strongly $\mathcal{B}$-constructive* or *B-ω-constructive* and $\mathcal{B}$-$\mathfrak{F}_n$-constructive systems are called *$\mathcal{B}$-n-constructive*.

We proceed to the second approach. Let $\mathfrak{A}$ be a system of a signature σ and its universe A be a subset of $\mathbb{N}$. We can discuss the effectiveness of various relations without handling numbers.

A system $\mathfrak{A}$ is called *$\mathcal{B}$-recursive* if its basic predicates and operations are of class $\mathcal{B}$. For many classes of computability $\mathcal{B}$, an abstract system is $\mathcal{B}$-constructivizable if and only if it is isomorphic to a $\mathcal{B}$-recursive system. From a $\mathcal{B}$-constructive system we can construct a $\mathcal{B}$-recursive system effectively with respect to $\mathcal{B}$ provided that it is possible to choose a single number in every set of numbers. The passage from a $\mathcal{B}$-recursive system to a $\mathcal{B}$-constructive one can usually be realized with the help of a $\mathcal{B}$-recursive function that enumerates all the universes of a $\mathcal{B}$-recursive model and defines the $\mathcal{B}$-constructivization of this system.

For systems that are $\mathfrak{F}$-constructive with respect to $\mathcal{B}$ we can define $\mathcal{B}$-$\mathfrak{F}$-recursive models in a similar way. If a model is isomorphic to a $\mathcal{B}$-$\mathfrak{F}_\omega$-recursive model, then it is $\mathcal{B}$-decidable. In the case $B \subseteq \mathbb{N}$, we use the notation $\mathcal{B}(B)$ for the class of B-recursive sets. $\mathcal{B}(B)$-$\mathfrak{F}$-recursive ($\mathcal{B}(B)$-$\mathfrak{F}$-constructive) systems are called $\mathfrak{F}$-recursive ($\mathfrak{F}$-constructive) with respect to B.

Proposition 3.1.8. *If systems $\langle \mathfrak{M}, \nu \rangle$ and $\langle \mathfrak{N}, \mu \rangle$ are n-constructive (strongly constructive) with respect to B, then the direct product $\mathfrak{M} \times \mathfrak{N}$ with the numbering $\nu \otimes \mu(n) = (\nu(l(n)), \mu(r(n)))$ is a model which is n-constructive (strongly constructive) with respect to B.*

PROOF. The assertion follows from the algorithm which reduces the question on the truth of formulas on the direct product to that on the truth of formulas with the same number of alternations of quantifiers but on the terms of the direct product [18]. □

Let $\langle \mathfrak{A}, \nu \rangle$ and $\langle \mathfrak{B}, \mu \rangle$ be enumerated models and let $\varphi \colon \mathfrak{A} \to \mathfrak{B}$ be a homomorphism. A homomorphism φ from $\langle \mathfrak{A}, \nu \rangle$ into $\langle \mathfrak{B}, \mu \rangle$ is called *C-recursive* if there exists a C-recursive function f such that $\varphi\nu = \mu f$, i.e., the diagram

$$\begin{array}{ccc} N & \stackrel{f}{\to} & N \\ \nu \downarrow & & \downarrow \mu \\ A & \stackrel{\varphi}{\to} & B \end{array}$$

is commutative. In this case, the function f represents φ and the latter is called a *C-homomorphism*.

If there exists a C-recursive isomorphism φ of $\langle \mathfrak{A}, \nu \rangle$ into $\langle \mathfrak{B}, \mu \rangle$, then $\langle \mathfrak{B}, \mu \rangle$ is called a *C-extension* of $\langle \mathfrak{A}, \nu \rangle$ *with respect to* φ. If $\mathfrak{A} \subseteq \mathfrak{B}$ and the identity embedding of $\mathfrak{A}$ into $\mathfrak{B}$ is C-recursive, then $\langle \mathfrak{B}, \mu \rangle$ is a C-extension of $\langle \mathfrak{A}, \nu \rangle$.

Proposition 3.1.9. *Let $\langle \mathfrak{B}, \mu \rangle$ be an enumerated system, $\mathcal{D}$ be a B-recursively enumerable subset of $\mathbb{N}$, and the set $\mu(\mathcal{D})$ be closed under the basic operations and contain the values of constants. Then a subsystem $\mathfrak{A}$ of the system $\mathfrak{B}$ with universe $\mu(\mathcal{D})$ has a numbering ν such that $\langle \mathfrak{B}, \mu \rangle$ is a B-extension of $\langle \mathfrak{A}, \nu \rangle$.*

PROOF. We obtain the required numbering if we put $\nu(n) \rightleftharpoons \mu f(n)$, where f is a B-recursive function enumerating the set $\mathcal{D}$. □

Proposition 3.1.10. *If $\langle \mathfrak{B}, \mu \rangle$ is a C-extension of $\langle \mathfrak{A}, \nu \rangle$ with respect to φ and $\langle \mathfrak{B}, \mu \rangle$ is constructive with respect to C, then $\langle \mathfrak{A}, \nu \rangle$ is also constructive with respect to C.*

PROOF. The assertion follows from the fact that the verification of the truth of all the predicates and equalities of $\langle \mathfrak{A}, \nu \rangle$ can be reduced to the verification of the truth of the corresponding predicates and equalities of $\langle \mathfrak{B}, \mu \rangle$ by means of a reducing function. □

Here and subsequently, we omit the symbol designating the class under consideration if for the class of recursive functions we take $\mathcal{B}$.

Proposition 3.1.11. *If a Boolean algebra $\langle \mathfrak{B}; \nu \rangle$ is $\mathfrak{F}_\alpha$-constructive with respect to B, then the Boolean ring $\langle R(\mathfrak{B}); \nu \rangle$ is $\mathfrak{F}_\alpha$-constructive with respect to B, where $\alpha \leqslant \omega$.*

3.1. Basic Notions

Since the basic operations of the same signature can be expressed in terms of quantifier-free formulas of some other signature, Proposition 3.1.11 is obvious. A similar reasoning leads to the following proposition.

Proposition 3.1.12. *If a Boolean algebra $\langle \mathfrak{B}; \nu \rangle$ is $\mathfrak{F}_\alpha$-constructive with respect to B, $\alpha \leqslant \omega$, then the Boolean lattice $\langle L(\mathfrak{B}); \nu \rangle$ is also $\mathfrak{F}_\alpha$-constructive.*

The operations in Boolean algebra $\mathfrak{B}$ can be expressed in terms of an order on the Boolean lattice $L(\mathfrak{B})$ with the help of ∀-formulas. Therefore, the following assertion holds.

Proposition 3.1.13. *If a Boolean lattice $\langle L(\mathfrak{B}); \nu \rangle$ is $\mathfrak{F}_{1+\alpha}$-constructive with respect to B, $\alpha \leqslant \omega$, then the Boolean algebra $\langle \mathfrak{B}, \nu \rangle$ is $\mathfrak{F}_\alpha$-constructive with respect to B.*

It is easy to see that if $\langle \mathfrak{A}, \nu \rangle$ is $\mathfrak{F}_\alpha$-constructive with respect to B, then $\langle \mathfrak{A}, \nu \rangle$ is $\mathfrak{F}_\beta$-constructive with respect to B for any $\beta < \alpha$. Generally speaking, the converse assertion is not true.

Proposition 3.1.14. *If $\langle \mathfrak{A}, \nu \rangle$ is constructive with respect to B, and the theory $\mathrm{Th}(\mathfrak{A}, a_1, \ldots, a_n)$ of the system $\mathfrak{A}$ in the enrichment by a finite number of constants is model-complete and B-decidable, then $\langle \mathfrak{A}, \nu \rangle$ is strongly constructive with respect to B.*

PROOF. It suffices to note that, in view of the B-decidability of $\mathrm{Th}(\mathfrak{A}, a_1, \ldots, a_n)$, it is possible to verify effectively the equivalence of formulas. However, any formula is equivalent (with respect to a model-complete theory) to an ∃-formula. Therefore, from any formula Φ we can find effectively two ∃-formulas $\exists \overline{y} \Psi_1$ and $\exists \overline{y} \Psi_2$ that are equivalent to Φ and $\neg \Phi$ respectively. To verify the truth of Φ on elements with the numbers $\overline{l} = (l_1, \ldots, l_n)$, it suffices to search through all possible finite sequences $\overline{c}$ and $\overline{b}$ of numbers of elements and, checking the satisfiability of $\Psi_1(\nu \overline{l}, \nu \overline{c})$ and $\Psi_2(\nu \overline{l}, \nu \overline{b})$, to find the first sequence such that $\Psi_1(\nu \overline{l}, \nu \overline{c})$ or $\Psi_2(\nu \overline{l}, \nu \overline{b})$ holds. If $\Psi_1(\nu \overline{l}, \nu \overline{c})$ holds, then, by the above equivalences, Φ is true on the sequences with numbers $\overline{l}$ and is false if $\Psi_2(\nu \overline{l}, \nu \overline{b})$ holds on $\nu \overline{b}$ for some $\overline{b}$. □

Proposition 3.1.14 and the model completeness of the theory of Boolean algebras with respect to the signature extended by the predicate symbols A_n, B_n, C_n, I_n, $n \in \mathbb{N}$ (cf. Sec. 2.2), lead to the following proposition.

Proposition 3.1.15. *If $\langle \mathfrak{B}, \nu \rangle$ is a B-constructive Boolean algebra of an extended signature of σ^* (cf. Sec. 2.6), then $\langle \mathfrak{B}, \nu \rangle$ is B-strongly constructive.*

The study of various nonequivalent effective representations is very important. There are several approaches to the condensation of equivalence of representations. We describe only the three main methods. Let ν and μ be numberings of the same system $\mathfrak{M}$.

The constructivizations ν and μ are called *B-recursively equivalent* if there are B-recursive functions f and g such that $\nu = \mu f$ and $\mu = \nu g$.

In the case of enumerated models, the B-recursive equivalence requires the existence of a pair of B-recursive functions f and g such that $\nu = \mu f$ and $\mu = \nu g$, but in the case of B-constructive models, the second function can be constructed from the first one.

It is easy to see that if, for a constructivizable system, the set of its automorphisms is of the cardinality of the continuum, then the set of nonrecursive equivalent constructivizations of this system also has the cardinality of the continuum. In particular, a countable atomless Boolean algebra has the continuum of nonrecursive equivalent constructivizations.

Constructivizations ν and μ of $\mathfrak{A}$ are called *B-autoequivalent* if there exists an automorphism φ of $\mathfrak{A}$ such that $\varphi \nu$ and μ are B-recursively equivalent. This definition is of algebraic character because algebra deals with the properties of systems up to an isomorphism. As we show below, a countable atomless Boolean algebra has a unique up to an autoequivalence constructivization. A system with such properties is called *autostable* and the maximal number of nonautoequivalent constructivizations of $\mathfrak{A}$ is called the *algorithmic dimension* and is denoted by $\dim_A(\mathfrak{A})$. A system $\mathfrak{A}$ is called *B-autostable* if any two of its B-constructivizations are B-autoequivalent. It is easy to prove that two constructivizations are autoequivalent if the same relations up to automorphisms are decidable on these constructivizations, i.e., the set of ν-numbers of $A_0 \subseteq A^n$ is recursive if and only if there exists an automorphism φ of $\mathfrak{A}$ such that the set of μ-numbers of $\varphi(A_0)$ is recursive.

We consider one more weakening of the notion of an equivalence. To this end, we turn to decidable relations that are definable in the model under our consideration. Namely, ν and μ are called *algebraically equivalent* if, for any set $M \subseteq A^\omega = \bigcup_{n<\omega} A^n$ stable under automorphisms, the set of sequences of μ-numbers of M is recursive if and only if the set of sequences of ν-numbers of M is recursive.

3.1. Basic Notions

It is clear that if ν and μ are recursively equivalent, then they are also autoequivalent and, consequently, they are algebraically equivalent. The converse assertions do not hold in general.

Corollary 3.1.1. *For any constructive Boolean algebra $\langle \mathfrak{A}, \nu \rangle$ and an element $a \in \mathfrak{A}$ there exist constructivizations ν_a and $\nu_{C(a)}$ of the algebras $\widehat{a}$ and $\widehat{C(a)}$ such that $\langle \mathfrak{A}, \nu \rangle$ and $(\widehat{a} \otimes \widehat{C(a)}, \nu_a \otimes \nu_{C(a)})$ are recursively isomorphic.*

We say that a subset $M \subseteq A^\omega$ is *decidable* (*enumerable*) in $\langle \mathfrak{A}, \nu \rangle$ if the set $\{\langle n_1, \ldots, n_k \rangle \mid (\nu n_1, \ldots, \nu n_k) \in M\}$ is recursive (recursively enumerable).

Proposition 3.1.16. *If $\langle \mathfrak{A}, \nu \rangle$ is a B-extension of $\langle \mathfrak{B}, \mu \rangle$ and of $\langle \mathfrak{B}, \gamma \rangle$ with respect to the same isomorphic embedding φ, then there exists a B-automorphism ψ of subsystem B such that $\psi\mu = \gamma$.*

The question on nonequivalent representations and their classifications is of great importance in the study of constructive systems. Two of the above-mentioned approaches to effective representations of systems allow us to investigate the same properties, choosing a language as a function of the situation. We intend to establish that these approaches are equivalent. To this end, we show the equivalence of the corresponding categories. Introducing notions concerning categories, we follow [13].

Let us consider a category Num of all enumerated models with homomorphisms as morphisms and a category Nat of all models whose universes are subsets of $\mathbb{N}$ with homomorphisms as morphisms. Let $\langle \mathfrak{M}, \nu \rangle$ be an enumerated model. We define the value of the functor Rec on $\langle \mathfrak{M}, \nu \rangle$ by setting

$$\text{Rec} \langle \mathfrak{M}, \nu \rangle \rightleftharpoons \langle N_\mathfrak{M}, \Sigma \rangle$$

where

$$N_\mathfrak{M} \rightleftharpoons \{n \mid n \text{ is the least element of } \nu n\}$$
$$P \rightleftharpoons \{\langle n_1, \ldots, n_k \rangle \in N_\mathfrak{M} \mid \mathfrak{M} \models P(\nu n_1, \ldots, \nu n_k)\}$$

for a predicate symbol $P \in \sigma$ and

$$F(n_1, \ldots, n_k) \rightleftharpoons \min\{m \mid F(\nu n_1, \ldots, \nu n_k) = \nu m\}$$

for a functional symbol $F \in \sigma$. We note that $\text{Rec}\langle \mathfrak{M}, \nu \rangle$ and $\mathfrak{M}$ are isomorphic. If φ is a homomorphism from $\langle \mathfrak{M}, \nu \rangle$ into $\langle \mathfrak{N}, \mu \rangle$, then

$$\text{Rec}(\varphi) \rightleftharpoons \{\langle n, m \rangle \mid n \in \mathbb{N}_\mathfrak{M}, m \in \mathbb{N}_\mathfrak{N} \text{ and } \varphi(\nu n) = \mu m\}$$

It is obvious that $\operatorname{Rec}(\varphi)$ is a homomorphism from $\operatorname{Rec}\langle\mathfrak{M},\nu\rangle$ into $\operatorname{Rec}\langle\mathfrak{N},\mu\rangle$.

REMARK 3.1.1. Models $\mathfrak{M}$ and $\operatorname{Rec}\langle\mathfrak{M},\nu\rangle$ are isomorphic.

REMARK 3.1.2. The value $\operatorname{Rec}(\varphi)$ of any isomorphism φ is also an isomorphism.

We define the functor Con from Nat into Num. To this end, we set $\operatorname{Con}(\mathfrak{M}) \rightleftharpoons \langle\mathfrak{M},\nu\rangle$, where ν is an enumeration of elements of $|\mathfrak{M}|$ in ascending order. If $|\mathfrak{M}|$ is finite, then all the numbers that do not appear in this enumeration are transformed into the last element of the enumeration of $\mathfrak{M}$. The numbering ν obtained in such a way will be denoted by ν_{Con}.

REMARK 3.1.3. A functor Rec determines the equivalence between categories Num and Nat.

We define subcategories Con^B of the category Num consisting of B-constructive models with B-recursive homomorphisms as morphisms and subcategories Rec^B of the category Nat consisting of B-recursive models with B-recursive homomorphisms as morphisms.

Theorem 3.1.4. *The restriction* Rec^B *of the functor* Rec *to the subcategory* Con^B *determines the equivalence between the categories* Con^B *and* Rec^B.

PROOF. It is easy to check that the model $\operatorname{Rec}\langle\mathfrak{M},\nu\rangle$ is B-recursive provided that $\langle\mathfrak{M},\nu\rangle$ is a B-constructive model. The B-recursivity of universes and the existence of a B-recursive function f such that $\varphi\nu = \mu f$ imply that $\operatorname{Rec}\varphi$ is a partially B-recursive function such that its graph is B-recursive.

We consider the restriction Rec^B of the functor Rec to the subcategory Con^B. It is a functor from Con^B into Rec^B. There exist functor isomorphisms $\varphi\colon 1_{\operatorname{Con}^B} \to \operatorname{Rec}\operatorname{Con}$ and $\psi\colon 1_{\operatorname{Rec}^B} \to \operatorname{Con}\operatorname{Rec}$ such that $\operatorname{Rec}\varphi = \psi\operatorname{Rec}$. □

Corollary 3.1.2. *Two enumerated models* $\langle\mathfrak{N},\nu\rangle$ *and* $\langle\mathfrak{M},\mu\rangle$ *are isomorphic if and only if* $\operatorname{Rec}\langle\mathfrak{N},\nu\rangle$ *and* $\operatorname{Rec}\langle\mathfrak{M},\mu\rangle$ *are isomorphic.*

Corollary 3.1.3. B-*constructive models* $\langle\mathfrak{N},\nu\rangle$ *and* $(\mathfrak{M},\mu)$ *are* B-*isomorphic if and only if* $\operatorname{Rec}\langle\mathfrak{M},\mu\rangle$ *and* $\operatorname{Rec}\langle\mathfrak{N},\nu\rangle$ *are* B-*isomorphic.*

3.1. Basic Notions

Thus, the study of constructivizations of a model $\mathfrak{M}$ up to an autoequivalence is equivalent to the study of recursive models up to a recursive isomorphism that are isomorphic to the model $\mathfrak{M}$. Namely,

if ν and μ are constructivizations of a model $\mathfrak{M}$, then $\mathrm{Rec}\,\langle\mathfrak{M},\nu\rangle$ and $\mathrm{Rec}\,\langle\mathfrak{M},\mu\rangle$ are recursive models isomorphic to $\mathfrak{M}$; moreover, ν and μ are autoequivalent if and only if $\mathrm{Rec}\,\langle\mathfrak{M},\nu\rangle$ and $\mathrm{Rec}\,\langle\mathfrak{M},\mu\rangle$ are recursively isomorphic and, for any recursive model $\mathfrak{N}$ isomorphic to $\mathfrak{M}$, there exists its constructivization ν such that $\mathrm{Rec}\,\langle\mathfrak{M},\nu\rangle$ and $\mathfrak{N}$ are recursively isomorphic.

Corollary 3.1.4. *Any infinite constructive algebra is recursively isomorphic to a constructive algebra with a single-valued numbering.*

We now consider numberings of quotient systems by congruences. If η is a congruence on an algebraic system $\mathfrak{A}$ and ν is a numbering of $\mathfrak{A}$, then we can define a quotient numbering ν/η of the quotient system $\mathfrak{A}/\eta$ as follows: $\nu/\eta(n) \rightleftharpoons \nu(n)/\eta$. If it is clear what congruence we mean, then we write simply ν instead of ν/η.

Proposition 3.1.17. *The following assertions hold:*

(a) *if η is a B-decidable strict congruence and $\langle\mathfrak{A},\nu\rangle$ is a B-constructive algebraic system, then $\langle\mathfrak{A}/\eta,\nu/\eta\rangle$ is a B-constructive algebraic system.*

(b) *if $\langle\mathfrak{A},\nu\rangle$ is a B-positive system and η is a B-enumerable congruence, then $\langle\mathfrak{A}/\eta,\nu/\eta\rangle$ is a B-positive system.*

Let $\mathfrak{M} = \langle M, P_0^{n_0}, \ldots, P_k^{n_k}, a_0, \ldots, a_s \rangle$ be a finite model of a finite signature σ without functional symbols and let ν be a mapping from $[0,n] = \{i \mid 0 \leqslant i \leqslant n\}$ onto M. A pair $\langle\mathfrak{M},\nu\rangle$ is called a *finitely enumerated* or *n-enumerated* model.

To every finite signature $\sigma = \langle P_0^{n_0}, \ldots, P_k^{n_k}, a_0, \ldots, a_s \rangle$ we assign the number $\langle\langle\langle 0,n_0\rangle,\ldots,\langle k,n_k\rangle\rangle,s\rangle$, $n_i \geqslant 1$. We extend the signature σ by constant symbols $c_0, \ldots, c_n, \ldots$ with natural indices. We define an $(n+1)$-diagram $\mathcal{D}\langle\mathfrak{M},\nu\rangle$ of a finite enumerated model $\langle\mathfrak{M},\nu\rangle$ by defining the enrichment $\mathfrak{M}^n$ of $\mathfrak{M}$ to the signature $\sigma_n = \sigma \cup \{c_0,\ldots,c_n\}$ so that the value of the constant c_i is equal to the element νi with the number $i \leqslant n$,

$$\mathcal{D}\langle\mathfrak{M},\nu\rangle = \{\varphi \mid \varphi \text{ is an atomic formula of the signature } \sigma_n$$
$$\text{without free variables}$$
$$\text{or the negation of such a formula and } \mathfrak{M}^n \vDash \varphi\}$$

Let $GD\langle\mathfrak{M},\nu\rangle$ be the set of the Gödel numbers of formulas of $\mathcal{D}\langle\mathfrak{M},\nu\rangle$. The sequence $\langle n, \langle\langle 0, n_0\rangle, \langle 1, n_1\rangle, \ldots, \langle k, n_k\rangle\rangle, s\rangle, u\rangle$, where u denotes the canonical number of the finite set $D_u = GD\langle\mathfrak{M},\nu\rangle$, is called the *Gödel number of an enumerated finite model of a finite signature* σ. Such models are also referred to as *finitely enumerated models* and their numbers will be called the *Gödel numbers*.

We note the following obvious properties of these numberings of finitely enumerated models.

⟨1⟩ *The set of the Gödel numbers of finitely enumerated models is recursive.*

⟨2⟩ *From the Gödel number of a finitely enumerated model $\langle\mathfrak{M},\nu\rangle$ it is possible to compute the number of elements of $|\mathfrak{M}|$.*

⟨3⟩ *From the Gödel number of a finitely enumerated model it is possible to compute how many predicate symbols and constant ones are contained in the signature σ and compute the number of places of all predicate symbols as well.*

⟨4⟩ *Given two Gödel numbers of finitely enumerated models, it is possible to recognize whether these models are considered with respect to the same signature.*

⟨5⟩ *The set of numbers of finite signatures is recursive.*

⟨6⟩ *If n is the number of a finitely enumerated model and m is the number of a signature, then it is possible to recognize whether a finitely enumerated model with the number n is a model of the signature with the number m.*

⟨7⟩ *From the number of a finitely enumerated model n and the number of a signature m it is possible to recognize whether a finitely enumerated model with the number n of the signature $\sigma = \langle P_0^{n_0}, \ldots, P_k^{n_k}, a_0, \ldots, a_s\rangle$ has some enrichment to a finitely enumerated model of the signature σ' with the number m.*

A finite n-enumerated model $\langle\mathfrak{M},\nu\rangle$ is called an *extension* of a k-enumerated model $\langle\mathfrak{N},\mu\rangle$ if $k < n$, the models $\mathfrak{M}$ and $\mathfrak{N}$ are considered in the same signature, and the set $\{\langle\nu(i),\mu(i)\rangle \mid i \leqslant k\}$ is an isomorphic embedding of $\mathfrak{M}$ into $\mathfrak{N}$.

⟨8⟩ *From any two Gödel numbers of finitely enumerated models n and m it is possible to recognize whether a finitely enumerated model*

3.1. Basic Notions

with the Gödel number m is an extension of a finitely enumerated model with the Gödel number n.

A finite enumerated model of the signature $\sigma = \langle P_0^{n_0}, \ldots, P_k^{n_k}, a_0, \ldots, a_s \rangle$ is called an $\langle \mathfrak{M}, \nu \rangle$-extension of a k-enumerated model $\langle \mathfrak{N}, \mu \rangle$ of the signature $\sigma' = \langle P_0^{m_0}, \ldots, P_{r'}^{m_{r'}}, a_0, \ldots, a_{s'} \rangle$ if $k < n$, $s' \leqslant s$, $r' \leqslant r$, for any $0 \leqslant i \leqslant r'$ the number of places m_i of the predicate P_i is equal to the number of places n_i of the predicate P_i, and $\{\langle \mu(i), \nu(i) \rangle \mid i \leqslant k\}$ is an isomorphism of $\mathfrak{N}$ into $\mathfrak{M} \upharpoonright \sigma$.

$\langle 9 \rangle$ *From any two Gödel numbers n and m of finitely enumerated models it is possible to recognize whether a model with the number n is an extension of a finitely enumerated model with the number m.*

Let $\langle \mathfrak{M}, \nu \rangle$ be an enumerated model of the signature

$$\sigma = \langle P_0^{n_0}, \ldots, P_k^{n_k}, a_0, \ldots, a_s, \ldots \rangle$$

Starting from $n \in \operatorname{dom} \nu$, we define $M_n = \{\nu i \mid i \leqslant n\}$ and consider the signature σ_n obtained from the signature σ in the following way: $\sigma_n = \langle P_0^{n_0}, \ldots, P_r^{n_r}, a_{i_1}, \ldots, a_{i_k} \rangle$, where r is equal to n if σ contains less than n predicates and r is equal to the number of predicates of σ if the latter contains less than n predicates. We note that the set $\{i_1, \ldots, i_k\}$ consists of all indices i of constants of σ such that $i \leqslant n$ and the value of the constant c_i belongs to M_n.

We now define the model $\mathfrak{M}_n$ as a submodel of the restriction $\mathfrak{M} \upharpoonright \sigma_n$ of the model $\mathfrak{M}$ to a model of the signature σ_n with the universe M_n. Finitely enumerated models $\langle \mathfrak{M}_n, \nu_n \rangle$, where $\nu_n(k) = \nu(k)$ for $k \leqslant n$, will be called *finitely enumerated submodels* of $\langle \mathfrak{M}, \nu \rangle$.

By a *representation* of an enumerated model $\langle \mathfrak{M}, \nu \rangle$ we mean the set $W \langle \mathfrak{M}, \nu \rangle$ of the Gödel numbers of all finitely enumerated submodels.

Proposition 3.1.18. *An enumerated model $\langle \mathfrak{M}, \nu \rangle$ is constructive if and only if its representation $W \langle \mathfrak{M}, \nu \rangle$ is recursively enumerable.*

To prove the assertion it suffices to compare the decidability problem for a model and that for its representation.

We adopt the convention that the empty model of the empty signature with the empty numbering and any finite enumerated model are *constructive*. Having the set $W \langle \mathfrak{M}, \nu \rangle$, we can define the model $\mathfrak{M}$ and its numbering ν as follows. We consider the set

$$M_W^0 = \{c_i \mid \text{there exists the Gödel number of a finite enumerated model in } W\langle \mathfrak{M}, \nu \rangle \text{ with the number } n+1 \text{ and } i \leqslant n\}$$

On M_W^0, we introduce the following equivalence relation: $c_i \sim_W c_j$ if $c_i = c_j$ is included in the diagram of some n-enumerated model with the number in $W\langle \mathfrak{M}, \nu \rangle$. Let M_W be equal to the quotient set $M_W^0/\sim_W$. We define the numbering ν_W by setting $\nu_W(i) = c_i/\sim_W$, where $i \in M_W^0$. We set $\nu_W(i) = a_j$ for a constant of the signature σ of the model $\mathfrak{M}$ if the equality $c_i = a_j$ is included in the diagram of some n-enumerated model with the number in W.

We define $P_i(\nu_W n_0, \ldots, \nu_W n_k)$ if the formula $P_i(c_{n_0}, \ldots, c_{n_k})$ is included in the diagram of some n-enumerated model with the number in W.

Thus, we define the model $\mathfrak{M}_W$ of the same signature that the model $\mathfrak{M}$ has and the numbering ν_W of $\mathfrak{M}_W$ as well. We also define $\varphi(c_i/\sim) \rightleftharpoons \nu i$ and find that φ is an isomorphism of an enumerated model $\langle \mathfrak{M}_W, \nu_W \rangle$ onto $\langle \mathfrak{M}, \nu \rangle$; moreover, on numbers it is represented by the identity function.

If $W\langle \mathfrak{M}, \nu \rangle$ is recursively enumerated, the model $\langle \mathfrak{M}_W, \nu_W \rangle$ is constructive. The converse assertion is obvious.

We now fix a finite signature σ without functional symbols and define a numbering $\varkappa^\sigma$ of all constructive models of the signature σ as follows. We take the principal numbering $\{W_n\}_{n \in \mathbb{N}}$ of all recursively enumerable subsets of $\mathbb{N}$. As usual, W_n^t is a part of the set W_n that is already enumerated to the step t in numbering. We recall that we enumerate only $x < t$ in W_n^t. From W_n we construct a new set V_n by setting $V_n^0 = \varnothing$ and verify the following conditions at the step $t+1$:

— any element of W_n^{t+1} is the Gödel number of some k-model of the signature σ,

— for any $x, y \in W_n^{t+1}$ one of the finitely enumerated models, the Gödel numbers of which are x and y, is an extension of other models.

We set $V_n^{t+1} \rightleftharpoons V_n^t$ if these conditions fail and $V_n^{t+1} \rightleftharpoons V_n^t \cup W_n^{t+1}$ otherwise. Let $V_n = \cup V_n^t$. It is obvious that the sequence of sets $\{V_n\}_{n \in \mathbb{N}}$ just constructed is computable. Consequently, there is a recursive function ρ such that $V_n = W_{\rho(n)}$ for any n. Furthermore, $W_{\rho(\rho(n))} = W_{\rho(n)}$ for any n.

It is easy to see that for any n the set V_n is recursively enumerable and is the representation of some constructive model. Using the construction of the last proposition, we restore the constructive model $\mathfrak{M}_{V_n}$ and its

3.1. Basic Notions

constructivization ν_{V_n} by V_n. We set $\varkappa^\sigma(n) \rightleftharpoons \langle \mathfrak{M}_{V_n}, \nu_{V_n} \rangle$. The model $\mathfrak{M}_{V_n}$ is denoted by $\mathfrak{M}_n^\varkappa$ and the numbering ν_{V_n} is written as $\nu_n^\varkappa$. From the characterization of recursive models via their representations we see that $\varkappa^\sigma(n)$ enumerates all the constructive models of the signature σ, including finitely enumerated models and the empty one. If σ is an infinite signature and a function $i \to n_i$ is recursive, where n_i denotes the number of places of the ith predicate symbol, then, acting in a similar way, we obtain $W\langle\mathfrak{M}, \nu\rangle$ and the numbering $\varkappa^\sigma$ enumerating all the constructive models of the signature σ and finite constructive models of finite parts of the signature σ provided that, constructing the sets V_n, we require that finitely enumerated models are models of the initial segments of the signature σ and take the condition of an extension.

The following notion of a computable sequence of constructive models is often used in the literature concerning constructive models. A sequence of constructive models $\langle \mathfrak{M}_0, \nu_0 \rangle, \dots, \langle \mathfrak{M}_n, \nu_n \rangle, \dots$ is called *computable* if these models are uniformly constructive, i.e., for necessary recursive characteristic functions of sets of numbers and recursive functions defining the operations, their indices can be computed for models $\langle \mathfrak{M}_n, \nu_n \rangle$ from the number n by means of recursive functions.

Proposition 3.1.19. *A sequence of constructive models $\langle \mathfrak{M}_0, \nu_0 \rangle, \dots, \langle \mathfrak{M}_n, \nu_n \rangle, \dots$ is computable if and only if there exist recursive functions f and g such that the models $\langle \mathfrak{M}_n, \nu_n \rangle$ and $\varkappa(f(n))$ are recursively isomorphic for any n and the number $g(n)$ of the recursive function $\varkappa_{g(n)}$ that defines this recursive isomorphism is computed from n by means of the function g, i.e., for $\varphi_n(\nu_n(m)) \rightleftharpoons \nu^\varkappa_{f(n)}(\varkappa_{g(n)}(m))$ we conclude that φ_n is an isomorphism of $\mathfrak{M}_n$ onto $\mathfrak{M}^\varkappa_{f(n)}$, where $\varkappa_n$ is the principal numbering of the set of all partially recursive functions.*

PROOF. The existence of functions f and g follows from the construction of recursively enumerable sets $W\langle\mathfrak{M}_n, \nu_n\rangle$ by $\langle\mathfrak{M}_n, \nu_n\rangle$ The computability of the sequence $\{\langle\mathfrak{M}_n, \nu_n\rangle \mid n \in \mathbb{N}\}$ implies the computability of the sequence $\{W\langle\mathfrak{M}_n, \nu_n\rangle \mid n \in \mathbb{N}\}$. Since W is principal, we obtain the reducing function f. The function g can be constructed from the diagram since it is the same for $\langle\mathfrak{M}_n, \nu_n\rangle$ and $\varkappa(n)$.

The converse assertion is true in view of the following simple propositions.

Proposition 3.1.20. *The sequence $\langle\mathfrak{M}_n^\varkappa, \nu_n^\varkappa\rangle$ of constructive models is computable.*

PROOF. The assertion follows from the construction and the definition of a computable sequence since for $\langle \mathfrak{M}_n^\varkappa, \nu_n^\varkappa \rangle$ from its number n we can find its diagram that is defined from $V_n = W\langle \mathfrak{M}_n^\varkappa, \nu_n^\varkappa \rangle$. □

Let $\alpha = \{\langle \mathfrak{M}_n, \nu_n \rangle\}$ and $\beta = \{\langle \mathfrak{N}_n, \mu_n \rangle\}$ be two sequences of enumerated models. The sequence α *is reduced* to the sequence β if there exists a recursive function f such that the constructive models $\langle \mathfrak{M}_n, \nu_n \rangle$ and $\langle \mathfrak{N}_{f(n)}, \mu_{f(n)} \rangle$ are recursively isomorphic (in the sense of constructive models). The fact that α is reduced to β is denoted by $\alpha \leqslant \beta$. We say that the sequence α *is effectively reduced* to the sequence β if there exist recursive functions f and g such that for any n the function $\varkappa_{g(n)}$ defines an isomorphism of the constructive model $\langle \mathfrak{M}_n, \nu_n \rangle$ onto the model $\langle \mathfrak{N}_{f(n)}, \mu_{f(n)} \rangle$, i.e., the mapping $\varphi_n(\nu_n(m)) \rightleftharpoons \mu_{f(n)}(\varkappa_{g(n)}(m))$ is well defined and is an isomorphism between $\mathfrak{M}_n$ and $\mathfrak{N}_{f(n)}$.

From definitions we immediately obtain the following claim.

Proposition 3.1.21. *If a sequence $\{\langle \mathfrak{M}_n, \nu_n \rangle \mid n \in \mathbb{N}\}$ of enumerated models is effectively reduced to a computable sequence $\{\langle \mathfrak{N}_n, \mu_n \rangle \mid n \in \mathbb{N}\}$ of constructive models, then $\{\langle \mathfrak{M}_n, \nu_n \rangle \mid n \in \mathbb{N}\}$ is a computable sequence of constructive models.*

Let $\mathcal{S}$ be the class of constructive models of a signature σ and let K be some construction defining some finite sequence of constructive models and homomorphisms on them by constructive models of $\mathcal{S}$. We say that a construction K is *uniformly effective* on the class $\mathcal{S}$ if there exists a partially recursive function g_K such that the fact that the constructive model $\varkappa^\sigma(n)$ belongs to $\mathcal{S}$ implies that $g_K(n)$ is defined and $g_K(n)$ is equal to the number of the pair of two finite sequences, where the first sequence yields the numbers $\varkappa^\sigma$ of constructive models that are constructed by K and the second one gives the numbers of recursive functions that define the required recursive homomorphisms with respect to the numbering $\varkappa$.

Proposition 3.1.22. *Let K be a uniformly effective construction on $\mathcal{S}$ that, from a constructive model $\langle \mathfrak{N}, \nu \rangle$ of $\mathcal{S}$, constructs a constructive model $\langle \mathfrak{M}, \mu \rangle$ of $\mathcal{S}$ and a homomorphism (φ, f) from $\langle \mathfrak{N}, \nu \rangle$ into $\langle \mathfrak{M}, \mu \rangle$ ($K\langle \mathfrak{N}, \nu \rangle = \langle \langle \mathfrak{M}, \mu \rangle, (\varphi, f) \rangle$). Then, for any constructive model $\langle \mathfrak{N}, \nu \rangle$ of $\mathcal{S}$, there is a computable sequence $\langle \mathfrak{N}_0, \nu_0 \rangle, \ldots, \langle \mathfrak{N}_n, \nu_n \rangle, \ldots$ of constructive models of $\mathcal{S}$ and a computable sequence $g_0, g_1, \ldots, g_n, \ldots$ of recursive functions such that*

(1) $\langle \mathfrak{N}_0, \nu_0 \rangle = \langle \mathfrak{N}, \nu \rangle$,

3.1. Basic Notions

(2) *for any n the function g_n defines a homomorphism φ_n from $\mathfrak{N}_n$ into $\mathfrak{N}_{n+1}$, where $\varphi_n(\nu_n(m)) = \nu_{n+1}g_n(m)$ for any n,*

(3) *for any n the construction K defines the model $\langle \mathfrak{N}_{n+1}, \nu_{n+1} \rangle$ and the homomorphism (φ_n, g_n) from $\langle \mathfrak{N}_n, \nu_n \rangle$.*

PROOF. The assertion follows from the recursive definition of the required indices by means of the function g_K representing the construction K on the class $\mathcal{S}$. $\square$

By the corollary of the theorem on reducing restricted theories to $\exists$-formulas and $\forall$-formulas of the signature of Boolean algebras extended by predicates of σ_n, we obtain the following claim.

Proposition 3.1.23. *An enumerated Boolean algebra $\langle \mathfrak{A}, \nu \rangle$ is n-constructive if and only if ν is a constructivization of the enrichment $\mathfrak{A}^{\sigma_n}$ of $\mathfrak{A}$ to the signature σ_n of the theory BA_n^*.*

PROOF. The necessity follows from the possibility of describing predicates of the signature σ_n in terms of Σ_n-formulas or Π_n-formulas of the signature of Boolean algebras. The sufficiency is obtained from the fact that any Σ_n-formula is equivalent to a $\forall$-formula and an $\exists$-formula of the signature σ_n of the theory BA_n^*.

Exercises

1. Prove that a countable atomless Boolean algebra is constructivizable.

2. Prove that any constructivization of an atomless Boolean algebra is strong.

3. Construct an example of a Boolean algebra having nonautoequivalent constructivizations.

4. Prove that there exist nonconstructivizable Boolean algebras.

5. Prove that a B-constructive atomic Boolean algebra such that the set of its atoms is B-recursive is B-strongly constructive.

6. Prove that all constructivizations of a Boolean algebra are autoequivalent.

7. Prove that if ν is a constructivization of $\mathfrak{B}$, I is an ideal of $\mathfrak{B}$, and the set $\nu^{-1}(I)$ is recursively enumerable, then the quotient algebra $\langle \mathfrak{B}/I, \nu/I \rangle$ is positive, where $\nu/I(n) \rightleftharpoons \nu(n)/I$.

8. Prove that if ν is a constructivization of $\mathfrak{B}$ and I is a ν-decidable ideal of $\mathfrak{B}$, then $\langle \mathfrak{B}/I, \nu/I \rangle$ is a constructive quotient algebra.

9. Prove that Rec^B is a functor from Con^B into Rec^B and Con^B is a functor from Rec^B into Con^B.

10. Prove that if $\mathfrak{M}$ is a B-recursive model, then $\mathrm{RecCon}\,(\mathfrak{M})$ and $\mathfrak{M}$ are B-isomorphic.

11. Prove that if $\langle \mathfrak{M}, \nu \rangle$ is a B-constructive model, then $\mathrm{ConRec}\,\langle \mathfrak{M}, \nu \rangle$ and $\langle \mathfrak{M}, \nu \rangle$ are B-isomorphic.

12. Prove that if $\langle \mathfrak{M}, \nu \rangle$ is a strongly constructive model, then $\mathrm{Rec}\,\langle \mathfrak{M}, \nu \rangle$ is a decidable model.

13. Prove that if $\mathfrak{M}$ is a decidable model, then $\mathrm{Con}\,(\mathfrak{M})$ is a strongly constructive model.

14. Prove that the functors Con^B and Rec^B define the equivalence of the categories Con^B and Rec^B.

15. Formulate and prove an "effective" version of the Vaught theorem (cf. Corollary 1.5.3).

16. Prove that two recursive atomless Boolean algebras are recursively isomorphic.

17. Prove that any two constructivizations of the Boolean algebra $\mathfrak{B}_\omega$ with recursive sets of numbers of atoms are isomorphic.

3.2. Constructibility in Linear Orders and Boolean Algebras

The problem of the existence of constructive models with given properties is one of the principal problems of the theory of constructive models. In constructing a Boolean algebra with some prescribed properties, the construction by a linear order on a Boolean algebra is of importance, which has an explicit effective character. In this section, we consider an effectivization of such a construction and apply it to study the existence of Boolean algebras with required algorithmic properties.

3.2. Linear Orders

Given a numbering ν of a linear order L with the least element, we construct a numbering γ_L of a Boolean algebra $\mathfrak{B}_L$ which inherits a series of algorithmic properties of the enumerated order $\langle L; \nu \rangle$.

Proposition 3.2.1. *If $L = \langle L; \nu \rangle$ is an enumerated linear order, then the Boolean algebra $\mathfrak{B}_L$ has the numbering γ_L and the recursive functions $f_\wedge$, $f_\vee$, f_C, i, and k such that*

(1) $\gamma_L(n) \cup \gamma_L(m) = \gamma_L(f_\vee(n, m))$,

(2) $\gamma_L(n) \cap \gamma_L(m) = \gamma_L(f_\wedge(n, m))$,

(3) $\gamma_L(f_C(n)) = C(\gamma_L(n))$,

(4) *if $\nu(n) \leqslant \nu(m)$, then $\gamma_L(i(n, m)) = [\nu(n), \nu(m)[$,*

(5) $\gamma_L(k(n)) = [\nu n, \infty[$;

moreover, $\langle \mathfrak{B}_L, \gamma_L \rangle$ is B-constructive if $\langle L; \nu \rangle$ is a B-constructive linear order.

PROOF. Let $\gamma(n)$ denote the number of a sequence of natural numbers $(m_0, \ldots, m_k)$ that can be found effectively from n. Every number m_i is considered as the number of the pair $(l(m_i), r(m_i))$. We set

$$\gamma_L(n) = \bigcup_{i=1}^{\tilde{\kappa}} [\nu l(m_i), \nu r(m_i)[\cup [a, b[$$

where $a = \nu l(m_0)$ and $b = +\infty$ if $r(m_0) = 0$, $a = b = \nu 0$ if $r(m_0) > 0$.

It is easy to obtain the γ_L-numbers of union, intersection, and complement, which allows us to construct the functions $f_\vee$, $f_\wedge$, and f_C. The method of constructing the functions i and k is obvious. Note that the question on the equality of elements with the γ_L-numbers n and m is reduced to the question whether the element with the γ_L-number $f_\vee(f_\wedge(n, f_C(m)), f_\wedge(f_C(n), m))$ is zero, and, in turn, this question is reduced to the verification of the condition $\nu x < \nu y$ in the linear order L. Consequently, the Boolean algebra $\langle \mathfrak{B}_L, \gamma_L \rangle$ is B-constructive if the linear order is B-constructive. □

If P is a preorder on L, i.e., P is reflexive and transitive, then we can define an order P^* on the quotient set $L^* \rightleftharpoons L/\sim_P$ by the relation $\sim_P$ as follows: $a \sim_P b \Leftrightarrow (a, b) \in P, (b, a) \in P, a/\sim_P \leqslant b/\sim_P \Leftrightarrow (a, b) \in P$. We say that P is a *linear preorder* on L if the order P^* on L^* induced by P is linear. To obtain constructive linear orders we construct recursive relations

P on recursively enumerable subsets L of the set of natural numbers. They define linear preorders on L. Starting from any recursive enumeration of L, we can construct a canonical constructivization of the linear order on L; moreover, the constructivizations obtained by means of different enumerations are recursively equivalent. More exactly, taking a recursive enumeration $f\colon \mathbb{N} \overset{\text{onto}}{\to} L$, we set $\nu_f(n) \rightleftharpoons f(n)/{\sim_P}$, where the relations $\nu n = \nu m$ and $\nu n \leqslant \nu m$ are recursive with respect to P and simply recursive if P is recursive. If g is some other enumeration of L, i.e., $g\colon \mathbb{N} \overset{\text{onto}}{\to} L$ is a recursive function, then the recursive function $\xi(n) = \mu m(g(m) \sim_P f(n))$ satisfies the equality $\nu_f(\xi(n)) = \nu_g(n)$, i.e., the numberings ν_f and ν_g are recursively equivalent. Consequently, up to a recursive equivalence, the method of constructing a constructivization ν_f of the linear order $\langle L^*; \leqslant_{P^*} \rangle$ by a recursive preorder P on an enumerable set L is independent of the choice of the function enumerating L.

Proposition 3.2.2. *For a B-constructive Boolean algebra $\langle \mathfrak{B}, \nu \rangle$ there exists a B-constructive linear order $\boldsymbol{L} = \langle L; \mu \rangle$ such that $\langle \mathfrak{B}, \nu \rangle$ and $\langle \mathfrak{B}_{\boldsymbol{L}}, \gamma_{\boldsymbol{L}} \rangle$ are B-isomorphic.*

PROOF. Acting step-by-step, we construct a B-enumerable set L of ν-numbers of a linearly ordered set with the largest element and the least element such that L generates a constructive Boolean algebra $\langle \mathfrak{B}, \nu \rangle$. At the step t, we construct a finite set L_t. A preorder is obtained by restriction of the set $P = \{(n, m) \mid \nu n \leqslant \nu m\}$ to L_t^2.

Let $f_\wedge$ and $f_\vee$ be B-recursive functions such that $\nu f_\wedge(n, m) = \nu n \wedge \nu m$, $\nu f_\vee(n, m) = \nu n \vee \nu m$, and $a, b \in \mathbb{N}$ are such that $\nu a = \boldsymbol{0}$, $\nu b = \boldsymbol{1}$.

STEP 0. We set $L_0 = \{a, b\}$.

STEP $t+1$. We have defined the finite set L_t consisting of the elements $n_0, n_1, \ldots, n_k$; moreover, $\boldsymbol{0} = \nu n_0 < \nu n_1 < \nu n_2 < \ldots < \nu n_k = \boldsymbol{1}$. We consider an element νt and set

$$L_{t+1} \rightleftharpoons L_t \cup \{f_\vee(f_\wedge(n_{i+1}, t), n_i) \mid i < k,\ \nu t \wedge \nu n_{i+1} \neq \nu n_{i+1},\ \text{and}$$
$$\nu n_i \neq (\nu n_{i+1} \wedge \nu t) \vee \nu n_i\}$$

It is clear that $\bigcup_{t \geqslant 0} L_t$ is exactly the set of numbers of linearly ordered generating sets constructed in Theorem 1.6.1. Since L_{t+1} is constructed from L_t effectively with respect to B, $L = \nu\bigl(\bigcup_{t \geqslant 0} L_t\bigr)$ is B-enumerable; moreover, every element of $\nu\bigl(\bigcup_{t \geqslant 0} L_t\bigr)$ obtains a unique number. Taking the restriction of the relation $\leqslant$ to L, we obtain a B-recursive linear order

3.2. Linear Orders

on L. Since L is B-enumerable, $L' \rightleftharpoons L\setminus\{1\}$ is also a B-enumerable set. From a B-enumerating function f we find a B-constructivization μ of the linear order $\langle L'; \leqslant \rangle$.

Let γ_μ be the B-constructivization of $\mathfrak{B}_{L'}$ constructed in Proposition 3.2.1. Since L' is a linearly ordered generating set without 1 in $\mathfrak{B}_{L'}$ and $\mathfrak{B}$, we see that $\mathfrak{B}_{L'}$ and $\mathfrak{B}$ are isomorphic; moreover, there exists an isomorphism $\varphi \colon \mathfrak{B} \to \mathfrak{B}_{L'}$ such that $\varphi(b) = [-\infty, b[$ for any element $b \in L'$ and from the ν-number of an element $b \in L'$ we can find effectively the γ_μ-number of an element $[-\infty, b[$. Since any element is generated by unions of elements $a' \setminus b'$, where $a', b' \in \nu L' \cup \{1\}$, there exists a B-recursive function f such that $\varphi \nu n = \gamma_\mu(f(n))$ for any n. □

Proposition 3.2.3. *If $\mathfrak{B}_L$ is an atomic Boolean algebra and, in the B-constructive linearly ordered set $\boldsymbol{L} = \langle L; \mu \rangle$, the set of pairs of numbers of neighboring elements is B-recursive (i.e., the set*

$$S_L \rightleftharpoons \{(a,b) \mid a < b, \text{ there is no } c \in L \text{ such that } a < c < b\}$$

has a B-recursive set of numbers), then $\langle \mathfrak{B}_L, \nu_L \rangle$ is a strongly B-constructive Boolean algebra.

PROOF. By Proposition 3.2.1, $(\mathfrak{B}_L, \gamma_L)$ is a B-constructive Boolean algebra. It is easy to see that an element $\boldsymbol{a} \in \mathfrak{B}_L$ is an atom if and only if $\boldsymbol{a} = [x, y[$, where $x < y$ and x, y are "neighbors," i.e., there is no z such that $z < y$ and $x < z$, or $\boldsymbol{a} = [b, \infty[$ if b is the greatest element of L. By the facts that the least element and the greatest element of $L \setminus \{0\}$ are found in a unique way, the set of their numbers is B-recursive, and the property of being "neighbors" is B-decidable by condition, we seek such a representation of elements of $\mathfrak{B}_L$. If it exists, then the element under consideration is an atom. Consequently, the set of the numbers of atoms is B-enumerable. But the set of numbers of elements that are not atoms is also B-enumerable since n is the γ-number of not atom provided that there is m such that $0 < \gamma m < \gamma n$. The last relation is effectively recognized with respect to B in view of the B-constructibility condition. Therefore, for a B-constructive Boolean algebra $\langle \mathfrak{B}_L, \gamma_L \rangle$ the set of its atoms is B-decidable. By Proposition 3.1.23, $\langle \mathfrak{B}_L, \gamma_L \rangle$ is a strongly B-constructive Boolean algebra. □

Theorem 3.2.1. *For any countable Boolean algebra $\mathfrak{B}$ there exists a unique atomic Boolean algebra $\mathfrak{A}(\mathfrak{B})$ such that*

$$\mathfrak{A}(\mathfrak{B}) / F(\mathfrak{A}(\mathfrak{B})) \cong \mathfrak{B}$$

and $\mathfrak{A}(\mathfrak{B})$ *is strongly B-constructivizable if $\mathfrak{B}$ is a B-constructivizable Boolean algebra.*

PROOF. By Proposition 1.6.1, there exists a linearly ordered basis L of $\mathfrak{B}_L$; moreover, $\mathfrak{B} \cong \mathfrak{B}_L$. If $\mathfrak{B}$ is B-constructivizable, then, by Proposition 3.2.2, the basis L can be chosen to be B-constructivizable. We now consider the linear order $\omega \times L$, where ω is the linear order of type of the natural order on natural numbers. Let $\mathfrak{A}(\mathfrak{B})$ be a Boolean algebra $\mathfrak{B}_{\omega \times L}$. Let us show that $\mathfrak{A}(\mathfrak{B})$ is the required algebra. It is clear that $\omega \times L$ is B-constructivizable provided that L is B-constructivizable. Hence $\mathfrak{A}(\mathfrak{B})$ is B-constructivizable if $\mathfrak{B}$ is constructivizable. We introduce a mapping φ from $\mathfrak{B}_L$ into $\mathfrak{A}(\mathfrak{B})/F$, where F denotes the Frechét ideal of the Boolean algebra $\mathfrak{A}(\mathfrak{B})$, by setting

$$\varphi\left(\bigcup_{i=1}^{n}[\alpha_i, \beta_i[\right) \rightleftharpoons \bigcup_{i=1}^{n}[\alpha'_i, \beta'_i[/F$$

here

$$\delta' \rightleftharpoons \begin{cases} \langle 0, \delta \rangle & \text{if } \delta \in L \\ -\infty & \text{if } \delta = -\infty \\ \infty & \text{if } \delta = \infty \end{cases}$$

By definition, φ preserves all the basic operations of Boolean algebras. Thus, φ is a homomorphism from $\mathfrak{B}_L$ into $\mathfrak{A}(\mathfrak{B})/F$. We note that elements of $\mathfrak{B}_{\omega \times L}$ of the form $[(n, \delta), (m, \delta)[$ and only such elements are unions of a finite number of atoms of $\mathfrak{B}_{\omega \times L}$, and elements of the form $[(n, \delta), (n+1, \delta)[$ are atoms. In this case, $\mathfrak{B}_{\omega \times L}$ is an atomic Boolean algebra and any of its elements of the form $[(n, \delta_1), (m, \delta_2)[$ differs from $[(0, \delta_1), (0, \delta_2)[$ by a finite number of atoms. Then φ maps $\mathfrak{B}_L$ onto $\mathfrak{B}_{\omega \times L}/F$. It remains to show that φ is an isomorphism. This fact immediately follows from the equivalence $\varphi(a) = \varnothing/F \Leftrightarrow a = 0$. The uniqueness of $\mathfrak{A}(\mathfrak{B})$ follows from Proposition 1.5.2. The strong B-constructibility follows from the fact that the set of atoms is recursive. □

Theorem 3.2.2. *If $\mathfrak{B}$ is an ω-atomic B-constructivizable Boolean algebra, then $\mathfrak{B}$ is strongly B-constructivizable.*

PROOF. We consider two cases. If there exists $n < \omega$ such that $\mathfrak{B}/F_{n+1}(\mathfrak{B})$ is a singleton, then $\mathfrak{B}$ is a superatomic Boolean algebra and $\mathfrak{B}$ is isomorphic to $\mathfrak{B}_{\omega^n \times m}$, where m is the number of atoms of the Boolean algebra $\mathfrak{B}/F_n(\mathfrak{B})$, and the latter contains more than one element. In this case, $\omega^n \times m$ has a numbering with respect to which the set of pairs of neigh-

3.2. Linear Orders

boring elements is decidable and, consequently, $\mathfrak{B}$ is even strongly constructivizable. If for any natural number n the Boolean algebra $\mathfrak{B}/F_n(\mathfrak{B})$ is not a singleton, then we consider the Boolean algebra $\mathfrak{A}(\mathfrak{B})$ from Theorem 3.2.1. Then the mapping

$$\psi\left(a/F_\omega(\mathfrak{B})\right) \rightleftharpoons (\varphi(a))/F_\omega\left(\mathfrak{A}(\mathfrak{B})/F(\mathfrak{A}(\mathfrak{B}))\right),$$

where φ is an isomorphism of $\mathfrak{B}$ onto $\mathfrak{A}(\mathfrak{B})/F(\mathfrak{A}(\mathfrak{B}))$, is also an isomorphism of $\mathfrak{B}/F_\omega(\mathfrak{B})$ onto $\left(\mathfrak{A}(\mathfrak{B})/F(\mathfrak{A}(\mathfrak{B}))\right)/F_\omega\left(\mathfrak{A}(\mathfrak{B})/F(\mathfrak{A}(\mathfrak{B}))\right)$. However, the mapping

$$\Delta \colon \mathfrak{A}(\mathfrak{B}) \to \left(\mathfrak{A}(\mathfrak{B})/F(\mathfrak{A}(\mathfrak{B}))\right)/F_\omega\left(\mathfrak{A}(\mathfrak{B})/F(\mathfrak{A}(\mathfrak{B}))\right)$$

such that $\Delta(a) \rightleftharpoons \left(a/F(\mathfrak{A}(\mathfrak{B}))\right)/F_\omega\left(\mathfrak{A}(\mathfrak{B})/F(\mathfrak{A}(\mathfrak{B}))\right)$ is a homomorphism of $\mathfrak{A}(\mathfrak{B})$ onto $\left(\mathfrak{A}(\mathfrak{B})/F(\mathfrak{A}(\mathfrak{B}))\right)/F_\omega\left(\mathfrak{A}(\mathfrak{B})/F(\mathfrak{A}(\mathfrak{B}))\right)$. But,

$$F_n(\mathfrak{A}(\mathfrak{B})/F(\mathfrak{A}(\mathfrak{B}))) = F_{n+1}(\mathfrak{A}(\mathfrak{B})).$$

Consequently, the kernel of the homomorphism Δ is exactly $F_\omega(\mathfrak{A}(\mathfrak{B}))$. Then $\mathfrak{A}(\mathfrak{B})/F_\omega(\mathfrak{A}(\mathfrak{B}))$ and $\left(\mathfrak{A}(\mathfrak{B})/F(\mathfrak{A}(\mathfrak{B}))\right)/F_\omega\left(\mathfrak{A}(\mathfrak{B})/F(\mathfrak{A}(\mathfrak{B}))\right)$ are isomorphic. Hence Boolean algebras $\mathfrak{A}(\mathfrak{B})/F_\omega(\mathfrak{A}(\mathfrak{B}))$ and $\mathfrak{B}/F_\omega(\mathfrak{B})$ are isomorphic too. Since $\mathfrak{B}/F_n(\mathfrak{B})$ is not a singleton for all $n < \omega$, we have $\mathfrak{B}/F_\omega(\mathfrak{B})$ is not a singleton either. Since $\mathfrak{A}(\mathfrak{B})/F(\mathfrak{A}(\mathfrak{B}))$ is isomorphic to $\mathfrak{B}$, the Boolean algebras $\mathfrak{A}(\mathfrak{B})$ and $\mathfrak{B}$ are ω-atomic. Then (cf. Exercise 3 of Sec. 1.4) $\mathfrak{B}$ and $\mathfrak{A}(\mathfrak{B})$ are isomorphic and $\mathfrak{A}(\mathfrak{B})$ is strongly B-constructivizable (by Theorem 3.2.1). Consequently, $\mathfrak{B}$ is strongly B-constructivizable. □

Corollary 3.2.1. *If a superatomic Boolean algebra $\mathfrak{B}$ is B-constructivizable, then it is strongly B-constructivizable.*

Exercises

1. Prove that if ν and μ are two autoequivalent constructivizations of L, then constructivizations γ_ν and γ_μ of the Boolean algebra $\mathfrak{B}_L$ are autoequivalent.

2. Prove that if φ is a recursive isomorphic embedding of an enumerated linear order $\langle L_0; \nu_0 \rangle$ into a constructive linear order $\langle L_1; \nu_1 \rangle$, then there exists a recursive isomorphic embedding of $\langle \mathfrak{B}_{L_0}, \gamma_{\nu_0} \rangle$ into $\langle \mathfrak{B}_{L_1}, \gamma_{\nu_1} \rangle$.

180   3. Constructive Boolean Algebras

3. Prove that the Boolean algebras $\mathfrak{B}_\omega$ and $\mathfrak{B}_{\omega \times \eta}$ have constructivizations with undecidable sets of atoms.

4. Prove that the Frechét ideal $F(\mathfrak{B})$ of a strongly constructive Boolean algebra $\langle \mathfrak{B}, \nu \rangle$ is enumerable.

5. Prove that the set $\nu^{-1}(F(\mathfrak{B}))$ is a Σ^0_2-set and $\nu^{-1}(\text{Atomless}\,(\mathfrak{B}))$ is a Π^0_2-set in a constructive Boolean algebra $\langle \mathfrak{B}, \nu \rangle$.

6. Prove that a superatomic Boolean algebra is constructivizable if and only if its ordinal type is a recursive ordinal.

3.3. Trees Generating Constructive Boolean Algebras

A construction of the Boolean algebra $\mathfrak{B}_D$ generated by a tree D is effective. In this section, we apply the tree method to study if constructivizations and strong constructivizations exist and investigate the dependence of the existence problem on a chosen signature.

Let $D \subseteq \mathbb{N}$ be a tree and let $\mathfrak{K}(D)$ be the set of all finite subsets of D. If D is a B-enumerable set, then, in the strongly computable numbering $\{\gamma(n) \mid n \in \mathbb{N}\}$ of finite subsets of $\mathbb{N}$, the set of numbers of elements of $\mathfrak{K}(D)$ is B-enumerable. We consider a function f enumerating all the numbers of elements of $\mathfrak{K}(D)$ and define the numbering

$$\nu_f(n) \rightleftharpoons \bigcup_{i \in \gamma(f(n))} A_i$$

of all elements of $\mathfrak{B}_D$. We note that for any finite subset $H \subseteq D$ the construction of $\bigcup_{i \in H} A_i$ of the canonical representation H^* by H is effective in view of Lemma 1.7.1, and the canonical representations of union, intersection, and complement can be effectively found with the help of constructions in Lemma 1.7.4. By the B-enumerability of the numbers of $\mathfrak{K}(D)$ and the effectiveness of finding all elements of a set from its number, we conclude that $\langle \mathfrak{B}_D, \nu_f \rangle$ is a B-constructive Boolean algebra. We can introduce partially B-recursive functions u and v such that

$$(\forall n)(n \in D \Rightarrow \nu_f(u(n)) = A_n), \quad (\forall n)\big(\nu_f(n) = \bigcup_{i \in \gamma(v(n))} A_i\big)$$

3.3. Generating Trees

i.e., for $n \in D$ we can find effectively with respect to B the ν_f-number of the corresponding element A_n of the Boolean algebra $\mathfrak{B}_D$ and from the ν_f-number n we can find the γ-number of a finite subset $K \subseteq D$ which defines the element $\nu_f(n) \in \mathfrak{B}_D$. Thus, we have proved the following theorem.

Theorem 3.3.1. *If D is a B-enumerable tree, then there exists a B-constructivization ν of the Boolean algebra $\mathfrak{B}_D$ and two partially B-recursive functions u and v such that*

(1) $(\forall n)(n \in D \Rightarrow \nu(u(n)) = A_n)$,

(2) $(\forall n)\left(\nu n = \bigcup_{i \in \gamma(v(n))} A_i\right)$;

moreover, if ν and μ are two B-constructivizations subject to the above-mentioned properties, then there exists a B-recursive function g such that $\nu = \mu g$, i.e., ν and μ are recursively B-equivalent.

The last property of recursive B-equivalences follows from the construction of the function g: for n we take the set with the γ-number $v(n)$ and from $i \in \gamma(v(n))$ we find the μ-numbers of A_i by means of the function u. If we know the number of the union of all these elements, then we can obtain the value of the function g at n.

The constructivization presented in Theorem 3.3.1 is denoted by ν_D and is called the *constructivization constructed by the tree D*.

The following question is associated with the universality of our construction of a B-constructive Boolean algebra $\langle \mathfrak{B}_D, \nu_D \rangle$ from a recursively B-enumerable tree D. The complete answer to this question is contained in the following theorem.

Theorem 3.3.2. *For any B-constructive Boolean algebra there exists a recursively B-enumerable tree D such that $\langle \mathfrak{B}, \nu \rangle$ and $\langle \mathfrak{B}_D, \nu_D \rangle$ are B-isomorphic.*

PROOF. It is required only to check the B-effectiveness of the construction of the tree $\langle D; \varphi \rangle$ generating a countable Boolean algebra in Theorem 1.7.1. More exactly, as an enumeration of elements of $\mathfrak{B}$ we take the sequence $b_0 \leftrightharpoons \nu 0$, $b_1 \leftrightharpoons \nu 1$, $b_2 \leftrightharpoons \nu 2, \ldots$ and construct a recursively B-enumerable tree D and a partially B-recursive function f such that $\langle D; \varphi \rangle$, where $\varphi = \nu f$, is a tree generating $\mathfrak{B}$. With the help of the functions u and v for ν_D, we construct a B-recursive function g and an isomorphism

$$\psi(\bigcup_{i\in H} A_i) = \bigvee_{i\in H} \nu f(i)$$

of the Boolean algebra $\mathfrak{B}_D$ onto the Boolean algebra $\mathfrak{B}$ such that the equality $\psi(\nu_D(n)) = \nu g(n)$ holds. This means that $\langle \mathfrak{B}, \nu \rangle$ and $\langle \mathfrak{B}_D, \nu_D \rangle$ are B-isomorphic. □

We say that $\langle D; f \rangle$ is a B-tree *generating* a B-constructive Boolean algebra $\langle \mathfrak{B}, \nu \rangle$ if $f\colon D \to \mathbb{N}$ is a partially B-recursive function and the tree $\langle D; \nu f \rangle$ generates $\mathfrak{B}$. As was already proved, such a generating B-tree exists for any B-constructive Boolean algebra.

Any B-constructive Boolean algebra $\langle \mathfrak{B}, \nu \rangle$ can be isomorphically B-embedded into the atomless constructive Boolean algebra $\langle \mathfrak{B}_\mathbb{N}, \nu_\mathbb{N} \rangle$. Indeed, since $\langle \mathfrak{B}, \nu \rangle$ is B-isomorphic to $\langle \mathfrak{B}_D, \nu_D \rangle$, it suffices to construct a B-isomorphism of $\langle \mathfrak{B}_D, \nu_D \rangle$ into $\langle \mathfrak{B}_\mathbb{N}, \nu_\mathbb{N} \rangle$. We take the identity embedding id: $\mathfrak{B}_D \to \mathfrak{B}_\mathbb{N}$ and a B-recursive function g so that we find the value $g(n)$ as follows: using the function ν_D, we find the γ-number of some representation of $\nu_D(n)$ in the generating tree from n and the $\nu_\mathbb{N}$-number of A_i from $i \in \nu_D(n)$. For $g(n)$ we take the number of the union $\bigcup_{i\in\gamma(\nu_D(n))} \nu_\mathbb{N}(u_\mathbb{N}(i))$.
We achieve much success if we embed $\langle \mathfrak{B}, \nu \rangle$ into $\langle \mathfrak{B}_\mathbb{N}, \nu_\mathbb{N} \rangle$ so that the image of $\mathfrak{B}$ in $\mathfrak{B}_\mathbb{N}$ turns out to be B-recursive, but this requires a more delicate construction than that considered here.

Lemma 3.3.1. *For any B-recursive Boolean algebra $\mathfrak{A}$ there exists a B-computable sequence $\{\mathfrak{A}_i\}_{i\in\mathbb{N}}$ of finite subalgebras such that $\cup \mathfrak{A}_i = \mathfrak{A}$, $\mathfrak{A}_{i+1} = \mathrm{gr}(\mathfrak{A}_i \cup \{a_i\})$, where a_i is an atom of $\mathfrak{A}_{i+1}$.*

PROOF. We define $\mathfrak{A}_0 = \{\mathbf{0}, \mathbf{1}\}$. We further suppose that $\mathbf{0}$ and $\mathbf{1}$ are the null element and the unit in each of the recursive Boolean algebras under consideration. Let $\mathfrak{A}_i$ have been constructed. We consider the least element $k \in \mathfrak{A}$ of $\mathbb{N}$ such that k does not belong to the subalgebra $\mathfrak{A}_i$ and there exists the least atom $b \in \mathfrak{A}_i$ of $\mathbb{N}$ such that $b \wedge k \neq \mathbf{0}$ and $b \wedge C(k) \neq \mathbf{0}$. Let $\mathfrak{A}_{i+1} = \mathrm{gr}(\mathfrak{A}_i \cup \{b \wedge k\})$ be a subalgebra of $\mathfrak{A}$ generated by $\mathfrak{A}_i \cup \{b \wedge k\}$. It is easy to check that the sequence constructed in such a way satisfies the required conditions. □

Theorem 3.3.3. *For any B-constructive Boolean algebra $\langle \mathfrak{B}, \nu \rangle$ there exists an isomorphism ψ of $\mathfrak{B}$ into $\mathfrak{B}_\mathbb{N}$ and a B-recursive function f such that $\psi(\mathfrak{B})$ is B-recursive and $\psi\nu = \nu_\mathbb{N} f$.*

3.3. Generating Trees

PROOF. Without loss of generality, we can assume that ν is a single-vauled numbering. We use the simple fact that if a set is enumerated by a monotone B-recursive function, then it is B-recursive. To construct the required isomorphism is to construct step-by-step the corresponding B-recursive function f. At the step $t+1$, we determine the approximation f^{t+1} of the function f so that $\bigcup_{t \geq 0} f^t = f$ and the characteristic function χ of the image $\mathfrak{B}$ in $\mathfrak{B}_{\mathbb{N}}$ at t. The value f^{t+1} is defined exactly on the numbers of elements of the subalgebra $\mathfrak{B}^t$ of $\mathfrak{B}$ generated by $\{\nu 0, \ldots, \nu(t)\}$ and χ is defined at the step t for $n < t$. It is clear that atoms of this subalgebra are presented by nonzero elements of the form $\nu 0^{\varepsilon_0} \wedge \nu 1^{\varepsilon_1} \wedge \ldots \wedge \nu t^{\varepsilon_t}$, where $\varepsilon_i \in \{0,1\}$, $a^0 \rightleftharpoons C(a)$, and $a^1 \rightleftharpoons a$.

STEP 0. Let n_0, n_1, m_0, and m_1 be numbers such that $\nu n_0 = 0$, $\nu n_1 = 1$, $\nu_{\mathbb{N}}(m_0) = 0$, and $\nu_{\mathbb{N}}(m_1) = 1$. We define $f^0(n_0) = m_0$, $f^0(n_1) = m_1$,

$$\chi(0) \rightleftharpoons \begin{cases} 1 & \text{if } \nu_{\mathbb{N}}0 = 1 \text{ or } \nu_{\mathbb{N}}0 = 0 \\ 0 & \text{otherwise} \end{cases}$$

and f^0 and χ are undefined on the rest of the values.

STEP $t+1$. Let $a_0, \ldots, a_n$ be the numbers of all atoms of the subalgebra $\mathfrak{B}^t$ of $\mathfrak{B}$ generated by $\{\nu 0, \ldots, \nu(t)\}$ and $b_0 \rightleftharpoons f^t(a_0), \ldots, b_n = f^t(a_n)$. By the inductive hypothesis, the mapping $\psi^t \rightleftharpoons \nu_{\mathbb{N}} f^t \nu^{-1}$ yields an isomorphism of $\mathfrak{B}^t$ into $\mathfrak{B}_{\mathbb{N}}$. For m such that $\chi(m) = 1$ the element νm belongs to $\psi^t(\mathfrak{B}^t)$ and for all $m < t$ such that $\nu_{\mathbb{N}}(m)$ is in $\psi^t(\mathfrak{B}^t)$ the value of χ is 1. For every $i \leq n$ such that $\nu a_i \wedge \nu(t) \neq 0$ and $\nu a_i \wedge C(\nu t) \neq 0$ we find the numbers z_i of nonzero elements such that $\nu_{\mathbb{N}}(z_i)$ do not belong to the subalgebra of $\mathfrak{B}_{\mathbb{N}}$ generated by the elements

$$\{\nu_{\mathbb{N}} b_0, \ldots, \nu_{\mathbb{N}} b_n\} \cup \{\nu_{\mathbb{N}} k \mid \chi(k) = 0 \,\&\, k \leq t\}, \quad \nu_{\mathbb{N}} z_i < \nu_{\mathbb{N}} b_i$$

For $i \leq n$ such that $\nu a_i \leq C(\nu(t))$, as z_i we take the $\nu_{\mathbb{N}}$-number of zero in $\mathfrak{B}_{\mathbb{N}}$ and set $z_i = b_i$ if $\nu a_i \leq \nu(t)$. Then nonzero elements of the form $\nu a_i \wedge \nu t$ and $\nu a_i \backslash \nu(t)$ are atoms of the Boolean algebra $\mathfrak{B}^{t+1}$ generated by elements $\{\nu 0, \ldots, \nu t\}$. Defining ψ^{t+1} on atoms,

$$\psi^{t+1}(\nu a_i \wedge \nu t) = \nu_{\mathbb{N}} z_i$$
$$\psi^{t+1}(\nu a_i \wedge C(\nu t)) = \nu_{\mathbb{N}} b_i \backslash \nu_{\mathbb{N}} z_i$$

we can define ψ^{t+1} as an isomorphism of the subalgebra $\mathfrak{B}^{t+1}$ into $\mathfrak{B}_{\mathbb{N}}$ which continues ψ^t. For $n \in \nu^{-1}(\mathfrak{B}^{t+1})$, as $f^{t+1}(n)$ we take the least

number of
$$\bigvee_{i \in K} \nu_{\mathbb{N}} z_i \vee \bigvee_{i \in L} \nu_{\mathbb{N}} b_i \setminus \nu_{\mathbb{N}} z_i$$
where $K \rightleftharpoons \{i \mid \nu a_i \wedge \nu t \leqslant \nu n\}$ and $L \rightleftharpoons \{i \mid \nu a_i \wedge C(\nu t) \leqslant \nu n\}$. It is easy to see that $\psi^{t+1} = \nu_{\mathbb{N}} f^{t+1} \nu^{-1}$. We set $\chi(t) = 1$ if $\nu_{\mathbb{N}}(t) \in \psi^{t+1}(\mathfrak{B}^{t+1})$ and $\chi(t) = 0$ otherwise.

We now have $f = \bigcup_{t \leqslant 0} f^t$ and can easily verify that $\psi \rightleftharpoons \nu_{\mathbb{N}} f \nu^{-1}$ is an isomorphic embedding of $\mathfrak{B}$ into $\mathfrak{B}_{\mathbb{N}}$ and that χ is the characteristic function of $\nu_{\mathbb{N}}^{-1}(\psi(\mathfrak{B}))$. □

Using the tree technique, we study connections between the constructivizabilities of Boolean algebras with respect to different signatures. We note that the strong constructibility of Boolean algebras is independent of the choice of the signature since the basic operations and predicates of these signatures can be formally expressed through one another. In the signatures $\langle \wedge^2, \vee^2, C^1, \mathbf{0}, \mathbf{1} \rangle$ and $\langle +, \cdot, \mathbf{0}, \mathbf{1} \rangle$ the basic operations are expressed throught one another by means of terms. Therefore, the constructibility property is also independent of the choice of the signature. But the situation changes if we consider a Boolean algebra of the signature of Boolean lattices. Since $x \leqslant y$ if and only if $x \wedge y = x$, the constructibility of a system as a Boolean lattice follows from its constructibility as a Boolean algebra. We show below that the converse assertion is false. However, if we do not fix a numbering, then, as was shown by Dzgoev [33], the existence problems are equivalent for different signatures.

Theorem 3.3.4. *For any Boolean algebra $\mathfrak{B}$ the following assertions are equivalent*:

(a) $\mathfrak{B}$ *is constructivizable,*

(b) *the Boolean ring $R(\mathfrak{B})$ is constructivizable,*

(c) *the Boolean lattice $L(\mathfrak{B})$ is constructivizable.*

PROOF. The implications (a)⇒(b) and (b)⇒(c) follow from the aforesaid. To prove the implication (c)⇒(a) we find a constructivization ν of the Boolean lattice $L(\mathfrak{B})$ with an infinite number of atoms of a recursively enumerable tree D such that $\mathfrak{B}_D \cong \mathfrak{B}$. This is sufficient in order to complete the proof in view of Theorem 3.3.1 and the constructivizability of a Boolean algebra with a finite number of atoms. We will construct step-by-step a strongly computable sequence of finite trees $\{D^t\}_{t \geqslant 0}$, a function

3.3. Generating Trees

$\lambda tn(t)$, subtrees D_k^t with subsets of their end vertices S_k^t, $k \leqslant n(t)$, and a function $\varphi^t \colon D_{n(t)}^t \to \mathbb{N}$ such that $D = \cup D^t$. Our purpose is to define, starting from a constructive Boolean lattice $L(\mathfrak{B})$ of the signature $\langle \leqslant \rangle$, a tree $D^* = \lim\limits_{t \to \infty} D_{n(t)}^t$ and a function $\varphi^* = \nu \lim\limits_{t \to \infty} \varphi^t$ so that the tree $\langle D^*; \varphi^* \rangle$ generates the Boolean algebra $\mathfrak{B}$, $D \supseteq D^*$, and $\mathfrak{B}_{D^*}$ is a subalgebra of $\mathfrak{B}_D$, where

$$\mathfrak{B}_D = \mathrm{gr}\{\mathfrak{B}_{D^*} \cup \{e_i^a \mid \bigvee_{i=1}^{n_a} e_i^a = a, e_i^a \wedge e_j^a = 0, i \neq j, a \text{ is an atom of } \mathfrak{B}_{D^*}\}$$

By Proposition 1.7.2, the Boolean algebra $\mathfrak{B}_D$ is isomorphic to the Boolean algebra $\mathfrak{B}_{D^*}$ and the latter is isomorphic to $\mathfrak{B}$ (cf. Theorem 1.7.2). By construction,

$$D_0^t \subseteq D_1^t \subseteq \ldots \subseteq D_{n(t)}^t \subseteq D^t$$
$$\varphi^{t+1} \upharpoonright D_k^t = \varphi^t \upharpoonright D_k^t, \quad k < n(t+1)$$

We intend to prove that for a sufficiently large t and any $n \in D_k^t$, $k < n(t)$ the following conditions hold:

$$\nu\varphi^t(n) \wedge \nu\varphi^t(S(n)) = 0$$
$$\nu\varphi^t(n) \vee \nu\varphi^t(S(n)) = \nu\varphi^t(H(n))$$
$$\nu k = \bigvee_{l \in S_k^t} \nu\varphi^t(l)$$

To this end, we try to show that for any t and $k < n(t)$ the following conditions hold.

(T_k^t) If $a \in S_k^t$, then there is no $m \leqslant t+1$ such that $\nu\varphi^t(a) < \nu m$, $\nu m \leqslant \nu\varphi^t(H(a))$, $\nu m \leqslant \nu k$. If a is an end vertex of D_{k+1}^t and $a \notin S_k^t$, then there is no $m \leqslant t+1$ such that $\nu m \leqslant \nu k$, $\nu m \leqslant \nu a$, and $\nu m \neq \mathbf{0}$.

(D_k^t) If a is an end vertex of the tree D_{k+1}^t, then there is no $m \leqslant t+1$ such that $\nu m \leqslant \nu\varphi^t(a)$ & $\nu m \leqslant \nu\varphi^t(S(a))$ & $\nu m \neq \mathbf{0}$ or $\nu\varphi^t(a) \leqslant \nu m$ & $\nu\varphi^t(S(a)) \leqslant \nu m$ & $\nu m < \nu\varphi^t(H(a))$.

(T_k^t) approximates the property $\nu\varphi^t(a) = \nu k \wedge \nu\varphi^t(H(a))$ for $a \in S_k^t$ and the property $\nu\varphi^t(a) \wedge \nu k = 0$ for an end vertex of the tree D_{k+1}^t which in not in S_k^t, i.e., it approximates the property of the choice of the greatest lower bound $\nu k \wedge \nu\varphi^t(H(a))$.

(D_k^t) approximates the property $\nu\varphi^t(a) \vee \nu\varphi^t(S(a)) = \nu\varphi^t(H(a))$, $\nu\varphi^t(a) \wedge \nu\varphi^t(S(a)) = \mathbf{0}$, i.e., it approximates the choice of a complement to an element $\nu\varphi^t(a)$ located under the element $\nu\varphi^t(H(a))$ for an end vertex of the tree D^t_{k+1}.

We choose $a, b \in \mathbb{N}$ so that $\nu a = \mathbf{0}$ and $\nu b = \mathbf{1}$.

STEP 0. We define $D^0_0 \rightleftharpoons \{0\}$, $\varphi^0(0) \rightleftharpoons b$, $n(0) \rightleftharpoons 0$, and $D^0 \rightleftharpoons D^0_0$.

STEP $t+1$. We find the greatest $n \leqslant n(t)$ such that the conditions (T^t_k) and (D^t_k) hold for $k < n$. We define the set

$$S^{t+1}_n \rightleftharpoons \{s \mid \nu\varphi^t(s) \leqslant \nu n,\ s \text{ is an end vertex of } D^t_n$$

or there exists a number $m \leqslant t+1$ such that

$$\nu m \neq \mathbf{0},\ \nu\varphi^t(H(s)) \not\leqslant \nu n,\ \nu m \leqslant \nu\varphi^t(H(s)),\ \nu m \leqslant \nu n,$$

$H(s)$ is an end vertex of D^t_n and $s = LH(s)\}$

$$D^{t+1}_{n+1} \rightleftharpoons D^t_n \cup \{S(a) \mid a \in S^{t+1}_n\} \cup S^{t+1}_n$$

and, setting $n(t+1) = n+1$,

$$D^{t+1}_m \rightleftharpoons D^t_m \text{ for } m \leqslant n, \quad D^{t+1} \rightleftharpoons D^t \cup D^{t+1}_{n+1}$$

It remains to define the function $\varphi^{t+1}\colon D^{t+1}_{n+1} \to \mathbb{N}$. For $m \in D^t_n$ we set $\varphi^{t+1}(m) \rightleftharpoons \varphi^t(m)$ and for $m \in S^{t+1}_n \setminus D^t_n$ let $\varphi^{t+1}(m)$ be equal to the least $k \leqslant t+1$ such that $\nu k \leqslant \nu\varphi^t(H(m))$ and $\nu k \leqslant \nu n$ and there is no $l \leqslant t+1$ such that $\nu k < \nu l$, $\nu l \leqslant \nu\varphi^t(H(m))$, and $\nu l \leqslant \nu n$. For $m \in \{S(a) \mid a \in S^{t+1}_n\} \setminus D^t_n$ it is required to find the least $k \in \mathbb{N}$ such that $\nu k \leqslant \nu\varphi^t(H(m))$, and there is no $l \leqslant t+1$ such that $\nu l \neq \mathbf{0}\ \&\ \nu l \leqslant \nu k\ \&\ \nu l \leqslant \nu\varphi^t S(m)$ or $\nu k \leqslant \nu l\ \&\ \nu\varphi^{t+1}(S(m)) \leqslant \nu l\ \&\ \nu l < \nu\varphi^t(H(m))$, and put $\varphi^{t+1}(m)$ equal to k. After that we pass to the next step.

We show that the construction just presented satisfies all the desired conditions. To this end, we prove several of its auxiliary properties.

⟨1⟩ *For any $x, y \in D^t_n$ and $n < n(t)$, the relation $x \prec y$ implies $\nu\varphi^t(x) < \nu\varphi^t(y)$.*

⟨2⟩ *If $k < n(t)$ for all $t \geqslant t_0$, then*

    (a) $D^t_k = D^{t_0}_k$ *and* $D^t_0 = D^{t_0}_0$ *for any* $t \geqslant t_0$,

    (b) $\varphi^t \upharpoonright D^{t_0}_k = \varphi^{t_0} \upharpoonright D^{t_0}_k$ *for any* $t \geqslant t_0$,

    (c) $D^t_{k+1} = D^t_k \cup \{L(s), R(s) \mid s \in A \text{ and } s \in D^t_k\}$ *for some set* A,

    (d) $\nu\varphi^t(L(s)) \vee \nu\varphi^t(R(s)) = \nu\varphi^t(s)$ *and* $\nu\varphi^t(L(s)) \wedge \nu\varphi^t(R(s)) = \mathbf{0}$ *for any s such that* $L(s) \in D^t_k$,

3.3. Generating Trees

(e) *if $L(s)$ is an end vertex of D_k^t and $L(s) \notin D_{k-1}^t$, then*

$$\nu\varphi^{t_0}(L(s)) = \nu\varphi^{t_0}(s) \wedge \nu(k-1)$$
$$\nu\varphi^{t_0}(R(s)) = \nu\varphi^{t_0}(s) \wedge C(\nu(k-1))$$
$$D_k^t \subseteq D_{k-1}^t \cup \{L(s), R(s) \mid \nu\varphi^{t_0}(s) \not\leqslant \nu(k-1)$$
$$\nu\varphi^{t_0}(s) \wedge \nu(k-1) \neq \mathbf{0}\}$$

The assertions can be easily established by induction on t from the construction.

⟨3⟩ *For any m there exists T such that for all $t \geqslant T$ the value $n(t)$ is at most m.*

We assume the contrary and consider the least m for which the assertion fails. It is clear that $m > 0$. Consequently, there exists T such that for all $t \geqslant T$ the value $n(t)$ is at most $m-1$ and there exist infinitely many t for which $n(t)$ coincides with $m-1$. In this case, the conditions (T_k^t) and (D_k^t) hold for all $k < m-1$ for $t \geqslant T$. However, the inequality $k < m-1 \,\&\, m-1 \leqslant n(t)$ for $t \geqslant T$ implies $D_k^t = D_k^T$ for $t \geqslant T$ while φ^t on D_k^t remains the same after the step T. We consider the tree D_{m-1}^T and all its end vertices $a_0, \ldots, a_n$. Let

$$\nu y_i = \nu(m-1) \wedge \nu\varphi^T(a_i), \quad \nu z_i = C(\nu(m-1)) \wedge \nu\varphi^T(a_i)$$

We consider the step $t \geqslant T$ such that $t \geqslant \max(\{y_i \mid i \leqslant n\} \cup \{z_i \mid i \leqslant n\})$, $n(t) = m-1$. At the step $t+1$,

$$S_{m-1}^{t+1} = \{L(a_i) \mid \nu\varphi^t(a_i) \not\leqslant \nu(m-1), \nu\varphi^t(a_i) \wedge \nu(m-1) \neq \mathbf{0}\}$$
$$\cup \{a_i \mid \nu\varphi^t(a_i) \leqslant \nu(m-1)\}$$
$$D_m^{t+1} = D_{m-1}^t \cup S_{m-1}^{t+1} \cup S(S_{m-1}^{t+1})$$

For a_i such that $\nu\varphi^T(a_i) \not\leqslant \nu(m-1)$ and $\nu\varphi^T(a_i) \wedge \nu(m-1) \neq \mathbf{0}$, from the construction it follows that φ^{t+1} is defined so that $\nu\varphi^{t+1}(L(a_i)) = \nu y_i$ and $\nu\varphi^{t+1}(R(a_i)) = \nu z_i$. Proceeding by induction, for any $s \geqslant t+1$ we find that the conditions (T_k^s) and (D_k^s) are satisfied for $k \leqslant m-1$. Hence $n(s) \geqslant m$ for all $s \geqslant t+1$, which contradicts the assumption.

⟨4⟩ *For any $k \in \mathbb{N}$ there exists a finite tree $D_k = \lim_{t \to \infty} D_k^t$, $D_0 \subseteq D_1 \subseteq \ldots \subseteq D_k \subseteq \ldots$, and a function $\varphi = \lim_{t \to \infty} \varphi^t$ such that*

$\varphi \colon D^* \to \mathbb{N}$, $D^* = \bigcup_{k \geqslant 0} D_k$, and $\langle D^*; \nu\varphi \rangle$ is a tree generating the Boolean algebra $\mathfrak{B}$.

The proof follows from the above claims since every element νk can be represented as a union of elements $\nu\varphi(D_{k+1})$; moreover, the elements $\nu\varphi(D_{k+1})$ and $\nu\varphi$ satisfy all the required relations.

⟨5⟩ *For any end element $a \in D^*$ there exists only a finite number of elements of $D = \bigcup_{t \geqslant 0} D^t$ located below a.*

Indeed, we consider a number k and a step t_0 after which $a \in D_k^t$, $n(t_0) = k$, the values $\lambda t D_n^t$, $n \leqslant k$, and $\lambda t \varphi^t \restriction D_k^t$ are unchanged. Then a is an end element of $D_k^{t_0}$ because $D_k = D_k^{t_0}$, $D^* \supseteq D_k$, and a is an end element of D^*. Since the tree $\langle D^*; \nu\varphi \rangle$ generates $\mathfrak{B}$, we conclude that $\nu\varphi(a)$ is an atom of the Boolean algebra. Therefore, after the step t_0 there is no element s such that $0 < \nu(s) < \nu\varphi(a)$. Proceeding by induction, it is easy to show that for all $t \geqslant t_0$ and $n(t) \geqslant n \geqslant k$ the element a is an end one of D_n^t. In this case, after the step t_0 there is no step t at which in D_n^t there are no elements located below a. Consequently, after the step t_0 in $D^{t+1} = D^t \cup D_{n(t+1)}^{t+1}$ there are no elements located below a except those elements that are already in D^{t_0}.

On the basis of ⟨1⟩–⟨4⟩, we obtain $\mathfrak{B} = \mathrm{gr}\bigl(\mathfrak{B}_{D^*} \cup \bigcup_{a \in T} \{e_1^a, \ldots, e_{n_a}^a\}\bigr)$, where T denotes the set of end vertices of D^* and the elements $e_1^a, \ldots, e_{n_a}^a$ of $\mathfrak{B}_D$ are pairwise disjoint so that $\bigvee_{i=1}^{n_a} e_i^a = A_a$. By Proposition 1.7.2, we conclude that $\mathfrak{B}_D$ and $\mathfrak{B}_{D^*}$ are isomorphic to Boolean algebras. Since D is a recursively enumerable tree, $\mathfrak{B}$ is a constructive Boolean algebra in view of Theorem 3.3.1. □

Exercises

1. Construct trees generating the Boolean algebras $\mathfrak{B}_{\omega^n}$, $\mathfrak{B}_{(\omega+n)\times\eta}$, and $\mathfrak{B}_{(\omega+\eta)\times\eta}$.

2. Prove that there exists a constructivization of $\mathfrak{B}_\omega$ which contains only principal ultrafilters.

3. Prove that a superatomic Ershov algebra $\mathfrak{B}$ is constructive if and only if its ordinal type $o(\mathfrak{B})$ is a recursive ordinal [63].

3.3. Generating Trees

4. Prove that a Boolean algebra $\mathfrak{B}$ has a constructivization in which all its ultrafilters are recursive if and only if $\mathfrak{B}$ is superatomic and constructivizable.

5. Prove that an atomic Boolean algebra $\mathfrak{B}$ is strongly constructivizable if and only if there exists its constructivization with decidable set of atoms [92].

6. Using the tree technique, prove that an atomic Boolean algebra $\mathfrak{B}$ is strongly constructivizable if the quotient algebra $\mathfrak{B}/F(\mathfrak{B})$ of $\mathfrak{B}$ by the Fréchet filter is constructive.

7. Prove that an atomic Boolean algebra $\mathfrak{B}$ is strongly constructivizable if there exists a constructivization ν of $\mathfrak{B}$ in which the complexity of the Frechét ideal is Δ_2^0 with respect to the arithmetic hierarchy [34].

8. Prove that an ultrafilter of a constructive Boolean algebra is enumerable if and only if it is decidable.

9. A constructive Boolean algebra $\langle \mathfrak{B}, \nu \rangle$ and a constructivization ν of $\langle \mathfrak{B}, \nu \rangle$ are called *effectively atomic* if there exists a recursive function $f(n,m)$ such that for any n and m

    (a) $\nu f(n,0) = \nu n$,

    (b) $\nu f(n, m+1) \leqslant \nu f(n,m)$,

    (c) there exists $\lim_{m \to \infty} f(n,m)$, and $\nu \lim_{m \to \infty} f(n,m)$ is an atom of the Boolean algebra $\mathfrak{B}$.

    Prove that any constructivization of a Boolean algebra $\mathfrak{B}_\omega$ is effectively atomic [64].

10. Prove that for any constructivizable infinite atomic Boolean algebra there exists an effectively atomic constructivization [64, 65].

11. Construct constructivizations of $\mathfrak{B}_{\omega^2}$ and $\mathfrak{B}_{\omega \times \eta}$ that are not effectively atomic.

12. Prove that any infinite constructive Boolean algebra admits a constructive proper elementary extension.

13. Prove that a constructive Boolean algebra containing an atomless element has a nonconstructivizable group of recursive automorphisms [130].

14. Prove that the group of all recursive automorphisms of a constructive infinite atomic Boolean algebra is not constructivizable [130, 143].

15. Prove that any infinite strongly constructivizable Boolean algebra has a constructivization such that any of its automorphisms moves only a finite number of atoms [131].

3.4. Decidable Boolean Algebras

The notion of a strongly constructive model $\langle \mathfrak{M}, \nu \rangle$ is the most natural one for an effectively prescribed model in studies of model-theoretic properties of decidable theories. A model $\mathfrak{M}$ is called *decidable* if there exists a numbering ν of $\mathfrak{M}$ such that the model $\langle \mathfrak{M}, \nu \rangle$ is strongly constructive. Considering a decidable model, we always assume that some of its constructivizations are fixed. We begin with some general facts concerning strongly constructive models and then proceed to the decidability of algorithmic properties of strong constructivizations.

In this section, we consider the existence problem for decidable models (in particular, Boolean algebras) with given model-theoretic properties. It is easy to see that a theory $\text{Th}(\mathfrak{M})$ of a decidable model $\mathfrak{M}$ is decidable. The converse assertion (which can be regarded as a "constructive" analog of an existence theorem for models of consistent theories) also holds.

Theorem 3.4.1 [existence theorem for decidable models]. *Any decidable theory has a decidable model.*

If a type $p(x_1, \ldots, x_n)$ is realized in a decidable model, then it is decidable, i.e., there exists an algorithm determining whether a formula $\varphi(x_1, \ldots, x_n)$ belongs to $p(x_1, \ldots, x_n)$. Hence any undecidable type is omitted in any decidable model. By Theorem 3.4.1, any decidable type is realized in some decidable model.

Theorem 3.4.2 [omitting-decidable-types theorem]. *If T is a decidable theory and $\{p_i(\overline{x}_i) \mid i \in \mathbb{N}\}$ is a computable family of decidable nonpricipal types, then there exists a decidable model of T omitting all the types $p_i(\overline{x}_i)$, $i \in \mathbb{N}$.*

Theorems 3.4.1 and 3.4.2 can be proved by constructing step-by-step a Henkin theory with the required properties. We extend the signature by new constants $c_0, c_1, \ldots, c_k, \ldots$ and construct step-by-step finite parts

3.4. Decidable Boolean Algebras

$\mathcal{D}^t$ of the Henkin theory by adding φ or $\neg\varphi$ at even steps in order to keep consistency with T. The consistency condition is effectively verified since T is decidable. Adding a formula φ of the form $(\exists y)\psi(y)$, we take a constant c_k that has not been used and add the formula $\psi(c_k)$, which does not violate the consistency. At odd steps, for every type $p_i(\overline{x})$ and a sequence of constants $\overline{c}$ we find $\varphi(\overline{x}) \in p_i(\overline{x})$ such that $\neg\varphi(\overline{c})$ is consistent with the part $\mathcal{D}^t$ constructed above and add the result to $\mathcal{D}^t$. We obtain $\mathcal{D} = \cup \mathcal{D}^t$ which is consistent with T. Thus, we obtain the Henkin theory T and the corresponding Henkin model $\mathfrak{M}_T$, which is required. The model is decidable since, for every sentence φ, we can consider a step on which φ or $\neg\varphi$ is added and determine what is true on $\mathfrak{M}_T$.

From the above construction we obtain the following claim.

Corollary 3.4.1. *If T_i is a computable family of complete decidable theories, then there exists a computable sequence of strongly constructive models $\langle \mathfrak{M}_n, \nu_n \rangle$ such that $\mathfrak{M}_n \vDash T_n$.*

As was shown in Chapter 2, the family of types $S_{\mathfrak{M}}$ realized in a homogeneous model and the homogeneity condition define such a countable homogeneous model uniquely up to an isomorphism. Therefore, it is natural to try to find the decidability condition in terms of types realized in a homogeneous model. First of all, we present the following simple condition.

Proposition 3.4.1. *If $\mathfrak{M}$ is a decidable model, then the family of types $S_{\mathfrak{M}}$ realized in $\mathfrak{M}$ is computable.*

PROOF. Taking any strong constructivization of $\mathfrak{M}$ and enumerating all the finite sequences of numbers of elements $\overline{n} = (n_0, \ldots, n_k)$, we define the type $p_{\overline{n}}(x_0, \ldots, x_k)$ consisting of those formulas that are true on elements with given numbers. Thereby, we define a computable family $S_{\mathfrak{M}}$ of types. □

In the general case, the above condition is necessary but not sufficient for decidability. Let us consider a computable principal numbering of all partial enumerable types $p_0(\overline{x}_0), p_1(\overline{x}_1), \ldots$ that are consistent with a decidable theory T and are such that, from a number n, it is possible to compute a sequence of variables $\overline{x}_n$ of a partial type with number n constructed in Sec. 3.1.

A family of types S is called a *computable family with effective extension* if it is computable and there exists a partially recursive function $R(n, m)$ such that if a type $p_n(\overline{x}_n) \in S$ and a formula $\varphi_m(\overline{x}_n, \overline{x})$ with the

Gödel number m are consistent, then $R(n,m)$ is defined, $p_{R(n,m)} \supseteq p_n$, $p_{R(n,m)} \in S$, and $\varphi_m(\overline{x}_n, \overline{x}) \in p_{R(n,m)}$.

Without loss of generality, we can assume that, for a computable family S with effective extension, there exists a recursively enumerable set W such that $S = \{p_n \mid n \in W\}$ and the value $R(n,m)$ belongs to W for any $n \in W$ and m such that φ_m is consistent with p_n.

Theorem 3.4.3 [the Goncharov–Peretyat'kin criterion for decidability of homogeneous models]. *A countable homogeneous model $\mathfrak{M}$ with the family $S_\mathfrak{M}$ of types realized in $\mathfrak{M}$ is decidable if and only if $S_\mathfrak{M}$ is a computable family with effective extension.*

PROOF. If $\mathfrak{M}$ is a decidable homogeneous model, then $S_\mathfrak{M}$ is a computable family of types in view of Proposition 3.4.1. To prove the effectiveness of extensions we define step-by-step the type $d_{c(n,m)}(\overline{x}_n, \overline{x}) = \cup d^t(\overline{x}_n, \overline{x})$. At every step $t+1$, we find the least sequence $\overline{m}_n$ such that p_n^t is true, where p_n^t is that part of p_n that is already computed at the step t. We consider the conjunction $\& d^{t+1}(\overline{x}_n, \overline{x}) \& \varphi_m(\overline{x}_n, \overline{x})$. If a formula $(\exists \overline{x})(\& d^{t+1}(\overline{x}_n, \overline{x}) \& \varphi_m(\overline{x}_n, \overline{x}))$ belongs to p_n^t, then we find the least sequence $\overline{m}$ such that this formula holds on a sequence of elements with numbers $\overline{m}_n, \overline{m}$. We consider a formula $\varphi(\overline{x}_n, \overline{x})$ with the least number among formulas that do not belong to d^t, but hold on the sequence of elements with numbers $(\overline{m}_n, \overline{m})$. If $(\exists \overline{x})(\& d^t(\overline{x}_n, \overline{x}) \& \varphi_m(\overline{x}_n, \overline{x}) \& \varphi(\overline{x}_n, \overline{x}))$ belongs to $p_n(\overline{x}_n)$, then we add $\varphi(\overline{x}_n, \overline{x})$ to d^t and pass to the following step; otherwise, we set $d^{t+1} = d^t$. Acting in such a way, we arrive at a computable sequence of partial types. Since p_n is universal, there is a recursive function $\alpha(n,m)$ such that $d_{c(n,m)} = p_{\alpha(n,m)}$, $n, m \in \mathbb{N}$. It is easy to see that α possesses the required properties.

The proof of the sufficiency is more complicated. We construct a Henkin theory $\mathcal{D}$ of the enrichment of the signature σ of T by new constant symbols $c_0, c_1, \ldots, c_n, \ldots$ so that the following conditions on types hold:

(1) for any type $p(\overline{x})$ of S there exists a sequence of constants $\overline{c}$ such that $p(\overline{c}) \subseteq \mathcal{D}$,

(2) for any sequence of constants $\overline{c}$ there exists a type $p(\overline{x})$ of S such that $p(\overline{c}) \subseteq \mathcal{D}$,

(3) for any types $p(\overline{x})$ and $q(\overline{x}, y)$ of S and a sequence of constants $\overline{c}$ there is a constant c such that $q(\overline{c}, c) \subseteq \mathcal{D}$ provided that $p(\overline{x}) \subseteq q(\overline{x}, y)$ and $p(\overline{c}) \subseteq \mathcal{D}$.

3.4. Decidable Boolean Algebras

By (1) and (2), all the types of S (and only they) are realized in the Henkin model $\mathfrak{M}_\mathcal{D}$ of the theory $\mathcal{D}$. By (3), $\mathfrak{M}_\mathcal{D}$ is homogeneous in the signature σ. Furthermore, (3) implies (1). Thus, to complete the construction it suffices to require the validity of (2) and (3). We consider an enumeration of all the types

$$p_{s_0}(\overline{x}'_0), \ldots, p_{s_n}(\overline{x}'_n), \ldots, p_{s_m}(\overline{x}'_m), \ldots$$

of S, where $W = \{s_0, s_1, s_2, \ldots\}$ is a recursively enumerable set. Starting from it, we construct computable sequences

$$\Delta_0(\overline{c}_0), \ldots, \Delta_n(\overline{c}_n), \ldots, \quad \Gamma_0(\overline{c}'_0, v), \ldots, \Gamma_n(\overline{c}'_n, v), \ldots$$

such that

— every element of Δ_n is obtained from a suitable p_{s_k} by replacing the variables $\overline{x}'_k$ by some sequence of constants $\overline{c}_n$ which are effectively found from n.

— for any $p_{s_n}(\overline{x}'_n)$ and a sequence $\overline{c}'$ of the same length as the sequence $\overline{x}'_n$, there exists k such that $\Delta_k = [p_{s_n}(\overline{x}'_n)]_{\overline{c}'}^{\overline{x}'_n}$; moreover, k is effectively found from s_n and $\overline{c}'$,

— for any element of $\Gamma_n(\overline{c}'_n, v)$ there exists a type $p_{s_k}(\overline{x}'_k)$ such that the number of variables of the sequence $\overline{x}'_k$ is equal to the number of constants of the sequence $\overline{c}'_n$ plus unit, Γ_n is a result of replacing the first variable by v and the rest of the variables by the corresponding constants of $\overline{c}'_n$, and s_k can be effectively found from n,

— for any type $p_{s_n}(\overline{x}'_n)$ in $m+1$ variables and a sequence of constant symbols $\overline{c}$ consisting of m elements, we can find effectively k such that $\Gamma_k(\overline{c}, v)$ is obtained from $p_{s_n}(\overline{x}'_n)$ by substitution of v for the first variable and the corresponding constants of $\overline{c}$ for the rest of the variables.

Let Δ_0^t and Γ_0^t denote those parts of the sets Δ_0 and Γ_0 that are already computed at the step t. We also assume that the type Δ_0 contains no constant and is equal to T. At every step t, we define a finite part of the Henkin theory $\mathcal{D}^t$, a finite sequence of types

$$\Delta_0, \Gamma_{s(1,t)}, \Delta_{m(1,t)}, \ldots, \Gamma_{s(k(t),t)}, \Delta_{m(k(t),t)}$$

and constant symbols $c_{d(1,t)}, \ldots, c_{d(k(t),t)}$ which will be inserted instead of the variable v in the corresponding types $\Gamma_{s(1,t)}, \ldots, \Gamma_{s(k(t),t)}$.

A sequence $p_{m_1}(\overline{c}_1), \ldots, p_{m_k}(\overline{c}_k)$ obtained from the types of S by substituting the sequences of the corresponding constants is called t-consistent if for any $i \leqslant k$ and formulas $\varphi(\bar{c}) \in p_{m_i}^t$ the formula $\varphi(\bar{c}_i) \upharpoonright \bar{c}_j$, obtained from $\varphi(\bar{c}_i)$ by replacing all the constant symbols that are in $\bar{c}_i$ but not in $\bar{c}_j$ by new variables $\bar{x}$, and the quantification of the existential quantifier belong to $p_{m_j}(\bar{c}_j)$, $j \neq i$. Formulas $\varphi(\bar{c}_i) \upharpoonright \bar{c}_j$ defined in such a way are called the $\exists$-*restriction* of φ to the set of constants $\bar{c}_j$. Formulas obtained by replacing only variables without quantification of quantifiers are called a *restriction* and are denoted by $\varphi(\bar{c}_i) \upharpoonright \bar{c}_j, \bar{x}$. By the completeness of types and the computability of S, the t-compatibility condition is effectively verified.

At every step, the following conditions hold:

— $\mathcal{D}^t$ is consistent with T,

— a sequence

$$\Delta_0, [\Gamma_{s(1,t)}]^v_{c_{d(1,t)}}, \Delta_{m(1,t)}, \ldots, [\Gamma_{s(k(t),t)}]^v_{c_{d(k(t),t)}}, \Delta_{m(k(t),t)}$$

is t-consistent and every $\Delta_{m(i,t)}$ contains those parts of sets of this sequence that are already computed at the step t,

— for the conjunction $\&\mathcal{D}^t$ and any type of this sequence the $\exists$-restriction $\&\mathcal{D}^t \upharpoonright \bar{c}$ to the sequence corresponding to this type belongs to the same type.

STEP 0. We define $\mathcal{D}^0 = \varnothing$. The sequence consists only of Δ_0.

STEP $t+1$. We find $s \geqslant t$ such that one of the following conditions holds:

⟨1⟩ there exists $i > s(k(t), t)$ and a new constant c_k that do not occur in formulas of $\mathcal{D}^t$ and in the formulas of the types considered; moreover, if we add the type $[\Gamma_j]^v_{c_k}$, with the least j for which the s-consistence condition holds, to the sequence under consideration, we obtain an s-consistent sequence for some j, $s(k(t), i) < j < i$,

⟨2⟩ the sequence of types constructed at the step t is not s-consistent.

If ⟨1⟩ holds, then we set $k(t+1) = k(t) + 1$. For the least j satisfying the s-consistency condition we add $[\Gamma_j]^v_{c_k}$, to the sequence under consideration and obtain an s-consistent sequence. Therefore, $\exists$-restrictions of formulas to the corresponding constants of $\mathcal{D}^t$ belong to this type. Furthermore, the following conditions hold:

⟨3⟩ there exists the least m' such that $\Delta_{m'}$ contains all the formulas that are already computed at the step $t+1$ for the members of our sequences and the type $[\Gamma_j]^v_{c_k}$,

3.4. Decidable Boolean Algebras

⟨4⟩ ∃-restrictions of conjunctions of formulas of $\mathcal{D}^t$ to the set of constants of the type $\Delta_{m'}$ belong $\Delta_{m'}$,

⟨5⟩ the type Δ_m possesses the same properties as the type $\Delta_{m'}$ and can be obtained as follows: replacing constants by variables, from the type $\Delta_{m'}$ we find the type $p_{m''}(\overline{x}_k)$ of S, in the conjunction $\&\mathcal{D}^t$ we replace constants belonging to $\Delta_{m'}$ by $\overline{x}_k$, and the remaining constants by $\overline{x}$, and take the Gödel number l of the resulting formula.

With the help of an effectively extending function R, we obtain the type $p_{R(m'',l)}(\overline{x}_k, \overline{x})$, in which we substitute variables for constants and obtain Δ_m as a result. It is easy to see that, for Δ_m, an ∃-restriction of $\&\mathcal{D}^t$ keeps this construction unchanged.

We set $s(k(t)+1, t+1) = i$, $m(k(t)+1, t+1) = m$, $d(k(t+1), t+1) = k$, $d(j, t+1) = d(j,t)$, $s(j, t+1) = s(j,t)$, and $m(j, t+1) = m(j,t)$, $j \leqslant k(t)$. Thereby, we have extended the sequence. We now consider a sentence φ with the least number such that $\varphi \notin \mathcal{D}^t$ and $\neg\varphi \notin \mathcal{D}^t$. We set $\varphi_0 \rightleftharpoons \varphi \& \& \mathcal{D}^t$ and $\varphi_1 \rightleftharpoons \neg\varphi \& \& \mathcal{D}^t$. If for any type of our sequences ∃-restrictions of φ_ε belong to this type, then to $\mathcal{D}^t$ we add the sentence φ in the case $\varepsilon = 0$, and $\neg\varphi$ in the case $\varepsilon = 1$. If we add a sentence φ of the form $(\exists y)\psi(y)$, then we consider the first constant c_r among those constants that are not used in formulas of $\mathcal{D}^t$ and in sequences of constants of our types. Then we add $\psi(c_r)$ to $\mathcal{D}^t$. If ⟨2⟩ holds, then we consider the maximal part of the form

$$\Delta_0, [\Gamma_{s(1,t)}]^v_{c_{d(1,t)}}, \Delta_{m(s,t)}, \ldots, [\Gamma_{s(i,t)}]^v_{c_{d(s,t)}}, \Delta_{m(s,t)}$$

which is $(t+1)$-consistent. In this case, we set $k(t+1) = s$, $s(j, t+1) = s(j,t)$, $d(j, t+1) = d(j,t)$, and $m(j, t+1) = m(j,t)$, $j \leqslant s$.

It is easy to check that $\lim_{t \to \infty} k(t) = \infty$. This means that the sequence of the types considered becomes stable and the stabilizing part increases. Consequently, $\bigcup_t \mathcal{D}^t$ is a Henkin theory and satisfies the conditions (2) and (3) on types. □

Corollary 3.4.2 [the Morley criterion for decidability of countably saturated models]. *A countably saturated model of a decidable theory T is decidable if and only if the family of all types of the theory T is computable.*

PROOF. Since a countably saturated model is homogeneous, it suffices to note that the effective extension condition also holds in the case in which the family of all types is computable.

3. Constructive Boolean Algebras

Let us consider a type $p_n(\overline{x})$ and a formula $\varphi(\overline{x}, \overline{y})$ with number m. We define the following procedure of constructing a new type $d_{c(m,n)}$.

STEP 0. We set $d^0_{c(m,n)} = \varnothing$.

STEP $t+1$. We consider the conjunction $\& d^t_{c(m,n)} \& \varphi(\overline{x}, \overline{y})$. Closing it by existential quantifiers in variables of $\overline{y}$, we verify whether

$$(\exists \overline{y})\big(\& d^t_{c(m,n)} \& \varphi(\overline{x}, \overline{y})\big)$$

belongs to $p^t_n(\overline{x})$. If not, then $d^{t+1}_{c(m,n)} = \varnothing$. Otherwise, we consider the least Gödel number m of a formula of the form $\psi(\overline{x}, \overline{y})$ such that $\psi \notin d^t_{c(m,n)}$ and $\neg \psi \notin d^t_{c(m,n)}$. We check the following conditions:

$$(\exists \overline{y})\big(\& d^t_{c(m,n)} \& \varphi(\overline{x}, \overline{y}) \& \psi(\overline{x}, \overline{y})\big) \in p^t_n(\overline{x}) \tag{3.4.1}$$

$$(\exists \overline{y})\big(\& d^t_{c(m,n)} \& \varphi(\overline{x}, \overline{y}) \& \neg \psi(\overline{x}, \overline{y})\big) \in p^t_n(\overline{x}) \tag{3.4.2}$$

In the case of (3.4.1), we set $d^{t+1}_{c(m,n)} = d^t_{c(m,n)} \cup \{\psi\}$. If (3.4.1) fails, but (3.4.2) is valid, then we set $d^{t+1}_{c(m,n)} = d^t_{c(m,n)} \cup \{\neg \psi\}$. If both conditions fail, then we change nothing: $d^{t+1}_{c(m,n)} = d^t_{c(m,n)}$. Thereby, we obtain a computable sequence of partial types. Moreover, if $p_n \in S_T$ and $\varphi(\overline{x}, \overline{y})$ is consistent with p_n, then the complete type can be constructed such that it is consistent with the theory T and, consequently, belongs to S_T, where S_T denotes the family of all decidable types consistent with T. Taking the function $\alpha(m, n)$ which reduces the sequence just constructed to $\{p_n \mid n \in \mathbb{N}\}$, we obtain an extending function. □

Corollary 3.4.3 [the Goncharov–Harrington criterion for decidability of prime models]. *A prime model of a decidable theory T is decidable if and only if the family of principal types of the theory T is computable.*

PROOF. It is required to verify the effective extension condition for the family of principal types. Let $p_{s_0}(\overline{x}_0), \ldots, p_{s_n}(\overline{x}_n), \ldots$ be an enumeration of all principal types. For a type $p_n(\overline{x})$ and a formula $\varphi(\overline{x}, \overline{y})$ with number m we consider the following procedure.

STEP 0. We set $d^0_{c(n,m)} = \varnothing$.

STEP t. We check whether the formula $(\exists y)\big(\& d^t_{c(n,m)} \& \varphi(\overline{x}, \overline{y})\big)$ belongs to $p^t_n(\overline{x})$. If not, then we set $d^{t+1}_{c(n,m)} = d^t_{c(n,m)}$. Otherwise, we find the least i such that

$$p^t_n(\overline{x}) \subseteq p_{s_i}(\overline{x}, \overline{y}) \tag{3.4.3}$$

3.4. Decidable Boolean Algebras

and the formula $\psi(\overline{x}, \overline{y})$ with the least number such that $\psi \notin d^t_{c(n,m)}$ and $\neg \psi \notin d^t_{c(n,m)}$ satisfies the conditions

$$\psi(\overline{x}, \overline{y}) \& \& d^t_{c(n,m)} \& \varphi(\overline{x}, \overline{y}) \in p_{s_i} \qquad (3.4.4)$$

$$(\exists \overline{y})(\psi(\overline{x}, \overline{y}) \& \& d^t_{c(n,m)} \& \varphi(\overline{x}, \overline{y})) \in p_n \qquad (3.4.5)$$

or

$$\neg \psi(\overline{x}, \overline{y}) \& \& d^t_{c(n,m)} \& \varphi(\overline{x}, \overline{y}) \in p_{s_i} \qquad (3.4.6)$$

$$(\exists \overline{y})(\neg \psi(\overline{x}, \overline{y}) \& \& d^t_{c(n,m)} \& \varphi(\overline{x}, \overline{y})) \in p_n \qquad (3.4.7)$$

By the completeness of the types p_{s_i} and p_n, the conditions (3.4.3), (3.4.4), and (3.4.7) are effectively verified. If these conditions are satisfied, then we set $d^{t+1}_{c(n,m)} = d^t_{c(n,m)} \cup \{\psi\}$ under the conditions (3.4.4), (3.4.5) and $d^{t+1}_{c(n,m)} = d^t_{c(n,m)} \cup \{\neg \psi\}$ otherwise. If the type p_n is principal and consistent with $\varphi(\overline{x}, \overline{y})$, then, after the appearance of a complete formula in p^t_n, we constantly choose the same i. In this case, $d_{c(n,m)} = \cup d^t_{c(n,m)}$ is equal to p_{s_i} and, consequently, is principal. As usual, the reduction of this sequence to $\{p_n \mid n \in \mathbb{N}\}$ yields a function that is necessary for an effective extension. $\square$

Theorem 3.4.4. *If the family S_T of all decidable types consistent with a theory T is computable, then there is a decidable prime model of the theory T.*

PROOF. It is required to show that the theory T is atomic and the family of all principal types is computable. To this end, for any formula $\varphi(\overline{x})$ consistent with T we construct a computable sequence of formulas $\varphi_n(\overline{x})$ such that $\varphi_0(\overline{x}) = \varphi(\overline{x})$ and for any n the set $T \cup \{\varphi_n(\overline{x})\}$ is consistent and $T \vdash (\forall \overline{x})(\varphi_{n+1}(\overline{x}) \to \varphi_n(\overline{x}))$. Furthermore, for any formula $\varphi(\overline{x})$ there exists n such that $\varphi_n(\overline{x})$ is complete. Consequently,

$$d_n(\overline{x}) \rightleftharpoons \{\psi(\overline{x}) \mid \text{there exists } n \text{ such that } T \vdash \varphi_n(\overline{x}) \to \psi(\overline{x})\}$$

is a principal type containing a formula $\varphi(\overline{x})$ with the Gödel number n.

Let $p_{s_0}(\overline{x}), \ldots, p_{s_n}(\overline{x}), \ldots$ be a computable sequence of all decidable types consistent with the theory T.

STEP 0. We set $\varphi_0(\overline{x}) = \varphi(\overline{x})$ if $\varphi(\overline{x})$ is consistent with T and $\varphi_0(\overline{x}) \rightleftharpoons x_0 = x_0$ otherwise.

198   3. Constructive Boolean Algebras

STEP $t+1$. We consider a formula ψ with the Gödel number t. If it has free variables different from the variables occurring in the formula $\varphi(\overline{x})$, then we set $\varphi_{t+1} = \varphi_t$. Otherwise, we consider two cases.

CASE 1: $\psi(\overline{x}) \& \varphi_t$ or $\neg\psi(\overline{x}) \& \varphi_t$ is not consistent with T. Then we set $\varphi_{t+1} = \varphi_t$.

CASE 2: $\psi(\overline{x}) \& \varphi_t$, as well as $\neg\psi(\overline{x}) \& \varphi_t$, is consistent with T. We find the least numbers i, j such that $\psi(\overline{x}) \& \varphi_t \in p_{s_i}$ and $\neg\psi(\overline{x}) \& \varphi_t \in p_{s_j}$. We set $\varphi_{t+1} = \varphi_t \& \neg\psi(\overline{x})$ if $i < j$ and $\varphi_{t+1} = \varphi_t \& \psi(\overline{x})$ if $j < i$.

If Case 2 holds infinitely many times, then the sequence $\{\varphi_t \mid t \in \mathbb{N}\}$ defines a decidable type which is consistent with the theory T and is different from each of the types in the enumeration $p_{s_0}, p_{s_1}, \ldots, p_{s_n}, \ldots$. This contradicts the assumption, which provides the validity of the conditions imposed on φ_t. □

Corollary 3.4.4. *If the family of all decidable types consistent with a theory T is computable, then T has a prime model $\mathfrak{N}$ and the family $S_\mathfrak{N}$ of types realized in $\mathfrak{N}$ is computable.*

Theorem 3.4.5. *If the family of all decidable types of a theory T is computable and $\mathfrak{N}$ is a homogeneous countable model of the theory T with the computable family of types $S_\mathfrak{N}$, then the model $\mathfrak{N}$ is decidable.*

PROOF. By the criterion for a homogeneous model to be decidable, it suffices to show that the family $S_\mathfrak{N}$ has an effectively extending function. From a type $p_n(\overline{x})$ and a formula $\varphi(\overline{x}, \overline{y})$ we construct the required type step-by-step. Let $p_{s_0}(\overline{x}_0), \ldots, p_{s_n}(\overline{x}_n), \ldots$ denote the computable family of all decidable types consistent with the theory T. We construct $d_{k(n,m)} = \bigcup_t d^t_{c(n,m)}$ from a partial type $p_n(\overline{x})$ and a formula $\varphi(\overline{x}, \overline{y})$ with the Gödel number m.

STEP 0. We set $d^0_{c(n,m)} = \{\varphi(\overline{x}, \overline{y})\}$.

STEP $t+1$. If the formula $(\exists \overline{y})(\& d^t \& \varphi(\overline{x}, \overline{y}))$ is not in p^t_n, we set $d^{t+1}_{c(n,m)} = d^t_{c(n,m)}$. Otherwise, we consider a formula $\psi(\overline{x}, \overline{y})$ with the Gödel number $l(t)$.

CASE 1: T satisfies the conditions

$$T \vdash \& p^t_n(\overline{x}) \to (\exists \overline{y})(\psi(\overline{x}, \overline{y}) \& \& d^t_{c(n,m)})$$

$$T \vdash \& p^t_n(\overline{x}) \to (\exists \overline{y})(\neg\psi(\overline{x}, \overline{y}) \& \& d^t_{c(n,m)})$$

3.4. Decidable Boolean Algebras

and the formula ψ contains only those free variables that occur in $\overline{x}, \overline{y}$. We find the least i and j such that

$$\psi(\overline{x},\overline{y}) \& \& d^t_{c(n,m)} \in p_{s_i}, \quad \neg\psi(\overline{x},\overline{y}) \& \& d^t_{c(n,m)} \in p_{s_j}$$

We set $d^{t+1}_{c(n,m)} = d^t_{c(n,m)} \cup \{\neg\psi(\overline{x},\overline{y})\}$ if $i < j$ and $d^{t+1}_{c(n,m)} = d^t_{c(n,m)} \cup \{\psi(\overline{x},\overline{y})\}$ if $j < i$.

CASE 2: T satisfies the condition

$$T \vdash \& p^t_n(\overline{x}) \to (\exists \overline{y})\psi(\overline{x},\overline{y}) \& \neg(\exists \overline{y})\neg\psi(\overline{x},\overline{y})$$

We set $d^{t+1}_{c(n,m)} = d^t_{c(n,m)} \cup \{\psi\}$.

CASE 3: T satisfies the condition

$$T \vdash \& p^t_n(\overline{x}) \to (\exists \overline{y})\neg\psi(\overline{x},\overline{y}) \& \neg(\exists \overline{y})\psi(\overline{x},\overline{y})$$

We set $d^{t+1}_{c(n,m)} = d^t_{c(n,m)} \cup \{\neg\psi(\overline{x},\overline{y})\}$.

CASE 4: Cases 1–3 fail. We set $d^{t+1}_{c(n,m)} = d^t_{c(n,m)}$.

The type $p_n(\overline{x}) \in S_{\mathfrak{N}}$ is complete. Therefore, if $\varphi(\overline{x},\overline{y})$ is consistent with $p_n(\overline{x})$, then the partial type $d_{c(n,m)} = \cup d^t_{c(n,m)}$ is also complete. If Case 1 happens infinitely many times, then the type just constructed is decidable and different from each of the types $p_{s_0}, p_{s_1}, \ldots, p_{s_n}, \ldots$. However, the last sequence contains all decidable types. Consequently, Case 4 can take place only a finite number of times. Then the type $d_{c(n,m)}(\overline{x},\overline{y})$ is principal over $p(\overline{x})$. Consequently, this type is realized in every model in which $p_n(x)$ is realized. Therefore, $d_{c(n,m)}(\overline{x},\overline{y}) \in S_{\mathfrak{N}}$. The function reducing the sequence of partial types to $\{p_n \mid n \in \mathbb{N}\}$ gives an effectively extending function.

Corollary 3.4.5. *If the family of all decidable types of a decidable theory T is computable, then a prime model $\mathfrak{N}$ of T exists and is decidable; if the family $S_{\mathfrak{N}}$ is computable, then a homogeneous countable model of T exists and is decidable.*

Corollary 3.4.6. *If some universal model of a theory T is decidable, then a prime model of T exists and is decidable; in addition, the family of all decidable types consistent with T is computable.*

We now study decidable and strongly constructive Boolean algebras. The existence theorem for decidable models and the decidability of complete elementary theories of all elementary characteristics imply the following claim.

Proposition 3.4.2. *For an elementary characteristic* $\chi = (\alpha, \beta, \gamma)$ *of Boolean algebras there exists a strongly constructive Boolean algebra* $\langle \mathcal{L}_\chi, \nu_\chi \rangle$ *such that* $\mathrm{ch}(\mathcal{L}_\chi) = \chi$ *which can be constructed uniformly in* χ.

PROOF. As is known (cf. Chapter 2), the elementary theory $\mathrm{Th}(\mathcal{L})$ of a Boolean algebra $\mathcal{L}$ is decidable and has a prime countably saturated model.

Theorem 3.4.6. *A countably saturated model of any elementary characteristic is decidable.*

PROOF. In view of the Morley criterion, it suffices to show the computability of the family of all types of the theory T_χ of a given elementary characteristic χ. Since any type can be reduced to a partition type, it suffices to define all partition types. The partition type

$$p(x_1, \ldots, x_n)\left(\bigvee_{i=1}^{n} x_i = 1 \in p(x_1, \ldots, x_n), \right.$$

$$\left. x_i \wedge x_j = 0 \in p(x_1, \ldots, x_n),\ i \neq j \right)$$

is determined by elementary characteristics of the element x_i. It is only required to describe all admissible characteristics of partitions for a saturated Boolean algebra of characteristic χ. Such a description follows from the density condition.

1. If $\chi = (\infty, 0, 0)$, then any partition with any characteristic of x_i is admissible provided that one of the x_i has the characteristic $(\infty, 0, 0)$, which defines all partition types.

2. If $\chi = (n, \infty, 0)$, then a partition type $p(x_1, \ldots, x_n)$ such that $\mathrm{ch}_1(x_i) < n \vee (\mathrm{ch}_3(x_i) = 0\ \&\ \mathrm{ch}_1(x_i) = n)$ is admissible and there exists i such that $\mathrm{ch}_1(x_i) = (n, \infty, 0)$.

3. If $\chi = (n, \infty, 1)$, then a partition type $p(x_1, \ldots, x_n)$ such that $\mathrm{ch}_1(x_i) \leqslant n$ is admissible and there exist i and j such that $\mathrm{ch}_1(x_i) = \mathrm{ch}_1(x_j) = n$, $\mathrm{ch}_3(x_i) = 1$, and $\mathrm{ch}_2(x_j) = \infty$.

4. If $\chi = (n, k, \varepsilon)$, then a partition type $p(x_1, \ldots, x_n)$ such that $\mathrm{ch}_1(x_i) \leqslant n$, $\mathrm{sg}\left(\sum_{i \in K} \mathrm{ch}_3(x_i) \right) = \varepsilon$ is admissible and $\sum_{i \in K} \mathrm{ch}_2(x_i) = k$ for $K \rightleftharpoons \{i \mid \mathrm{ch}_1(x_i) = n\}$, where $\mathrm{sg}\,(0) = 0$ and $\mathrm{sg}\,(x + 1) = 1$ for all x.

It is easy to see that all admissible partitions are realized in dense Boolean algebras of the corresponding characteristics and any partition re-

3.4. Decidable Boolean Algebras

alized in some Boolean algebra of a given characteristic is admissible. The above description of partition types and the possibility of reducing any type to some partition type imply that the constructed family is computable and coincides with the set of types realized in a countably saturated model of characteristic χ. □

Corollary 3.4.7. *A prime model of the theory of Boolean algebras of any elementary characteristic is decidable.*

PROOF. The assertion follows from Theorem 3.4.4 and the existence and decidability of countably saturated models. □

Corollary 3.4.8. *Any countable homogeneous Boolean algebra $\mathfrak{A}$ such that $\mathrm{ch}_1(\mathfrak{A}) < \infty$ is decidable.*

PROOF. By the decidability of a countably saturated model of the theory $\mathrm{Th}(\mathfrak{A})$ and Corollary 3.4.5, it suffices to show that the family of types realized in $\mathfrak{A}$ is computable. As in the case of a saturated model, it suffices to describe only partition types, while the rest of the types can be constructed from the partition types in a well-known way.

We now describe all partition types in the case of an algebra $\mathfrak{A}$ of the homogeneity type $\langle 0, \langle p_0, \ldots, p_n \rangle \rangle$.

1. Let $\mathrm{ch}(\mathfrak{A}) = \langle n, 1, 0 \rangle$ and $p_{n-1} = 1$. A partition $x_1, \ldots, x_m$ is realized in $\mathfrak{A}$ if the following conditions hold:

    (1) there exists a unique i such that $\mathrm{ch}(x_i) = n$ and $\mathrm{ch}(x_j) < n$, $j \neq i$,
    (2) there exists at most one j such that $\mathrm{ch}(x_j) = \langle n-1, \infty, \varepsilon \rangle$,
    (3) there exists j such that $\mathrm{ch}(x_j) = \langle k, \infty, \varepsilon \rangle$, $k < n-1$, if and only if $p_k = 2$.

    It is easy to see that such partition types (and only they) are realized in $\mathfrak{A}$ of the homogeneity type $\langle 0, \langle p_0, \ldots, p_n \rangle \rangle$. It is clear that the family of such types is computable.

2. Let $\mathrm{ch}(\mathfrak{A}) = \langle n, 1, 0 \rangle$ and $p_{n-1} \neq 1$. Only partitions subject to (1) and (3) with $k \leq n-1$ can be realized in $\mathfrak{A}$. Therefore, all such partition types can be enumerated.

3. Let $\mathrm{ch}(\mathfrak{A}) = \langle n, \alpha, \beta \rangle$, where $\alpha \in \mathbb{N}$. Only those partitions $x_1, \ldots, x_m$ can be realized in $\mathfrak{A}$ that are subject to the following conditions:

(1) there exists i such that $\mathrm{ch}_1(x_i) = n$ and the set $K \leftrightharpoons \{i \mid \mathrm{ch}_1(x_i) = n\}$ satisfies the equalities

$$\sum_{i \in K} \mathrm{ch}_2(x_i) = \alpha, \quad \mathrm{sg}\left(\sum_{i \in K} \mathrm{ch}_3(x_i)\right) = \beta$$

(2) $\mathrm{ch}_1(x_i) \leqslant n$ for any i,

(3) for any $k < n$ an element x_i may exist such that $\mathrm{ch}(x_i) = (k, \infty, 0)$ for $p_k = 2$.

Thereby, we have defined an enumeration of all possible partition types.

4. Let $\mathrm{ch}(\mathfrak{A}) = \langle n, \infty, \beta \rangle$. Then one of the following possibilities takes place: $p_n = \omega$, $p_n = \omega + \omega$, and $p_n = \omega \times \eta$. The first two conditions on a partition $x_1, \ldots, x_n$ are the same while the third one is different for all these three cases:

(1) there exists i such that $\mathrm{ch}(x_i) = (n, \infty, \varepsilon)$, and

$$\mathrm{sg}\left(\sum_{i \in K} \mathrm{ch}_3(x_i)\right) = \beta$$

for the set $K = \{i \mid \mathrm{ch}_1(x_i) = n\}$,

(2) $\mathrm{ch}_1(x_i) \leqslant n$ for any i, and j may exist such that $\mathrm{ch}(x_j) = (k, \infty, \varepsilon)$, $k < n$, only if $p_k = 2$,

(3a) (if $p_n = \omega$) there exists a unique i_0 such that $\mathrm{ch}(x_{i_0}) = (n, \infty, \varepsilon)$, $\varepsilon \leqslant \beta$,

(3b) (if $p_n = \omega + \omega$) the set $\{i \mid \mathrm{ch}_1(x_i) = n \,\&\, \mathrm{ch}_2(x_i) = \infty\}$ consists of at most two elements,

(3c) (if $p_n = \omega \times \eta$) except for (1) and (2), no additional conditions are imposed.

By the description of partitions realized in $\mathfrak{A}$, we see that these partitions are enumerable in the cases just listed. Consequently, the family of types realized in $\mathfrak{A}$ is computable. □

In the case of Boolean algebras of the first infinite characteristic, there is a continuum of the homogeneous models and only part of homogeneous models can be decidable. As is known, at least a prime model and a countably saturated model are decidable. However, there are other decidable homogeneous models.

3.4. Decidable Boolean Algebras

Theorem 3.4.7 [132]. *If $\mathfrak{A}$ is a countable homogeneous Boolean algebra of elementary characteristic $\langle \infty, 0, 0 \rangle$ and of the homogeneity type $\langle \alpha, \langle p_n \mid n \in \mathbb{N} \rangle \rangle$, then $\mathfrak{A}$ is a decidable model if and only if $\{n \mid p_n = 0\} \in \Pi_2^0$.*

PROOF. Let $\mathfrak{A}$ be decidable. By definition,

$p_n = 0 \Leftrightarrow$ there exists no element x such that $\mathrm{ch}(x) = (n, \infty, 0)$

$\Leftrightarrow$ any element x such that $\mathrm{ch}_1(x) = n$ has a finite second characteristic

$\Leftrightarrow$ for any $m \in \mathbb{N}$ if $\mathfrak{A} \vDash \mathrm{ch}_1(\nu m) = n$, then there exists k and $x_1, \ldots, x_{k+1}$ such that

$$\mathfrak{A} \vDash \underset{i=1}{\overset{k}{\&}} \mathrm{Atom}_n(\nu(x_i)) \& \mathrm{Atomless}_n(\nu(x_{k+1})) \& \bigvee_{i=1}^{k+1} \nu(x_i) = \nu m$$

where ν denotes some strong constructivization of $\mathfrak{A}$. By the strong constructibility of $\langle \mathfrak{A}, \nu \rangle$, the conditions

$\mathfrak{A} \vDash \mathrm{ch}_1(\nu m) = n$

$\mathfrak{A} \vDash (\underset{i=1}{\overset{k}{\&}} \mathrm{Atom}_n(\nu(x_i)) \& \mathrm{Atomless}_n(\nu(x_{k+1})) \& \bigvee_{i=1}^{k+1} \nu(x_i) = \nu m)$

are decidable. By the Tarski–Kuratowski algorithm, $\{n \mid p_n = 0\} \in \Pi_2^0$.

To prove the converse assertion we enumerate all those partitions that can be realized in a decidable model $\mathfrak{A}$. Let $\{n \mid p_n = 0\} \in \Pi_2^0$. By definition, there exists a recursive relation A such that $n \in \{n \mid p_n = 0\} \Leftrightarrow (\forall x)(\exists y)(A(n, x, y))$. Hence $n \in \{n \mid p_n = 2\} \Leftrightarrow (\exists x)(\forall y) \neg A(n, x, y)$.

For $\mathfrak{A}$ we consider the homogeneity type $\langle \alpha, \langle p_n \mid n \in \mathbb{N} \rangle \rangle$. A partition $x_1, \ldots, x_n$ is called α-*admissible* if the following conditions hold:

— for $\alpha = 1$ there exists exactly one element x_i such that $\mathrm{ch}(x_i) = (\infty, 0, 0)$,

— for $\alpha = 2$ there exist at most two elements, but for at least one of them, say x_i, $\mathrm{ch}(x_i) = (\infty, 0, 0)$,

— for $\alpha = 3$ there exists at least one element x_i such that $\mathrm{ch}(x_i) = (\infty, 0, 0)$.

For $\gamma = \langle \langle n_1, l_1, \delta_1 \rangle, \ldots, \langle n_s, l_s \delta_s \rangle \rangle$ we consider a family S_γ of partition types satisfying the following conditions:

(1) $x_1, \ldots, x_n$ is an α-admissible partition,

(2) there exist $x_{i_1}, \ldots, x_{i_s}$ such that $\mathrm{ch}_1(x_{i_j}) = n_{i_j}$, $\mathrm{ch}_3(x_{i_j}) = \delta_{i_j}$, $n_{i_j} \in \{n_1, \ldots, n_s\}$, while the second characteristic is progressively defined depending on $n_1, l_1, n_2, l_2, \ldots, n_s, l_s$ and the step t,

(3) for the rest of the x_i we set $\mathrm{ch}_1(x_i) = \infty$ or $\mathrm{ch}(x_i) = (m_i, \alpha_i, \beta_i)$, where $\alpha_i < \infty$,

At the step t, we find that $\mathrm{ch}_2(x_{i_j}) \geqslant t$ or we set $\mathrm{ch}_2(x_{i_j}) = t$. Such a system of inequalities shows that $\mathrm{ch}_2(x_{i_j}) = \infty$ or $\mathrm{ch}_2(x_{i_j}) = t$ at some step t, which completely defines the elementary characteristic of x_{ij}. If $\mathrm{ch}_1(x_{i_j}) = n_{i_j}$ is undefined at the step $t + 1$, then we consider n_i such that $n_i = n_{i_j}$ and verify the formula $(\forall y \leqslant t)\neg A(n_i, l_i, y)$. If the latter is true, then we define $\mathrm{ch}_2(x_{i_j}) \geqslant t + 1$. If the formula $(\exists y \leqslant t) A(n_i, l_i, y)$ is true, then we set $\mathrm{ch}_2(x_{i_j}) = t + 1$ and do not redefine characteristics in the sequel. It is clear that for any γ the family S_γ is enumerated and partitions of S_γ are realized in $\mathfrak{A}$. Therefore, the family $S = \bigcup_\gamma S_\gamma$ is enumerable and all partition types of S are realized in $\mathfrak{A}$. Since a partition is unique, there is only a finite number of elements such that the first characteristic is finite and the second one is infinite. For such a partition, it is easy to pick up γ so that it has been enumerated in S_γ.

Thus, we have established that the family $S_\mathfrak{A}$ is enumerable and a countably saturated model of this characteristic is decidable. Consequently, a homogeneous model $\mathfrak{A}$ with computable family of types is decidable by Theorem 3.4.5. □

The following simple theorem (but having many applications) provides an effective method of establishing the decidability and strong constructibility of models.

Theorem 3.4.8. *If a complete theory T is model-complete and decidable, then any of its constructive models is strongly constructive.*

PROOF. The assertion follows from Proposition 3.1.14. □

We now examine the existence problem for strongly constructive elementary extensions of strongly constructive models. In particular, we discuss the question of whether any strongly constructive model has a constructive elementary extension to a strongly constructive homogeneous model.

Proposition 3.4.3. *For any constructive atomless Boolean algebra there exists a proper constructive elementary extension to a strongly constructive Boolean algebra.*

3.4. Decidable Boolean Algebras

PROOF. In view of the model completeness of the theory of atomless Boolean algebras in the signature of Boolean algebras, it suffices to find a constructive isomorphic embedding of a constructive atomless Boolean algebra $\langle \mathfrak{A}, \nu \rangle$ into a constructive atomless Boolean algebra $\langle \mathfrak{B}, \mu \rangle$ so that $\varphi \colon \mathfrak{A} \to \mathfrak{B}$ is an isomorphic embedding but not an epimorphism and there exists a recursive function f such that $\varphi\nu(n) = \mu f(n)$ for any $n \in \mathbb{N}$. By Proposition 3.2.2, there exists a constructive linear order L such that $\langle \mathfrak{B}_L, \nu_L \rangle \cong \langle \mathfrak{A}, \nu \rangle$. The linear order $\eta \times L$ is also constructive. The Boolean algebra $\langle \mathfrak{B}_{\eta \times L}, \nu_{\eta \times L} \rangle$ is constructive and atomless.

For η we take the set of rational numbers $\mathbb{Q}$ with the natural order. We define an embedding φ of $\mathfrak{B}_L$ into $\mathfrak{B}_{\eta \times L}$ as follows:

$$\varphi(\cup [a_i, b_i[) \rightleftharpoons \bigcup_i [\alpha_i, \beta_i[$$

where $\alpha_i = (0, a_i)$ if $a_i \in L$, $\alpha_i = -\infty$ if $a_i = -\infty$, $\beta_i = (0, b_i)$ if $b_i \in L$, and $\beta_i = \infty$ if $b_i = \infty$. The composition of a constructive isomorphism of $\langle \mathfrak{A}, \nu \rangle$ onto $\langle \mathfrak{B}_L, \nu_L \rangle$ and the constructive embedding φ define the required proper embedding in an obvious way. □

Theorem 3.4.9. *For any strongly constructive atomic Boolean algebra $\langle \mathfrak{A}, \nu \rangle$ there exists a constructive elementary extension φ to a strongly constructive Boolean algebra $\langle \mathfrak{B}, \mu \rangle$ such that there exists an element $b \in \mathfrak{B}$ such that for any $a \in \mathfrak{A}$, $\mathrm{ch}_2(a) = \infty$, we have $\mathrm{ch}_2(a \wedge b) = \infty$, $\mathrm{ch}_2(a \wedge C(b)) = \infty$, and b does not belong to $\mathfrak{A}$ (if $\mathfrak{A}$ is infinite); moreover, $\langle \mathfrak{B}, \mu \rangle$ is uniformly constructed from $\langle \mathfrak{A}, \nu \rangle$ and for an infinite Boolean algebra this extension is proper.*

PROOF. We consider the enrichment of a signature of a Boolean algebra $\mathfrak{A}$ by a unary predicate A distinguishing atoms. In this enrichment $\mathfrak{A}'$, the theory of the Boolean algebra $\mathfrak{A}'$ is model-complete. We consider a tree $\mathcal{D}$ such that $\langle \mathfrak{B}_\mathcal{D}, \mu_\mathcal{D} \rangle \cong \langle \mathfrak{A}, \nu \rangle$. Such a tree exists by Theorem 3.3.2. Since $\langle \mathfrak{A}, \nu \rangle$ is strongly constructive, the recursive and isomorphic Boolean algebra $\langle \mathfrak{B}_\mathcal{D}, \mu_\mathcal{D} \rangle$ is also strongly constructive. Therefore, the set of numbers of its atoms, as well as the set of end vertices of the tree $\mathcal{D}$, is recursive.

We identify $\mathcal{D}$ with a tree generating the Boolean algebra $\mathfrak{A}$ and consider elements of $\mathcal{D}$ as elements of $\mathfrak{A}$, keeping in mind the standard embedding. We introduce a new constant c and constants $c_0, c_1, \ldots, c_n, \ldots$ with respect to which we consider the diagram of the Boolean algebra $\mathfrak{A}$ under the assumption that the value of the constant c_n is νn. We define all relations between c and all elements of the generating tree. Thereby, we define all relations between c and all elements of $\mathfrak{A}$. Let $c_{g(s)}$ denote the constant corresponding to an element s of $\mathcal{D}$.

With the help of an algorithm that decides all formulas, from any strongly constructive Boolean algebra $\langle \mathfrak{A}, \nu \rangle$ we can uniformly construct a tree $\mathcal{D}$ and a recursive embedding g of $\mathcal{D}$ into $\mathbb{N}$ so that the tree $\langle \mathcal{D}, \nu g \rangle$ generates $\mathfrak{A}$. In view of the strong constructibility of $\langle \mathfrak{A}, \nu \rangle$, the tree $\mathcal{D}$, as well as the set of endpoints of $\mathcal{D}$, is also recursive. At the step t, we define relations between c, its complement $C(c)$, and elements of $\mathcal{D}$ of level t if no relations are defined for them. Perhaps such relations are also defined for a finite number of points of higher level $s > t$, but only for those elements of level t under which at most four atoms are located. The set of relations constructed at the step t is denoted by S^t. Constructing a partition of c by elements of a Boolean algebra, we use counters $r(x,t)$ defining a part of the partition of c (or $C(c)$) the intersection of which with $\nu(x)$ is finite. By a relation between c and an element n of a tree, we mean an expression of one of the following forms:

(1) $c \wedge c_{g(n)} = c_{g(n)}$, or $c \wedge c_{g(n)} \neq c_{g(n)}$, or $c \wedge c_{g(n)} = \mathbf{0}$, or $c \wedge c_{g(n)} \neq \mathbf{0}$,

(2) $c \wedge c_{g(n)} \neq c$,

(3) $A(c \wedge c_{g(n)})$ or $\neg A(c \wedge c_{g(n)})$.

Similar relations are defined for the complement $C(c)$ to an element c.

STEP 0. We verify if at most four atoms are located under $\nu g(0)$. By the strong constructibility of $\langle \mathfrak{A}, \nu \rangle$, this condition is decidable. If the condition fails, then $\nu g(0) = \bigvee_{i=1}^{s} \nu x_i$, where $x_i \in \mathcal{D}$, x_i is an endpoint of $\mathcal{D}$, and $s \leqslant 4$. We set $S^0 = \{C(c) = c_{g(0)}\}$ and $r(0,0) = 0$. If there are more than four atoms, then $S^0 = \{c \wedge c_{g(0)} \neq c_{g(0)}, C(c) \wedge c_{g(0)} \neq c_{g(0)}, c \wedge c_{g(0)} \neq \mathbf{0}, C(c) \wedge c_{g(0)} \neq \mathbf{0}, \neg A(c \wedge c_{g(0)}), \neg A(C(c) \wedge c_{g(0)})\}$ and $r(0,0) = 0$.

STEP $t+1$. For all elements $x_1, \ldots, x_n$ of $\mathcal{D}$ of level $t+1$ we define all necessary relations if they are not already defined. If, for some $y \preccurlyeq x_i$, the relation $c \wedge c_{g(y)} = c_{g(y)}$ or $C(c) \wedge c_{g(y)} = c_{g(y)}$ is already defined, then the relations between $c_{g(x_i)}$ and c are defined or not depending on what was added to S^t. We add similar equalities to the relations connecting $c_{g(x_i)}$ with c and $C(c)$. It is not necessary to define A on an intersection since it is already defined in the diagram of $\mathfrak{A}$. If the previous condition fails, then, for $y = H(x_i) \in \mathcal{D}$, the relations $c \wedge c_{g(y)} \neq c_{g(y)}$ and $C(c) \wedge c_{g(y)} \neq c_{g(y)}$ are already defined and both intersections are marked as not atoms. We also consider the neighboring element $S(x_i)$ that belongs to $\mathcal{D}$ and is located

3.4. Decidable Boolean Algebras

under y. We define all the relations and all the counters simultaneously. The following cases are possible.

CASE 1: there are at most four atoms located under $\nu g(x_i)$ and under $\nu g(S(x_i))$. Since at most four atoms of $\mathfrak{A}$ are located under $g(y)$, at least four atoms are located under $\nu g(x_i)$ and under $\nu g(S(x_i))$ as well. Let $y_1, \ldots, y_m$, $m \geqslant 4$, be those end vertices of $\mathcal{D}$ that are situated under x_i and under $S(x_i)$. We define $c \wedge c_{g(y_i)} = c_{g(y_i)}$, $1 \leqslant i \leqslant 2$, and $C(c) \wedge c_{g(y_i)} = c_{g(y_i)}$, $2 < i \leqslant m$. For the rest of the $z \in \mathcal{D}$ such that $z \preccurlyeq x_i$ or $z \preccurlyeq S(x_i)$ we can obtain relations for $c_{g(z)}$ from the above relations since $\nu g(z) = \vee \{\nu g(y_i) \mid i \in K_z\}$, where $K_z \subseteq \{1, \ldots, m\}$.

CASE 2: there are at least four atoms located under $\nu g(x_i)$ and under $\nu g(S(x_i))$. We define S^{t+1} by adding the following relations to S^t:

$$c_{g(x_i)} \wedge c \neq \mathbf{0} \& c_{g(x_i)} \wedge c \neq c_{g(x_i)} \& \neg A(c_{g(x_i)} \wedge c)$$

We add the same relations to S^t for $S(x_i)$:

$$c_{g(S(x_i))} \wedge c \neq \mathbf{0} \& c_{g(S(x_i))} \wedge c \neq c_{g(S(x_i))} \& \neg A(c_{g(S(x_i))} \wedge c)$$

As for a complement, we add both these collections of relations after the replacement of c by $C(c)$. We set $r(x_i, t+1) = r(S(x_i), t+1) = r(H(x_i), t)$ in the first two cases.

CASE 3: under one of the elements of $\{\nu g(x_i), \nu g(S(x_i))\}$ there are at most four atoms while under another element there are more than four atoms. Let $a \in \{x_i, S(x_i)\}$ be such that at most four atoms are located under $\nu g(a)$, and more than four atoms are located under $\nu g(b)$, where $b \in \{x_i, S(x_i)\}$. If $r(H(x_i), t)$ is even, then we assume that, in S^{t+1}, S^t is completed by the relations $c_{g(a)} \wedge c = c_{g(a)}$ and $c_{g(a)} \wedge C(c) = \mathbf{0}$. If $r(H(x_i), t)$ is odd, then we assume that S^{t+1} contains $c_{g(a)} \wedge C(c) = c_{g(a)}$ and $c_{g(a)} \wedge c = \mathbf{0}$. For b we add the following relations to S^{t+1}:

$$c \wedge c_{g(b)} \neq c_{g(b)} \& c \wedge c_{g(b)} \neq \mathbf{0} \& \neg A(c \wedge c_{g(b)})$$
$$C(c) \wedge c_{g(b)} \neq c_{g(b)} \& C(c) \wedge c_{g(b)} \neq \mathbf{0} \& \neg A(C(c) \wedge c_{g(b)})$$

Naturally, we add all relations of S^t to S^{t+1}. We set $r(a, t+1) = r(H(a), t)$ and $r(b, t+1) = r(H(b), t) + 1$.

We consider the theory $T_c \leftrightharpoons \mathrm{At}_1^* \cup \mathcal{D}_1\langle \mathfrak{A}, \nu \rangle \cup \bigcup_t S^t$, where $\mathcal{D}_1\langle \mathfrak{A}, \nu \rangle$ is the diagram of the Boolean algebra $\mathfrak{A}$ of the signature $\sigma_1 = \sigma \cup \{I_0, A_1\}$, where the value of the constant c_n is assumed to be equal to νn and At_1^* is the theory of atomic infinite Boolean algebras of the signature σ_1. For any t we can choose $b \in \mathfrak{A}$ such that, provided that the value of c is assumed to be equal to b, all relations of S^t are satisfied by c in the enrichment $\langle \mathfrak{A}, b \rangle$ of $\mathfrak{A}$ by b for the value of the constant symbol c. By the Mal'tsev compactness theorem, (cf. Theorem 1.1.2) the set of formulas T_c is consistent and there is an enrichment $\mathfrak{B}'$ of some Boolean algebra $\mathfrak{B}$ such that $\mathfrak{B}' \vDash T_c$. Hence $\mathfrak{B}$ is atomic. Since $\mathfrak{B}' \vDash \mathcal{D}_1 \langle \mathfrak{A}, \nu \rangle$, we can assume that $\mathfrak{B}'$ is an extension of the enrichment $\mathfrak{A}_1$ of $\mathfrak{A}$ to the signature σ_1. The theory of atomic Boolean algebras of the signature σ_1 is model-complete. Consequently, $\mathfrak{A}_1 \preccurlyeq \mathfrak{B}'$ and $\mathfrak{A} \preccurlyeq \mathfrak{B}$.

If there are infinitely many elements of $\mathcal{D}$ under $n \in \mathcal{D}$, then $c \wedge c_{g(n)} \neq c_{g(n)}$ and $C(c) \wedge c_{g(n)} \neq c_{g(n)}$. Consequently, for any element n such that $\mathrm{ch}_2(\nu(n)) = \infty$ we have $c \neq c_n$. Therefore, any element $\nu(n) \in \mathfrak{A}$ such that $\mathrm{ch}_2(\nu(n)) = \infty$ is divided by c into two parts in such a way that, under $c \wedge c_n$ and under $C(c) \wedge c_n$ as well, there are infinitely many c_m such that νm is an atom of $\mathfrak{A}$ (that is the reason why the counters $r(x,t)$ in x were introduced). Therefore, the value of c in the Boolean algebra $\mathfrak{B}'$ differs from the values of c_n and, consequently, it does not belong to $\mathfrak{A}$. Thus, $\mathfrak{B}$ is a proper extension of $\mathfrak{A}$ subject to the required conditions.

We consider the subalgebra $\mathfrak{B}_0$ of $\mathfrak{B}$ generated by $\{\nu g(n) \wedge (c)_{\mathfrak{B}} \mid n \in \mathcal{D}\} \cup \{\nu g(n) \wedge C((c)_{\mathfrak{B}}) \mid n \in \mathcal{D}\}$, where $(c)_{\mathfrak{B}}$ denotes the value of the constant c in $\mathfrak{B}'$. It is obvious that $\mathfrak{A} \preccurlyeq \mathfrak{B}_0$ since atoms of $\mathfrak{A}$ go to atoms of $\mathfrak{B}_0$ and any element that is not an atom goes into such an element. The Boolean algebra $\mathfrak{B}_0$ is atomic, which follows from the construction of relations S^t for atoms and the fact that $\mathfrak{A}$ is atomic. However, elements of $\mathfrak{B}_0$ are as follows:

$$\bigvee_{i=1}^{k} (g(n_i) \wedge c) \vee \bigvee_{i=1}^{l} (g(m_i) \wedge C(c))$$

where $\{n_i \mid 1 \leqslant i \leqslant k\} \cup \{m_i \mid 1 \leqslant i \leqslant l\} \subseteq \mathcal{D}$. We regard x as the number of the pair $c(n,m)$ and n, m as numbers of finite sequences $n = \langle n_1, \ldots, n_k \rangle$ and $m = \langle m_1, \ldots, m_l \rangle$. Then

$$\nu_0(x) \rightleftharpoons \left(\bigvee_{i=1}^{k} g(n_i) \wedge c \right) \vee \left(\bigvee_{i=1}^{l} g(m_i) \wedge C(c) \right)$$

is a numbering of the Boolean algebra $\mathfrak{B}_0$. By the properties of the tree $\mathcal{D}$, it is easy to obtain the existence of recursive functions f, g, h construction of which depends only on the structure of the tree. Consequently, $f, g,$

3.4. Decidable Boolean Algebras

and h are uniformly determined by the algebra $\langle \mathfrak{A}, \nu \rangle$ so that

$$\nu_0(x) \lor \nu_0(y) = \nu_0 f(x,y)$$
$$\nu_0(x) \land \nu_0(y) = \nu_0 g(x,y)$$
$$C(\nu_0(x)) = \nu_0 h(x)$$

Having the description of S^t, we are always able to determine if $\nu_0(x)$ is an atom. Starting from x and y, we can also define whether the values $\nu_0(x)$ and $\nu_0(y)$ are equal. Thus, ν_0 is a strong constructivization of the atomic Boolean algebra $\mathfrak{B}_0$. The embedding φ of $\langle \mathfrak{A}, \nu \rangle$ into $\langle \mathfrak{B}_0, \nu_0 \rangle$ is recursive. Hence the element $\nu g(n)$, $n \in \mathcal{D}$, is $(\nu g(n) \land c) \lor (\nu g(n) \land C(c))$. From n we find $x(n)$ such that $\nu_0(x(n)) = \nu g(n)$. However, any element of $\mathfrak{A}$ is effectively defined as a union of elements of the generating tree. But the images of such elements are effectively defined. Consequently, the embedding of $\langle \mathfrak{A}, \nu \rangle$ into $\langle \mathfrak{B}_0, \nu_0 \rangle$ is constructive and a function $f_\mathfrak{A}$ providing the constructibility of the embedding ($\varphi\nu = \nu_0 f_\mathfrak{A}$) is uniformly constructed from $\mathfrak{A}$. To construct this function we need only information about the tree $\mathcal{D}$. □

Corollary 3.4.9 [43]. *For any infinite strongly constructive Boolean algebra $\langle \mathfrak{A}, \nu \rangle$ there exists a proper strongly constructive elementary extension.*

PROOF. If $\mathfrak{A}$ is atomic, Theorem 3.4.9 immediately yields the required conclusion. Otherwise, $\mathfrak{A}$ contains a nonzero atomless element a and the decomposition $\mathfrak{A} \cong \widehat{a} \times \widehat{C(a)}$ holds. Since $\langle \mathfrak{A}, \nu \rangle$ is a constructive algebra, ν induces the constructivizations ν_a and $\nu_{C(a)}$ such that $\langle \mathfrak{A}, \nu \rangle \cong \langle \widehat{a}, \nu_a \rangle \times \langle \widehat{C(a)}, \nu_{C(a)} \rangle$. In this case, ν_a and $\nu_{C(a)}$ are strong constructivizations because ν is a strong constructivization and all the predicates of the signature σ^* that are necessary for model completeness are defined on $x \leqslant a$ in $\widehat{a}$ in the same manner as in $\mathfrak{A}$. Using Proposition 3.4.3, we consider a proper elementary constructive extension $\langle \mathfrak{B}, \mu \rangle$ of $\langle \widehat{a}, \nu_a \rangle$. We note that $\langle \mathfrak{B}, \mu \rangle \times \langle \widehat{C(a)}, \nu_{C(a)} \rangle$ is a proper constructive elementary extension of $\langle \mathfrak{A}, \nu \rangle$. The strong constructibility of the product follows from the strong constructibility of its members. □

Corollary 3.4.10 [on constructive saturation of atomic Boolean algebras]. *For any infinite atomic strongly constructive Boolean algebra $\langle \mathfrak{A}, \nu \rangle$ there exists its constructive elementary extension to a strongly constructive saturated Boolean algebra.*

PROOF. We consider a strongly constructive infinite Boolean algebra $\langle \mathfrak{A}, \nu \rangle$. We now construct uniformly a computable sequence of strongly constructive Boolean algebras $\langle \mathfrak{A}_n, \nu_n \rangle$ and effective elementary embeddings $\varphi_n \colon \mathfrak{A}_n \preccurlyeq \mathfrak{A}_{n+1}$ expressed in terms of recursive functions f_n, i.e., $\varphi_n \nu_n = \nu_{n+1} f_n$, where $\langle \mathfrak{A}_0, \nu_0 \rangle = \langle \mathfrak{A}, \nu \rangle$. To construct the above sequence we use the construction of the theorem on elementary proper extensions of strongly constructive atomic Boolean algebras. In this case, for any n and $a \in \mathfrak{A}_n$ such that $\mathrm{ch}_2(a) = \infty$ there is an element $b \in \mathfrak{A}_{n+1}$ such that $\mathrm{ch}_2(a \setminus b) = \mathrm{ch}_2(a \wedge b) = \infty$. We define a model $\mathfrak{A}_\infty$ as the union of models of this elementary chain and define a numbering ν_∞ by setting $\nu_\infty c(n,m) \rightleftharpoons \nu_n(m)$. In view of the computability of elementary embeddings and the uniform strong constructibility of algebras $\langle \mathfrak{A}_n, \nu_n \rangle$, we conclude that $\langle \mathfrak{A}_\infty, \nu_\infty \rangle$ is a strongly constructive algebra. Since $\mathfrak{A}_{n+1}$ is a condensation of $\mathfrak{A}_n$ for any n, we conclude that $\bigcup \mathfrak{A}_n = \mathfrak{A}_\infty$ is a dense Boolean algebra. Consequently, $\mathfrak{A}_\infty$ is a saturated Boolean algebra. Thus, $\langle \mathfrak{A}_\infty, \nu_\infty \rangle$ is a strongly constructive saturated atomic Boolean algebra extending $\langle \mathfrak{A}_0, \nu_0 \rangle$, and consequently $\langle \mathfrak{A}, \nu \rangle$, in a constructive and elementary way.

Proposition 3.4.4. *From any elementary characteristic α of a Boolean algebra it is possible to uniformly construct a strongly constructive prime model $\langle \mathfrak{A}_\alpha, \nu_\alpha \rangle$ and a strongly constructive saturated model $\langle \mathfrak{B}_\alpha, \nu_\alpha \rangle$ of the theory of Boolean algebras of characteristic α.*

PROOF. We can obtain this assertion in view of the uniformity of constructing the corresponding families of types from characteristics and the uniformity of constructing a simple model and a saturated model from a family of types. □

Proposition 3.4.5. *If $\langle \mathfrak{A}, \nu \rangle$ is a strongly constructive Boolean algebra, $\mathrm{ch}(\mathfrak{A}) = (n, \infty, 0)$, $\langle \mathfrak{B}, \mu \rangle$ is a strongly constructive Boolean algebra, and $\mathrm{ch}(\mathfrak{B}) = (n+1, 1, 0)$, then a constructive elementary embedding of $\langle \mathfrak{B}, \mu \rangle$ into the product $\langle \mathfrak{A}, \nu \rangle \times \langle \mathfrak{B}, \mu \rangle$ can be uniformly constructed from $\langle \mathfrak{A}, \nu \rangle$, $\langle \mathfrak{B}, \mu \rangle$, and n.*

PROOF. By the strong constructibility of the model $\langle \mathfrak{B}, \mu \rangle$, from a number m of the element of $\langle \mathfrak{B}, \mu \rangle$ we can effectively recognize if $\mathrm{ch}_1(\mu m) < n+1$ or $\mathrm{ch}_1(\mu m) = n+1$ holds. The embedding

$$\varphi(\mu m) = \begin{cases} \langle 0, \mu m \rangle, & \mathrm{ch}_1(\mu m) < n+1 \\ \langle 1, \mu m \rangle, & \mathrm{ch}_1(\mu m) = n+1 \end{cases}$$

3.4. Decidable Boolean Algebras

of $\mathfrak{B}$ into $\mathfrak{A} \times \mathfrak{B}$ becomes an isomorphic embedding in the enrichments of these algebras to the signature σ_{4n+1}. Since $\mathfrak{B}$ and $\mathfrak{A} \times \mathfrak{B}$ have characteristics $(n+1,1,0)$, the model completeness of these enrichments means that φ is an elementary embedding. Its constructibility is obvious in view of the aforesaid. The strong constructibility of the product follows from the same property of its members. □

It is not so easy to construct an embedding in the case of characteristic $(n+1,0,1)$.

Proposition 3.4.6. *From n and a strongly constructive Boolean algebra $\langle \mathfrak{B}, \mu \rangle$ of characteristic $(n+1,0,1)$ it is possible to construct effectively a strongly constructive prime Boolean algebra $\langle \mathfrak{A}, \nu \rangle$ of characteristic $(n, \infty, 0)$ and a constructive elementary embedding of $\langle \mathfrak{B}, \mu \rangle$ into the product $\langle \mathfrak{A}, \nu \rangle \times \langle \mathfrak{B}, \mu \rangle$.*

PROOF. In view of the uniformity of constructing a strongly constructive prime model from its elementary characteristic appearing in Proposition 3.4.4, we obtain the existence and construction of $\langle \mathfrak{A}, \nu \rangle$ uniformly in n; moreover, $\mathrm{ch}\,(\nu f(R^l(0))) = (n, \infty, 0)$ for $l \in \mathbb{N}$, where $(\mathcal{D}, \nu f)$ is the tree generating $\mathfrak{A}$ and f is a recursive function.

For a strongly constructive Boolean algebra $\langle \mathfrak{B}, \mu \rangle$ it is not hard to construct its generating tree $(\mathcal{D}', \varphi')$ and a recursive function g such that $\varphi'(n) = \nu g(n)$ for any $n \in \mathcal{D}$; moreover, the elements $\nu g(R^k(0))$, $k \in \mathbb{N}$, have characteristic $(n+1,0,1)$. By the assumptions on the strong constructibility of $\langle \mathfrak{B}, \mu \rangle$ and $\mathrm{ch}(\mathfrak{B}) = (n+1,0,1)$, such a generating tree is uniformly constructed from $\langle \mathfrak{B}, \mu \rangle$. We now construct the required embedding. Since the tree $(\mathcal{D}', \varphi')$ generates $\mathfrak{B}$, from any number m we can find uniformly elements of the canonical representation $n_0, \ldots, n_k \in \mathcal{D}'$ such that $n_0, \ldots, n_k$ are pairwise inconsistent in $\mathcal{D}'$ and $\vee \mu g(n_i) = \mu m$. If, among $\{n_i \mid i \leqslant k\}$, there are no elements of the form $R^l(0)$, $l \in \mathbb{N}$, we set $\varphi(\mu m) = \langle 0, \mu m \rangle$. If such elements exist, we define φ-images for all $\mu g(n_i)$ in such a way that with μm the union of images is associated:

$$\varphi(\mu g(n_i)) = \begin{cases} \langle 0, \mu g(n_i) \rangle & \text{if } n_i \text{ is not of the form } R^l(0) \\ \langle \nu f R^l(0), \mu g(n_i) \rangle & \text{if } n_i = R^l(0) \end{cases}$$

It is easy to see that φ is a constructive embedding constructed uniformly and φ is an isomorphic embedding in the enrichment to the signature σ_{4n+4}. In view of the model completeness of this enrichment, the embedding φ is elementary. □

Proposition 3.4.7. *From any n and a strongly constructive Boolean algebra $\langle \mathfrak{A}, \nu \rangle$ such that $\mathrm{ch}_1(\mathfrak{A}) = n$ it is possible to construct effectively an elementary extension $\langle \mathfrak{B}, \mu \rangle$ and an elementary embedding φ such that $\mathrm{ch}_2(\mathfrak{A}) = \infty$ implies the existence of an element $b \in \mathfrak{B}$ such that $\mathrm{ch}_1(b) = \mathrm{ch}_1(C(b)) = n$ and $\mathrm{ch}_2(b) = \mathrm{ch}_2(C(b)) = \infty$.*

PROOF. The assertion is proved by a slight modification of the proof of Theorem 3.4.9 in the case $n = 1$. The difference consists of using counters (that are similar to those for atoms) not only for atoms of the nth quotient algebra by some Ershov–Tarski ideal I_n, but for characteristics $k < n$, where elements of characteristic $\mathrm{ch}_1(a) = k$ are included in c or $C(c)$ respectively. □

Propositions 3.4.5–3.4.7 yield the following claim.

Proposition 3.4.8. *From any n and a strongly constructive Boolean algebra $\langle \mathfrak{A}, \nu \rangle$ it is possible to construct effectively its constructive elementary extension to a strongly constructive Boolean algebra $\langle \mathfrak{B}, \mu \rangle$; moreover, if $\mathrm{ch}_1(\mathfrak{A}) = n$, then $\mathfrak{B}$ is a condensation of $\mathfrak{A}$.*

PROOF. We consider the set of sequences

$$A_n \leftrightharpoons \{(m, k, (k+1, 1, 0)) \mid \mathrm{ch}(\nu m) = (k+1, 1, 0) \text{ and } k < n\}$$
$$\cup \{(m, k, (k+1, 0\ 1)) \mid \mathrm{ch}(\nu m) = (k+1, 0, 1) \text{ and } k < n\}$$
$$\cup \{(m, k, (k, \infty, 0)) \mid \mathrm{ch}_1(\nu m) = k \text{ and } k \leqslant n\}$$

Since $\langle \mathfrak{A}, \nu \rangle$ is a strongly constructive system, the set A_n is recursive. We can construct an enumeration $\alpha_0, \ldots, \alpha_n, \ldots$ of all sequences of A_n. Using the above constructions, we construct uniformly a sequence of strongly constructive models and constructive embeddings

$$\langle \mathfrak{A}, \nu \rangle = \langle \mathfrak{A}_0, \nu_0 \rangle \preccurlyeq \langle \mathfrak{A}_1, \nu_1 \rangle \preccurlyeq \ldots \preccurlyeq \langle \mathfrak{A}_n, \nu_n \rangle \preccurlyeq \ldots$$

so that the embeddings $\varphi \colon \mathfrak{A}_n \to \mathfrak{A}_{n+1}$ and $\varphi_n \colon \mathfrak{A}_0 \to \mathfrak{A}_n$ are represented in terms of the recursive functions g_n and f_n as follows. Let $\langle \mathfrak{A}_0, \nu_0 \rangle \leftrightharpoons \langle \mathfrak{A}, \nu \rangle$. We assume that

$$\langle \mathfrak{A}_0, \nu_0 \rangle \preccurlyeq \langle \mathfrak{A}_1, \nu_1 \rangle \preccurlyeq \ldots \preccurlyeq \langle \mathfrak{A}_n, \nu_n \rangle$$

are already constructed and consider the sequence α_n. If $\alpha_n = (m, k, (k+1, 1, 0))$, then we apply the construction of Proposition 3.4.5 to the Boolean algebra $\langle \widehat{\nu_n f_n(m)}, \nu_n \upharpoonright \nu f_n(m) \rangle$ and obtain the required elementary extension. Multiplying it by $(\widehat{C(\nu m)}, \nu \upharpoonright \widehat{C(\nu m)})$, we arrive at the desired extension.

3.4. Decidable Boolean Algebras

If $\alpha_n = (m, k, (k+1, 0, 1))$, then the construction is similar to that in the previous case. But, in this case, we apply the construction of Proposition 3.4.6 to $\widehat{\nu(m)}$.

If $\alpha_n = (m, k, (k, \infty, 0))$, then we apply the construction of Proposition 3.4.7 to $(\widehat{\nu m}, \nu \upharpoonright \widehat{\nu(m)})$ and multiply the strongly constructive model obtained as a result by $(\widehat{C(\nu m)}, \nu \upharpoonright \widehat{C(\nu m)})$.

Taking $\langle \mathfrak{A}_\infty, \nu_\infty \rangle$ as the union of this computable chain of elementary constructive extensions, we conclude that $\mathfrak{A}_\infty$ is a condensation of $\mathfrak{A}$ and $\langle \mathfrak{A}_\infty, \nu_\infty \rangle$ is a constructive elementary extension of $\mathfrak{A}$. □

Applying the construction of Proposition 3.4.8 ω times, from any strongly constructive model $\langle \mathfrak{A}, \nu \rangle$ we obtain its constructive elementary extension to a dense strongly constructive Boolean algebra. However, dense Boolean algebras are countably saturated. Thus, we have proved the following theorem.

Theorem 3.4.10. *Any strongly constructive Boolean algebra $\langle \mathfrak{A}, \nu \rangle$ such that $\mathrm{ch}_1(\mathfrak{A}) < \infty$ has a constructive elementary extension to a strongly constructive saturated Boolean algebra.*

We now consider the case in which the first characteristic is infinite. For elements of a finite characteristic and Boolean algebras of an infinite characteristic the density conditions can be realized in extensions with the help of the constructions described in the above propositions. In the case of infinite characteristics, in order to obtain the density conditions, we prove the following proposition using the same method as above.

Proposition 3.4.9. *From any strongly constructive Boolean algebra $\langle \mathfrak{A}, \nu \rangle$ such that $\mathrm{ch}_1(\mathfrak{A}) = \infty$ it is possible to construct effectively its constructive elementary extension $\langle \mathfrak{B}, \mu \rangle$ containing $c \in |\mathfrak{B}|$ such that for any $a \in \mathfrak{A}$ such that $\mathrm{ch}_1(a) = \infty$ the elementary characteristics of $a \wedge c$ and $a \wedge C(c)$ coincide and are equal to $(\infty, 0, 0)$.*

PROOF. We use a modification of the construction of Theorem 3.4.9 on elementary extensions of atomic Boolean algebras. But instead of the existence condition for four atoms, we verify the condition that $\mathrm{ch}_1(x) > t_n + 4$. In addition, the counter indicates the minimal characteristic of elements added under c and $C(c)$ but not the number of elements. □

Proposition 3.4.9 and Theorem 3.4.10 imply the following claim.

Theorem 3.4.11 [on strongly constructive extensions]. *Any strongly constructive Boolean algebra has a constructive elementary extension to a strongly constructive saturated Boolean algebra.*

PROOF. We construct effectively a chain of condensations $\langle \mathfrak{A}_0, \nu_0 \rangle \preccurlyeq \langle \mathfrak{A}_1, \nu_1 \rangle \preccurlyeq \ldots \preccurlyeq \langle \mathfrak{A}_i, \nu_i \rangle \preccurlyeq \ldots$ so that $\langle \mathfrak{A}_i, \nu_i \rangle$ is a computable sequence of constructive models, f_n is a computable sequence of recursive functions, and $\varphi_n \colon \mathfrak{A}_n \preccurlyeq \mathfrak{A}_{n+1}$ is an elementary embedding of $\mathfrak{A}_n$ into $\mathfrak{A}_{n+1}$ such that $\mathfrak{A}_{n+1}$ is a condensation of $\mathfrak{A}_n$ and $\varphi_n \nu_n(m) = \nu_{n+1} f_n(m)$ for any $m, n \in \mathbb{N}$. Defining $\langle \mathfrak{A}_\infty, \nu_\infty \rangle$ as the union of this chain, we obtain the required strongly constructive elementary extension of $\langle \mathfrak{A}, \nu \rangle$, where $\langle \mathfrak{A}_0, \nu_0 \rangle = \langle \mathfrak{A}, \nu \rangle$, $\mathfrak{A}_\infty = \varinjlim \mathfrak{A}_n$, and $\nu_\infty(c(n,m)) \leftrightharpoons \nu_n(m)$. The saturation of $\mathfrak{A}_\infty$ follows from its density. □

Corollary 3.4.11. *Any strongly constructive Boolean algebra $\langle \mathfrak{A}, \nu \rangle$ has a constructive elementary extension to a strongly constructive homogeneous Boolean algebra.*

3.5. Restricted Fragments of the Theory of Boolean Algebras and Decidable Algebras

The decidability problem for various classes of algorithmic problems on effectively prescribed systems is one of the main purposes of the theory of constructive models. Therefore, studying constructive models, it is important to investigate correlations between the decidability of different algorithmic problems in a given constructive model as well as in its various constructivizations. We study the interconnection between the decidability of all elementary properties of a constructive Boolean algebra and that of all atomic properties of enrichments of the Boolean algebra, i.e., properties that can be expressed in terms of quantifier-free formulas of the corresponding extended signatures.

From the corollary to the theorem on reducing formulas of restricted theories to ∃-formulas and ∀-formulas but in the enrichment of the signature of Boolean algebras with predicates of σ_n we obtain the following proposition.

Proposition 3.5.1. *An enumerated Boolean algebra $\langle \mathfrak{A}, \nu \rangle$ is n-constructive if and only if ν is a constructivization of the enrichment $\mathfrak{A}^{\sigma_n}$ of a Boolean algebra $\mathfrak{A}$ to a system of the signature σ_n of the theory BA_n^*.*

3.5. Restricted Fragments of the Theory of Boolean Algebras

PROOF. The necessity follows from the possibility of describing a predicate of the signature σ_n in terms of Σ_n-formulas or Π_n-formulas of the signature of Boolean algebras. The sufficiency follows from the fact that any Σ_n-formula is equivalent to a $\forall$-formula and an $\exists$-formula of the signature σ_n. □

Answering the question whether a Boolean algebra is decidable, it is important to study the existence of constructivizations of enrichments of a Boolean algebra $\mathfrak{B}$ to algebraic systems of the restricted signatures σ_n, $n \geqslant 0$. We first study the question concerning the decidability of a Boolean algebra if its first characteristic is 0 or 1.

Proposition 3.5.2. *If a Boolean algebra $\mathfrak{A}$ such that $\mathrm{ch}_1(\mathfrak{A}) = 0$ admits a constructivization with recursive set of numbers of atoms, then the constructivization is strong and the algebra $\mathfrak{A}$ itself is decidable.*

PROOF. If the first characteristic of $\mathfrak{A}$ is zero, then $\mathfrak{A}$ can be represented as the product of the atomic part and the atomless part: $\mathfrak{A} = \widehat{a} \times \widehat{C(a)}$. Any constructivization of an atomless Boolean algebra is strong since its theory is model-complete and decidable. Any constructivization of an atomic Boolean algebra with recursive set of numbers of atoms can be regarded as a constructivization of an enrichment to a system of the signature σ_1. An atomic Boolean algebra of the signature σ_1 has a model-complete and decidable theory. Consequently, the constructivizations of $\mathfrak{A}$ of the signature σ_1 are strong and the Boolean algebra $\mathfrak{A}$ is decidable. □

We now study the existence problem for Boolean algebras such that the set of numbers of atoms of any of its constructivizations is not recursive. We note that such algebras are not decidable. To clarify the connections between the notions of constructivizability and strong constructivizability, with every Boolean algebra $\mathfrak{B}$ we associate a subset $\mathbb{N}(\mathfrak{B}) \subseteq \mathbb{N}$ invariant under isomorphisms. Having estimated its algorithmic complexity, we arrive at the necessary condition for the strong constructivizability. Let $\mathfrak{B}$ be a Boolean algebra and let $F_n(\mathfrak{B})$, $n \in \mathbb{N}$, be a sequence of its Frechét ideal. On $\mathfrak{B}$ we introduce the following relations:

(1) $\alpha_n(x) \rightleftharpoons x \in F_n(\mathfrak{B})$,

(2) $\lambda_n(x) \rightleftharpoons x \notin F_n(\mathfrak{B})$,

(3) $\mathrm{Atomistic}_n(x) \rightleftharpoons x/F_n(\mathfrak{B})$ is an atomic element of $\mathfrak{B}/F_n(\mathfrak{B})$,

(4) $\gamma_n^k(x) \rightleftharpoons x \notin F_{n+k}(\mathfrak{B})$ & $(\forall y)$ $(y \leqslant x \Rightarrow (y/F_k(\mathfrak{B})$ is not an atomic element of $\mathfrak{B}/F_k(\mathfrak{B})$ or $y \in F_{n+k}(\mathfrak{B}))$ and there exist infinitely many pairwise disjoint elements z located under x belonging to $F_{n+k}(\mathfrak{B})$ but not to $F_{n+k-1}(\mathfrak{B})$,

(5) $\Phi_n^k \rightleftharpoons (\exists x)\gamma_n^k(x)$.

The following lemma is obvious.

Lemma 3.5.1. *The following assertions hold*:

(a) *if* $\mathfrak{A} \cong \mathfrak{B}$, *then* $\mathfrak{A} \vDash \Phi_n^k \Rightarrow \mathfrak{B} \vDash \Phi_n^k$,

(b) $\mathfrak{B}_{\omega^{n+k}+\omega^k \times \eta} \vDash \Phi_n^k$, $n + k \geqslant 1$,

(c) *if* $k \geqslant 1$, *then* $\mathfrak{B}_{\sum_{m \in X}(\omega^{m+k}+\omega^k \times \eta)} \vDash \Phi_n^k \Leftrightarrow n \in X$

The following lemma is easily proved by induction.

Lemma 3.5.2. *The following assertions hold*:

(a) $\alpha_0(x) \Leftrightarrow x = 0$

$$\alpha_{n+1}(x) \Leftrightarrow (\exists m \in \mathbb{N})(\exists x_1 \ldots x_m)\left(x = \bigvee_{i=1}^{m} x_i \ \& \underset{i \neq j}{\&} x_i \wedge x_j = 0 \right.$$
$$\left. \& \underset{i=1}{\overset{m}{\&}} (\forall y)(y \leqslant x_i \Rightarrow (\alpha_n(y) \vee \alpha_n(C(y) \wedge x_i)))\right),$$

(b) $\lambda_n(x) \Leftrightarrow \neg \alpha_n(x)$,

(c) $\text{Atomistic}_n(x) \Leftrightarrow (\forall y)(y \leqslant x \Rightarrow (\alpha_n(y) \vee (\exists z)(z \leqslant y \ \& \ (\forall a)(a \leqslant z \Rightarrow (\alpha_n(a) \vee \alpha_n(z \wedge C(a)))) \ \& \ \neg \alpha_n(z))))$,

(d) $\gamma_n^k(x) \Leftrightarrow \lambda_{n+k}(x) \ \& \ (\forall m \in \mathbb{N})(\exists x_1 \ldots x_m) \ (\underset{i \neq j}{\&} x_i \wedge x_j = 0 \ \&$

$\underset{i=1}{\overset{m}{\&}}(\lambda_{n+k-1}(x_i) \ \& \ \alpha_{n+k}(x_i) \ \& \ x_i \leqslant x)) \ \& \ (\forall y)(y \leqslant x$

$\rightarrow (\neg \text{Atomistic}_k(y) \vee \alpha_{n+k}(y)))$.

For desired invariants we take the sets $\mathbb{N}_k(\mathfrak{B}) \rightleftharpoons \{m \mid \mathfrak{B} \vDash \Phi_m^k\}$ and estimate their algorithmic complexity in the case of strongly constructivizable Boolean algebras.

Proposition 3.5.3. *If $\mathfrak{B}$ is a strongly constructivizable Boolean algebra, then* $\mathbb{N}_k(\mathfrak{B}) \in \Phi(2k+1, 2)$.

3.5. Restricted Fragments of the Theory of Boolean Algebras 217

PROOF. Let ν be a strong constructivization of $\mathfrak{B}$. From the strong constructibility of $\langle \mathfrak{B}, \nu \rangle$ it follows that α_1 can be represented as a Σ_1^0-form.

Applying by induction the Tarski–Kuratowski algorithm [178] to α_n, Atomistic$_n$, and Φ_n^k and using Lemma 3.5.2, we obtain the representations of $\alpha_{n+1}(x)$ as a Σ_{2n+1}^0-form and of λ_{n+1} as a Π_{2n+1}^0-form. In the representation of Φ_n^k as a $\Sigma_{2n+2k+1}^0$-form, we replace the recursive part P by a recursive function f such that

$$P(k,n,z_1,\ldots,z_{2n+2k+1}) \Leftrightarrow f(\langle z_1,\ldots,z_{2n+2k+1}\rangle) = 1$$

It is clear that we can choose the same function f for all Φ_n^k. In view of this representation, $\mathbb{N}_k(\mathfrak{B}) \in \mathfrak{D}(2k+1, 2)$. □

We proceed to constructing a constructivizable atomic Boolean algebra which is not strongly constructivizable. Let O stand for the set of all countable ordinals.

Proposition 3.5.4. *If* $\Delta \colon \mathbb{N}^3 \to O$, $\omega^{2m+k-1} \leqslant \Delta(m,i,h) \leqslant \omega^{2m+k}$ *then*

$$\mathfrak{B} \rightleftharpoons \mathfrak{B}_{\sum_n \sum_i \sum_h (\Delta(n,i,h)+1+\omega^k \times \eta)} \vDash \Phi_{2m}^k \Leftrightarrow (\exists i)^{\neg}(\exists^{\omega} h)(\Delta(m,i,h) = \omega^{2m+k})$$

PROOF. Let $\mathfrak{B} \vDash \Phi_{2m}^k$, but $(\forall i)(\exists^{\omega} h)(\Delta(m,i,h) = \omega^{2m+k})$. If $\mathfrak{B} \vDash \Phi_{2m}^k$, then there exists an element $b = [\alpha_1, \alpha_2[\cup \ldots \cup [\alpha_n, \alpha_{n+1}[$ such that $\mathfrak{B} \vDash \gamma_{2m}^k(b)$. By definition, b contains no elements of $F_{2m+k+1}(\mathfrak{B})$, but it contains ω pairwise disjoint elements of $F_{2m+k} \setminus F_{2m+k-1}$. For one of the intervals $[\alpha_i, \alpha_{i+1}[$, say $[\alpha_1, \alpha_2[$, this assertion is also true. Then $[\alpha_1, \alpha_2[$ contains ω intervals $[\alpha, \beta[\subseteq [\alpha_1, \alpha_2[$ that are pairwise disjoint and are well ordered by a type which exceeds ω^{2m+k-1} but is strictly less than ω^{2m+k}. However, this is possible only if $[\alpha_1, \alpha_2[$ contains an interval of the form

$$[\gamma + \sum_{h \leqslant h_0}(\Delta(m,i_0,h)+1+\omega^k \times \eta+1), \gamma + \sum_{h \in \omega}(\Delta(m,i_0,h)+1+\omega^k \times \eta)+1[$$

for some γ, i_0, and h_0. In view of the equality $(\forall i)(\exists^{\omega} h)(\Delta(m,i,h) = \omega^{2m+k})$, each such interval contains an interval $[\alpha, \beta[$ which is a well-ordered set of type ω^{2m+k}. Therefore, there is an element contained in $[\alpha_1, \alpha_2[$ (consequently, in b) and in $F_{2m+k+1} \setminus F_{2m+k}$. The contradiction obtained shows that $(\exists i)^{\neg}(\exists^{\omega} h)(\Delta(m,i,h) = \omega^{2m+k})$ in the case $\mathfrak{B} \vDash \Phi_{2m}^k$.

We now prove the inverse implication. Let $(\exists i)^{\neg}(\exists^{\omega} h)(\Delta(m,i,h) = \omega^{2m+k})$. This means that there is i_0 such that there exists only a finite number of $h_1, \ldots, h_n$ such that $\Delta(m,i_0,h_i) = \omega^{2m+k}$ and $\Delta(m,i_0,h) <$

ω^{2m+k} for the remaining h. We define

$$h_0 = \max\{h_i \mid i \in \{1,\ldots,n\}\} + 1$$
$$b \rightleftharpoons [\gamma + \sum_{h \leqslant h_0}(\Delta(m,i_0,h), +1 + \omega^k \times \eta) + 1, \ \gamma$$
$$+ \sum_{h \in \omega}(\Delta(m,i_0,h) + 1 + \omega^k \times \eta) + 1[$$

All of the intervals $[\alpha,\beta[$ from b, being well-ordered sets, have the ordering type strictly less than ω^{2m+k}. Therefore, b contains no elements of $F_{2m+k+1} \setminus F_{2m+k}$. We note that $[\alpha,\beta[$ is an atomic element of the kth quotient provided that it contains no interval of type $\omega^k \times \eta$. By the definition of the order just introduced, this interval is a well-ordered set. By the properties of Boolean algebras constructed from a linearly ordered set, it belongs to F_{2m+k}. Furthermore, any element

$$[\gamma + \sum_{h \leqslant t}(\Delta(m,i_0,h) + 1 + \omega^k \times \eta) + 1, \ \gamma + \sum_{h \leqslant t}(\Delta(m,i_0,h) + 1 + \omega^k \times \eta) + 1[$$

for $t \geqslant h_0$ is contained in b and belongs to $F_{2m+k} \setminus F_{2m+k-1}$. Therefore, $\mathfrak{B} \models \Phi_{2m}^k$. □

We first construct a series of constructive linear orders and, using them, obtain the required constructivizable but not strongly constructivizable Boolean algebra. To this end, we consider a recursive linear order $(\mathbb{N}, \subset_k)$ of the type $1 + \omega^k \times \eta$ and a series of recursive linear orders $(M_k, <_k)$ of type

$$\sum_m \sum_i \sum_h (\Delta(m,i,h) + 1 + \omega^k \times \eta)$$

where $\Delta(m,i,h) = \omega^{2m+k}$ and

$$\begin{aligned}M_k = \{&(m,i,h,z_1,\ldots,z_{2m+k},l,q) \mid (l = 0 \Rightarrow (q = -1 \\ & \& \ z_1,\ldots,z_{2m+k} \in \mathbb{N})) \\ & \& \ (l = 1 \Rightarrow ((z_1,\ldots,z_{2m+k}) = (-1,\ldots,-1) \ \& \ q \in \mathbb{N})) \\ & \& \ i,m,h \in \mathbb{N} \ \& \ l \in \{0,1\}\}\end{aligned}$$

3.5. Restricted Fragments of the Theory of Boolean Algebras

The order $<_k$ is defined as follows:

$$(m, i, h, z_1, \ldots, z_{2m+k}, l, q) <_k (\widehat{m}, \widehat{i}, \widehat{h}, \widehat{z}_1, \ldots, \widehat{z}_{2\widehat{m}+k}, \widehat{l}, \widehat{q})$$
$$\Leftrightarrow \langle m, i, h, l\rangle <_{\text{lex}} \langle \widehat{m}, \widehat{i}, \widehat{h}, \widehat{l}\rangle \vee (\langle m, i, h, l\rangle$$
$$= \langle \widehat{m}, \widehat{i}, \widehat{h}, \widehat{l}\rangle \,\&\, ((l = 1 \,\&\, q \subset_k \widehat{q})$$
$$\vee (l = 0 \,\&\, (z_1, \ldots, z_{2m+k}) <_{\text{lex}} (\widehat{z}_1, \ldots, \widehat{z}_{2m+k}))))$$

Taking the quotient of the order $\langle M_k, <_k\rangle$, we obtain a recursive order of type $\sum_m \sum_i \sum_h (\Delta^k(m, i, h) + 1 + \omega^k \times \eta)$, where $\omega^{2m+k-1} \leqslant \Delta^k(m, i, h) \leqslant \omega^{2m+k}$. To this end, we consider a recursive function I with large amplitude [109] (i.e., $(\forall x)(\exists^\omega y)(I(y) = x)$) and set

$$\langle m, i, h, z_1, \ldots, z_{2m+k}, l, q\rangle \leqslant^k \langle \widehat{m}, \widehat{i}, \widehat{h}, \widehat{z}_1, \ldots, \widehat{z}_{2\widehat{m}+k}, \widehat{l}, \widehat{q}\rangle$$

if

$$\langle m, i, h, z_1, \ldots, z_{2m+k}, l, q\rangle <_k \langle \widehat{m}, \widehat{i}, \widehat{h}, \widehat{z}_1, \ldots, \widehat{z}_{2\widehat{m}+k}, \widehat{l}, \widehat{q}\rangle$$
$$\vee (m = \widehat{m} \,\&\, i = \widehat{i} \,\&\, h = \widehat{h} \,\&\, l = \widehat{l} = 0 \,\&\, q = \widehat{q} = -1 \,\&$$
$$\underset{i=1}{\overset{2m+k-1}{\&}} z_i = \widehat{z}_i \,\&\, \&\, (\forall n)(\widehat{z}_{2m+k} \leqslant n < z_{2m+k}$$
$$\Rightarrow r_k(\langle i, I(h), z_1, \ldots, z_{2m+k-1}, n\rangle) \neq 1))$$

where r_k is the recursive function from Proposition 3.1.4.

Let $\eta_k \rightleftharpoons \{\langle x, y\rangle \mid x \leqslant^k y \,\&\, y \leqslant^k x\}$. Then (all the constructions are effective) $\langle M_k/\eta_k, \leqslant^k\rangle$ is a constructive linear order with respect to an effective numbering of all elements of M_k. Introduce the notation

$$\Delta^k(m, i, h) \rightleftharpoons$$
$$\langle \{\langle m, i, h, z_1, \ldots, z_{2m+k}, 0, -1\rangle \mid \langle z_1, \ldots, z_{2m+k}\rangle \in \mathbb{N}^{2m+k}\}/\eta_k, \leqslant^k\rangle$$

Then

$$\mathfrak{M}_k \rightleftharpoons \langle M_k/\eta_k, \leqslant^k\rangle \cong \sum_m \sum_i \sum_h (\Delta^k(m, i, h) + 1 + \omega^k \times \eta)$$

Lemma 3.5.3. *The constructed linear orders possess the following properties*:

(1) $\omega^{2m+k-1} \leqslant \Delta^k(m, i, h) \leqslant \omega^{2m+k}$,

(2) $\Delta^k(m, i, h) = \omega^{2m+k} \Leftrightarrow (\exists^\omega z_1) \ldots (\exists^\omega z_{2m+k}) r_k(\langle i, I(h), z_1, \ldots, z_{2m+k}\rangle) = 1$,

(3) $\Delta^k(m,i,h) = \Delta^k(m,i,\widehat{h})$ if $I(h) = I(\widehat{h})$.

PROOF. After taking a quotient algebra, only elements different on the $(2m+k+3)$-coordinate can be different. Hence property (1) is obvious. Property (3) follows from the fact that the quotient of the summands $\Delta(m,i,h)$ and $\Delta(m,i,\widehat{h})$ for $I(h) = I(\widehat{h})$ is defined in the same way. Let us prove (2). Fix m, i, and h. Let

$$\mu_1(z_1, \ldots, z_{2m+k-1})$$
$$\rightleftharpoons \langle \{\langle m, i, h, z_1, \ldots, z_{2m+k}, 0, -1\rangle \mid z_{2m+k} \in \mathbb{N}^{2m+k}\}/\eta_k, \leqslant^k\rangle.$$

For $r = 2, 3, \ldots, 2m+k$ we set by induction

$$\mu_r(z_1, \ldots, z_{2m+k-r}) \rightleftharpoons \sum_{z \in \mathbb{N}} \mu_{r-1}(z_1, \ldots, z_{2m+k-r}, z)$$

For $r = 1, \ldots, 2m+k$ we introduce the notation

$$A_r(z_1, \ldots, z_{2m+k-r}) \rightleftharpoons (\exists^\omega z_{2m+k-r+1}) \cdots$$
$$(\exists^\omega z_{2m+k})(r_k(\langle i, I(h), z_1, \ldots, z_{2m+k}\rangle) = 1)$$

We show that for r ($1 \leqslant r \leqslant 2m+k$) the following relations hold:

$\langle 1 \rangle$ $(\forall z_1) \ldots (\forall z_{2m+k-r})(A_r(z_1, \ldots, z_{2m+k-r}) \Rightarrow \mu_r(z_1, \ldots, z_{2m+k-r}) = \omega^r)$,

$\langle 2 \rangle$ $(\forall z_1) \ldots (\forall z_{2m+k-r})(\neg A_r(z_1, \ldots, z_{2m+k-r})) \Rightarrow \omega^{r-1} \leqslant \mu_r(z_1, \ldots, z_{2m+k-r}) < \omega^r$.

Let $r = 1$. We fix $m, i, h, z_1, \ldots, z_{2m+k-1}$. If

$$\neg(\exists^\omega z_{2m+k})(r_k(\langle i, I(h), z_1, \ldots, z_{2m+k}\rangle) = 1)$$

then there exists only a finite number of z_{2m+k} such that

$$r_k(\langle i, I(h), z_1, \ldots, z_{2m+k}\rangle) = 1$$

By definition, almost all of the elements $\langle m, i, h, z_1, \ldots, z_{2m+k}, l, q\rangle$, where $z_{2m+k} \in \mathbb{N}$, are identified and $1 \leqslant \mu_1 < \omega$. If

$$(\exists^\omega z_{2m+k})(r_k(\langle i, I(h), z_1, \ldots, z_{2m+k}\rangle) = 1)$$

then ω elements are not identified and $\mu_1(z_1, \ldots, z_{2m+k-1}) = \omega$. Thereby, we have shown that the basis of induction holds for the relations $\langle 1 \rangle$ and $\langle 2 \rangle$. Let $\langle 1 \rangle$ and $\langle 2 \rangle$ hold for $r = s-1$. Prove them for $r = s$. It is clear that

$$\sum_z \mu_s(z_1, \ldots, z_{2m+k-s}, z) = \omega^s$$

3.5. Restricted Fragments of the Theory of Boolean Algebras

if and only if there exist ω elements $z_{2m+k-s+1} \in \mathbb{N}$ such that

$$\mu_{s-1}(z_1, \ldots, z_{2m+k-s+1}) = \omega^{s-1}$$

The step of induction is proved.

Putting $r = 2m + k$, we complete the proof. □

Lemma 3.5.4. *The following relation holds*:

$$\mathfrak{B}_{\mathfrak{M}_k} \models \Phi_{2m}^k \Leftrightarrow (\exists i)(\forall j)^{\neg}(\exists^\omega z_1) \ldots (\exists^\omega z_{2m+k})$$
$$r_k(\langle i, j, z_1, \ldots, z_{2m+k}\rangle) = 1.$$

PROOF. To establish necessity we assume the contrary:

$$(\forall i)(\exists j)(\exists^\omega z_1) \ldots (\exists^\omega z_{2m+k}) \, r_k(\langle i, j, z_1, \ldots, z_{2m+k}\rangle) = 1$$

Then $(\forall i)(\exists j)\Delta^k(m, i, j) = \omega^{2m+k}$ by Lemma 3.5.3. Since

$$(\forall j_1)(I(j_1) = I(j) \Rightarrow \Delta^k(m, i, j_1) = \Delta^k(m, i, j))$$

we have $(\forall i)(\exists^\omega j)(\Delta^k(m, i, j) = \omega^{2m+k})$ and $\mathfrak{B}_{\mathfrak{M}_k} \models^{\neg} \Phi_{2m}^k$ by Proposition 3.5.4, which yields a contradiction.

To prove the sufficiency we note that

$$(\exists i)(\forall j)^{\neg}(\exists^\omega z_1) \ldots (\exists^\omega z_{2m+k}) r_k(\langle i, j, z_1, \ldots, z_{2m+k}\rangle) = 1$$

implies $(\exists i)(\forall j)\Delta(m, i, j) < \omega^{2m+k}$ by Lemma 3.5.3. Consequently,

$$(\exists i)^{\neg}(\exists^\omega j)\Delta(m, i, j) = \omega^{2m+k}$$

and $\mathfrak{B}_{\mathfrak{M}_k} \models \Phi_{2m}^k$ by Proposition 3.5.4. □

Lemma 3.5.5. $\{n \mid \mathfrak{B}_{\mathfrak{M}_k} \models \Phi_n^k\} \dot\notin \Phi(2k+1, 2)$.

PROOF. By Lemma 3.5.4 and Proposition 3.1.4, for m the following equivalences are true:

$$2m \in \{n \mid \mathfrak{B}_{\mathfrak{M}_k} \models \Phi_n^k\} \Leftrightarrow (\exists i)(\forall j)^{\neg}(\exists^\omega z_1) \ldots$$
$$(\exists^\omega z_{2m+k}) \, r_k(\langle i, j, z_1, \ldots, z_{2m+k}\rangle) = 1$$
$$\Leftrightarrow 2m \in X(2k+1, 2)$$

By Proposition 3.1.3, $\{n \mid \mathfrak{B}_{\mathfrak{M}_k} \models \Phi_n^k\} \dot\notin \Phi(2k+1, 2)$. □

Theorem 3.5.1. *For any $k \in \mathbb{N}$ there exists a k-atomic, but not a $(k+1)$-atomic Boolean algebra which is constructivizable but not strongly constructivizable.*

PROOF. It is easy to see that the Boolean algebra $\mathfrak{B}_{\mathfrak{M}_k}$ is k-atomic but not $(k+1)$-atomic. By Proposition 3.2.1, $\mathfrak{B}_{\mathfrak{M}_k}$ is constructivizable. However, by Lemma 3.5.5 and Proposition 3.5.3, it is not strongly constructivizable. □

We now study the question on the decidability of a Boolean algebra $\mathfrak{A}$ such that $\mathrm{ch}_1(\mathfrak{A}) = 1$.

Theorem 3.5.2. *If a Boolean algebra $\mathfrak{A}$ admits a constructivization with recursive set of numbers of atoms and $\mathrm{ch}(\mathfrak{A}) = (1, 1, 0)$, then $\mathfrak{A}$ is decidable.*

PROOF. For a Boolean algebra $\mathfrak{A}$ we consider three possible cases.

CASE 1: there exists an atomic element a such that under $C(a)$ there is no atomic element containing infinitely many atoms.

CASE 2: for any atomic element $a \in \mathfrak{A}$ there exists an atomic element $b \leqslant C(a)$ such that $\widehat{b} \cong \mathfrak{B}_\omega$.

CASE 3: Cases 1, 2 fail and for any atomic element $a \in \mathfrak{A}$ there exists an atomic element $b \leqslant C(a)$ such that $\widehat{b} \cong \mathfrak{B}_{\omega \times \eta}$.

We show that for any countable Boolean algebra of characteristic (1,1,0) one of these cases holds. We note that any two of them cannot hold simultaneously.

We assume that there exists a Boolean algebra $\mathfrak{A}$ such that none of the above-mentioned cases holds and $\mathrm{ch}(\mathfrak{A}) = (1, 1, 0)$. In particular, for $\mathfrak{A}$ Case 2 fails. Hence there is an atomic element $a_0 \in \mathfrak{A}$ such that under $C(a_0)$ there is no atomic element $b \leqslant C(a_0)$ such that $\widehat{b} \cong \mathfrak{B}_\omega$.

Since Case 3 does not hold for $\mathfrak{A}$, there is an atomic element a_1 such that under $C(a_1)$ there is no element $b \leqslant C(a_1)$ such that $\widehat{b} \cong \mathfrak{B}_{\omega \times \eta}$.

We now consider $a = a_0 \vee a_1$. Since a_0 and a_1 are atomic, a is also atomic. Since Case 1 fails for $\mathfrak{A}$, there is an element $b \leqslant C(a)$ such that b contains infinitely many atoms. We consider the quotient algebra $\widehat{b}/F(\widehat{b})$ by the Frechét ideal. Under b there are infinitely many atoms, hence $b \notin F(\widehat{b})$ and the Boolean algebra $\widehat{b}/F(\widehat{b})$ is not trivial. If there exists $b_0 \leqslant b$ such that $b_0/F(\widehat{b})$ is an atom of the Boolean algebra $\widehat{b}/F(\widehat{b})$, then $\widehat{b}_0$ is a superatomic Boolean algebra of type (1,1) and $\widehat{b}_0 \cong \mathfrak{B}_\omega$. However, $b_0 \leqslant b \leqslant C(a) \leqslant C(a_0)$, which contradicts the choice of a_0. Consequently, $\widehat{b}/F(\widehat{b})$ is an atomless Boolean algebra. By Corollary 1.5.4, $\widehat{b} \cong \mathfrak{B}_{\omega \times \eta}$, but $b \leqslant C(a) \leqslant C(a_1)$, which contradicts the choice of a_1. The contradiction

3.5. Restricted Fragments of the Theory of Boolean Algebras 223

obtained shows that one of the above cases takes place for the Boolean algebra $\mathfrak{A}$.

If Case 1 holds for $\mathfrak{A}$, then there exists an atomic element a such that under $C(a)$ there is no atomic element containing infinitely many atoms. Hence any element $x \in \widehat{C(a)}$ is of one of the following types: either $x \notin I(\mathfrak{A})$ and ch $(x) = (1,1,0)$ or $x \in I(\mathfrak{A})$ and ch $(x) = (0,n,\varepsilon)$. But each of these characteristics is described by a single formula. Consequently, the Boolean algebra $\widehat{C(a)}$ is a prime model of characteristic $(1,1,0)$. A prime model is decidable (cf. Corollary 3.4.7) and $\widehat{a}$ is atomic. Furthermore, it has a constructivization with decidable set of atoms. As has already been indicated, it is also decidable. Since the product of decidable models is decidable, $\mathfrak{A} = \widehat{a} \times \widehat{C(a)}$ is a decidable model.

If Case 3 holds for $\mathfrak{A}$, then we consider an atomic element a_0 such that there is no atomic element $b \leqslant C(a_0)$ such that $\widehat{b} \cong \mathfrak{B}_\omega$. Such an element exists because, in view of the condition of Case 3, Case 2 fails for $\mathfrak{A}$. As in the previous case, we conclude that $\widehat{a}$ is a decidable Boolean algebra since it has a constructivization with recursive set of atoms.

We consider the Boolean algebra $C(a)$. The conditions of Case 3 imply that $C(a)$ is a dense Boolean algebra. Consequently, it is countable, saturated, and, by Theorem 3.4.6, decidable. Therefore, the Boolean algebra $\mathfrak{A}$, being the product of decidable Boolean algebras, is also decidable.

It remains to consider Case 2. A Boolean algebra $\mathfrak{A}$ of type $(1,1,0)$ satisfying the conditions of Case 2 is called ω-limit.

Lemma 3.5.6. *If a Boolean algebra $\mathfrak{A}$ is ω-limit and is constructivizable with respect to the extended signature σ_1, then there exists its constructivization μ of the same signature with $0'$-recursive Ershov–Tarski ideal $I(\mathfrak{A})$.*

PROOF. For a constructive Boolean algebra $\langle \mathfrak{A}, \nu \rangle$ of the signature σ_1 we consider a recursive tree $\mathcal{D}$ such that $\langle \mathfrak{A}, \nu \rangle \cong \langle \mathfrak{B}_\mathcal{D}, \nu_\mathcal{D} \rangle$. We fix a mapping φ and a partially recursive function g such that $\langle \mathcal{D}, \varphi \rangle$ is a tree generating $\mathfrak{A}$, and $\varphi(n) = \nu g(n)$ for any $n \in \mathcal{D}$.

The recursivity of the set of numbers of atoms implies the recursivity of the tree $\mathcal{D}$. Consequently, the set of end vertices of the tree $\mathcal{D}$ is recursive. We note that the recursivity of the tree $\mathcal{D}$ is a necessary and sufficient condition for the recursivity of the set of numbers of atoms of the Boolean algebra $\mathfrak{B}_\mathcal{D}$ with respect to the numbering $\nu_\mathcal{D}$.

We now consider the representation of $\mathcal{D}$ as a computable sequence of finite subtrees $\mathcal{D}_0 \subseteq \mathcal{D}_1 \subseteq \ldots \subseteq \mathcal{D}_n \subseteq \ldots$, where $\bigcup_n \mathcal{D}_n = \mathcal{D}$, $\mathcal{D}_0 = \{0\}$

and $\mathcal{D}_{n+1} \rightleftharpoons \mathcal{D} \cap \{x \mid h(x) \leqslant n+1\}$. An element $x \in \mathcal{D}_n$ is called *t-atomless* if $y \in \mathcal{D}$ for any $y \preccurlyeq x$ and $h(y) \leqslant t + h(x)$.

We note that an element $\varphi(x)$ of a Boolean algebra is atomless if and only if $x \in \mathcal{D}$ is a t-atomless element for any $t \in \mathbb{N}$. If $x \in \mathcal{D}$, then the number $i > h(x)$ is called the *t-atomlessness level* of x if there exists an element $y \prec x$ such that y is t-atomless and $h(y) = i$, and the element $H(y)$ is not t-atomless. An element $x \in \mathcal{D}$ is called *t-principal* if the element x is not t-atomless and, for any $y \in \mathcal{D}$ such that $y \neq x$ and $h(y) = h(x)$, the following conditions hold :

there exists $k > h(x)$ such that k is the t-atomlessness level of x but not of y;

a number i, $k < i \leqslant h(x)$, is the t-atomlessness level of x if and only if i is the t-atomlessness level of y.

We construct step-by-step a new tree $M = \bigcup_{t \geqslant 0} M^t$ such that $\mathfrak{B}_M \cong \mathfrak{B}_\mathcal{D}$. A unique chain of elements of M defining a nonzero element of $\mathfrak{B}_M / I(\mathfrak{B}_M)$ is a $0'$-recursive set. It is easy to see that a function f is $0'$-recursive if and only if there exists a computable sequence of functions f_n such that $f(x) = \lim_{n \to \infty} f_n(x)$ for any $x \in \mathbb{N}$. Therefore, to define such a chain we construct a computable sequence of finite characteristic functions χ_n, which yields the required $0'$-recursive function as a limit.

STEP 0. We set $M^0 = \{0\}$, $i(0) = 0$, $\varphi^0(0) = 0$, $n(0) = 0$, $\widehat{0} = 0$, $m^0(0) = 0$, and $K^0 = \varnothing$. No marks are fixed.

STEP $t+1$. We find the least $t' > i(t)$ such that for any $i \leqslant n(t) + 1$ there exists a t'-prime element x_i such that under x_i there exists an endpoint $y \preccurlyeq x$ of $\mathcal{D}_{t+1}$, which is not t'-atomless and is not an endpoint of $\mathcal{D}_{t+1}$. The existence of such t' follows from the fact that the characteristic of the Boolean algebra $\mathfrak{A}$ under consideration is $(1,1,0)$. The effectiveness of verification is a consequence of the recursivity of the tree $\mathcal{D}$.

If for any $i \leqslant n(t)$ the equality $x_i = m^t(i)$ holds, then $x_{n(t)+1} \prec m^t(n(t))$ or $x_{n(t)+1} \not\prec m^t(n(t))$.

For $x_{n(t)+1} \prec m^t(n(t))$ we set

$n(t+1) \rightleftharpoons n(t) + 1$, $\quad i(t+1) \rightleftharpoons t'$

$m^{t+1}(i) = m^t(i)$, $\; i \leqslant n(t)$, $\quad m^{t+1}(n(t)+1) = x_{n(t)+1}$

$M^{t+1} = \{x \mid H(x) \in M^t,\; x \text{ is marked by some tag } \boxed{k}$

$\qquad$ or there exists $y \in \mathcal{D}^{t+1} \setminus \mathcal{D}^t$ and $H(x) = \varphi^t(y)\}$

$K^{t+1} = \{x \mid x \in M^{t+1} \setminus M^t,\; H(x) \text{ is marked by some tag } \boxed{k}\}$

3.5. Restricted Fragments of the Theory of Boolean Algebras 225

If x was marked by the tag $\boxed{k}$, then we move $\boxed{k}$ by $R(x)$.

In the case $x_{n(t)+1} \not\prec m^t(n(t))$ we set $n(t+1) = n(t)+1$, $i(t+1) = t'$, and $m^{t+1}(i) = m^t(i)$ for $i \leqslant n(t)$, $m^{t+1}(n(t) + 1) = x_{n(t)+1}$. We find an end element y of M^t, which is not in K^t, $y_0 \prec m^t(n(t))$, and an end element z_0 of $\mathcal{D}_{t+1}$ such that $\varphi^t(z_0) = y_0$.

For all $s \in \mathcal{D}^t$ such that $s \precsim x_{n(t)+1}$ fails and $s \neq z_0$ we define $\varphi^{t+1}(s) = \varphi^t(s)$ and $\varphi^{t+1}(z_0) = L(y_0)$. Under the element $R(y_0)$ we locate a subtree $\mathcal{D}''$ isomorphic to $\mathcal{D}' = \{z \in \mathcal{D}^{t+1} \mid z \precsim x_{n(t)+1}\}$. Defining φ^{t+1} on $\mathcal{D}'$ so that φ^{t+1} is an isomorphism of $\mathcal{D}'$ onto $\mathcal{D}''$, we add elements of $\mathcal{D}''$ to M^{t+1}. For any element $y \in M^{t+1}$ marked by $\boxed{k}$, we add two elements $R(y)$ and $L(y)$ to M^{t+1}, move $\boxed{k}$ by the $R(y)$, and add $L(y)$ to K^{t+1}. We add $L(y_0)$ to M^{t+1}. We also add two elements $L(\varphi^t(H(x_{n(t)+1})))$ and $R(\varphi^t(H(x_{n(t)+1})))$ to M^{t+1}, setting $\varphi^t(L(H(x_{n(t)+1}))) = L\varphi^t(H(x_{n(t)+1}))$, and mark $R(\varphi^t(H(x_{n(t)+1})))$ by $\boxed{x_{n(t)+1}}$. The elements $R(y), L(y) \in \mathcal{D}^{t+1}\setminus \mathcal{D}^t$ that are end ones of $\mathcal{D}^{t+1}$ and are not located under $x_{n(t)+1}$ are added to M^{t+1}. We set $\varphi^{t+1}(L(y)) = L\varphi^{t+1}(y)$ and $\varphi^{t+1}(R(y)) = R\varphi^{t+1}(y)$.

If there exists $i \leqslant n(t)$, $x_i \neq m^t(i)$, then we consider the largest $i_0 < n(t)$ such that $(\forall j \leqslant i_0) x_j = m^t(j)$. We define $n(t+1) \leftrightharpoons i_0$. For every $k \in \mathcal{D}^t$ such that $h(k) > i_0$ and there is the mark $\boxed{k}$, we consider an element y marked by $\boxed{k}$ and add the elements $L(y)$ and $R(y)$ to M^{t+1}. For every k, $h(k) > i_0$, we choose the greatest element $x \in \mathcal{D}^t$ such that $k \prec x$ and $h(x) \geqslant i_0$, but, on $z \succ x$, there is no $\boxed{k'}$ such that $h(k') > i_0$. Under $\varphi^t(x)$ we locate a subtree isomorphic to $\{z \mid z \in \mathcal{D}^{t+1}, z \precsim x\}$ and add the corresponding isomorphism to the mapping φ^{t+1} if the set $\{z \in \mathcal{D}^t \mid h(z) \leqslant i_0\}$ is nonempty, i.e., there exists at least one x with the above-mentioned properties. Otherwise, φ^{t+1} coincides with φ^t on this element x. We continue the constructed mapping to an isomorphic embedding $\mathcal{D}^{t+1}$ into M^{t+1} with respect to the order $\precsim$. After that the φ^{t+1}-images of end elements of $\mathcal{D}$ belonging to $\mathcal{D}^{t+1}$ are added to K^t. As a result, we obtain K^{t+1}.

From the construction and the fact that the characteristic of $\mathfrak{A}$ is $(1,1,0)$, it follows that $\lim\limits_{t\to\infty} n(t) = \infty$, $\varphi(x) \leftrightharpoons \lim\limits_{t\to\infty} \varphi^t(x)$ exists for any x, and $\lim\limits_{t\to\infty} m^t(i)$ exists for any i. Let $\langle \mathcal{D}, I \rangle$ be a tree generating $\mathfrak{B}_\mathcal{D}$. From the above properties we obtain that the quotient algebras $\widehat{I(x)}/F(\widehat{I(x)})$ and $\widehat{\varphi(I(x))}/F(\widehat{\varphi(I(x))})$ are isomorphic for $x \in \mathcal{D}$ such that $\widehat{I(x)}$ is an atomic Boolean algebra. Since elements of $\bigcup\limits_{t \geqslant 0} M^t$ determining atomic elements of the Boolean algebra $\mathfrak{B}_M$ define a finite Boolean algebra or $\mathfrak{B}_\omega$, or differ from some summands of $\varphi(I(x))$ by a finite summands for some $x \in \mathcal{D}$ such that $I(x)$ is an atomic element of $\mathfrak{B}_\mathcal{D}$, it is easy to construct an isomorphism between $\mathfrak{B}_M$ and $\mathfrak{B}_\mathcal{D}$. However, in $\mathfrak{B}_M$, for any i the elements $\lim\limits_{t\to\infty} \varphi(m^t(i))$ determine an infinite chain of elements of $\mathfrak{B}_M$ of type $(1,1,0)$. From the construction it follows that $\mathfrak{B}_M$ is a recursive Boolean algebra with recursive set of atoms and $0'$-recursive Ershov–Tarski ideal.

The next stage of the proof of the theorem is based on the following lemma.

Lemma 3.5.7. *If a recursive Boolean algebra $\mathfrak{B}$ of characteristic $(1,1,0)$ with recursive set of atoms and the $0'$-recursive Ershov–Tarski ideal $I(\mathfrak{B})$ is ω-limit one, then there exists a recursive Boolean algebra $\mathfrak{A}$ isomorphic to $\mathfrak{B}$ with recursive set of atoms and recursive Ershov–Tarski ideal.*

PROOF. If the Ershov–Tarski ideal is $0'$-recursive, then for any recursive tree $\mathcal{D}$ such that the Boolean algebra $\mathfrak{B}_\mathcal{D}$ is recursively isomorphic to $\mathfrak{B}$ there exists a recursive function f such that there exists $\overline{f}(n) = \lim\limits_{t\to\infty} f(n,t)$ for any n; moreover, $\overline{f}(x) = 0 \Leftrightarrow I(x) \in I(\mathfrak{B}_\mathcal{D})$ and $\overline{f}(x) = 1 \Leftrightarrow I(x) \notin I(\mathfrak{B}_\mathcal{D})$, where $\langle \mathcal{D}, I \rangle$ is a tree generating $\mathfrak{B}_\mathcal{D}$. In this case, we stretch the tree $\mathcal{D}$ and construct a new tree $\mathcal{D}'$ and embeddings φ'_t of $\mathcal{D}$ into $\mathcal{D}'$ so that φ'_t becomes stable whenever the function $f(n,t)$ stabilizes. In this case, we keep the isomorphism type unchanged. Thereby, we attain the recursivity of the Ershov–Tarski ideal without disturbing the recursivity of the set of atoms and without changing the isomorphism type.

Our next aim is to analyze a recursive Boolean algebra $\mathfrak{A}$ with recursive set of atoms and recursive Ershov–Tarski ideal $I(\mathfrak{A})$.

We note the following simple fact.

Proposition 3.5.5. *If a recursive Boolean algebra $\mathfrak{A}$ of characteristic $(1,1,0)$ has a recursive set atoms and the recursive Ershov–Tarski ideal $I(\mathfrak{A})$, then there exists a computable sequence $\mathfrak{A}_i$ of recursive Boolean algebras of characteristic $\mathrm{ch}_1(\mathfrak{A}_i) = 0$ such that the sets of their atoms are uniformly recursive and $\sum\limits_{i\in\mathbb{N}} {}_{\{0,1\}}\mathfrak{A}_i \cong \mathfrak{A}$.*

PROOF. We proceed by sequentially detaching the direct summands. The effective realization of this procedure is justified by the recursivity of the Ershov–Tarski ideal and the fact that, for any x and y such that $x \wedge y =$

3.5. Restricted Fragments of the Theory of Boolean Algebras 227

0 and $x \vee y \notin I(\mathfrak{A})$, the relation $x \in I(\mathfrak{A}) \& y \notin I(\mathfrak{A})$ or $x \notin I(\mathfrak{A}) \& y \in I(\mathfrak{A})$ holds.

We say that a constructive Boolean algebra $\langle \mathfrak{A}, \nu \rangle$ is *effectively represented as the direct sum of constructive Boolean algebras* $\langle \mathfrak{A}_i, \nu_i \rangle$ (denoted by $\langle \mathfrak{A}, \nu \rangle \cong_{\text{rec}} \sum_{i \in \mathbb{N}} {}_{\{0,1\}} \langle \mathfrak{A}_i, \nu_i \rangle$) if $\mathfrak{A} \cong \sum_{i \in \mathbb{N}} {}_{\{0,1\}} \mathfrak{A}_i$ and there exist recursive functions $f \colon \mathbb{N} \times \mathbb{N} \to \mathbb{N}$ and $g \colon \mathbb{N} \to \mathbb{N}$ such that $\forall n \in \mathbb{N}\, \nu(n) = \sum_{i \in \mathbb{N}} \nu_i f(i, n)$, where $\forall i \in \mathbb{N}\ \nu_i f(i, n) \in \mathfrak{A}_i$; moreover, $\forall i > g(n)\ \nu_i f(i, n) = 0$ or $\forall i > g(n)\ \nu_i f(i, n) = 1$.

Proposition 3.5.6. *Let a Boolean algebra $\mathfrak{B}$ is represented as the direct sum of a computable sequence of recursive Boolean algebras $\mathfrak{A}_i$ of the signature σ_1, where $\forall i \in \mathbb{N}\ \mathfrak{A}_i \cong \mathfrak{A}'_i \times \mathfrak{B}'_i$, $\mathfrak{A}'_i$ is an atomic Boolean algebra, and $|\mathfrak{A}'_i| = \infty$, $\mathfrak{B}'_i$ is a nontrivial atomless Boolean algebra. Then $\mathfrak{B}$ is decidable.*

PROOF. It suffices to find a constructivization such that $\operatorname{Atom}(\mathfrak{B})$, $\operatorname{Atomless}(\mathfrak{B})$, $\operatorname{Atomic}(\mathfrak{B})$, and $I(\mathfrak{B})$ are decidable. Then the assertion follows in accordance with Theorem 3.4.8.

Let $\mathcal{D}_i$ be a tree that generates $\mathfrak{A}_i$ and is constructed uniformly on i. Without loss of generality, we can assume that, at every step of constructing the tree, only two new end elements are added to the ready part of the tree, i.e., $\mathcal{D}_i^0 = \{0\}$, $\mathcal{D}_i^{t+1} = \mathcal{D}_i^t \cup \{a, b\}$, $\mathcal{D}_i = \bigcup_{i \in \mathbb{N}} \mathcal{D}_i^t$, where $a = L(c)$, $b = R(c)$, and c is some end element of $\mathcal{D}_i^t$. We consider a mapping $\delta_i \colon \mathcal{D}_i \stackrel{1-1}{\to} \mathfrak{A}_i$. We construct step-by-step a new tree M_i so that $\mathfrak{B}_{M_i} \cong \mathfrak{A}'_i$ and $\mathfrak{B}_{M_i} \times \mathfrak{B}_\eta \cong \mathfrak{A}_i$ are isomorphic.

We give an informal description of the procedure. We introduce the notion of a complete element of a tree (an analog of an atomless element of the generated Boolean algebra) and that of an n-complete element so that an element is complete if and only if it is n-complete for any natural number n. Constructing step-by-step, we keep track of n-complete elements of $\mathcal{D}_i^{t+1}$ so that the growth of n is observed by a counter. The latter allows the element to be unchanged until the value of the counter is not greater than some fixed one. If the counter exceeds the fixed value, then we move the element along the tree M_i^t, "pasting" it as well as its elements located under it at the step $t+1$ to the least free end element which is not marked as an end element for the whole tree. Thus, complete elements go into atom elements due to the constant movement of n-complete fragments in view of the constant growth of n. Atomic and finitely complete elements are stabilized and, beginning from some step, are no longer moved.

228 **3. Constructive Boolean Algebras**

Although we add an infinite number of end elements, the new tree generates a Boolean algebra isomorphic to the atomic part of the initial Boolean algebra. This can be explained by the proposition on preservation of a type of isomorphism of a Boolean algebra: the above construction preserves a type of isomorphism after pasting finite subtrees to end vertices. We now proceed to the precise description of a construction. At the step t, we define a part of the tree M_i^t, a function $\varphi_i^t : \mathcal{D}_i^t \to M_i^t$, and a counter $\lambda x\, r(x,t) : \mathcal{D}_i^t \to \mathbb{N}$.

STEP 0. We define $M_i^0 \rightleftharpoons \{0\}$, $r(0,0) \rightleftharpoons 0$, and $\varphi_i^0(0) \rightleftharpoons 0$.

STEP $t+1$. We set $\mathcal{D}_i^{t+1} = \mathcal{D}_i^t \cup \{a,b\}$, where $a = L(c)$, $b = R(c)$, and c is an end element of $\mathcal{D}_i^t$. We assume that, at the step $t+1$, the tree M_i^t, the function $\varphi_i^t : \mathcal{D}_i^t \to M_i^t$, and the values $r(x,t)$ for $x \in \mathcal{D}_i^t$ have been defined.

We set $\widehat{x} \rightleftharpoons \{y \in \mathbb{N} \mid x \succcurlyeq y\}$. An element $a \in \mathcal{D}$ is called *complete in* $\mathcal{D}$ if $\widehat{a} \cap \mathcal{D} = \widehat{a}$ and n-*complete in* $\mathcal{D}$ if $\mathcal{D} \supseteq \{x \mid H(x,i) = a\, \&\, x \preccurlyeq a\, \&\, i \leqslant n\}$. If the value n is clear, then we talk about finitely complete elements. At this step, we introduce the "atomization" procedure for complete elements of the tree $\mathcal{D}_i$.

CASE 1: there exists $x \in \mathcal{D}_i^t$ such that the following condition holds:

(∗) $\neg(\exists y \in \mathcal{D}_i^{t+1}(y \preccurlyeq x\, \&\, \delta_i(y) \in \mathrm{Atom}(\mathfrak{A}_i)))$ and x is not n-complete in $\mathcal{D}_i^t$, but is n-complete in $\mathcal{D}_i^{t+1}$, where $n > 0$ and $n \geqslant r(x,t)$.

Let $C(\delta_i(x)) \neq \bigvee\limits_{i=1}^{k} b_i$ for any $k \in \mathbb{N}$, where $b_i \in \mathfrak{B}_{\mathcal{D}_i^t}$ and b_i is an atom of $\mathfrak{A}_i$. Let x_0 be the greatest element among those $x \in \mathcal{D}_i^t$ that satisfy the condition (∗). Such an element x_0 exists since, by condition, it is not n-complete in $\mathcal{D}_i^t$, but is n-complete in $\mathcal{D}_i^{t+1}$, i.e., an element appears under x_0. At the step $t+1$, we add only a and b such that $H(a) = H(b)$. Hence all the elements subject to (∗) already belong to the set $\{x \mid x_0 \succcurlyeq x\}$ for some x_0 satisfying (∗).

We consider $y \succcurlyeq x_0$ satisfying the following condition:

(∗∗) under $S(y)$ there are end elements z of $\mathcal{D}_i^t$ and $\delta_i(z) \notin \mathrm{Atom}(\mathfrak{A}_i)$.

We consider the least y_0 satisfying the condition (∗∗) and the least (with respect to the ordinary order on the set of natural numbers) z_0 such that $z_0 \preccurlyeq S(y_0)$, z_0 is an end element of $\mathcal{D}_i^t$, and $\delta_i(z_0) \notin \mathrm{Atom}(\mathfrak{A}_i)$. Since there are infinitely many atoms in the Boolean algebra $\mathfrak{A}_i$, in view of the last part of (∗) such y_0 and z_0 can be found. We set $r(y,t+1) \rightleftharpoons r(y,t)+1$ for $y \preccurlyeq x_0$, $y \in \mathcal{D}_i^{t+1}$, and leave the previous values for the rest of the

3.5. Restricted Fragments of the Theory of Boolean Algebras 229

arguments at $t+1$. We consider $d \rightleftharpoons \varphi_i^t(z_0)$. Let ψ be an isomorphism mapping $\mathcal{D}_i^{t+1} \cap \widehat{x}_0$ into a subtree with the vertex $L(d)$, i.e., $\psi : (\mathcal{D}_i^{t+1} \cap \widehat{x}_0) \to \widehat{L(d)}$. Then $M_i^{t+1} \rightleftharpoons M_i^t \cup \operatorname{range} \psi \cup \{R(d)\}$, $\varphi_i^{t+1}|_{(\mathcal{D}_i^t \setminus \widehat{x}_0) \setminus \{z_0\}} \rightleftharpoons \varphi_i^t|_{(\mathcal{D}_i^t \setminus \widehat{x}_0) \setminus \{z_0\}}$, $\varphi_i^{t+1}|_{\mathcal{D}_i^{t+1} \cap \widehat{x}_0} \rightleftharpoons \psi|_{\mathcal{D}_i^{t+1} \cap \widehat{x}_0}$, $\varphi_i^{t+1}(z_0) \rightleftharpoons R(d)$.

CASE 2: at the step $t+1$, there is no x satisfying $(*)$. At the step $t+1$, we already have $\mathcal{D}_i^{t+1} = \mathcal{D}_i^t \cup \{a,b\}$, where $a = L(c)$, $b = R(c)$, and c is an end element of $\mathcal{D}_i^t$. We set $c' \rightleftharpoons \varphi_i^t(c)$ and define $\varphi_i^{t+1}|_{\mathcal{D}_i^t} \equiv \varphi_i^t$, $\varphi_i^{t+1}(a) \rightleftharpoons L(c')$, $\varphi_i^{t+1}(b) \rightleftharpoons R(c')$, $M_i^{t+1} \rightleftharpoons M_i^t \cup \{L(c'), R(c')\}$.

Let $M_i \rightleftharpoons \bigcup_{t \in \mathbb{N}} M_i^t$. For any $x \in \{y \in \mathcal{D}_i \mid \delta_i(y) \notin \operatorname{Atom}(\mathfrak{A}_i)\}$ there exists a number $t^x \in \mathbb{N}$ such that $\forall t \geqslant t^x (\varphi_i^t(x) = \varphi_i^{t^x}(x))$. For the above set of arguments we introduce a function φ_i as follows: $\varphi_i(x) \rightleftharpoons \lim_{t \to \infty} \varphi_i^t(x)$. Let $\mathfrak{B}_{M_i}$ be a Boolean algebra generated by the tree M_i and let $\mu_i : M_i \overset{1-1}{\to} \mathfrak{B}_{M_i}$ be the generating mapping. By construction, $\mathfrak{B}_{M_i}$ is atomic and the set of its atoms is recursive.

Lemma 3.5.8. *The embedding $\mathfrak{A}_i' \to \mathfrak{B}_{M_i}$ holds.*

Without loss of generality, we can assume that $\mathfrak{A}_i'$ is a subalgebra of the algebra $\mathfrak{B}_{M_i}$. Indeed, let $p_1, \ldots, p_N$ list all the maximal complete elements of the tree $\mathcal{D}_i$. From the tree $\mathcal{D}_i$ we cut off all the elements $H(p_1), \ldots, H(p_N)$ and all the elements $\widehat{p}_1, \ldots, \widehat{p}_N$. For subtrees $\widehat{S(p_1)} \cap \mathcal{D}_i, \ldots, \widehat{S(p_n)} \cap \mathcal{D}_i$ we move their vertices and place them under elements $H(p_1), \ldots, H(p_N)$ with the help of the function H (Fig. 3.5.1).

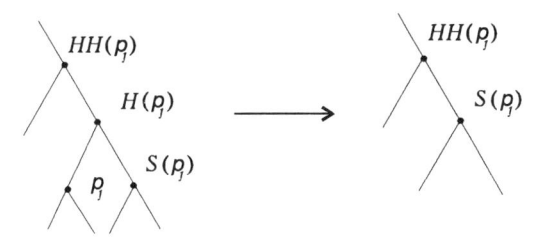

Fig. 3.5.1

The tree obtained will be denoted by $\mathcal{D}_i'$. It is obvious that $\mathfrak{B}_{\mathcal{D}_i'} \cong \mathfrak{A}_i'$. We consider the composition $\mu_i \varphi_i : \mathcal{D}_i' \overset{1-1}{\to} \mathfrak{B}_{M_i}$ as a generating mapping onto its image, i.e., onto $\mu_i \varphi_i(\mathcal{D}'_i) \subseteq \mathfrak{B}_{M_i}$. In other words, we can assume that the Boolean algebra $\mathfrak{A}'_i$ is isomorphically embedded into the Boolean algebra $\mathfrak{B}_{M_i}$.

Lemma 3.5.9. *For all i the Boolean algebras $\mathfrak{A}'_i$ and $\mathfrak{B}_{M_i}$ are isomorphic.*

PROOF. The assertion follows from the Remmel theorem and the following two properties of the construction:

⟨1⟩ $\forall x \in M_i((\exists^\infty y \in M_i(y \text{ is an end element of } M_i \& y \preccurlyeq x) \rightarrow (\exists^\infty z \in \mathcal{D}_i)(z \text{ is an end element of } \mathcal{D}_i \& \varphi_i(z) \preccurlyeq x)))$,

⟨2⟩ $\forall x \in M_i(\exists y \in \mathcal{D}_i)(x \succcurlyeq \varphi_i(y)$ and, between x and $\varphi_i(y)$, there are only a finite number of finitely complete elements obtained by the "atomization" procedure and corresponding to n-complete elements of the tree $\mathcal{D}_i$).

Thus, we have shown that the Boolean algebras $\mathfrak{A}'_i$ and $\mathfrak{B}_{M_i}$ are isomorphic. Hence we have isomorphisms $\mathfrak{A}_i \cong \mathfrak{A}'_i \times \mathfrak{B}_\eta \cong \mathfrak{B}_{M_i} \times \mathfrak{B}_\eta$. Since the Boolean algebra $\mathfrak{B}_\eta$ is constructivizable uniformly on i and the set of atoms of the Boolean algebra $\mathfrak{B}_{M_i}$ is decidable uniformly on i, the Boolean algebra $\sum_{i \in \mathbb{N}} {}_{\{0,1\}} (\mathfrak{B}_{M_i} \times \mathfrak{B}_\eta)$ can be constructivizable and has a decidable set of atoms and that of atomless elements. The Boolean algebra $\sum_{i \in \mathbb{N}} {}_{\{0,1\}} (\mathfrak{B}_{M_i} \times \mathfrak{B}_\eta)$ is isomorphic to the Boolean algebra $\mathfrak{B} \cong \sum_{i \in \mathbb{N}} {}_{\{0,1\}} (\mathfrak{A}'_i \times \mathfrak{B}_\eta)$ since all the summands are isomorphic respectively. Since the elementary characteristic is $(1,1,0)$ and, in the constructed Boolean algebra $\mathfrak{B}_{M_i}$, all elements are atomic, $\sum_{i \in \mathbb{N}} {}_{\{0,1\}} \mathfrak{B}_{M_i} \times \mathfrak{B}_\eta$ is recursive in the enrichments of Atom, Atomless, Atomic, and I; moreover, it is decidable in view of the model completeness of its theory.

We now complete the proof of the theorem. We have shown that a recursive Boolean algebra $\mathfrak{B}$ of characteristic $(1,1,0)$ with recursive set of atoms admits a constructivization ν such that $\langle \mathfrak{B}, \nu \rangle$ with decidable set of atoms and the Ershov–Tarski characteristic $(1,1,0)$ can be effectively represented as the direct sum of constructive Boolean algebras $\langle \mathfrak{A}_i, \nu_i \rangle$ of characteristic $\mathrm{ch}_1 \mathfrak{A}_i = 0$ for any i (i.e., the first Ershov–Tarski ideal is recursive). Let us show that, in this case, the Boolean algebra $\mathfrak{B}$ is strongly constructivizable and decidable. Indeed, let the Boolean algebra $\langle \mathfrak{B}, \nu \rangle$ admit the effective representation $\langle \mathfrak{B}, \nu \rangle \cong_{\mathrm{rec}} \sum_{i \in \mathbb{N}} {}_{\{0,1\}} \langle \mathfrak{B}_i, \nu_i \rangle$, where for

3.5. Restricted Fragments of the Theory of Boolean Algebras

any $i \in \mathbb{N}$ the constructive Boolean algebra $\mathfrak{A}_i \cong \mathfrak{A}'_i \times \mathfrak{B}'_i$, $\mathfrak{A}'_i$ is atomic and $\mathfrak{B}'_i$ is atomless or 0. The main nontrivial case was discussed above. We note two simple algebraic claims.

Lemma 3.5.10. *If $\mathfrak{B}$ is a Boolean algebra, $\mathfrak{B} \cong \sum_{n \in \mathbb{N}} {}_{\{0,1\}} \mathfrak{B}_i$, and $f: \mathbb{N} \to \mathbb{N}$ is strictly monotone such that $f(0) = 0$, then the Boolean algebras $\sum_{i \in \mathbb{N}} {}_{\{0,1\}} \left(\sum_{f(n) \leqslant i < f(n+1)} \mathfrak{B}_i \right)$ and $\mathfrak{B}$ are isomorphic.*

Indeed, these Boolean algebras are isomorphic since all the finite fragments of the direct sums of type $\mathfrak{B}_0 \oplus \ldots \oplus \mathfrak{B}_{f(n+1)}$ and $(\mathfrak{B}_0 \oplus \ldots \oplus \mathfrak{B}_{f(t)}) \oplus \ldots \oplus (\mathfrak{B}_{f(n)} \oplus \ldots \oplus \mathfrak{B}_{f(n+1)})$ are isomorphic. □

The following effective analog of Lemma 3.5.10 holds.

Lemma 3.5.11. *Let a constructive Boolean algebra $\langle \mathfrak{B}, \nu \rangle$ be effectively represented in the form $\langle \mathfrak{B}, \nu \rangle \cong_{\text{rec}} \sum_{i \in \mathbb{N}} {}_{\{0,1\}} \langle \mathfrak{B}_i, \nu_i \rangle$ and let $f: \mathbb{N} \to \mathbb{N}$ be a strictly monotone recursive function such that $f(0) = 0$. If $\langle \mathfrak{A}, \mu \rangle \cong_{\text{rec}} \sum_{i \in \mathbb{N}} {}_{\{0,1\}} \left(\sum_{f(n) \leqslant i < f(n+1)} \langle \mathfrak{B}_i, \nu_i \rangle \right)$, then $\langle \mathfrak{B}, \nu \rangle$ and $\langle \mathfrak{A}, \mu \rangle$ are recursively isomorphic.*

REMARK 3.5.1. Let $\mathfrak{B} \cong \sum_{i \in \mathbb{N}} {}_{\{0,1\}} \mathfrak{A}_i$, where $\mathfrak{A}_i \cong \mathfrak{A}'_i \times \mathfrak{B}'_i$, $\mathfrak{A}'_i$ is atomic, $\mathfrak{B}'_i \cong \mathfrak{B}_\eta$ or $\mathfrak{B}'_i \cong 0$. If $\exists^\infty i \mathfrak{B}'_i \cong 0$, then $\exists^\infty i \mathfrak{B}'_i \cong \mathfrak{B}_\eta$. Otherwise, $\text{ch} \mathfrak{B} = 0$, which contradicts the condition.

CASE 1: $\exists^\infty i \mathfrak{B}'_i \cong 0$. By the above remark and Lemma 3.5.10, it is possible to construct a new constructivization of $\mathfrak{B}$ with recursive set of atoms and the decomposition $\mathfrak{B} \cong \sum_{i \in \mathbb{N}} {}_{\{0,1\}} \mathfrak{A}_i \times \mathfrak{B}''_i$ such that $\mathfrak{B}''_i \cong \mathfrak{B}_\eta$ for any $i \in \mathbb{N}$. Without loss of generality, we can assume that $\mathfrak{B}'_i \not\cong 0$ for any i.

CASE 2: $\exists^\infty i (|\mathfrak{A}'_i| < \infty) \,\&\, \exists^\infty i (|\mathfrak{A}'_i| = \infty)$.

CASE 2.1: there exist infinitely many i such that there is $\mathfrak{A}'_i$ of type $\mathfrak{B}_\omega$. We consider the Boolean algebra $\langle \mathfrak{B}', \nu' \rangle \cong_{\text{rec}} \sum_{i \in \mathbb{N}} {}_{\{0,1\}} \langle \mathfrak{A}'_i \times \mathfrak{B}_\omega \times \mathfrak{B}'_i, \nu_i \times \nu_\omega \rangle$, where ν_i is a constructivization of $\mathfrak{A}'_i \times \mathfrak{B}'_i$, ν_ω is a strong constructivization of $\mathfrak{B}_\omega$, and $\langle \mathfrak{B}, \nu \rangle \cong_{\text{rec}} \sum_{i \in \mathbb{N}} {}_{\{0,1\}} \langle \mathfrak{A}'_i \times \mathfrak{B}'_i, \nu_i \rangle$. Using the Vaught criterion, it is easy to show that $\mathfrak{B}'$ and $\mathfrak{B}$ are isomorphic. Furthermore, in $\mathfrak{B}'$ the set of atoms is decidable if it is decidable in $\langle \mathfrak{B}, \nu \rangle$. Thus, this case is reduced to Proposition 3.5.6.

CASE 2.2: among i such that $\mathfrak{A}'_i$ is infinite, there are only a finite number of i such that $\mathfrak{B}_\omega$ is isomorphic to the direct summands $\mathfrak{A}'_i$. Then the quotient algebra $\mathfrak{A}'_i$ by the Frechét filter is atomless for almost all $\mathfrak{A}'_i$. Consequently, almost all $\mathfrak{A}'_i$ such that $|\mathfrak{A}'_i| = \infty$, are isomorphic to $\mathfrak{B}_{\omega\times\eta}$. As in Case 2.1, it suffices to consider the algebra $\langle \mathfrak{B}', \nu'\rangle \cong_{\text{rec}} \sum_{i\in\mathbb{N}} {}_{\{0,1\}} \langle \mathfrak{A}_i \times \mathfrak{B}_{\omega\times\eta}, \nu_i \otimes \nu_{\omega\times\eta}\rangle$ which is isomorphic to $\mathfrak{B}$ and satisfies the hypotheses of Proposition 3.5.6, where $\nu_{\omega\times\eta}$ is a strong constructivization of $\mathfrak{B}_{\omega\times\eta}$.

CASE 3: $\exists^{<\infty} i |\mathfrak{A}'_i| < \infty$. This case is reduced to Proposition 3.5.6 by "pasting together" the first summands containing all the numbers subject to the condition $|\mathfrak{A}'_i| < \infty$ into a common summand.

CASE 4: $|\mathfrak{A}'_i| < \infty$ for any $i \in \mathbb{N}$. Then the Boolean algebra $\mathfrak{B}$ is isomorphic to $\mathfrak{B}_{\omega+\eta}$ and is decidable.

CASE 5: $\exists^{<\infty} i |\mathfrak{A}'_i| = \infty$. We separate summands with infinite part $\mathfrak{A}'_i$ into one summand which is strongly constructive because it is atomic and the set of its atoms is recursive. The remaining part satisfies Case 4. Hence it is also strongly constructivizable. Since the product of strongly constructive Boolean algebras is strongly constructive, the conclusion of the theorem is valid. □

Corollary 3.5.1. *If $\mathfrak{B}$ is a Boolean algebra of characteristic $(1,1,0)$ with recursive sets* Atom$(\mathfrak{B})$ *and* Atomless$(\mathfrak{B})$, *then $\mathfrak{B}$ is decidable.*

In the case of characteristic $(1,0,1)$, the complete answer is not obtained. The question on the decidability of a Boolean algebra admitting a recursive representation in its enrichment with the predicates Atom and Atomless distinguishing the sets of atoms and that of atomless elements is still open.

Odintsov [150] establishes the following criterion for decidability.

Theorem 3.5.3. [150] *Let $\mathfrak{B}$ be a recursive Boolean algebra, $\text{ch}\mathfrak{B} = (1,0,1)$, and the sets* Atom$(\mathfrak{B})$, Atomless$(\mathfrak{B})$, *and* Atomic$(\mathfrak{B})$ *be recursive. Then there exists a decidable Boolean algebra $\mathfrak{D}$ such that $\mathfrak{D} \cong \mathfrak{B}$.*

Before proving this theorem we establish the criterion for isomorphism of Boolean algebras of characteristic $(1,0,1)$. Let $\mathfrak{A}$ be a Boolean

3.5. Restricted Fragments of the Theory of Boolean Algebras

algebra of characteristic $(1,0,1)$. A sequence of its nonzero atomic elements $\{a_0, a_1, \ldots\}$ is called *principal* if

(1) $a_i \wedge a_j = 0$, $i \neq j$,

(2) any atomic element of $\mathfrak{A}$ is located under the union of a finite number of elements of the sequence A,

(3) if $d \in A$, then for all, except a finite number, $i \in \mathbb{N}$ either $d \wedge a_i = 0$, or $a_i \leqslant d$.

Let $\mathcal{P}_A$ stand for a subset of $\mathcal{P}(\mathbb{N})$ defined as follows. A subset $I \subset \mathbb{N}$ belongs to $\mathcal{P}_A$ if and only if there is an element $d \in \mathfrak{A}$ such that $a_i \leqslant d$ for all $i \in I$ and $a_i \wedge d = 0$ for all $i \notin I$.

Let $I, J \subset \mathbb{N}$. We set $I \sim_f J$ if and only if $(I \backslash J) \cup (J \backslash I)$ is a finite set. The following properties are obvious:

(1) if $I, J \in \mathcal{P}_A$, then $I \cup J \in \mathcal{P}_A$ and $I \cap J \in \mathcal{P}_A$,

(2) if $I \in \mathcal{P}_A$, then $\mathbb{N} \backslash I \in \mathcal{P}_A$,

(3) if $I \in \mathcal{P}_A$ and $I \sim_f J$, then $J \in \mathcal{P}_A$.

Lemma 3.5.12. *Boolean algebras $\mathfrak{B}$ and $\mathfrak{D}$ of characteristic $(1,0,1)$ are isomorphic if and only if there exist principal sequences $A = \{a_0, a_1, \ldots, \}$ of elements of $\mathfrak{B}$ and $C = \{c_0, c_1, \ldots\}$ of elements of $\mathfrak{D}$ such that $(\widehat{a}_i)_\mathfrak{B} \cong (\widehat{c}_i)_\mathfrak{D}$, $i \in \omega$, and $\mathcal{P}_A = \mathcal{P}_C$.*

Let A and C be principal sequences of atomic elements of the algebras $\mathfrak{B}$ and $\mathfrak{D}$ satisfying the hypotheses of the lemma. We show that $\mathfrak{B}$ and $\mathfrak{D}$ are isomorphic. Let $b \in \mathfrak{B}$. We set $J^+(b) = \{i \mid a_i \leqslant b\}$, $J^-(b) = \{i \mid a_i \wedge b = 0\}$, and $J^0(b) = \mathbb{N} \backslash (J^+(b) \cup J(b))$. It is clear that $J^0(b)$ is a finite set for any $b \in \mathfrak{B}$. Similarly, we define $J^+(d)$, $J^-(d)$, and $J^0(d)$ for an arbitrary element $d \in \mathfrak{D}$. For $i \in \omega$ we fix an isomorphism $f_i : \widehat{(a_i)}_\mathfrak{B} \to \widehat{(c_i)}_\mathfrak{D}$ and use the Vaught criterion for isomorphism of Boolean algebras. The isomorphism condition S is defined as the set of pairs $\langle b, d \rangle \in \mathfrak{B} \times \mathfrak{D}$ such that $J^+(b) = J^+(d)$, $J^-(b) = J^-(d)$, and $f_i(a_i \wedge b) = c_i \wedge d$ for any $i \in J^0(d)$. We show that S is the isomorphism condition. It is obvious that the pairs $\langle 0, 0 \rangle$ and $\langle 1, 1 \rangle$ are in S. We assume that $\langle a, b \rangle \in S$ and consider the pair $\langle C(a), C(b) \rangle$. It is clear that $J^+(C(a)) = J^-(a)$, $J^-(C(a)) = J^+(a)$. Analogous assertions are valid for b. Therefore, $J^+(C(a)) = J^+(C(b))$ and $J^-(C(a)) = J^-(C(b))$. For any $i \in J^0(C(a))$ we have $f_i(C(a) \wedge a_i) = C_{\widehat{a}_i}(f_i(a \wedge a_i)) = C_{\widehat{c}_i}(b \wedge c_i) = C(b) \wedge c_i$. Consequently, $\langle C(a), C(b) \rangle \in S$. Let $\langle a, b \rangle \in S$ and $d \in \mathfrak{B}$. Consider the elements $a \wedge d$ and $a\backslash d$. From the condition $\mathcal{P}_A = \mathcal{P}_C$ it follows that there exists an element $e' \in \mathfrak{D}$ such that $e' \leqslant b$, $J^+(e') = J^+(a \wedge d)$, and $J^-(e') = \mathbb{N}\backslash J^+(e')$. Hence $J^+(b\backslash e') = J^+(a\backslash d) \cup (J^0(a \wedge d) \cap J^+(a))$ and $J^-(b\backslash e') = J^-(a\backslash d) \cup (J^-(a \wedge d) \cap J^0(a)) \cup F$, where F is a subset of $J^0(a)$ such that $d \wedge a_i = a \wedge a_i$ for any $i \in F$. Let $G = (J^0(a \wedge d) \cap J^+(a)) \cup (J^-(a \wedge d) \cap J^0(a))$. For $i \in G$ we set $g_i = f_i(a \wedge d)$. Let $e = e' \vee \bigvee_{i \in G} g_i \vee \bigvee_{i \in F} (b \wedge c_i)$. It is clear that $\langle a \wedge d, b \wedge e \rangle$ and $\langle a\backslash d, b\backslash e \rangle \in S$. Acting by analogy, we can prove that $\langle a, b \rangle \in S$ and $e \in \mathfrak{D}$ imply that there is $d \in \mathfrak{B}$ such that $\langle a \wedge d, b \wedge c \rangle$ and $\langle a\backslash d, b\backslash e \rangle \in S$. Consequently, S is the isomorphism condition. The inverse implication follows from the proof of Lemma 3.5.13 (below).

Lemma 3.5.13. *Let $\mathfrak{B}$ be an infinite Boolean algebra and let the set of its atomic elements be recursive. Then there exists a recursively enumerable principal sequence of atomic elements of $\mathfrak{B}$.*

PROOF. Let $1, b_0, b_1, \ldots$ be an arbitrary recursively enumerable sequence generating the algebra $\mathfrak{B}$, and $\mathfrak{B}^s = \mathrm{gr}(\{b_0, b_1, \ldots, b_s\})^*$, where b_{s+1} is an atom of the Boolean algebra $\mathfrak{B}^{s+1}$ and $\mathfrak{B}^s \neq \mathfrak{B}^{s+1}$. Denote by A^s the set of elements enumerated in A at the step s.

STEP 0. We set $A^0 = \{\varnothing\}$.

STEP $s + 1$. The set A^s is already defined. Let a be an atom of the algebra $\mathfrak{B}^s$ such that $b_{s+1} < a$. If there is an atomic element $c \in \{b_{s+1}, a\backslash b_{s+1}\}$ such that for any $d \in A^s$ it is not true that $c \leqslant d$, then we set $A^{s+1} = A^s \cup \{c\}$. Otherwise, we set $A^{s+1} = A^s$. The proof is completed by an immediate verification of the required conditions. □

We denote by $\mathfrak{C}$ the set of finite sequences consisting of 0 and 1. If u and v are such sequences, then $u * v$ denotes the sequence obtained from u by assigning v from the right. The notation $u \leqslant v$ means that u is the initial segment of the sequence v. A sequence of sequences $\overline{r} = \langle r_0, \ldots, r_n \rangle$ is called a *cross section* of $\mathfrak{C}$ if none of the sequences $r_0, \ldots, r_s$ is the initial segment of another sequence and for any sequence $u \in \mathfrak{C}$ either $r_i \leqslant u$ or $u \leqslant r_i$ for some $i \leqslant n$. Denote by $\mathfrak{C}_{\overline{r}}$ the set of all initial segments (not necessarily proper) of sequences of the sequence $\overline{r}$. Let $\mathfrak{B}$ be a Boolean algebra satisfying the hypotheses of Theorem 3.5.3. We fix a recursively enumerable sequence $1 = b_0, b_1, \ldots$ generating $\mathfrak{B}$. Let $R(\mathfrak{B})$ denote the set $\bigcup_{s \in \omega} \mathrm{Atom}(\mathfrak{B}^s)$, where b_{s+1} is an atom of the algebra $\mathfrak{B}_{s+1}$ and $\mathfrak{B}_s = \mathrm{gr}\{b_0, \ldots, b_s\}$.

3.5. Restricted Fragments of the Theory of Boolean Algebras 235

Lemma 3.5.14. *There exists a recursive function* $g : \mathbb{N} \times \mathfrak{C} \to \mathfrak{B}$ *subject to the following conditions:*

(a) *for any sequence* $r \in \mathfrak{C}$ *there exists* $\lim_s g(s, r) = h(r)$ *and the function* h *realizes an injection of the set* $\mathfrak{C}$ *into the set* $R(\mathfrak{B}) \setminus I(\mathfrak{B})$,

(b) $\operatorname{gr}(\{h(r)/I(\mathfrak{B}) \mid r \in \mathfrak{C}\}) = \mathfrak{B}/I(\mathfrak{B})$,

(c) $h(r)/I(\mathfrak{B}) = (h(r*0) \vee h(r*1))/I(\mathfrak{B})$; *moreover*, $h(r*0)/I(\mathfrak{B}) \wedge h(r*1)I(\mathfrak{B}) = 0_{\mathfrak{B}}/I(\mathfrak{B})$.

The function $g(s, r)$ will be determined in parallel with the construction of the algebra $\mathfrak{B}^s$ and the enumeration of the Ershov–Tarski ideal $I(\mathfrak{B})$ (the ideal $I(\mathfrak{B})$ is recursively enumerable since the sets Atomless $(\mathfrak{B})$ and Atomic $(\mathfrak{B})$ are recursive). At every step, either one of the atoms of the constructed finite subalgebra of $\mathfrak{B}$ is divided or the next element of $I(\mathfrak{B})$ is enumerated. We denote by $\mathfrak{B}^s$ a subalgebra of $\mathfrak{B}$ that is already constructed at the step s and by I^s the finite set of elements of the ideal $I(\mathfrak{B})$ that are already enumerated at the step s. We suppose that $I^s \subset \mathfrak{B}^s$.

STEP 0. We set $\mathfrak{B}^0 = \{\mathbf{1}, \mathbf{0}\}$, $I^0 = \{\varnothing\}$, and $g(0, r) = \mathbf{1}$ for all $r \in \mathfrak{C}$.

STEP $s + 1$. For all $r \in \mathfrak{C}$ we already have $\mathfrak{B}^s$, I^s, and $g(s, r)$. The function $\lambda r g(s, r)$ possesses the following property. Let $\{a_0, \ldots, a_n\} = \operatorname{Atom}(\mathfrak{B}^s) \setminus I^s$. Then there exists a sequence $\bar{r} = \langle r_0, \ldots, r_n \rangle$ that forms a cross section of $\mathfrak{C}$, satisfying the condition $g(s, r_i) = a_i$, and the function $\lambda r g(s, r)$ is an injection on the set $\mathfrak{C}_{\bar{r}}$. Furthermore, if $r_i \leqslant u$, then $g(s, r_i) = g(s, u)$. Let the atom $a \in \operatorname{Atom}(\mathfrak{B}^s)$ be divided at this step. Two cases are possible.

CASE 1: $a \in I^s(\mathfrak{B})$. We set $g(s+1, r) = g(s, r)$ for all $r \in \mathfrak{C}$.

CASE 2: $a \notin I^s(\mathfrak{B})$ and the atom a is divided into two parts: b and c: $a = b \vee c$, $b \wedge c = 0$. Let $a = g(s, r)$. We set $g(s+1, r*0) = b$ and $g(s+1, r*1) = c$. If $r*0 \leqslant u$, then $g(s+1, u) = g(s+1, r*0)$ and if $r*1 \leqslant u$, then $g(s+1, u) = g(s+1, r*1)$. The values of the functions remain the same on the rest of the sequences.

Let the next element $a \in I(\mathfrak{B})$ be enumerated at the step $s+1$. If $a \notin \operatorname{rang} \lambda r g(s, r)$, then the function g remains unchanged. Let $a = g(s, r_0)$ and $r_0 = r*1$. We set $g(s+1, r) = g(s, r*(1-i))$. For any sequence v we set $g(s+1, r*v) = g(s, r*(1-i)*v)$. In the rest of the cases, the function remains unchanged. The description of how to construct the function $\lambda s \lambda r g(s, r)$ is completed. It is clear that the constructed functions are recursive.

We prove that for any sequence $r \in \mathfrak{C}$ there exists $\lim_s g(s, r) = h(r)$, and if $s_0 = \min\{s \mid g(s, r) = h(r) \& g(s, r) \neq g(s, r*0)\}$, then $g(s, r*0) \notin I(\mathfrak{B})$ and $g(s, r*1) \notin I(\mathfrak{B})$. To this end, we consider an empty sequence Λ. We have $g(0, \Lambda) = \mathbf{1}$. Let s' be the greatest number of a step at which the algebra $\mathfrak{B}^{s'}$ contains exactly one atom $a \in A(\mathfrak{B}^{s'})$ that does not belong to the ideal $I(\mathfrak{B})$. Then, at the step $s'+1$, this atom is divided into two parts b and c: $a = b \vee c$, $b \wedge c = 0$; moreover, $b \notin I(\mathfrak{B})$ and $c \notin I(\mathfrak{B})$. From the description of the construction, it follows that there is a number of step $s'' \geqslant s'+1$ such that $g(s'', \Lambda) = a$, $g(s'', \Lambda*0) = b$, and $g(s'', \Lambda*1) = c$. It is clear that $g(s'', \Lambda) = h(\Lambda)$, $g(s'', \Lambda*0) \notin I(\mathfrak{B})$, and $g(s'', \Lambda*1) \notin I(\mathfrak{B})$. We assume that the assertion is valid for sequences of length less than the length of the sequence r and $r = u*i$. By assumption, there is s such that $g(s, u) = h(u)$ and $g(s, r) \notin I(\mathfrak{B})$. The proof is completed as in the case of the sequence Λ. From what has been proved, we obtain all the assertions of the lemma in a standard way. □

We now describe a construction of a recursive Boolean algebra $\mathfrak{D}$ of which existence is required for the proof of the theorem. As usual, the algebra $\mathfrak{D}$ is the union of an infinite sequence of finite Boolean algebras $\mathfrak{D}_0 \subset \mathfrak{D}_1 \subset \ldots$, subslgebras of a recursive atomless Boolean algebras. At the step s, several atoms of the algebra $\mathfrak{D}^s$ are divided into parts. In the construction, elements of the set $R(\mathfrak{D}) \rightleftharpoons \bigcup_s \mathrm{Atom}(\mathfrak{D}^s)$ can be marked by marks of the following set:

$$\{\mathrm{Atomic},\ \mathrm{Atomless}\} \cup \{\boxed{r} \mid r \in \mathfrak{C}\} \cup \{\boxed{\boxed{a}} \mid a \in \mathrm{Atomic}(\mathfrak{B})\}$$

Let A be a recursively enumerable principal sequence of atomic elements of the algebra $\mathfrak{B}$. We construct $\mathfrak{D}$ in parallel with the algebra $\mathfrak{B}$ and the enumeration A. As usual, $\mathfrak{B}^s$ denotes a finite subalgebra of $\mathfrak{B}$ constructed at the step s and A^s denotes the set of elements enumerated in A at the step s.

STEP 0. We set $\mathfrak{D}^0 = \{\mathbf{0}, \mathbf{1}\}$. We mark the unit by $\boxed{\Lambda}$.

STEP $s+1$. The subalgebra $\mathfrak{D}^s$ is already constructed. We consider all the atoms $\mathrm{Atom}(\mathfrak{D}^s)$ having marks from $\mathfrak{C}$. We assume that for any sequence r of length s there is an element of $\mathrm{Atom}(\mathfrak{D}^s)$ which is marked by $\boxed{r}$. We divide an atom marked by $\boxed{r}$ into four parts. The parts are marked by Atomic, Atomless, $\boxed{r*0}$, $\boxed{r*1}$ respectively. All elements of $\mathrm{Atom}(\mathfrak{D}^s)$ marked by Atomless are divided into two parts; each of them is marked by Atomless. If, at the step $s+1$, an element $a \in \mathrm{Atom}(\mathfrak{B}^s)$ is divided into two parts b and c and there is an element of $\mathrm{Atom}(\mathfrak{D}^s)$

3.5. Restricted Fragments of the Theory of Boolean Algebras

marked by $\boxed{a}$, then we divide this element into two parts. One of them is marked by $\boxed{b}$ and the other is marked by $\boxed{c}$. We assume that there exists $d \in A^s$ which is not a mark of some element of the algebra $\mathfrak{D}^s$. We find a sequence r such that $d \leqslant g(s,r)$ and d is incomparable with $g(s,r*0)$ and $g(s,r*1)$. It is easy to see that such a sequence always exists. We seek an element of Atom $(\mathfrak{D}^s)$ that is marked by Atomic, is not marked by any element of $\mathfrak{B}$, and is located under an element marked by $\boxed{r}$. If such an element d_0 exists, then we mark it by $\boxed{d}$ and divide it in such a way that it is possible to establish an isomorphism between Atom $\widehat{(d_0)}_{\mathfrak{D}^{s+1}}$ and $\widehat{(d)}_{\mathfrak{B}^{s+1}}$. Atoms of Atom $([d_0]_{\mathfrak{D}^{s+1}})$ are marked by the corresponding atoms of the algebra $\widehat{(d)}_{\mathfrak{B}^{s+1}}$. This completes the description of the construction.

We emphasize some properties of the constructed algebra $\mathfrak{D}$. We denote by a_r an element marked by $\boxed{r}$. It is obvious that for all r the element a_r does not belong to $I(\mathfrak{D})$ and $a_r/I(\mathfrak{D}) = (a_{r*0} \vee a_{r*1})/I(\mathfrak{D})$, $a_{r*0} \wedge a_{r*1} = 0$. Therefore, $\mathfrak{D}/I(\mathfrak{D})$ is an atomless Boolean algebra. Consequently, $\mathfrak{D}$ has elementary characteristic $(1,0,1)$.

An element of $\mathfrak{D}$ belongs to the ideal $I(\mathfrak{D})$ if and only if it is located under the union of a finite number of elements of $\mathfrak{D}$, each of them marked by Atomic or Atomless. Therefore, the ideal $I(\mathfrak{D})$ is recursive. An element of $\mathfrak{D}$ is atomic if and only if it is located under the union of a finite number of elements of $\mathfrak{D}$ that are marked by Atomic. Therefore, the set Atomic $(\mathfrak{B})$ is recursive. Finally, an element of $\mathfrak{D}$ is atomless if and only if it is the union of a finite number of elements marked by Atomless. Therefore, the ideal of atomless elements is also recursive.

To establish the recursivity of the set of atoms, we make the following remark. In constructing $\mathfrak{D}$, with each element $a \in \mathfrak{D}$ marked by Atomic we associate (as a mark) an element $b \in A$. The further construction is realized so that a recursive isomorphism can be established between $\widehat{(a)}_{\mathfrak{D}}$ and $\widehat{(b)}_{\mathfrak{B}}$. Therefore, the recursivity of the set of atoms of $\mathfrak{D}$ follows from the recursivity of the set of atoms of the algebra $\mathfrak{B}$. The recursivity of the sets Atom $(\mathfrak{D})$, Atomless $(\mathfrak{D})$, Atomic $(\mathfrak{D})$, and $I(\mathfrak{D})$ implies the decidability of the algebra $\mathfrak{D}$ since ch $(\mathfrak{D}) = (1,0,1)$. It remains to show that $\mathfrak{D} \cong \mathfrak{B}$. We note that not only will any element marked by Atomic also be marked by an element of A, but any element of A appears as a mark of an element of $\mathfrak{D}$. Therefore, elements of $\mathfrak{D}$ marked by Atomic can be enumerated so that we obtain a recursively enumerable principal sequence $C = \{c_0, c_1 \ldots\}$ of atomic elements of the algebra $\mathfrak{D}$ such that

$(\widehat{c}_i)_{\mathfrak{D}} \cong (\widehat{a}_i)_{\mathfrak{B}}$. Using Lemma 3.5.12, it is not hard to establish the equality $\mathcal{P}_B = \mathcal{P}_D$. Therefore, $\mathfrak{B} \cong \mathfrak{D}$. □

Proposition 3.5.2 and Theorem 3.5.1 lead to the following claim.

Corollary 3.5.2. *The following assertions hold*:

(a) *a Boolean algebra $\mathfrak{A}$ with $\operatorname{ch}(\mathfrak{A}) = 0$ is decidable if and only if its enrichment to a system of the signature σ_1 is constructivizable.*

(b) *there exists an undecidable not constructivizable Boolean algebra $\mathfrak{A}$ with $\operatorname{ch}_1(\mathfrak{A}) = 0$ and elementary characteristics $(0, \infty, 0)$ and $(0, \infty, 1)$, and only such ones,*

(c) *for the elementary theory of any Boolean algebra $\mathfrak{A}$ with $\operatorname{ch}_1(\mathfrak{A}) > 0$ there exists a Boolean algebra of this theory that is undecidable but constructivizable.*

Thus, even for the signatures σ and σ_1 the classes of constructive algebras and algebras constructive in their enrichments as well as the classes of Boolean algebras constructivizable in the signatures σ and σ_1 are different.

We now study the decidability of restricted theories of constructive Boolean algebras in other signatures. For recursive linear orders $L_1 = (\mathbb{N}, <_1)$ and $L_2 = (\mathbb{N}, <_2)$ we define $L_1 + L_2 = (\mathbb{N}, <)$, where

$$x < y \leftrightharpoons G(x) < G(y) \vee \bigl(G(x) = G(y) = 0 \& [x/2] <_1 [y/2]\bigr)$$
$$\vee \bigl(G(x) = G(y) = 1 \& [x/2] <_2 [y/2]\bigr)$$

and $G(x) = 0$ if 2 divides x and $G(x) = 1$ otherwise. It is clear that $L_1 + L_2$ determines the linear order equal to the sum of the linear orders L_1 and L_2. For recursive linear orders L_i, $i = 1, \ldots, k$, we can define $L_1 + L_2 + \ldots + L_k \leftrightharpoons (\ldots (L_1 + L_2) + \ldots + L_k)$. Let $L = (\mathbb{N}, <)$ be a recursive linear order with the least element. Consider the Boolean algebra $\mathfrak{B}_L$ and its constructivization ν_L.

Let η be a linear order of the type of rational numbers, let ν_η be its single-valued constructivization, and let $<_\eta$ be the corresponding recursive order. We construct step-by-step a recursive set B such that

$(\forall xy)(x <_\eta y) \Rightarrow (\exists z_1, z_2)(z_1 \in B \& z_2 \notin B \& x <_\eta z_1 <_\eta y \& x <_\eta z_2 <_\eta y)$

STEP 0. We set $B_0 = \varnothing$ and $B'_0 = \varnothing$.

STEP $2n + 1$. For $r(n) <_\eta l(n)$, since η is dense and B'_{2n} is finite, there exists $x \notin B'_{2n}$ such that $r(n) <_\eta x <_\eta l(n)$. We consider the least

3.5. Restricted Fragments of the Theory of Boolean Algebras 239

such x and set $B_{2n+1} \leftrightharpoons B_{2n} \cup \{x\}$, $B'_{2n+1} \leftrightharpoons B'_{2n}$. If $r(n) \geqslant_\eta l(n)$ we change nothing, i.e., we define $B_{2n+1} \leftrightharpoons B_{2n}$ and $B'_{2n+1} \leftrightharpoons B'_{2n}$.

STEP $2n + 2$. For $r(n) <_\eta l(n)$, since η is dense and B_{2n+1} is finite, there exists $x \notin B_{2n+1}$ such that $r(n) <_\eta x <_\eta l(n)$. We define $B_{2n+2} \leftrightharpoons B_{2n+1}$ and $B'_{2n+2} \leftrightharpoons B'_{2n+1} \cup \{x\}$. For $r(n) \geqslant_\eta l(n)$ we set $B_{2n+2} \leftrightharpoons (B_{2n+1} \cup \{0, 1, \ldots, n\}) \setminus B'_{2n+1}$, $B'_{2n+2} \leftrightharpoons B'_{2n+1}$. It follows from the construction that the sets $B = \bigcup B_n$ and $B' = \bigcup B'_n$ are recursively enumerable and $B \cap B' = \emptyset$, $B \cup B' = \mathbb{N}$. Therefore, B is recursive and, by construction, possesses the required property. □

Let C_B be a general recursive characteristic function of B and G be a set of odd numbers such that $G(x) = 1$ for odd x and $G(x) = 0$ for even x. For every $n \geqslant 0$ we define effectively on n a recursive strict linear preorder $L_n = \langle \mathbb{N}, <_n \rangle$ and two auxiliary recursive strict linear preorders $L_n^0 = \langle \mathbb{N}, <_n^0 \rangle$ and $L_n^1 = \langle \mathbb{N}, <_n^1 \rangle$. We note that if P is an order on A, then $P' \leftrightharpoons \{(x, y) \mid P(x, y) \,\&\, \neg P(y, x)\}$ is a strict order. The order is uniquely defined from P'. Let

$$x \leqslant_0 y \Leftrightarrow x, y \in \mathbb{N}, \quad x <_0^0 y \Leftrightarrow x < y$$
$$x <_0^1 y \Leftrightarrow (x = 0 \,\&\, y \neq 0) \vee (x > 0 \,\&\, y > 0 \,\&\, x - 1 <_\eta y - 1)$$
$$x <_1 y \Leftrightarrow r(x) < r(y) \vee \Big(r(x) = r(y) \,\&\, \Big((G(r(x)) = 0$$
$$\&\, l(x) <_0 l(y)) \vee (G(r(x)) = 1 \,\&\, l(x) <_0^1 l(y))\Big)\Big).$$

If $<_n^0$, $<_n^1$, $<_n$ $<_{n+1}$ are already constructed, then we define $<_{n+1}^0$, $<_{n+1}^1$, and $<_{n+2}$ as follows:

$$x <_{n+1}^0 y \Leftrightarrow r(x) < r(y) \vee (r(x) = r(y) \,\&\, l(x) <_n l(y))$$
$$x <_{n+1}^1 y \Leftrightarrow (x = 0 \,\&\, y > 0) \vee \Big(x > 0 \,\&\, y > 0$$
$$\&\, \Big(r(x - 1) <_\eta r(y - 1) \vee \Big(r(x - 1) = r(y - 1)$$
$$\&\, (r(x - 1) \in B \,\&\, l(x - 1) <_n^0 l(y - 1))$$
$$\vee (r(x - 1) \notin B \,\&\, l(x - 1) <_n^1 l(y - 1))\Big)\Big)\Big)$$
$$x <_{n+2} y \Leftrightarrow r(x) < r(y) \vee \Big(r(x) = r(y) \,\&\, \big((G(r(x)) = 0$$
$$\&\, l(x) <_{n+1}^0 l(y)\big) \vee (G(r(x)) = 1 \,\&\, l(x) <_{n+1}^1 l(y))\big)\Big)$$

240   3. Constructive Boolean Algebras

We note that 0 is the least element of all of these recursive orders. Starting from the recursive orders L_n, L_n^0, and L_n^1, we effectively define the constructive Boolean algebras $(\mathfrak{B}_{L_n}, \nu_{L_n})$, $(\mathfrak{B}_{L_n^0}, \nu_{L_n^0})$, and $(\mathfrak{B}_{L_n^1}, \nu_{L_n^1})$. The Boolean algebra $\mathfrak{B}_{L_n}$, together with the constructivization obtained from the recursive linear order L_n, is denoted by $(\mathfrak{B}_n, \nu_n)$. By construction, we can prove by induction the following lemma.

Lemma 3.5.15. *The elementary characteristics of* $\mathfrak{B}_n$, $\mathfrak{B}_{L_n^1}$, *and* $\mathfrak{B}_{L_n^0}$ *are equal to* $(n, 1, 0)$, $(n, 0, 1)$, *and* $(n, \infty, 0)$ *respectively.*

Lemma 3.5.16. *Let* $n > 1$ *and* $[x, y[\in \mathfrak{B}_{L_n}$. *Then*

(a) *if* $r(x) = r(y)$ *and* $G(r(x)) = 0$, *then* $(\mathfrak{B}_{L_n})_{[x,y[} \cong (\mathfrak{B}_{L_{n-1}})_{[l(x),l(y)[}$,

(b) *if* $r(x) = r(y)$ *and* $G(r(x)) = 1$, *then* $(\mathfrak{B}_{L_n})_{[x,y[} \cong (\mathfrak{B}_{L_{n-1}^1})_{[l(x),l(y)[}$,

(c) *if* $r(x) < r(y)$, $k_1 \leftrightharpoons |\{i \mid G(i) = 0, r(x) \leqslant i < r(y)\}|$, $k_2 \leftrightharpoons 1$ *if* $\{i \mid G(i) = 1, r(x) < i < r(y)\} \neq \emptyset$ *or* $G(r(y)) = 1 \& l(y) > 0$, *and* $k_2 \leftrightharpoons 0$ *otherwise, then the elementary characteristic of* $[x, y[$ *is* (n, k_1, k_2),

(d) *if* $[x, \infty[\in \mathfrak{B}_{L_n}$, *then the elementary characteristic of* $[x, \infty[$ *is* $(n, 1, 0)$.

Lemma 3.5.17. *Let* $0 \neq [x, y[\in \mathfrak{B}_{L_n^1}$ *and* $n > 0$. *Then*

(a) *if* $(x = 0 \& y > 0) \vee (x > 0 \& y > 0 \& r(x - 1) <_\eta r(y - 1))$, *then the elementary characteristic of* $[x, y[$ *is* $(n, 0, 1)$.

(b) *if* $x > 0 \& y > 0 \& r(x - 1) = r(y - 1) \in B$, *then* $(\mathfrak{B}_{L_n^1})_{[x,y[} \cong (\mathfrak{B}_{L_{n-1}^0})_{[l(x-1),l(y-1)[}$,

(c) *if* $x > 0 \& y > 0 \& r(x - 1) = r(y - 1) \notin B$, *then* $(\mathfrak{B}_{L_n^1})_{[x,y[} \cong (\mathfrak{B}_{L_{n-1}^1})_{[l(x-1),l(y-1)[}$.

Lemma 3.5.18. *The algebras* $\mathfrak{B}_{L_1}$ *and* $\mathfrak{B}_{\omega+\eta}$ *are isomorphic and in the case* $0 \neq [x, y[\in \mathfrak{B}_{L_1}$

(a) *if* $r(x) < r(y)$, $k_1 \leftrightharpoons |\{i \mid G(i) = 0 \& r(x) \leqslant i < r(y)\}|$, $k_2 = 1$ *for* $\{i \mid G(i) = 1 \& r(x) < i < r(y)\} \neq \emptyset$ *or* $G(r(y)) = 1 \& l(y) > 0$ *and* $k_2 = 0$ *otherwise, then the elementary characteristic of* $[x, y[$ *is* $(0, k_1, k_2)$,

3.5. Restricted Fragments of the Theory of Boolean Algebras

(b) *if $r(x) = r(y) \& G(r(x)) = 1$, then the elementary characteristic of $[x, y[$ is $(0, 0, 1)$.*

Lemma 3.5.19. *The algebras $\mathfrak{B}_{L_0^1}$ and $\mathfrak{B}_{1+\eta}$ are isomorphic. If $0 \neq [x, y[\in \mathfrak{B}_{L_0^1}$, then the elementary characteristic of $[x, y[$ is $(0, 0, 1)$.*

In view of Lemmas 3.5.15–3.5.19 and the method of constructing $\mathfrak{B}_L$, we can effectively find the elementary characteristic of an element from its number in the constructivization of the Boolean algebra $(\mathfrak{B}_n, \nu_n)$. The following two lemmas are an immediate consequence of definitions.

Lemma 3.5.20. *If $a \in |\mathfrak{B}_n|$, then its elementary characteristic (n_1, n_2, n_3) is such that $n_1 \leqslant n \& n_2 < \infty \& (n_1 = n \to (n_2 = 1 \& n_3 = 0))$.*

Lemma 3.5.21. *Let the elementary characteristic of a Boolean algebra $\mathfrak{B}$ be $(n, 1, 0)$ and let $a_i \wedge a_j = 0 \& \bigvee_{i=1}^{m} a_i = 1$ for $a_1, \ldots, a_m \in |\mathfrak{B}|$, $i \neq j$. Then there exists a unique j_0 such that for any $j \neq j_0$ the elementary characteristic (n_1^j, n_2^j, n_3^j) of a_j is such that $n_1^j < n$, and the elementary characteristic of a_{j_0} is $(n, 1, 0)$.*

Lemma 3.5.22. *The following assertions hold:*

(a) *if $a \in |\mathfrak{B}|$, then a has the elementary characteristic (m_1, m_2, m_3), where $m_1 < \infty$, $m_2 < \infty$, and $m_3 \in \{0, 1\}$, if and only if $\mathfrak{B} \models E^{\langle m_1, m_2, m_3 \rangle}(a)$,*

(b) *if the elementary characteristic of $\mathfrak{B}$ is (n_1, n_2, n_3), then the formula $(\exists x)(E^{\langle m_1, m_2, m_3 \rangle}(x))$ is true in $\mathfrak{B}$ for any triple (m_1, m_2, m_3) such that $(m_1 = n_1 \& m_2 \leqslant n_2 \& m_3 \leqslant n_3) \vee (m_1 < n_1 \& m_3 \in \{0, 1\} \& m_2 < \infty)$.*

Let $\mathfrak{B}$ be a Boolean algebra of characteristic (n_1, n_2, n_3).

Proposition 3.5.7. *If $n_2 < \infty$, then there exists a Boolean algebra $\mathfrak{B}'$ such that $\mathfrak{B} \equiv \mathfrak{B}'$ and $(\forall a)(a \in |\mathfrak{B}'| \Rightarrow \mathrm{ch}_2(a) < \infty)$.*

If $n_2 = \infty$, then there exists a Boolean algebra $\mathfrak{B}'$ such that $\mathfrak{B} \equiv \mathfrak{B}'$ and $(\forall a)\Big(a \in |\mathfrak{B}'| \Rightarrow (\mathrm{ch}_1(a) < n_1 \Rightarrow \mathrm{ch}_2(a) < \infty) \& \big(\mathrm{ch}_1(a) < n_1 \vee (\mathrm{ch}_1(a) = n_1 \& (\mathrm{ch}_2(a) < \infty \vee (\mathrm{ch}_1(C(a)) \leqslant n_1 \& \mathrm{ch}_2(C(a)) < \infty)))\big)\Big)$.

PROOF. In view of Lemmas 3.5.15–3.5.20, as $\mathfrak{B}'$ we can take the Boolean algebra $\mathfrak{B}_{L_{n_1}}^{n_2} \times \mathfrak{B}_{L_{n_1}^1}^{n_3}$ for $n_1 < \infty$, $n_2 < \infty$, the Boolean algebra defined by the order $\sum_{i\in\mathbb{N}} L_i$ for $n_1 = \infty$, and the Boolean algebra $\mathfrak{B}_{L_{n_1}^0} \times \mathfrak{B}_{L_{n_1}^1}^{n_3}$ for $n_2 = \infty$. □

We denote by $(\mathfrak{B}^n, \nu^n)$ the strongly constructive Boolean algebra of characteristic $(n, 1, 0)$ from Proposition 3.4.2.

Proposition 3.5.8. *For any n and $a_1, \ldots, a_k \in \mathbb{N}$ the numbers $b_1, \ldots, b_k \in \mathbb{N}$ can be effectively found so that*

$$(\mathfrak{B}_n, \nu_n a_1, \ldots, \nu_n a_k) \equiv (\mathfrak{B}^n, \nu^n b_1, \ldots, \nu^n b_k)$$

PROOF. We consider $\overline{a}_{(i_1,\ldots,i_k)} \leftrightharpoons \bigwedge_{j=1}^{k} \nu_n a_j^{i_j}$, where $(i_1, \ldots, i_k) \in \{0,1\}^k$, $a^0 \leftrightharpoons C(a)$, and $a^1 \leftrightharpoons a$. Since the elementary characteristic of an element of $\langle \mathfrak{B}_n, \nu_n \rangle$ is effectively found from its number in this constructivization, it is possible to find the elementary characteristics of the elements $\overline{a}_{(i_1,\ldots,i_k)}$. We denote by (n_1^a, n_2^a, n_3^a) the elementary characteristic of $\nu_n a$. We enumerate the set $\{\overline{a}_{(i_1,\ldots,i_k)}\}$ so that $i \geqslant j \Rightarrow n_1^{a_i} \geqslant n_1^{a_j}$, say $\{\overline{a}_i\}_{i=1}^{l}$. By Lemma 3.5.21, $n_1^{\overline{a}_l} = n+1$, $n_2^{\overline{a}_l} = 1$, and $n_3^{\overline{a}_l} = 0$. By Lemma 3.5.20, $(\forall i < l)(n_1^{\overline{a}_i} \leqslant n \,\&\, n_2^{\overline{a}_i} < \infty)$. We can assume that $l \geqslant 2$ since the case $l = 1$ is trivial. We apply the following process.

STEP 1. We consider $E^{\langle n_1^{\overline{a}_1}, n_2^{\overline{a}_1}, n_3^{\overline{a}_1} \rangle}(x)$. Inserting subsequently elements $\nu^n m$ of $\mathfrak{B}^n$ into the above formula and using the strong constructibility of $(\mathfrak{B}^n, \nu^n)$, we find the number of $\overline{b}_1$ such that

$$\mathfrak{B}^n \models E^{\langle n_1^{\overline{a}_1}, n_2^{\overline{a}_1}, n_3^{\overline{a}_1} \rangle}(\nu^n \overline{b}_1)$$

STEP i, $1 < i < l$. We choose an element $\overline{b}_i$ such that $\nu^n \overline{b}_i$ satisfies the formula

$$E^{\langle n_1^{\overline{a}_i}, n_2^{\overline{a}_i}, n_3^{\overline{a}_i} \rangle}(x) \,\&\, x \wedge \left(\bigvee_{j<i} \nu^n \overline{b}_j \right) = 0$$

By Lemma 3.5.22, at every step $i < l$, such $\overline{b}_i$ can be found. Defining $\overline{b}_i$, $i \leqslant l-1$, as $\overline{b}_l$ we take the number of element $\mathbf{1} \setminus \left(\bigvee_{j<l} \nu^n \overline{b}_j \right)$. By the strong constructibility of $(\mathfrak{B}^n, \nu^n)$, the numbers $\overline{b}_i$ are effectively found. By Lemmas 3.5.21 and 3.5.22 they are such that the elementary characteristic of $\nu^n \overline{b}_i$ for any $i \leqslant l$ in $\mathfrak{B}^n$ is $(n_1^{\overline{a}_i}, n_2^{\overline{a}_i}, n_3^{\overline{a}_i})$. Therefore, using Lemma 3.5.22 and setting $\nu b_i = \bigvee_{\overline{a}_j \leqslant \nu a_i} \nu \overline{b}_j$, we obtain the required elements. □

3.5. Restricted Fragments of the Theory of Boolean Algebras

Propositions 3.4.4 and 3.5.8 imply the following claim.

Corollary 3.5.3. $\langle \mathfrak{B}_n, \nu_n \rangle$ *is a strongly constructive Boolean algebra and the deciding procedure is effectively constructed from n.*

Acting as in the proof of Proposition 3.5.8, we can prove the following claim.

Proposition 3.5.9. *The following assertions hold:*

(a) $(\mathfrak{B}_{L_n^0}, \nu_{L_n^0})$ *is strongly constructive,*

(b) $(\mathfrak{B}_{L_n^1}, \nu_{L_n^1})$ *is strongly constructive.*

We consider the recursive linear order $L = (\mathbb{N}, \subset)$ constructed in Theorem 3.5.1 such that $\mathfrak{B}_L$ is an atomic Boolean algebra constructivizable but not strongly constructivizable. For any constructivization ν of the Boolean algebra $\mathfrak{B}_L$ $\forall$-theory $(\mathfrak{B}_L, \nu 0, \nu 1, \nu 2, \ldots)$ is not decidable; otherwise, the set of atoms is recursive (since a $\forall$-formula is distinguished); consequently, this constructivization must be strong in view of Theorem 3.4.8, which contradicts the assumption.

Let $L_n = (\mathbb{N}, <_n)$ be the recursive orders just defined. We introduce a new order $L^n = (\mathbb{N}, <^n)$ by setting

$$x <^n y \rightleftharpoons r(x) \subset r(y) \vee (r(x) = r(y) \mathbin{\&} l(x) <_n l(y))$$

Lemma 3.5.23. $(\mathfrak{B}_{L^n})/I_n(\mathfrak{B}_{L^n}) \cong \mathfrak{B}_L$.

PROOF. It is easy to check that if

$$\varphi\left(\bigcup_{i=1}^{k}[x_i, y_i[\right) \rightleftharpoons \bigcup_{i=1}^{k}[c(0, x_i), c(0, y_i)[/I_n(\mathfrak{B}_{L^n})$$

then φ is an isomorphism of $\mathfrak{B}_L$ onto $\mathfrak{B}_{L^n}/I_n(\mathfrak{B}_{L^n})$. □

Lemma 3.5.24. *If* $0 \neq [x, y[\in \mathfrak{B}_{L^n}$ *and* $r(x) \subset r(y)$, *then*

$$\mathrm{ch}_1([x, y[) \geqslant n$$

Lemma 3.5.25. *If* $0 \neq [x, y[\in \mathfrak{B}_{L^n}$ *and* $r(x) = r(y)$, *then*

$$\mathrm{ch}_{\mathfrak{B}_{L^n}}([x, y[) = \mathrm{ch}_{\mathfrak{B}_{L_n}}([l(x), l(y)[)$$

Proposition 3.5.10. *For any* $a_1, \ldots, a_k \in \mathbb{N}$ *it is possible to find effectively* $b_1, \ldots, b_k$ *such that*

$$(\mathfrak{B}_{L^n}, \nu_{L^n} a_1, \ldots, \nu_{L^n} a_k) \equiv_{4n} (\mathfrak{B}_{L^{n+1}}, \nu_{n+1} b_1, \ldots, \nu_{n+1} b_k)$$

PROOF. As in Proposition 3.5.8, we define elements $\overline{a}_{(i_1,\ldots,i_k)}$. In view of Lemmas 3.5.15–3.5.19, starting from the number of $\overline{a}_{(i_1,\ldots,i_k)}$, we can effectively recognize its elementary characteristic provided that $\mathrm{ch}_1(\overline{a}_{(i_1,\ldots,i_k)}) < n$. In the opposite case, which is also effectively recognized, we cannot find the elementary characteristic precisely, but we know that $\mathrm{ch}_1(\overline{a}_{(i_1,\ldots,i_k)}) \geqslant n$. We enumerate elements $\overline{a}_{(i_1,\ldots,i_k)}$ so that elements of the first type such that $\mathrm{ch}_1(\overline{a}_{(i_1,\ldots,i_k)}) < n$ are at the beginning, and those elements such that $\mathrm{ch}_1(\overline{a}_{(i_1,\ldots,i_k)}) \geqslant n$ are after them. For elements of the first type we define the corresponding elements $\overline{b}_{(i_1,\ldots,i_k)}$ of the same elementary characteristic as $\overline{a}_{(i_1,\ldots,i_k)}$. A similar procedure was applied in Proposition 3.5.8 for $i < l$. It is obvious that the complement a to the union of elements of the first type in $\mathfrak{B}_{L^n}$ and the complement b to the union of the corresponding $\overline{b}_{(i_1,\ldots,i_k)}$ in $\mathfrak{B}_{L^{n+1}}$ have elementary characteristics equal to elementary characteristics $\mathfrak{B}_{L^n}$ and $\mathfrak{B}_{L^{n+1}}$. Let $\overline{a}_{(i_1^1,\ldots,i_k^1)}, \ldots, \overline{a}_{(i_1^s,\ldots,i_k^s)}$ be the rest of the elements. We now choose elements $\nu_{n+1} x_1, \ldots, \nu_{n+1} x_s$ of $\mathfrak{B}_{L^{n+1}}$ such that for every i, $1 \leqslant i \leqslant s$, the elementary characteristic $\nu_{n+1}(x_i)$ is equal to $(n, 1, 0)$ and $\nu_{n+1} x_i \leqslant b$, $\nu_{n+1} x_i \wedge \nu_{n+1} x_j = 0$, $i \neq j$. Define

$$\overline{b}_{(i_1^1,\ldots,i_k^1)} = b \setminus \bigvee_{i=2}^{s} \nu_{n+1} x_i$$

and for $j \geqslant 2$ define $\overline{b}_{(i_1^j,\ldots,i_k^j)} \rightleftharpoons \nu_{n+1} x_i$. Then

$$(\mathfrak{B}_{L^n}, \nu_{L^n} a_1, \ldots, \nu_{L^n} a_k) \equiv_{4n} (\mathfrak{B}_{L^{n+1}}, \nu_{n+1} b_1, \ldots, \nu_{n+1} b_k)$$

where $\nu_{n+1} b_i \rightleftharpoons \bigvee_{\overline{a}_{(i_1,\ldots,i_k)} \leqslant \nu_{L^n} a} b_{(i_1,\ldots,i_k)}$. □

Theorem 3.5.4. *There exists a Boolean algebra* $\mathfrak{B}$ *such that the theory* $\mathrm{Th}_{4n+1}(\mathfrak{B}, \nu_0, \nu_1, \ldots)$ *is undecidable for any constructivization* ν *of* $\mathfrak{B}$, *but there exists a constructivization* μ *of* $\mathfrak{B}$ *such that the theory* $\mathrm{Th}_{4n}(\mathfrak{B}, \mu_0, \mu_1, \ldots)$ *is decidable.*

PROOF. We consider a Boolean algebra $\mathfrak{B}_{L^n}$. By Proposition 3.5.10, $\mathrm{Th}_{4n}(\mathfrak{B}_{L^n}, \nu_{L^n}(0), \nu_{L^n}(1))$ is decidable. Assume that there exists a constructivization μ of $\mathfrak{B}_{L^n}$ such that $\mathrm{Th}_{4n+1}(\mathfrak{B}_{L^n}, \mu_0, \mu_1, \ldots)$ is decidable. Then $\mathrm{Th}_1\left(\mathfrak{B}_{L^n}/I_n, \mu/I_n(0), \mu/I_n(1), \ldots\right) = \mathrm{Th}_1(\mathfrak{B}_L, \nu 0, \nu 1, \ldots)$ is de-

3.5. Restricted Fragments of the Theory of Boolean Algebras

cidable. In view of the choice of the order L and the constructivization ν of the Boolean algebra $\mathfrak{B}_L$ from Theorem 3.5.1, the theory

$$\Pi_1 \cap \operatorname{Th}_1(\mathfrak{B}_L, \nu_0, \nu_1, \ldots)$$

is undecidable and restricted. The contradiction obtained completes the proof of the theorem. □

We construct a Boolean algebra $\mathfrak{B}$ which is not strongly constructivizable, but there exists a constructivization ν such that all the restricted theories $\operatorname{Th}_n(\mathfrak{B}, \nu 0, \nu 1, \ldots)$ are decidable.

Lemma 3.5.26. *If $\langle \mathfrak{B}, \nu \rangle$ is a strongly constructive Boolean algebra, then $\omega_{\mathfrak{B}} \rightleftharpoons \{m \mid (\exists x \in |\mathfrak{B}|) \operatorname{ch}(x) = (m, \infty, 0)\} \in \Sigma_2^0$.*

PROOF. It is obvious that

$$m \in \omega_{\mathfrak{B}} \Leftrightarrow (\exists n)(\forall k)\Big[(\exists x \in |\mathfrak{B}|)\big(E^{(m,k,0)}(x) \& x \leqslant \nu n\big)$$

$$\& (\forall y)\big(y \leqslant \nu n \to \neg E^{(m,0,1)}(y)\big) \& I_{m+1}(\nu n) \& \neg I_m(\nu n)\Big]$$

Since $\langle \mathfrak{B}, \nu \rangle$ is a strongly constructive Boolean algebra, the above formula can be transformed as follows: $(\exists n)(\forall k) P'(n, k, m)$, where P' is a recursive predicate. □

We consider $A \in \Pi_2^0 \setminus \Sigma_2^0$. Applying the Kreisel–Shoenfield–Wang algorithm [178], we conclude that there exists a recursive predicate $P(x, y)$ such that $x \in A \Leftrightarrow \exists^\omega y\, P(x, y)$.

Since the set

$$L_A \rightleftharpoons \{c(c(m, i), 0) \mid m, i \in \mathbb{N}\} \cup \{c(c(m, i), z + 1) \mid i, z, m \in \mathbb{N} \,\&\, P(m, z)\}$$

is recursive, there exists a recursive function $f \colon \mathbb{N} \underset{\text{onto}}{\to} L_A$. We introduce a recursive linear order $<_A$ as follows:

$$x <_A y \rightleftharpoons f(x) = c(c(m_1, i_1), z_1) \,\&\, f(y) = c(c(m_2, i_2), z_2)$$
$$\&\, \big(m_1 < m_2 \vee (m_1 = m_2 \,\&\, z_1 < z_2) \vee (m_1 = m_2 \,\&\, z_1 = z_2$$
$$\&\, i_1 <_{m_1+1} i_2)\big)$$

where $<_m$ is the order just defined.

Lemma 3.5.27. *For any m there exists a partially recursive function f_m and a number k such that $f_m \colon \Delta_m \underset{\text{onto}}{\overset{1-1}{\to}} \mathbb{N}$ and $(\forall xy \in \Delta_m)(x <_A y \Leftrightarrow f_m(x) \prec_{m+1}^k f_m(y))$; moreover, $k = 0 \Leftrightarrow m \in A$ and $\prec_m^0 \rightleftharpoons <_{m+1}^0$ is the above order and for $k > 0$, $\prec_m^k$ is a recursive order*

on $L_m + L_m + \ldots + L_m$,
$$\Delta_m \rightleftharpoons \{x \mid f^{-1}(c(c(m,0),0)) \leqslant_A x <_A f^{-1}(c(c(m+1,0),0))\}$$

PROOF. If $m \notin A$, then there exists k_0 such that $\exists^{\leqslant k_0} z\, P(m,z)$. We consider $z_1 < z_2 < \ldots < z_{k_0}$ such that $P(m, z_i)$. By construction,
$$\Delta_m = \{x \mid f(x) = c(c(m,n), z)\ \&\ z \in \{0, z_1+1, \ldots, z_{k_0}+1\}\}$$
We define $f_m(x) = 2^{k-i}(2rlf(x)+1)$, where $rf(x) = z_i + 1$ if $x \in \Delta_m$ and $f_m(x) = 2^k rlf(x)$ if $rf(x) = 0$ and $x \in \Delta_m$, $k \rightleftharpoons k_0 + 1$. It is easy to check that $f_m(x)$ is required and, in this case, $k > 0$.

If $m \in A$, then $\exists^\omega z\, P(m,z)$ and the set $A_m \rightleftharpoons \{z \mid P(m,z)\}$ is infinite and recursive. Hence there exists a general recursive monotonically (strictly) increasing function g_m enumerating the set A_m. We introduce the function
$$f_m(x) \rightleftharpoons \begin{cases} \text{is not defined,} & x \notin \Delta_m \\ c(rlf(x), 0), & rf(x) = 0,\ x \in \Delta_m \\ c(rlf(x), \mu y(g_m(y) = rf(x)) + 1)) & \text{otherwise} \end{cases}$$

It is obvious that f_m realizes an isomorphic mapping of the interval Δ_m onto $(\mathbb{N}, \prec^0_{m+1})$. □

Let $\overline{\Delta}_m \rightleftharpoons \sum_{i>m} \Delta_i$, $\overline{\nu}_m$ be the restriction of ν_{L_A} to $(\mathfrak{B}_{L_A})_{\overline{\Delta}_m}$. For brevity, ν_{L_A} will be written as ν_A.

Lemma 3.5.28. If
$$x <_A y\ \&\ (llf(x) = llf(y) \vee (llf(x) < llf(y)\ \&\ rf(x) < rf(y)))$$
then $\mathrm{ch}_1([x, y[) \geqslant llf(x)$.

Lemma 3.5.29. If
$$x <_A y, llf(x) = llf(y), rf(x) = rf(y)$$
then
$$(\mathfrak{B}_{L_A})_{[x,y[} \cong (\mathfrak{B}_{L_{llf(x)+1}})_{[rlf(x), rlf(y)[}$$
If $[x, \infty[\in \mathfrak{B}_{L_A}$, then $\mathrm{ch}_1([x, \infty[) = \infty$.

These lemmas follow from Lemma 3.5.27 and the construction. It is easy to obtain the following claim.

3.5. Restricted Fragments of the Theory of Boolean Algebras 247

Corollary 3.5.4. *Starting from the number of an element of $((\mathfrak{B}_{L_A})_{\overline{\Delta}_m}, \overline{\nu}_m)$, it is possible to effectively recognize whether the first coordinate of the elementary characteristic of this element is greater than m or define it exactly.*

Lemma 3.5.30. *For any $a_1, \ldots, a_k \in \mathbb{N}$ it is possible to effectively find $b_1, \ldots, b_k \in \mathbb{N}$ such that*
$$((\mathfrak{B}_{L_A})_{\overline{\Delta}_m}, \overline{\nu}_m a_1, \ldots, \overline{\nu}_m a_k) \equiv_{4m} (\mathfrak{B}_{L_{m+2}}, \nu_{m+2} b_1, \ldots, \nu_{m+2} b_k)$$

PROOF. The reasoning is similar to that of the proof of Proposition 3.5.10 in view of Corollary 3.5.4. □

Corollary 3.5.5. $\mathrm{Th}_{4m}((\mathfrak{B}_{L_A})_{\overline{\Delta}_M}, \overline{\nu}_m 0, \overline{\nu}_m 1, \ldots)$ *is decidable.*

Lemmas 3.5.20 and 3.5.27 imply the following claim.

Lemma 3.5.31. $(\forall a \in \mathfrak{B}_{L_A})((a \subseteq \overline{\Delta}_m \,\&\, \mathrm{ch}_1(a) < m) \Rightarrow \mathrm{ch}_2(a) < \infty)$.

Theorem 3.5.5. *There exists a Boolean algebra $\mathfrak{B}$ of characteristic $(\infty, 0, 0)$ such that*

(a) *for any constructivization ν for $\mathfrak{B}$ the theory $\mathrm{Th}(\mathfrak{B}, \nu 0, \nu 1, \ldots)$ is undecidable,*

(b) *there exists a constructivization μ of $\mathfrak{B}$ such that for any m the theory $\mathrm{Th}_m(\mathfrak{B}, \mu 0, \mu 1, \ldots)$ is decidable.*

PROOF. Let us show that a Boolean algebra $\mathfrak{B} \rightleftharpoons \mathfrak{B}_{L_A}$ constructed from the recursive order L_A defined above satisfies the hypotheses of the theorem.

We assume that there exists a strong constructivization ν of $\mathfrak{B}$. By Lemma 3.5.26, $\omega_{\mathfrak{B}} \in \Sigma_2^0$, but by Lemma 3.5.27 and 3.5.31, $1 + m \in \omega_{\mathfrak{B}} \Leftrightarrow m \in A$. Consequently, $\omega_{\mathfrak{B}} \in \Pi_2^0 \setminus \Sigma_2^0$ in view of the choice of $A \in \Pi_2^0 \setminus \Sigma_2^0$. The contradiction obtained proves (a).

For $\mathfrak{B}_{L_A}$ we consider a constructivization constructed from the recursive order L_A and denote ν_{L_A} by ν. It is obvious that

$$(\mathfrak{B}_{L_A}, \nu) \cong \left(\prod_{i=1}^{m}(\mathfrak{B}_{L_A})_{\Delta_i} \times (\mathfrak{B}_{L_A})_{\overline{\Delta}_m}, \nu\right)$$

By Lemma 3.5.27, $((\mathfrak{B}_{L_A})_{\Delta_i}, \nu_{\Delta_i}) \cong (\mathfrak{B}_{i+1}, \nu_{i+1})^k$ or $((\mathfrak{B}_{L_A})_{\Delta_i}, \nu_{\Delta_i}) \cong (\mathfrak{B}_{L_{i+2}^0}, \nu_{L_{i+2}^0})$. By Propositions 3.5.8 and 3.5.9, ν_{Δ_i} is a strong constructivization. Therefore, $\mathrm{Th}_m((\mathfrak{B}_{L_A})_{\Delta_i}, \nu_{\Delta_i}0, \nu_{\Delta_i}1, \ldots)$ is decidable. By Corollary 3.5.5, $\mathrm{Th}_m((\mathfrak{B}_{L_A})_{\overline{\Delta}_m}, \nu_{\overline{\Delta}_m}0, \nu_{\overline{\Delta}_m}1, \ldots)$ is decidable. In view of Proposition 3.1.8, $\mathrm{Th}_m(\mathfrak{B}_{L_A}, \nu 0, \nu 1, \ldots)$ is decidable. □

Applying the above results and methods, one can prove the following theorem.

Theorem 3.5.6. *For any n and the elementary characteristic* (n_1, n_2, n_3), *where* $n_1 > n \vee (n_1 = n \& n_2 = \infty)$, *there exists a Boolean algebra* $\mathfrak{B}$ *of characteristic* (n_1, n_2, n_3) *such that for any constructivization* ν *of* $\mathfrak{B}$ *the theory* $\mathrm{Th}_{4n+1}(\mathfrak{B}, \nu 0, \nu 1, \ldots)$ *is undecidable, but there exists a constructivization* μ *of* $\mathfrak{B}$ *such that* $\mathrm{Th}_{4n}(\mathfrak{B}, \mu 0, \mu 1, \ldots)$ *is decidable.*

By comparison, if $\langle \mathfrak{B}, \nu \rangle$ is a constructive Boolean algebra, the theory $\mathrm{Th}_{4n+1}(\mathfrak{B}, \nu 0, \nu 1, \ldots)$ is decidable, and $n+1 = \mathrm{ch}_1(\mathfrak{B})$, then the theory $\mathrm{Th}(\mathfrak{B}, \nu 0, \nu 1, \ldots)$ is decidable and if $\mathrm{Th}_{4n+4}(\mathfrak{B}, \nu 0, \nu 1, \ldots)$ is decidable and $\mathrm{ch}_1(\mathfrak{B}) = n + 1 \& \mathrm{ch}_2(\mathfrak{B}) < \infty$, then the theory $\mathrm{Th}(\mathfrak{B}, \nu 0, \nu 1, \ldots)$ is decidable.

Exercises

1. Prove that an atomic constructive Boolean algebra $\langle \mathfrak{B}, \nu \rangle$ with the Frechét ideal of complexity Δ_2^0 in the arithmetic hierarchy is strongly constructivizable.

2. Prove that a Boolean algebra of characteristic (n, k, m) is constructive with respect to the signature
$$\langle \vee^2, \wedge^2, C^1, 0, 1; \mathrm{Atom}_0, \mathrm{Atomless}_0, \mathrm{Atomic}_0, I_0, \ldots, \mathrm{Atom}_n,$$
$$\mathrm{Atomless}_n, \mathrm{Atomic}_n, I_n \rangle$$

3. Prove that a Boolean algebra of characteristic $(0, \infty, 0)$ with decidable set of atoms is strongly constructive.

4. Prove that a constructive Boolean algebra of characteristic $(0, \infty, 1)$ wtih recursive set of atoms is strongly constructive.

5. Prove that a constructive Boolean algebra of characteristic $(1, 1, 0)$ with decidable sets of atoms, atomless, and atomic elements respectively is strongly constructive.

6. Prove that constructive Boolean algebras $\langle \mathfrak{B}, \nu \rangle$ with $\mathrm{ch}_1(\mathfrak{B}) = 1$ and decidable sets of atoms, atomless, and atomic elements as well as the first Ershov–Vaught ideal I_1 is strongly constructive.

7. Construct an example of a constructive Boolean algebra of characteristic $(2, 0, 1)$ with decidable sets of atoms and atomless elements which is not strongly constructivizable.

8. Construct examples of constructive Boolean algebras of characteristics $(2, 1, 0)$ and $(1, 0, 1)$ with decidable sets of atoms which are not strongly constructivizable.

   [HINT. Use the technique of the Feiner hierarchy.]

9. Prove that the Boolean algebras $\mathfrak{B}_\omega$, $\mathfrak{B}_{\omega+\eta}$, $\mathfrak{B}_{\omega\times\eta}$ have not strong constructivizations.

10. Prove that $\mathfrak{B}_\omega$, $\mathfrak{B}_{\omega+\eta}$, and $\mathfrak{B}_{\omega\times\eta}$ with the natural numberings constructed from the linear orders ω, $\omega + \eta$, and $\omega \times \eta$, are strongly constructive.

3.6. Algorithmic Dimension of Boolean Algebras

The questions on uniqueness and the number of (strong) constructivizations are the principal ones in studying the structures of constructivizable models. To clarify the dependence of algorithmic properties of models upon the choice of a constructivization, various equivalences are introduced into consideration and, with respect them, the possible spectrum of nonequivalent classes of constructivizations is estimated. The most significant success in this direction is achieved for recursive equivalences and autoequivalences.

It is obvious that a system can have at most a countable number of nonautoequivalent constructivizations. The number of classes of auto equivalent constructivizations is called an *algorithmic dimension* of the system $\mathfrak{M}$ and is denoted by $\dim_A(\mathfrak{M})$. The number of classes of recursively equivalent constructivizations is denoted by $\dim_r(\mathfrak{M})$. We note that $\dim_r(\mathfrak{M}) \leqslant 2^\omega$.

The investigations of constructive models show (cf. [59, 67]) that the spectra of possible dimensions $\dim_A(\mathfrak{M})$ and $\dim_r(\mathfrak{M})$ are exactly the sets $\mathbb{N} \cup \{\omega\}$ and $\mathbb{N} \cup \{\omega\} \cup \{2^\omega\}$ respectively. However, for partial classes of models these spectra get considerably narrow.

In this section, we are interested in the spectrum of algorithmic dimensions of Boolean algebras as well as the algebraic characterization of Boolean algebras of a given algorithmic dimension. All the results concerning algorithmic dimensions of Boolean algebras are based on a common criterion for an algorithmic dimension to be infinite, which is referred to as the branching theorem.

Let $\langle \mathfrak{M}, \nu \rangle$ be an enumerated model of a signature σ and let σ_n be a computable sequence of finite signatures such that $\cup \sigma_n = \sigma$ and $\sigma_n \subseteq \sigma_{n+1}$ for any n.

By a *representation* of an enumerated model $\langle \mathfrak{M}, \nu \rangle$ we mean a pair $\langle \{\mathfrak{M}_n \mid n \in \mathbb{N}\}, f \rangle$, where f is a recursive function and $\{\mathfrak{M}_n, n \in \mathbb{N}\}$ is a strictly computable sequence of finite models of the signatures σ_n, $n \in \mathbb{N}$, which satisfy the following conditions:

$$\mathfrak{M}_n \leqslant \mathfrak{M}_{n+1} \restriction \sigma_n, \quad \nu f \colon \mathfrak{M}_n \to \mathfrak{M} \text{ is an isomorphic embedding,}$$

$$\bigcup_{n \geqslant 0} |\mathfrak{M}_n| \text{ is a recursive set}, \quad \nu f \left(\bigcup_{n \geqslant 0} |\mathfrak{M}_n| \right) = |\mathfrak{M}|.$$

It is easy to see that an enumerated locally finite model has a representation if and only if it is constructive. Moreover, it is possible to construct a representation in such a way that the universes of the models $\mathfrak{M}_n$ are initial intervals of natural numbers.

Let $\psi(\overline{x}, \overline{y}) = \bigwedge_{i \geqslant 1} \psi_i(\overline{x}, \overline{y})$ be an infinite conjunction of $\forall$-formulas in variables $\overline{x}$ and $\overline{y}$, let $\Pi = \langle \{\mathfrak{M}_n \mid n \in \mathbb{N}\}, f \rangle$ be a representation of a constructive model $\langle \mathfrak{M}, \nu \rangle$, and let $\overline{a}$ be a sequence of elements of $|\mathfrak{M}|$. We suppose that

$$B_{\psi, \overline{a}}^m \rightleftharpoons \{ \langle m_0, \ldots, m_l \rangle \mid \mathfrak{M} \models \psi(\nu f(m_0), \ldots, \nu f(m_l), \nu f(\overline{a})) \}$$

and, for any finite sequences $\overline{m}_0, \ldots, \overline{m}_d$ such that $\overline{m}_d = (m_0, \ldots, m_l)$, there exist infinitely many t and the correspondning isomorphic embeddings $\varphi_t \colon \mathfrak{M}_t \to \mathfrak{M}_{t+1}$ are such that for some $i \leqslant d$ the following conditions hold:

$$\mathfrak{M}_{t+1} \models^{\neg} \psi^{t+1}(\varphi_t(m_0^i), \ldots, \varphi_t(m_t^i), \overline{a}),$$

where $\psi^{t+1} \rightleftharpoons \bigwedge_{i=1}^{t+1} \psi_i, \overline{m}_i = (m_0^i, \ldots, m_l^i)$,

φ_t is the identity mapping on $\mathfrak{M}_m$ and on $\{\overline{m}_j \mid j < i\}$.

Let $\{\psi_m^i(\overline{x}_m, \overline{y}_m) \mid m, i \in \mathbb{N}\}$ be a computable sequence of $\forall$-formulas, i.e., from m, i we can effectively find the Gödel number of $\psi_{m,i}$, and $\{\overline{a}_m \mid$

3.6. Algorithmic Dimension

$m \in \mathbb{N}\}$ is a computable enumeration of sequences of elements of $\bigcup_{n \geqslant 0} |\mathfrak{M}_n|$;
moreover, the elements of $\overline{a}_m$ are in $|\mathfrak{M}_m|$ for any $m \in \mathbb{N}$.

We say that the representation $\Pi_\nu \rightleftharpoons \langle \{\mathfrak{M}_n \mid n \in \mathbb{N}\}, f \rangle$ is *branching with respect to sequences of* $\{\psi_m \rightleftharpoons \bigwedge_{i \geqslant 0} \psi_{m,i}(\overline{x}_m, \overline{y}_m) \mid m \in \mathbb{N}\}$ $\forall$-*formulas and finite sequences* $\{\overline{a}_m, m \in \mathbb{N}\}$ if for any m the set $B^m_{\psi_m, \overline{a}_m}$ is nonempty and $\psi_m(\mathfrak{M}, \nu f(\overline{a}_m)) \setminus \nu f(B^m_{\psi_m, \overline{a}_m})$ is finite. In the sequel, we omit the subscripts $\overline{a}_m$ in formulas if their occurrence is clear.

A model $\mathfrak{M}$ is called *branching* if there exists a representation of its constructivization branching with respect to some computable sequences of $\forall$-formulas and sequences of elements.

A class of constructivizations of a model $\mathfrak{M}$ is called *effectively infinite* if from any computable class of constructivizations of this model we can effectively find a constructivization that is not autoequivalent to any constructivization of this class.

Theorem 3.6.1 [on branching models]. *If a constructivizable model $\mathfrak{M}$ has branchings, then the class of its constructivizations is effectively infinite.*

PROOF. Let (S, γ) be a computable class of constructive models isomorphic to $\mathfrak{M}$. We consider a constructivization ν of $\mathfrak{M}$, computable enumeration of infinite $\forall$-formulas and finite sequences, and the representation $\Pi_\nu = \langle \{\mathfrak{M}_n \mid n \in \mathbb{N}\}, f \rangle$ branching with respect to these sequences. We construct step-by-step a constructivization μ and its representation $\Pi_\mu = \langle \{\mathfrak{M}^\mu_n \mid n \in \mathbb{N}\}, f^\mu \rangle$, which is the required one. We also consider representations of (S, γ). Denote by $\langle \{\mathfrak{M}^m_n \mid n \in \mathbb{N}\}, f^m \rangle$ the representation of $\langle \mathfrak{M}^m \mid \lambda n \gamma(m, n) \rangle$ which is effectively constructed from m.

At every step t, we construct a finite model $\mathfrak{M}^\mu_t$, a partial numbering μ^t, a partial mapping $\overline{\mu}^t$, and an auxiliary function r. We use the marks $\langle m, k \rangle$, where $m, k \in \mathbb{N}$. Let $\overline{z}$ be variables for which the elements $\overline{a}_m$ are substituted and let $x_0, \ldots, x_l$ be the rest of the variables occurring in the formula ψ_m. We order the sequences of numbers $(m_0, \ldots, m_l)$ by their numbers with respect to some fixed computable numbering. We also put the marks $\langle m_0, \ldots, m_l \rangle$ on the marks $\langle m, k \rangle$.

When constructing the model, we use a partially recursive function $\lambda n x K(n, x)$ that is universal for the class of unary partially recursive functions. If a function has been computed in less than t steps, then $K^t(n, x)$ denotes its value. Otherwise, the value of the function is not defined.

We now describe the construction.

STEP 0. We set $\mathfrak{M}^\mu_0 \rightleftharpoons \varnothing, \overline{\mu^0} = \mu^0 = \varnothing, r(0, n) = 0$ for all n.

STEP $t+1$. We check if there exists a mark $\langle m, k\rangle \leqslant t+1$ such that $\lambda x K^t(k, x)$ is defined on $\{0, \ldots, r(t, \langle m, k\rangle)\}$ and $\gamma(m, \operatorname{rang}\lambda x K^t(k, x)) \supseteq \{\gamma(m, 0), \ldots, \gamma(m, r(t, \langle m, k\rangle))\}$, the mark $\langle m, k\rangle$ is located nowhere; otherwide, it has the mark $\langle m', m_0^*, \ldots m_l^*\rangle$; moreover, one of the following conditions holds:

(i) $\lambda n\gamma(m, K^{t+1}(k, n))$ is not an isomorphic embedding of $\mathfrak{M}_t^\mu \upharpoonright \operatorname{dom} \lambda n K^{t+1}(k, n)$ into $\mathfrak{M}_m$;

(ii) the condition (i) fails but there exist numbers $m', n_0, \ldots, n_l$ that are not greater than $t+1$ and an isomorphic embedding $\varphi \colon \mathfrak{M}_t \to \mathfrak{M}_{t+1}$ such that

$$\mathfrak{M}_t^m \models \underset{i=0}{\overset{t}{\&}} \psi_{m',i}((f^m)^{-1}K^t(k, (n_0, \ldots, n_l)), (f^m)^{-1}K^t(k, \overline{a}_{m'})$$

$$\mathfrak{M}_{t+1} \models \neg \underset{i=0}{\overset{t}{\&}} \psi_{m',i}(\varphi\overline{\mu}^t((n_0, \ldots, n_l)), \overline{a}_{m'})$$

the restriction of φ on the set G is identical, where $(\overline{\mu})^{-1}(G)$ contains the numbers with marks less than $\langle m, k\rangle$ and the numbers in $\overline{a}_m$; moreover, for all sequences $m_0, \ldots, m_l$ such that $\langle m_0, \ldots, m_l\rangle < \langle K^t(k, n_0), \ldots, K^t(k, n_l)\rangle$ the following fact holds: if $m_i, i \leqslant l$, belong to $\mathfrak{M}_{t+1}^m$ and in $\mathfrak{M}_{t+1}^m$ the formula $\underset{i=0}{\overset{t+1}{\&}} \psi_{m',i}$ is true on these sequences, then $(\lambda x K^t(k, x))^{-1}(m_i), i \leqslant l$, belong to G and

$$\langle m_0, \ldots, m_l\rangle < \langle K^t(k, n_0), \ldots, K^t(k, n_l)\rangle$$

or $\mathfrak{M}_{t+1}^m \models \neg \underset{i=0}{\overset{t}{\&}} \psi_{m',i}((f^m)^{(-1)}(m_0^*, \ldots, m_l^*, K^t(k, \overline{a}_m)))$.

If such a mark does not exist, then we pass to the stage A (below) which completes the step $t+1$. Otherwise, we choose such a mark with the least number. Let $\langle m, k\rangle$ be such a mark. If for $\langle m, k\rangle$ the condition (i) is satisfied, then we put this mark on every element of $\mathfrak{M}_t^\mu$. Then we take off all marks that are greater than $\langle m, k\rangle$ as well as all marks from marks that are at least $\langle m, k\rangle$. After that we pass to the stage A. If $\langle m, k\rangle$ satisfies the condition (ii), then we choose the least sequence $\langle n_0, \ldots, n_l\rangle$ having the properties indicated in (ii). We take off marks from those marks that are not less than $\langle m, k\rangle$ and take off all marks that are at least $\langle m, k\rangle$. We enumerate elements of $\mathfrak{M}_{t+1}$ in such a way that $0, 1, \ldots, k_t$ become the numbers of the elements $\overline{\mu}^t\varphi(0), \ldots, \overline{\mu}^t\varphi(k_t)$, where $[0, k_t] = |\mathfrak{M}_t|$. We fix this numbering $\overline{\mu}^{t+1} \colon |\mathfrak{M}_{t+1}| \to |\mathfrak{M}_{t+1}|$. On the set $|\mathfrak{M}_{t+1}|$, we consider the mapping $\overline{\mu}^{t+1}$ that induces predicates, and constants from $\mathfrak{M}_{t+1}$. The model obtained is denoted by $\mathfrak{M}_{t+1}^\mu$. We put the marks $\langle m, k\rangle$ on elements

3.6. Algorithmic Dimension

of the model $\mathfrak{M}_{t+1}^{\mu}$ and the marks $\langle m', n_0, \ldots, n_l \rangle$ on $\langle m, k \rangle$, where m' and $\overline{n}$ are the least elements satisfying (ii). We set

$$\mu^{t+1} \leftrightharpoons \nu f \overline{\mu}^{t+1}, \quad r(t+1, \langle m, k \rangle) \leftrightharpoons \|\mathfrak{M}_{t+1}\|$$
$$r(t+1, n) \leftrightharpoons r(t, n), \quad n \neq \langle m, k \rangle$$

where $\|\mathfrak{M}_{t+1}\|$ is the number of elements of $\mathfrak{M}_{t+1}$, and pass to the following step.

A. We extend the mapping $\overline{\mu}^t \colon |\mathfrak{M}_t^{\mu}| \to |\mathfrak{M}_{t+1}|$ to $\overline{\mu}^{t+1} \colon |\mathfrak{M}_{t+1}| \xrightarrow[\text{onto}]{1-1} |\mathfrak{M}_{t+1}|$, set $\mathfrak{M}_{t+1}^{\mu} \leftrightharpoons (\overline{\mu}^{t+1})^{-1}(\mathfrak{M}_{t+1})$, $\mu^{t+1} \leftrightharpoons \nu f \overline{\mu}^{t+1}$, $r(t+1, n) = r(t, n)$ for all n, and pass to the following step, defining on $\mathfrak{M}_{t+1}^{\mu}$ the mappings of $\mathfrak{M}_{t+1}$ that are induced by the mapping $\overline{\mu}^{t+1}$.

We consider the sequence of models $\mathfrak{M}_0^{\mu} \subseteq \ldots \subseteq \mathfrak{M}_n^{\mu} \subseteq \ldots$ and set $\mathfrak{M}^{\mu} \leftrightharpoons \cup \mathfrak{M}_n^{\mu}$. By Lemma 3.6.5 (below), the mapping $\overline{\mu}(n) = \lim_{n \to \infty} \overline{\mu}^t(n)$ is defined for all n. For any n there exists t_n such that $\overline{\mu}^t(n) = \overline{\mu}^{t_n}(n)$ if $t \geqslant t_n$. Therefore, the restriction $\overline{\mu}$ on any $\mathfrak{M}_n^{\mu}$ is an isomorphic embedding of $\mathfrak{M}_n^{\mu}$ into $\bigcup_{n \geqslant 0} \mathfrak{M}_n$. Taking into account the representation of ν, we conclude that $\nu f \overline{\mu}$ is an isomorphism of $\mathfrak{M}^{\mu}$ onto $\mathfrak{M}$. Let $\mu(n) \leftrightharpoons \nu f \overline{\mu}(n)$ for all $n \in \mathbb{N}$. By Lemma 3.6.6 (below), the numbering μ is a constructivization and, by Lemma 3.6.7 (below), it is not autoequivalent to any constructivization of S. The following simple remarks on construction and Lemmas 3.6.1–3.6.7 complete the proof. □

REMARK 3.6.1. We take off $\langle m', n_0, \ldots, n_l \rangle$ from $\langle m, k \rangle$ if the imposed mark is less than $\langle m', n_0, \ldots, n_l \rangle$ or $\mathfrak{M} \models^{\neg} \psi_{m'}(\gamma(m, K(k, n_0)), \ldots, \gamma(mk, (k, n_l)))$.

REMARK 3.6.2. If marks that are at most $\langle m, k \rangle$ are imposed only a finite number of times, then the mark $\langle m', n_0, \ldots, n_l \rangle$ is imposed on $\langle m, k \rangle$ only a finite number of times.

Lemma 3.6.1. *If $\langle m, k \rangle$ is imposed infinitely many times, then the function $\lambda x K(k, x)$ is defined everywhere and $\gamma(m, K(k, \mathbb{N})) = |\mathfrak{M}|$.*

PROOF. The assertion follows from the definition of r and the constructivization condition (ii). □

Lemma 3.6.2. *If all marks that are less than $\langle m, k \rangle$ are imposed only a finite number of times, then $\langle m, k \rangle$ is also imposed a finite number of times.*

PROOF. We assume the contrary. By Lemma 3.6.1, $\gamma(m, K(k, \mathbb{N})) = |\mathfrak{M}|$ and the function $\lambda x K(k, x)$ is defined everywhere. Let t_0 be a step after which no mark at most $\langle m, k \rangle$ is imposed. We choose m' so that $|\mathfrak{M}_{m'}^\mu|$ contains all numbers with marks that are less than $\langle m, k \rangle$ at the step $t_0 + 1$. Let $\psi_{m'}(\mathfrak{M}, \nu f \overline{a}_{m'}) \backslash \nu f B_{\psi_{m'} \overline{a}_{m'}}^{m'}$ contain d elements. Since $B_{\psi_{m'} \overline{a}_{m'}}^{m'}$ is not empty and $\psi_{m'}(\mathfrak{M}, \nu f \overline{a}_{m'}) \backslash \nu f B_{\psi_{m'}}^{m'}$ contains exactly d elements, in $\mathfrak{M}$ there exist at least $d+1$ sequences of elements $\overline{m}_0, \ldots, \overline{m}_d$ such that $\mathfrak{M} \models \psi_{m'}(\overline{m}_j, K(k, \overline{a}_{m'}))$, where $j \leqslant d$.

We now consider sequences of numbers $\overline{m}'_0, \ldots, \overline{m}'_d$ that are in range $\lambda x K(k, x)$ and $\gamma(m, \overline{m}'_0) = \overline{m}_0, \ldots, \gamma(m, \overline{m}'_d) = \overline{m}_d$. Consider a step $t_1 > t_0$ after which elements of the sequences are in range $\lambda x K^t(k, x)$. Among such sequences we choose $\overline{m}^*$ with the greatest number. After the step t_1, every sequence $\langle m', \overline{m} \rangle$ can be imposed on $\langle m, k \rangle$ only a finite number of times.

We consider a step $t_2 > t_1$ such that all the numbers of the indicated sequences are in range $\lambda x K^{t_2}(k, x)$, and in $\mathfrak{M}_{t_2}^m$ the formula $\overset{t_2}{\underset{i=0}{\&}} \text{range } \psi_{m', i}$ is false on those sequences $\overline{m}$ for which in $\mathfrak{M}$ the formula $\psi_{m'}$ is false and $\overline{m} < \overline{m}^*$. Thus, there exist sequences $\overline{m}'_0, \ldots, \overline{m}'_{d'}$ such that $\mathfrak{M} \models \psi_{m'}(\gamma(m, \overline{m}'_i), \gamma(m, K(k, \overline{a}_{m'})))$, $i \leqslant d'$, $d' \geqslant d$, and for all sequences $\overline{m}$ with numbers less than the numbers of $\overline{m}^*$ we have $\overline{m} = \overline{m}'_i$ for some $i \leqslant d'$ or $\mathfrak{M} \models \neg \psi_{m'}(\gamma(m, (\overline{m}, K(k, \overline{a}_{m'}))))$.

We consider a step $t_3 \geqslant t_2$ such that those numbers of sequences that are in $\cup \mathfrak{M}_t^m$ belong to $\mathfrak{M}_{t_3}^m$ and $\mathfrak{M} \models \neg \psi_{m'}(\gamma(m, \overline{m}), \overline{a}_{m'})$, $\overline{m} < \overline{m}^*$, implies $\mathfrak{M}_{t_3}^m \models \neg \overset{t_3}{\underset{i=0}{\&}} \psi_{m', i}((f^m)^{-1}(\overline{m}), f(\overline{a}_{m'}))$. We choose sequences of numbers $\overline{m}''_0, \ldots, \overline{m}''_{d'}$ such that $K(k, \overline{m}''_0) = \overline{m}'_0, \ldots, K(k, \overline{m}''_{d'}) = \overline{m}''_d$ and a step $t_4 > t_3$ under which all the numbers of these sequences are in $f^m(\mathfrak{M}_{t_4}^m)$. Since, after the step t_0, on $\langle m, k \rangle$ every mark $\overline{m}$ can be imposed only a finite number of times, there exists a step $t_5 > t_4$, after which $\overline{m}^*$ are imposed on $\langle m, k \rangle$. Then, after the step t_5, μ^t remains constant on the sequences $\overline{m}''_0, \ldots, \overline{m}''_{d'}$. If for some sequence $\overline{m}_j$ the formula $\neg \psi_{m'}(\mu^{t_5}(\overline{m}''_j))$ holds in $\mathfrak{M}$, then we consider a step $t_6 > t_5$ such that $\mathfrak{M}_{t_6} \models \neg \overset{t_6}{\underset{i=0}{\&}} \psi_{m', i}(\overline{\mu}^{t_6}(\overline{m}''_j), \overline{a}_{m'})$ and on $\langle m, k \rangle$ the mark $\overline{m}'_j$ is imposed and not taken off in the sequel. If $\mathfrak{M} \models \psi_{m'}(\overline{\mu}^{t_5}(\overline{m}''_j), \overline{a}_{m'})$ for all j, then at least one of these sequences belongs to $B_{\psi_{m'} \overline{a}_{m'}}^{m'}$. In this case, there is a step $t_6 > t_5$, at which the mark $\langle m, k \rangle$ of this sequence satisfies the conditions of the step and on $\langle m, k \rangle$ a mark with $\overline{m}'_j$ must be imposed, where $j \leqslant d'$, which yields a contradiction. □

3.6. Algorithmic Dimension

Corollary 3.6.1. *Every mark $\langle m, k \rangle$ is imposed only a finite number of times.*

Lemma 3.6.3. *There exist infinitely many marks $\langle m, k \rangle$ that are imposed at some step and are not taken off in the sequel.*

PROOF. We assume the contrary. Let there exist a step t_0 after which all the imposed marks are taken off. We consider the least mark imposed after the steps t_0 and the step $t_1 > t_0$ at which it happens. From the construction and the condition (ii) it follows that this mark can be taken off only if its mark changes. Moreover, just this mark must be imposed. All marks can be imposed only a finite number of times. Hence, beginning from a certain step, it is left constant. It remains to show that, after the step t_0, at least one mark must be imposed. We assume the contrary and consider a number m that is greater than each of the marks that are imposed before the step t_0 and at the step t_0. We also consider a recursive permutation $\lambda x K(m, x)$. If $\lambda x \gamma(0, K(m, x))$ is an isomorphic embedding of $\bigcup_{t \geqslant 0} \mathfrak{M}_t^m$ into $\mathfrak{M}_0$, then, acting as in Lemma 3.6.2, we can find a step at which the mark $\langle 0, m \rangle$ must be imposed. Otherwise, the mark $\langle 0, m \rangle$ satisfies the condition (ii) at some step. Consequently, this mark will be imposed, which contradicts the assumption. □

Lemma 3.6.4. *If after a step t_0, on an element n, there is a mark which is not taken off, then $\overline{\mu}^t(n) = \overline{\mu}^{t_0}(n)$ for all $t \geqslant t_0$*

PROOF. The assertion is obvious in view of the construction. □

Lemma 3.6.5. *The value $\overline{\mu}(n) \leftrightharpoons \lim_{t \to \infty} \overline{\mu}^t(n)$ is defined for all n and $\overline{\mu} \colon \mathbb{N} \underset{\text{onto}}{\to} \mathbb{N}$.*

PROOF. The assertion follows from Lemmas 3.6.3 and 3.6.4 as well as the fact that $\overline{\mu}^t$ permutates $|\mathfrak{M}_t|$ for any t. □

Lemma 3.6.6. *The numbering $\mu = \nu f \overline{\mu}$ is a constructivization of $\mathfrak{M}$.*

PROOF. The assertion follows from Lemma 3.6.5 and the definition of relations on $\mathfrak{M}$. □

Lemma 3.6.7. *The constructivization μ is not autoequivalent to any constructivization of S.*

PROOF. We assume the contrary. Let μ be autoequivalent to a constructivization $\lambda n \gamma(m, n)$. Let $\lambda x K(k, x)$ be a general recursive function such that $\mu(n) = \gamma(m, K(k, n))$, $\gamma(m, K(k, \mathbb{N})) = \mathfrak{M}^m$ for all $n \in \mathbb{N}$, and the function $\lambda n K(k, n)$ induces an autoequivalence between $\lambda n \gamma(m, n)$ and $\lambda n \mu(n)$. Consider the corresponding mark $\langle m, k \rangle$. Using the construction of Lemma 3.6.2 and Corollary 3.6.1, we can show that $\langle m, k \rangle$ is imposed at some step and will not be taken off; otherwise, it will be imposed infinitely many times, which contradicts Corollary 3.6.1. □

Using Theorem 3.6.1, we now characterize autostable Boolean algebras, i.e., algebras of unit algorithmic dimension and describe the spectrum of all possible algorithmic dimensions of Boolean algebras.

Theorem 3.6.2. *The class of constructivizations of a constructivizable Boolean algebra with infinite set of atoms is effectively infinite.*

PROOF. In order to use the theorem on branching models it suffices to define a ∀-formula and find a representation with respect to which it ramifies. Let $\langle \mathfrak{B}, \nu \rangle$ be a constructive Boolean algebra with infinite set of atoms. Without loss of generality, we can assume that ν is a one-to-one numbering since any constructivization of an infinite algebraic system is recursively equivalent to a one-to-one constructivization. Thus, the constructivization ν induces on $\mathbb{N}$ the structure of the recursive Boolean algebra $\mathfrak{B}^*$ isomorphic to $\mathfrak{B}$. Using Lemma 3.3.1, we choose a computable sequence $\{\mathfrak{A}_i \mid i \in \mathbb{N}\}$ of finite subalgebras of this recursive Boolean algebra such that $\cup \mathfrak{A}_i = \mathfrak{B}^*$, $\mathfrak{A}_{i+1} = \mathrm{gr}(\mathfrak{A}_i \cup \{a_i\})$, where a_i is an atom of $\mathfrak{A}_{i+1}$. Then $\{\mathfrak{A}_i \mid i \in \mathbb{N}\}$ and the identity function form a representation of ν. As a sequence of ∀-formulas we take the sequence formed by the same ∀-formula: $\psi(x) \rightleftharpoons (\forall y)(y < x \Rightarrow y = 0) \& x \neq 0$. We show that the sequence chosen in such a way satisfies all the conditions of the definition of branching. We first prove that the set B_ψ^m is not empty. We consider an atom b of the Boolean algebra $\mathfrak{A}_m$ under which there are infinitely many elements of $\mathfrak{B}^*$ and an atom a of the Boolean algebra $\mathfrak{B}^*$ lying under b. We show that $a \in B_\psi^m$. In accordance with the choice of a, in $\mathfrak{B}$ the formula ψ holds on νa. Let $a_1, \ldots, a_d = a$ be a finite sequence of elements of $\mathfrak{B}^*$. Let $\mathfrak{B} \models^\neg \psi(\nu a_i)$ for some i. Under the identity embedding of $\mathfrak{A}_t$ into $\mathfrak{A}_{t+1}$, we have $a \in B_\psi^m$ beginning from some t_0. If $\mathfrak{B} \models \psi(\nu a_i)$, for any $a_1, \ldots, a_d$, then we choose the least i such that a_i is located under b. Since there are infinitely many elements under b and, at each step t, there are only a finite number of elements of $\mathfrak{A}_t$ under b, there exist infinitely many t such that, at the step $t + 1$, under b there is an element of $\mathfrak{A}_{t+1}$ that does

3.6. Algorithmic Dimension

not belong to $\mathfrak{A}_t$. We consider a step $t > m$ at which $a_1, \ldots, a_d$ belong to $\mathfrak{A}_t$. In accordance with the above construction, $\mathfrak{A}_{t+1} = \mathrm{gr}(\mathfrak{A}_t \cup \{c\})$, where c is an atom of $\mathfrak{A}_{t+1}$. Let c_0 be an atom of $\mathfrak{A}_t$ under which c is located. We define the embedding φ_t of $\mathfrak{A}_t$ into $\mathfrak{A}_{t+1}$ by setting

$$\varphi_t(x) \leftrightharpoons \begin{cases} x \vee c, & a_i \leqslant x \\ x \setminus c, & a_i \wedge x = 0 \end{cases}$$

It is easy to see that φ_t is an isomorphic embedding of $\mathfrak{A}_t$ into $\mathfrak{A}_{t+1}$; moreover, φ_t keeps all elements of $\mathfrak{A}_m$ and all elements that are located under those atoms of $\mathfrak{A}_m$ that are different from b. We note that $\varphi_t(a_i) = a_i \vee c$. Thus, in $\mathfrak{A}_{t+1}$ the formula ψ is false on $\varphi_t(a_i)$ and for all $j < i$ the mapping φ_t leaves a_j, $j < i$, at their places. Consequently, all the conditions imposed on a are fulfilled and $a \in B_\psi^m$, i.e., $B_\psi^m \neq \varnothing$. The aforesaid means the following: if an atom a of $\mathfrak{B}^*$ is located under an atom b of $\mathfrak{A}_m$ under which, in turn, there are infinitely many elements of $\mathfrak{B}^*$, then a belongs to B_ψ^m. In this case, if a does not belong to B_ψ^m, then it is located under an atom of $\mathfrak{A}_m$ under which, in turn, there are only a finite number of elements of $\mathfrak{B}^*$. Therefore, there are only a finite number of such a and the set $\psi(\mathfrak{B}^*) \setminus B_\psi^m$ is finite. By Theorem 3.6.1 on branching models, the class of all constructivizations of the Boolean algebra $\mathfrak{B}$ is effectively infinite. □

We now describe all autostable Boolean algebras and state the algebraic criterion for autostability.

Theorem 3.6.3. *A constructivizable Boolean algebra is autostable if and only if it has a finite number of atoms.*

PROOF. The necessity follows from Theorem 3.6.2. We prove the sufficiency. Let $\mathfrak{B}$ be a constructivizable Boolean algebra with finite number of atoms and let $\mathfrak{B}$ have m atoms $a_1, \ldots, a_m$. Consider $a = \bigvee_{i=1}^{m} a_i$ and two arbitrary constructivizations ν and μ of $\mathfrak{B}$. We show that ν and μ are autoequivalent. If $\mathfrak{B}$ is a finite Boolean algebra, then the autoequivalence of ν and μ is obvious. Otherwise, we consider constructivizations ν_a, μ_a of the algebra $\widehat{a}$ and $\nu_{C(a)}$, $\mu_{C(a)}$ of the algebra $\widehat{C(a)}$ from Corollary 3.1.1. Let $\langle \mathcal{D}_1, \varphi_1 \rangle$ and $\langle \mathcal{D}_1, \varphi_2 \rangle$ be recursively enumerable generating trees such that $\langle \widehat{C(a)}, \nu_{C(a)} \rangle$ is isomorphic to $\langle \mathfrak{B}_{\mathcal{D}_1}, \nu_{\mathcal{D}_1} \rangle$ and $\langle \widehat{C(a)}, \mu_{C(a)} \rangle$ is isomorphic to $\langle \mathfrak{B}_{\mathcal{D}_2}, \nu_{\mathcal{D}_2} \rangle$. Such trees exist by Theorem 3.3.2. Since $\widehat{C(a)}$ is an atomless Boolean algebra, we conclude that $\mathcal{D}_1 = \mathcal{D}_2 = \mathbb{N}$ and $\langle \widehat{C(a)}, \nu_{C(a)} \rangle$, $\langle \widehat{C(a)}, \mu_{C(a)} \rangle$ are isomorphic to $\langle \mathfrak{B}_\mathbb{N}, \nu_\mathbb{N} \rangle$. Consequently, $\nu_{C(a)}$ and $\mu_{C(a)}$ are autoequivalent. Therefore, $\langle \mathfrak{A}, \nu \rangle \cong \langle \widehat{a}, \nu_a \rangle \times \langle \widehat{C(a)}, \nu_{C(a)} \rangle \cong \langle \widehat{a}, \mu_a \rangle \times \langle \widehat{C(a)}, \mu_{C(a)} \rangle \cong \langle \mathfrak{A}, \mu \rangle$. □

Corollary 3.6.2. *The algorithmic dimension of a Boolean algebra can assume only three values*: $0, 1, \omega$.

Corollary 3.6.3. *The class of constructivizations of any nonautostable Boolean algebra is effectively infinite.*

It is interesting to study relationships between algorithmic dimensions of Boolean algebras depending on the choice of a signature. We emphasize that any numbering ν of a Boolean algebra $\mathfrak{B}$ is a constructivization of $\mathfrak{B}$ if and only if ν is a constructivization of the Boolean ring $R(\mathfrak{B})$ and automorphisms of $\mathfrak{B}$ are also automorphisms of $R(\mathfrak{B})$. Therefore, the algorithmic dimension of $\mathfrak{B}$ as a Boolean algebra coincides with the algorithmic dimension of $\mathfrak{B}$ regarded as a ring. A different situation arises if we consider $\mathfrak{B}$ with respect to the signature $\langle \leqslant \rangle$. It is clear that any constructivization ν regarded as a Boolean algebra is also a constructivization with respect to the signature $\langle \leqslant \rangle$. The converse assertion is false although (cf. Theorem 3.3.4) $\dim_A(\mathfrak{B}) \neq 0$ if and only if $\dim_A(L(\mathfrak{B})) \neq 0$. If constructivizations ν and μ of a Boolean algebra $\mathfrak{B}$ are nonautoequivalent, then they are nonautoequivalent with respect to the signature $\langle \leqslant \rangle$ since the class of automorphisms of $\mathfrak{B}$ coincides with that of $L(\mathfrak{B})$. Therefore, $\dim_A(L(\mathfrak{B})) \geqslant \dim_A(\mathfrak{B})$. Consequently, for Boolean algebras with an infinite number of atoms we have $\dim_A(L(\mathfrak{B})) = \dim_A(\mathfrak{B})$. Thus, it remains to consider the algorithmic dimension of lattices $L(\mathfrak{B})$ with a finite number of atoms.

Theorem 3.6.4. *If $L(\mathfrak{B})$ is an infinite constructivizable Boolean lattice, then the class of all its constructivizations is effectively infinite.*

PROOF. Since $L(\mathfrak{B})$ is a constructivizable Boolean lattice, it is also constructivizable as a Boolean algebra by Theorem 3.3.4. Let ν be a one-to-one constructivization of $\mathfrak{B}$ with respect to the signature of Boolean algebras such that it induces a recursive Boolean algebra $\mathfrak{B}^*$ isomorphic to $\mathfrak{B}$. Taking into account Lemma 3.3.1 and Corollary 3.1.4, we consider a computable sequence $\{\mathfrak{A}_i\}_{i \in \mathbb{N}}$ of finite subalgebras $\mathfrak{B}^*$ such that $\cup \mathfrak{A}_i = \mathfrak{B}^*$ and $\mathfrak{A}_{i+1} = \mathrm{gr}(\mathfrak{A}_i \cup \{a_i\})$, where a_i is an atom of $\mathfrak{A}_{i+1}$. This sequence, together with the identity function, yields a representation of the constructivization ν. We consider two cases.

3.6. Algorithmic Dimension

CASE 1: A Boolean algebra $\mathfrak{B}$ has infinitely many atoms. As was proved in Theorem 3.6.2, we have branching with respect to the formula $(\forall y)(y < x \Rightarrow y = 0)\&x \neq 0$ that distinguishes atoms. This formula is also suitable in the case of the signature $\langle \leqslant \rangle$. The proof of the fact that it has branching is the same as that in Theorem 3.6.2. Therefore, the lattice $L(\mathfrak{B})$ has effectively infinite class of constructivizations.

CASE 2: A Boolean algebra $\mathfrak{B}$ has a finite number of atoms. Let νa be a complement to the union of all atoms. Without loss of generality, we can assume that $a \in \mathfrak{A}_i$ for any i. From $\{\mathfrak{A}_i\}_{i \in \mathbb{N}}$ we select a computable sequence $\{\mathfrak{A}_{f(i)}\}_{i \in \mathbb{N}}$ so that any atom of $\mathfrak{A}_{f(i)}$ that is located under a is already not an atom of $\mathfrak{A}_{f(i+1)}$. Since any element lying under a in $\mathfrak{B}^*$ is atomless, such a choice is possible. Further, we regard this sequence as a representation of $\mathfrak{B}^*$ and denote it by $\{\mathfrak{B}_n \mid n \in \mathbb{N}\}$. As the required $\forall$-formula we take a formula of the form

$$\psi_m(x,y,z) \rightleftharpoons (\forall t)((x \leqslant t \& y \leqslant t, \rightarrow z \leqslant t) \& x \leqslant z \& y \leqslant z \& x \not\leqslant y$$
$$\& y \not\leqslant x \& z \leqslant a \& (\bigvee_{i=1}^{k_m} z \leqslant a_i)$$

where $a_1, \ldots, a_{k_m}$ are those atoms of $\mathfrak{B}_m$ that are less than a. We verify that the branching conditions are fulfilled for the enumeration of formulas ψ_m and sequences $\bar{a}_m$ constructed in such a way. To this end, it suffices to derive $B^m_{\psi_m \bar{a}_m} = \psi_m \langle \mathfrak{B}^*, \bar{a}_m \rangle$. By definition, $B^m_{\psi_m \bar{a}_m} \subseteq \psi_m \langle \mathfrak{B}^*, \bar{a}_m \rangle$. Therefore, it suffices to prove the inclusion $\psi_m \langle \mathfrak{B}^*, \bar{a}_m \rangle \subseteq B^m_{\psi_m \bar{a}_m}$, i.e., if $(a, b, c) \in \psi_m \langle \mathfrak{B}^*, \bar{a}_m \rangle$, then $(a, b, c) \in B^m_{\psi_m \bar{a}_m}$. We consider a finite sequence $(a_0, b_0, c_0), \ldots, (a_n, b_n, c_n) = (a, b, c)$. If there exists i such that $\mathfrak{B}^* \models \neg \psi_m(a_i, b_i, c_i, \bar{a}_m)$, then the required condition holds under the identity embedding of $\mathfrak{B}_t$ into $\mathfrak{B}_{t+1}$ for sufficiently large t. If $\mathfrak{B}^* \models \psi_m(a_i, b_i, c_i, \bar{a}_m)$ for any i, then we consider the sequence a_0, b_0, c_0. Since ψ_m is true on this sequence of elements a_0, b_0, c_0, the Boolean algebra $\mathfrak{B}^*$ satisfies the following conditions: $a_0 \vee b_0 = c_0$, $a_0 \not\leqslant b_0$, $b_0 \not\leqslant a_0$, c_0 is less than a and is located under one of the atoms of the Boolean algebra $\mathfrak{B}_m$. Let $t > m$ and let $\mathfrak{B}_t$ contain all the elements constituting the above-mentioned sequences. We define an isomorphic embedding $\varphi_t \colon \mathfrak{B}_t \to \mathfrak{B}_{t+1}$ of the signature $\langle \leqslant \rangle$ so that

$$\mathfrak{B}_{t+1} \models \neg \psi_m(\varphi_t(a_0), \varphi_t(b_0), \varphi_t(c_0), \bar{a}_m)$$

and the isomorphism φ_t is the identity mapping on $\mathfrak{B}_m$. Let d be an atom of $\mathfrak{B}_t$ located under $a_0 \setminus b_0$. By construction, the element d of $\mathfrak{B}_{t+1}$ is not an atom. If $d_0 < d$ and d_0 is an atom of $\mathfrak{B}_{t+1}$, then we define an embedding φ_t of $\mathfrak{B}_t$ into $\mathfrak{B}_{t+1}$ as follows:

$$\varphi_t(x) \leftrightharpoons \begin{cases} x \setminus c_0 \vee ((x \wedge c_0) \setminus d_0), & c_0 \not\leqslant x,\ d \leqslant x \wedge c_0 \\ x & \text{otherwise} \end{cases}$$

It is easy to see that $\varphi_t(x) \leqslant \varphi_t(y) \Leftrightarrow x \leqslant y$. Hence φ_t is an isomorphic embedding of $\mathfrak{B}_t$ into $\mathfrak{B}_{t+1}$ with respect to the signature $\langle \leqslant \rangle$. We note that φ_t leaves elements of $\mathfrak{B}_m$ at their places and $\varphi_t(c_0) = c_0, \varphi_t(b_0) = b_0$, but $\varphi_t(a_0) = a_0 \setminus d_0$. Therefore, $\varphi_t(c_0) \neq \varphi_t(b_0) \vee \varphi_t(a_0)$. Furthermore, $\varphi_t(b_0) \vee \varphi_t(a_0)$ belongs to $\mathfrak{B}_{t+1}$. Consequently, $\mathfrak{B}_{t+1} \models \neg \psi_m(\varphi_t(a_0), \varphi_t(b_0), \varphi_t(c_0), \bar{a}_m)$. Thus, $(a, b, c) \in B^m_{\psi_m \bar{a}_m}$. □

Corollary 3.6.4. *A Boolean lattice is autostable if and only if it is finite.*

Exercises

1. Show that the mapping φ_t appearing in the proof of Theorem 3.6.2 is an isomorphic embedding.

2. Prove that the mapping φ from Theorem 3.6.4 is an isomorphism between $\mathfrak{B}_t$ and $\mathfrak{B}_{t+1}$.

3. Construct directly a recursive isomorphism between arbitrary constructivizations ν and μ of an atomless Boolean algebra.

4. Find an algebraic criterion for autostability for Boolean algebras of the signature extended by the predicate A distinguishing atoms.

5. Find an algebraic criterion for autostability for Boolean algebras of the signature extended by the predicate A distinguishing atoms and the predicate B distinguishing atomless elements.

6. Prove that, in the signature extended by predicates for formulas of $\{\text{Atom}_0,\ \text{Atomless}_0, \text{Atomic}_0, I_0, \ldots, \text{Atom}_n, \text{Atomless}_n, \text{Atomic}_n, I_n, \ldots\}$, a Boolean algebra is autostable if and only if it can be represented as a sum of prime models.

   [HINT: Use the Nurtazin theorem on characterization of models that are autostable relative to strong constructivizations. [144].]

7. Prove that a Boolean algebra $\mathfrak{B}$ is autostable if and only if the Boolean ring $R(\mathfrak{B})$ is autostable.

3.7. Algorithmic Properties

8. Construct a constructivization of an atomless lattice $\mathfrak{B}$ which is not a constructivization of the Boolean algebra $B(\mathfrak{B})$.

9. Prove that for an atomless Boolean algebra $\mathfrak{B}$ the classes of constructivizations of the signature $\langle \leqslant, \vee \rangle$ and $\langle \leqslant, \wedge \rangle$ are different.

10. Find a criterion for autostability for Boolean algebras of the signature extended by the predicate F distinguishing the Frechét ideal.

11. Find a criterion for Boolean algebras of the signature σ_k for a fixed n taken from Chapter 2.

3.7. Algorithmic Properties of Subalgebras and Quotient Algebras of Constructive Boolean Algebras

In this section, we are interested in computable algebras and ideals of the class of constructive Boolean algebras. Any recursively enumerable subalgebra of a constructive algebra is constructive. At the same time, such an assertion with respect to taking the quotient is not true. Quotient algebras of constructive Boolean algebras by recursively enumerable ideals form a new class of positive Boolean algebras. It was Feiner [45] who first showed that this class is wider than that of constructivizable Boolean algebras. The class of positive Boolean algebras is of interest because it is closed under taking quotient algebras by enumerable ideals, direct products, and enumerable subalgebras. Furthermore, it realizes the Lindenbaum algebras of recursively axiomatizable theories.

We first study universal algebras of special subclasses of the class of positive Boolean algebras. We recall that a numbering $\nu \colon \mathbb{N} \to B$ of a Boolean algebra $\mathfrak{B} = \langle B; \vee, \wedge, C \rangle$ is called positive if there are recursive functions $f_\vee$, $f_\wedge$, and f_C such that for any $x, y \in \mathbb{N}$

$$\nu f_\vee(x,y) = \nu x \vee \nu y, \quad \nu f_\wedge(x,y) = \nu x \wedge \nu y, \quad \nu f_C(x) = C\nu(x)$$

and the equality relation $\eta_\nu \rightleftharpoons \{(x,y) \mid \nu x = \nu y\}$ is recursively enumerable. If η_ν is recursive, then the numbering ν is a constructivization of the Boolean algebra $\mathfrak{B}$.

For a positive numbering ν of a Boolean algebra $\mathfrak{B}$ we can introduce the Boolean algebra $\mathfrak{B}_\nu \rightleftharpoons \langle \mathbb{N}/\eta_\nu, \vee_\nu, \wedge_\nu, C_\nu \rangle$, where $x/\eta_\nu \vee_\nu y/\eta_\nu \rightleftharpoons f_\vee(x,y)/\eta_\nu$, $x/\eta_\nu \wedge_\nu y/\eta_\nu \rightleftharpoons f_\wedge(x,y)/\eta_\nu$, $C_\nu(x/\eta_\nu) \rightleftharpoons f_C(x)/\eta_\nu$. The

Boolean algebra defined in such a way is called *positive*. It is isomorphic to a recursive algebra if η_ν is recursive. It is easy to see that the Boolean algebras $\mathfrak{B}_\nu$ and $\mathfrak{B}$ are isomorphic under the natural mapping $\varphi_\nu(x/\eta_\nu) \rightleftharpoons \nu(x)$.

A homomorphism $\varphi \colon \mathfrak{A}_\nu \to \mathfrak{B}_\mu$, where $\mathfrak{A}_\nu$ and $\mathfrak{B}_\mu$ are positive Boolean algebras, is called *recursive* if there exists a recursive function f such that $\varphi(x/\eta_\nu) = f(x)/\eta_\nu$. This condition is exactly the corresponding condition for recursivity of the homomorphism $\overline{\varphi}$ of enumerated Boolean algebras $\langle \mathfrak{A}, \nu \rangle$ and $\langle \mathfrak{B}, \mu \rangle$, where $\overline{\varphi} \rightleftharpoons \varphi_\mu^{-1} \varphi \varphi_\nu$,

$$\begin{array}{ccc} \mathfrak{A} & \xrightarrow{\overline{\varphi}} & \mathfrak{B} \\ \varphi_\nu \downarrow & & \downarrow \varphi_\mu \\ \mathfrak{A}_\nu & \xrightarrow{\varphi} & \mathfrak{B}_\mu \end{array}$$

We introduce the product $\langle \mathfrak{A} \times \mathfrak{B}, \nu \otimes \mu \rangle$ of two positively enumerated algebras $\langle \mathfrak{A}, \nu \rangle$ and $\langle \mathfrak{B}, \mu \rangle$, where $\mathfrak{A} \times \mathfrak{B}$ is the Cartesian product of Boolean algebras and $(\nu \otimes \mu)(n) \rightleftharpoons \langle \nu l(n), \mu r(n) \rangle$. It is obvious that $\langle \mathfrak{A} \times \mathfrak{B}, \nu \otimes \mu \rangle$ is a positive Boolean algebra. We can also introduce the direct product of recursively enumerable Boolean algebras $\mathfrak{A}_\nu$ and $\mathfrak{B}_\mu$ by setting

$$\mathfrak{B}_{\nu \otimes \mu} \rightleftharpoons \langle \mathbb{N}/\eta_{\nu \otimes \mu}, \vee_{\nu \otimes \mu}, \wedge_{\nu \otimes \mu}, C_{\nu \otimes \mu} \rangle$$

It is clear that $\mathfrak{A}_\nu \otimes \mathfrak{B}_\mu \cong \mathfrak{B}_{\nu \otimes \mu}$. The positive Boolean algebra $\mathfrak{B}_{\nu \otimes \mu}$ will be denoted by $\mathfrak{A}_\nu \times_r \mathfrak{B}_\mu$. Positive Boolean algebras $\mathfrak{A}_\nu$ and $\mathfrak{B}_\mu$ are called *recursively isomorphic* if there exists a recursive isomorphism $\varphi \colon \mathfrak{A}_\nu \xrightarrow[\text{onto}]{} \mathfrak{B}_\mu$. In this case, the isomorphism φ^{-1} is also recursive. We use the notation $\mathfrak{A}_\nu \cong_{\text{rec}} \mathfrak{B}_\mu$ if $\mathfrak{A}_\nu$ and $\mathfrak{B}_\mu$ are recursively isomorphic.

Let a class of positive Boolean algebras $\mathcal{K}$ be given. Following Hanf, we call a Boolean algebra $\mathfrak{A}_\nu$ *universal in* $\mathcal{K}$ if $\mathfrak{A}_\nu \in \mathcal{K}$ and for any recursively enumerable Boolean algebra $\mathfrak{B}_\mu \in \mathcal{K}$ there exists a positive Boolean algebra $\mathfrak{B}_\pi$ such that $\mathfrak{B}_\mu \times_r \mathfrak{B}_\pi \cong_{\text{rec}} \mathfrak{A}_\nu$. We discuss the question on the existence of universal Boolean algebras for the following three classes of positive Boolean algebras:

Atomless	(atomless positive Boolean algebras)
Atom	(atomic positive Boolean algebras)
Atom$_r$	(atomic recursive Boolean algebras)

For all these classes, negative answers are obtained.

3.7. Algorithmic Properties

Proposition 3.7.1. *If $\mathfrak{A}_\nu$, $\mathfrak{B}_\mu$, and $\mathfrak{B}_\pi$ are positive Boolean algebras and $\mathfrak{A}_\nu \cong_{\mathrm{rec}} \mathfrak{B}_\mu \times \mathfrak{B}_\pi$, then $\mathfrak{A}_\nu$ is a recursive Boolean algebra if and only if $\mathfrak{B}_\mu$ and $\mathfrak{B}_\pi$ are recursive Boolean algebras.*

PROOF. The assertion is established by an immediate verification with the help of a recursive isomorphism. □

An atomless Boolean algebra has exactly one constructivization up to an equivalence. Therefore, any two recursive atomless Boolean algebras are recursively isomorphic. We fix a constructivization ν of an atomless Boolean algebra $\mathfrak{B}$ and construct, with the help of ν, the recursive Boolean algebra B.

Proposition 3.7.2. *For any positive Boolean algebra $\mathfrak{A}_\mu$ there exists a recursively enumerable congruence η on B such that $\mathfrak{A}_\mu \cong_{\mathrm{rec}} B/\eta$.*

PROOF. We first construct a tree D generating $\mathfrak{A}_\mu$. We consider an element $n_0 \in \mathbb{N}$ such that $n_0/\eta_\nu = \mathbf{1}$ and put $g(0) = n_0$. Assuming that the value $g(x)$ is already defined for all $x \in \bigcup_{i \leqslant t} E_i$, we define g on elements of E_{t+1}. If $x \in E_{t+1}$, then $LH(x) = x$ or $RH(x) = x$, where the functions L, R, and H and the sets E_n were introduced at the beginning of Sec. 1.7. Let

$$g(LH(x)) \rightleftharpoons f_\wedge(g(H(x)), t), \quad g(RH(x)) \rightleftharpoons f_\wedge(gH(x), f_C(t))$$

The set $\{g(n)/\eta_\nu \mid n \in \mathbb{N}\}$ generates $\mathfrak{A}_\mu$. Using the function g, we define a recursive homomorphism of the recursive Boolean algebra $\mathfrak{B}_\mathbb{N}$ onto the Boolean algebra $\mathfrak{A}_\mu$. The kernel of this homomorphism is recursively enumerable. It is easy to verify that the congruence η by this kernel is the required congruence. □

The aforesaid shows that the study of positive Boolean algebras can be reduced to the study of quotient algebras of atomless Boolean algebras by recursively enumerable congruences.

We prove one more property of positive Boolean algebras. We note that $\mathfrak{B} \cong \mathfrak{B}_a \times \mathfrak{B}_{C(a)}$ for any $a \in \mathfrak{B}$, where $\mathfrak{B}_a$ is equal to $\langle B_a; \vee_a, \wedge_a, C_a, 0_a, 1_a \rangle$ and $B_a \rightleftharpoons \{b \in B \mid b \leqslant a\}$, $C_a \rightleftharpoons C(x) \wedge a$, $1_a \rightleftharpoons a$, while $\wedge_a$, $\vee_a$, and 0_a are the same as in the case of $\mathfrak{B}$. We derive a recursive analog of this claim.

Let $\langle \mathfrak{B}, \nu \rangle$ be a positive enumerated Boolean algebra and let $a \in |\mathfrak{B}|$. Then the set $\{x \mid \nu x \leqslant a\}$ is recursively enumerable. We consider an arbitrary recursive function $f \colon \mathbb{N} \underset{\mathrm{onto}}{\to} \{x \mid \nu x \leqslant a\}$. The numbering

$\nu_a^f(x) \rightleftharpoons \nu f(x)$ of the Boolean algebra $\mathfrak{B}_a$ is positive. Furthermore, up to a recursive isomorphism, it is independent of the choice of f. Thereby, we obtain the positive enumerated Boolean algebra $\langle \mathfrak{B}_a, \nu_a \rangle$ and the positive Boolean algebra $\langle \mathfrak{B}_a, \nu_a \rangle$. It is clear that $\mathfrak{B}_\nu \cong_{\text{rec}} (\mathfrak{B}_a)_{\nu_a} \times_r (\mathfrak{B}_{C(a)})_{\nu_{C(a)}}$. The converse assertion is also true.

Proposition 3.7.3. *If $\mathfrak{A}_\nu$, $\mathfrak{B}_\mu$, and $\mathfrak{B}_\pi$ are positive Boolean algebras and $\mathfrak{A}_\nu \cong_{\text{rec}} (\mathfrak{B}_\mu \times_r \mathfrak{B}_\pi)$, then there exists an element a of the Boolean algebra $\mathfrak{A}$ such that $\mathfrak{B}_\mu \cong_{\text{rec}} (\mathfrak{A}_a)_{\nu_a}$.*

Thus, the positive Boolean algebra $\mathfrak{B}_\mu$ is a direct summand of the positive Boolean algebra $\mathfrak{A}_\nu$ if and only if there exists a recursive isomorphism φ from $\mathfrak{B}_\mu$ onto $(\mathfrak{A}_a)_{\nu_a}$ for some a, which is equivalent to the existence of an isomorphism and a recursive function g such that $\varphi \mu(x) = \nu g(x)$ for $x \in \mathbb{N}$.

Let $\varphi_n(x)$ be a universal partially recursive function for the class of unary partially recursive functions. We set $\varphi_n^t(x) = \varphi_n(x)$ if $\varphi_n(x)$ can be computed in less than t steps. Otherwise, $\varphi_n^t(x)$ is not defined.

We now turn to the above-mentioned problem of the existence of universal Boolean algebras.

Theorem 3.7.1. *There is no universal positive Boolean algebra for the class* Atomless.

PROOF. We assume the contrary. Let $\mathfrak{A}_\pi$ be a universal positive Boolean algebra for the class Atomless, let f_Δ^π be the recursive functions for $\Delta \in \{\vee, \wedge, C\}$, and let $\text{Ord} = \bigcup_{t \geqslant 0} \text{Ord}^t$, where $\text{Ord}^t \subseteq \text{Ord}^{t+1}$, $\{\text{Ord}^t \mid t \in \mathbb{N}\}$ is a strongly computable sequence [41] of finite sets, and $\text{Ord} \rightleftharpoons \{\langle x, y \rangle \mid \pi x \leqslant \pi y\}$ is a recursively enumerable set. We construct a positive Boolean algebra $\mathfrak{B}_\mu$ of Atomless that cannot be represented as a direct summand of $\mathfrak{A}_\pi$. Let B be a recursive Boolean algebra constructed for $\langle \mathfrak{B}_\mathbb{N}, \nu_\mathbb{N} \rangle$, where $\nu_\mathbb{N}$ is associated with the function $f_\mathbb{N}$, let $f_\vee$, $f_\wedge$ be recursive functions corresponding to the operations on B, and let A_n be an element of $\mathfrak{B}_\mathbb{N}$ corresponding to the element n of the generating tree $\mathbb{N}$.

We consider recursive functions e, h, and g such that $e(n) = k$ if $\nu_\mathbb{N}(n) = A_k$, $\nu_\mathbb{N} g(n) = A_{LR^n(0)}$, and $\nu_\mathbb{N} h(n) = A_n$. We also introduce into consideration numbers n_0 and n_1 such that $\nu_\mathbb{N}(n_0) = \varnothing$ and $\nu_\mathbb{N}(n_1) = \mathbb{N}$. In the sequel, we omit the subscript $\mathbb{N}$. The process is to construct finite sets η^t, U_n^t and functions $r(n,t)$, ξ^t.

STEP 0. We set $\eta^0 \rightleftharpoons \varnothing$, $U_n^0 \rightleftharpoons \varnothing$, $r(n,0) = 0$ for all n, but ξ^0 is defined nowhere.

3.7. Algorithmic Properties

STEP $t = c(n,m) + 1$.

CASE 1: $\xi^{t-1}(n)$ is not defined and $\varphi_n^t(g(n))$ is defined. We set $\xi^t(n) \rightleftharpoons g(n)$ and keep everything else unchanged, i.e., $r(m,t) \rightleftharpoons r(m, t-1)$, $U_m^t \rightleftharpoons U_m^{t-1}$, and $\eta^t \rightleftharpoons \eta^{t-1}$ for all m.

CASE 2: $\xi^{t-1}(n)$ is defined, $\langle \varphi_n(g(n)), n\rangle \notin \mathrm{Ord}^t$,

$$\{x \leqslant r(n, t-1) \mid \langle x, \varphi_n(g(n))\rangle \in \mathrm{Ord}^t\} \subseteq \{x \mid (\exists y \leqslant t)(x = \varphi_n^t(y)$$
$$\vee (\langle x, \varphi_n^t(y)\rangle \in \mathrm{Ord}^t \wedge \langle \varphi_n^t(y), x\rangle \in \mathrm{Ord}^t))\}$$

and φ_n^t preserves the operations for x such that $x \leqslant r(n,t)$, $\nu x \leqslant \nu g(n)$ and $x \in \mathrm{dom}\,\varphi_n^t$ (i.e.,

$$\langle \varphi_n^t(z), f_\Delta^\pi(\varphi_n^t(x), \varphi_n^t(y))\rangle \in \mathrm{Ord}^t, \quad \langle f_\Delta^\pi(\varphi_n^t(x), \varphi_n^t(y)), \varphi_n^t(z)\rangle \in \mathrm{Ord}^t$$

where $\Delta \in \{\wedge, \vee\}$, for any x, y, z of this set if $\nu z = \nu x \Delta \nu y$),

$$(\exists \overline{y} \subseteq \eta^{t-1})(\nu x = \mathbf{0} \vee \nu x \leqslant \bigvee_{i=1}^{s} \nu h y_i) \Rightarrow \langle \varphi_n^t(x), n_0\rangle \in \mathrm{Ord}^t$$

$\langle x, n_0\rangle \in \mathrm{Ord}^t$ for any $x \in U_n^{t-1}$. We set

$$\xi^t(x) \rightleftharpoons hRe\xi^{t-1}(n), \quad U_n^t \rightleftharpoons U_n^{t-1} \cup \{hLe\xi^{t-1}(n)\}$$
$$\eta^t \rightleftharpoons \eta^{t-1} \cup \{hLe\xi^{t-1}(n)\}, \quad r(n,t) \rightleftharpoons r(n, t-1)$$

everything else is left constant.

If Cases 1 and 2 fail, then we change nothing and pass to the following step.

We denote by I a recursively enumerable ideal generated by the recursively enumerable set $\eta = \cup \eta^t$. Consider a recursively enumerable quotient algebra $\mathfrak{B}_\mathbb{N}/I$. It is easy to see that $\mathfrak{B}_\mathbb{N}/I$ is isomorphic to the direct sum

$$\sum_{n \in \mathbb{N}} {}_0\mathfrak{B}_{\nu g(n)}/I \cap |\mathfrak{B}_{\nu g(n)}|$$

We show that each of its summands is an atomless Boolean algebra. It is easy to see that for every n, Case 1 can be realized only once. We show that Case 2 holds only a finite number of times. We assume the contrary. Then $\langle \varphi_n g(n), n_0\rangle \notin \mathrm{Ord}^t$ for all t. Consequently, $\pi\varphi_n g(n) \neq \mathbf{0}$. We consider two elements a and b such that $\mathbf{0} < a, b < \pi\varphi_n g(n)$ and $a \wedge b = \mathbf{0}$. Let a have a π-number n_a and b have a π-number n_b. Since Case 2 holds infinitely many times, there is a step t_0 such that $n_a, n_b < r(n, t_0)$ and $\{n_a, n_b\} \subseteq \{x \mid \langle x, \varphi_n g(n)\rangle \in \mathrm{Ord}^{t_0}\}$. Therefore, after the step t_0 there is a step t_1 at which Case 2 is realized for n. Hence there are elements

$\hat{n}_a$ and $\hat{n}_b$ such that $\varphi_n^{t_1}(\hat{n}_a) \approx_{t_1} n_a$ and $\varphi_n^{t_1}(\hat{n}_b) \approx_{t_1} n_b$ (here $x \approx_t y \leftrightharpoons x = y \vee (\langle x,y \rangle, \langle y,x \rangle \in \mathrm{Ord}^t))$, where $\nu \hat{n}_a \leqslant \nu g(n)$ and $\nu \hat{n}_b \leqslant \nu g(n)$. Consequently, there exist sequences $p_0, \ldots, p_k$ and $m_0, \ldots, m_l$ such that $\nu \hat{n}_a = A_{p_0} \cup \ldots \cup A_{p_k}$ and $\nu \hat{n}_b = A_{m_0} \cup \ldots \cup A_{m_l}$ (since $\pi \hat{n}_a \neq 0$, $\pi \hat{n}_b \neq 0$, and $\pi \hat{n}_a \wedge \pi \hat{n}_b = 0$) and there exists a number e such that $E_e \supseteq \{m_0, \ldots, m_l, p_0, \ldots, p_k\}$, $p_i \preccurlyeq eg(n)$, $i \leqslant k$, $m_i \preccurlyeq eg(n)$, $i \leqslant l$, and $\{p_0, \ldots, p_k\} \cap \{m_0, \ldots, m_l\} = \varnothing$.

We consider a step t^* at which, for n, Case 2 has been realized e times. All the elements $h(x)$ for $x \preccurlyeq eg(n)$, $x \in E_e$ (perhaps, one element), belong to the ideal generated by η^{t^*} Then there exist $l_1, \ldots, l_p \in \eta^{t^*}$ such that

$$\left(\nu \hat{n}_a \leqslant \bigvee_{i=1}^{p} \nu h(l_i)\right) \vee \left(\nu \hat{n}_b \leqslant \bigvee_{i=1}^{p} \nu h(l_i)\right)$$

Therefore, at the following step $t_2 > t^*$ at which Case 2 is again realized for n, we have $\langle n_a, n_0 \rangle \in \mathrm{Ord}^{t_2}$ or $\langle n_b, n_0 \rangle \in \mathrm{Ord}^{t_2}$, but this is impossible in view of the choice of $a \neq 0$ and $b \neq 0$.

We note that the ideal I is generated by $\{\nu n \mid n \in \cup \eta^t\}$ and, for an element having nonempty intersection with $\nu g(n)$, its number is contained in η^t only if Case 2 for n holds. Hence only a finite number of elements (denote them by $a_1, \ldots, a_n$ and their union by $\vee a_i$) having nonempty intersection with $\nu g(n)$ get into η; moreover, $a_1, \ldots, a_k < \nu g(n)$. Therefore, $\mathfrak{B}_{\nu g(n)} \cong \prod_{i=1}^{n} \mathfrak{B}_{\nu a_i} \times \mathfrak{B}_{\nu g(n) \setminus (\vee a_i)}$ and $\vee a_i \neq \nu g(n)$. Thus, $\mathfrak{B}_{\nu g(n)}/I \cap |\mathfrak{B}_{\nu g(n)}| \cong \mathfrak{B}_{\nu g(n) \setminus (\vee a_i)}$ and $\mathfrak{B}_\mathbb{N}/I$ is an atomless Boolean algebra because $\mathfrak{B}_{\nu g(n) \setminus (\vee a_i)}$ is an atomless Boolean algebra.

We show that $\mathfrak{B}_\mathbb{N}/I$ cannot be represented as a direct summand of $\mathfrak{A}_\pi$. We assume the contrary. Then there exists a recursive function φ_n realizing an isomorphism of $\mathfrak{B}_\mathbb{N}/I$ onto $(\mathfrak{A}_\pi)_a$ for some a, i.e.,

$$(\forall xy)(\pi \varphi_n f_\vee(x,y) = (\pi \varphi_n(x) \vee \pi \varphi_n(y)) \,\&\, \pi \varphi_n f_\wedge(x,y)$$
$$= (\pi \varphi_n(x) \wedge \pi \varphi_n(y)) \,\&\, (\nu x/I = \mathbf{0} \Leftrightarrow \pi \varphi_n(x) = \mathbf{0}))$$

Moreover, if $\nu x/I = \mathbf{1}$, then $(\forall y)(\pi y \leqslant \pi \varphi_n(x) \Rightarrow (\exists z)(\pi \varphi_n(z) = \pi y))$.

Let t_1 be a step after which Cases 1 and 2 for n are not realized. For all $t \geqslant t_1$ we have $r(n,t) = r(n,t_1)$. Since φ_n is defined everywhere, Case 1 for n is realized. Consequently, $\xi^{t_1}(n)$ is defined. Since φ_n realizes an isomorphism onto $(\mathfrak{A}_\pi)_a$ for some a and φ_n is defined everywhere, there is a step $t_2 = c(n,m) + 1 > t_1$ at which all the conditions of Case 2 hold for n. Hence Case 2 for n is realized, which is impossible. □

3.7. Algorithmic Properties

For Atom and Atom$_r$, results similar to Theorem 3.7.1 can be derived from the following claim.

Theorem 3.7.2. *For any atomic positive Boolean algebra $\langle \mathfrak{A}, \mu \rangle$ there exists a recursive atomic Boolean algebra which cannot be represented as a direct summand of $\langle \mathfrak{A}, \mu \rangle$.*

PROOF. As in Theorem 3.7.1, we consider a Boolean algebra $\langle \mathfrak{B}_\mathbb{N}, \nu_\mathbb{N} \rangle$ and partially recursive functions g, h, and e such that $\nu g(n) = A_{LR^n(0)}$, $\nu h(n) = A_n$ for all n and $e(n) = k$ if and only if $\nu n = A_k$. We define step-by-step a subalgebra $\mathfrak{B}$ of the algebra $\mathfrak{B}_\mathbb{N}$. At the step t, we construct a finite set $B^t \subseteq \mathbb{N}$ and define the values $r(n,t)$ for all n. This construction consolidates all the ideas of the proofs of the previous assertions.

STEP 0. We set $B^0 \rightleftharpoons \{\nu^{-1}(\varnothing), \nu^{-1}(A_0), \nu^{-1}(A_1), \nu^{-1}(A_2)\}$ and $r(n,t) = 0$ for all n.

STEP $t+1$. Let $t = c(n,l)$. We first assume that for n the following conditions hold:

(1) φ_n^t is defined on $g(n)$,

(2) φ_n^t is defined on $B^t \cap \{i \mid i \leqslant r(n,t)\}$,

(3) $\varphi_n^t(B^t) \supseteq \{i \mid i \leqslant r(n,t) \,\&\, \langle i, \varphi_n g(n) \rangle \in \mathrm{Ord}^t\}$,

(4) φ_n^t preserves all the operations for x such that $x \leqslant r(n,t)$ and $x \in B^t$, i.e., for any $x, y, z \in B^t \cap \{i \mid i \leqslant r(n,t)\}$ for $\nu x \Delta \nu y = \nu z$

$$\langle \varphi_n(z), f_\Delta^\mu(\varphi_n(x), \varphi_n(y)) \rangle \in \mathrm{Ord}^t, \ \langle f_\Delta^\mu(\varphi_n(x), \varphi_n(y)), \varphi_n(z) \rangle \in \mathrm{Ord}^t$$

where $\Delta \in \{\vee, \wedge\}$; if $\nu x \neq \mathbf{0}$, then $\langle \varphi_n(x), n \rangle \notin \mathrm{Ord}^t$; if $\nu x = \mathbf{0}$, then $\langle \varphi_n(x), n_0 \rangle \in \mathrm{Ord}^t$; if $\nu x \neq \nu y$, then

$$\langle \varphi_n(x), \varphi_n(y) \rangle \notin \mathrm{Ord}^t \vee \langle \varphi_n(y), \varphi_n(x) \rangle \notin \mathrm{Ord}^t$$

(5) for any $x \in \{x \leqslant r(n,t) \mid \langle x, \varphi_n g(n) \rangle \in \mathrm{Ord}^t\}$, there exists $z \in B^t$ such that $\langle \varphi_n(z), x \rangle \in \mathrm{Ord}^t$ and $\langle x, \varphi_n(z) \rangle \in \mathrm{Ord}^t$. Then $r(n, t+1) \rightleftharpoons r(n,t) + 1$, $r(m, t+1) \rightleftharpoons r(m,t)$ for all $m \neq n$

$$B^{t+1} \rightleftharpoons \nu^{-1}(\mathrm{gr}(\nu B^t \cup \{\nu g(t+1)\} \cup \{\nu h(k) \mid k \preccurlyeq g(n) \,\&\, k \in E_{r(n,t)+1}\}))$$

If the conditions (1)–(5) fail, then we set $r(m, t+1) = r(m,t)$ for all m and $B^{t+1} \rightleftharpoons \nu^{-1}(\mathrm{gr}(\nu B^t \cup \{\nu g(t+1)\}))$ and pass to the following step.

Let $\mathfrak{B} \subset \mathfrak{B}_\mathbb{N}$. The universe has the form $B = \bigcup_{t \geqslant 0} B^t$. For any $a \in \mathfrak{B}_\mathbb{N}$ there exist elements $n_1, \ldots, n_k$ such that $a \leqslant \nu g(n_1) \vee \ldots \vee \nu g(n_k)$

or $C(a) \leqslant \nu g(n_1) \vee \ldots \vee \nu g(n_k)$. Therefore, $\mathfrak{B} = \sum_n {}_{0,1}(|\mathfrak{B}| \cap \mathfrak{B}_{\nu g(n)})$.
To prove that $\mathfrak{B}$ is an atomic Boolean algebra, it suffices to note that the algebra $|\mathfrak{B}| \cap \mathfrak{B}_{\nu g(n)}$ is finite for any n. At the step $c(n, l) + 1$, only element $\nu g(n)$ and elements less than $\nu g(n)$ can be added to $\mathfrak{B}$ while elements of $|\mathfrak{B}| \cap \mathfrak{B}_{\nu g(n)}$ are added to $\mathfrak{B}$ only at the steps $c(n, l) + 1$ provided that all the conditions (1)–(5) of the corresponding step are fulfilled. We show that, for n, the conditions (1)–(5) can be satisfied only a finite number of times. We assume the contrary. Then $\lim_{t \to \infty} r(n, t) = \infty$. Therefore, all elements of $\mathfrak{B}_\mathbb{N}$ that are less than $\nu g(n)$ are ultimately added to $\mathfrak{B}$. Consequently, $\nu g(n)$ is an atomless element of $\mathfrak{B}$. Since $\lim r(n, t) = \infty$ and, for n, the conditions (1)–(5) hold infinitely many times, $\varphi_n k$ is defined for all k such that $\nu k \leqslant \nu g(n)$. For all νk satisfying the inequality $\nu k \leqslant \nu g(n)$ we define a mapping ψ by setting $\psi(\nu k) \leftrightharpoons \mu \varphi_n(k')$, where k' is the least number such that $\nu k' = \nu k$. By definition, ψ is well defined. From the condition (4) of the steps $c(n, l) + 1$, which is satisfied infinitely many times, it follows that ψ preserves the operations $\vee$ and $\wedge$; moreover, $\psi(x) = \mathbf{0} \Leftrightarrow x = \mathbf{0}$, i.e., ψ is an isomorphic embedding. From the condition (3) of the same steps it follows that ψ is an isomorphism onto $|\mathfrak{A}_{\mu \varphi_n g(n)}|$. Therefore, ψ is isomorphism of an atomless Boolean algebra onto a direct summand of $\mathfrak{A}$, which contradicts the fact that the Boolean algebra $\mathfrak{A}$ is atomic. Consequently, the Boolean algebra $\mathfrak{B}$ is atomic. We consider a constructivization π of the Boolean algebra $\mathfrak{B}$ taking a recursive function $f \colon \mathbb{N} \overset{1-1}{\underset{\text{onto}}{\to}} B$, which exists since B is recursively enumerated, and by setting $\pi n \leftrightharpoons \nu f(n)$. Since ν is a constructivization, π is a constructivization of a subalgebra $\mathfrak{B}$ of the algebra $\mathfrak{B}_\mathbb{N}$.

It remains to show that $\langle \mathfrak{B}, \pi \rangle$ cannot be represented as a direct summand of $\langle \mathfrak{A}, \mu \rangle$. We assume the contrary. Then there exists a recursive function φ such that $\mu \varphi \pi^{-1}$ is an isomorphism of $\mathfrak{B}$ onto $\mathfrak{A}_{\mu \varphi \pi^{-1}(1)}$. We consider a function φ_n such that $\varphi_n(x) = \varphi(\mu k(f(k) = x))$. The mapping $\mu \varphi_n \nu^{-1}$ is also an isomorphism of $\mathfrak{B}$ onto $\mathfrak{A}_{\mu \varphi \pi^{-1}(1)}$. Consequently, $\mu \varphi_n \nu^{-1} \upharpoonright (\mathfrak{B})_{\nu g(n)}$ is an isomorphism of $\mathfrak{B}_{\nu g(n)}$ onto $\mathfrak{A}_{\mu \varphi_n g(n)}$. We now consider a step t_0 after which for n the conditions (1)–(5) fail at the steps $c(n, l) + 1$. After the step t_0, the value $r(n, t)$ remains constant. Since $\mu \varphi_n \nu^{-1}(\mathfrak{B})_{\nu g(n)}$ is an isomorphism of $(\mathfrak{B})_{\nu g(n)}$ onto $\mathfrak{A}_{\mu \varphi_n g(n)}$, at some step $t_1 > t_0$, all the conditions (1)–(5) hold. For l such that $c(n, l) + 1 \geqslant t_1$ the conditions (1)–(5) hold for n, which yields a contradiction. □

Corollary 3.7.1. *There is no universal recursive Boolean algebra for the class* Atom_r.

3.7. Algorithmic Properties

Corollary 3.7.2. *There is no universal positive Boolean algebra for the class* Atom.

We now study the connection between the constructivizability of a Boolean algebra and that of its quotient algebras. As was mentioned, Feiner [45] constructed an example of an enumerated ideal of a recursive atomless Boolean algebra such that the quotient algebra by this ideal is a nonconstructivizable positive Boolean algebra. We are interested in the cases of special ideals: the Frechét ideal, the atomless ideal, and the Ershov–Tarski ideal.

Proposition 3.7.4. *There is a constructivizable atomic Boolean algebra such that its quotient algebra by the Frechét ideal is not constructivizable.*

PROOF. We consider an atomic constructivizable but not strongly constructivizable Boolean algebra constructed in Theorem 3.5.1. We show that its quotient algebra by the Frechét ideal is not constructivizable. Indeed, if $\mathfrak{B}/F(\mathfrak{B})$ is a constructivizable Boolean algebra, then, by Theorem 3.2.1, there exists an atomic strongly constructivizable Boolean algebra $\mathfrak{A}$ such that $\mathfrak{A}/F(\mathfrak{A}) \cong \mathfrak{B}/F(\mathfrak{B})$. However, in view of Proposition 1.5.2, the quotient algebras of Boolean algebras by the Frechét ideals are isomorphic provided that the Boolean algebras are isomorphic. Therefore, $\mathfrak{B}$ is strongly constructivizable, which contradicts the choice of $\mathfrak{B}$. □

By Theorem 3.2.1, the constructivizability of the quotient algebras by the Frechét ideals implies even the strong constructivizability of the corresponding atomic Boolean algebra, but the converse assertion is not true in general.

It is not hard to construct an example of a Boolean algebra that is not atomic and is not constructivizable while its quotient algebra by the Frechét ideal is constructivizable.

Proposition 3.7.5. *There exists a constructivizable Boolean algebra whose quotient algebra by the Ershov–Tarski ideal is not constructivizable.*

PROOF. We consider a constructivizable atomic Boolean algebra having the nonconstructivizable quotient algebra by the Frechét ideal from Proposition 3.7.4. Let L be a recursive linear order such that $\mathfrak{B}_L \cong \mathfrak{B}$. Such an order exists by Proposition 3.2.2. We consider the linear order $(2 + \eta) \times L$ and the Boolean algebra $\mathfrak{A} \cong \mathfrak{B}_{(2+\eta) \times L}$. The element $[(\alpha_1, \beta_1), (\alpha_2, \beta_2)[$ belongs to the Ershov–Tarski ideal $I(\mathfrak{A})$ if and only if

$[\beta_1, \beta_2[$ belongs to the Frechét ideal of the algebra $\mathfrak{B}_L$. Therefore, the quotient algebras $\mathfrak{B}_L/F(\mathfrak{B}_L)$ and $\mathfrak{A}/I(\mathfrak{A})$ are isomorphic. Hence $\mathfrak{A}/I(\mathfrak{A})$ is not constructivizable. □

As an example of a nonconstructivizable algebra with the constructivizable quotient algebra by the Ershov–Tarski ideal we indicate a Boolean algebra with a linear basis obtained as a result of multiplication of the order $\omega \times \eta$ and the generating linear order on the quotient algebra of a nonconstructivizable 2-atomic Boolean algebra by the Frechét ideal.

It remains to consider the case of the atomless ideal $\text{Atomless}(\mathfrak{B}) \leftrightharpoons \{a \in |\mathfrak{B}| \mid a \text{ is atomless in } \mathfrak{B}\}$. By the cardinality reasoning, the constructivizability of the quotient algebra by the ideal $\text{Atomless}(\mathfrak{B})$ does not imply the constructivizability of $\mathfrak{B}$. Namely, for a countable atomic Boolean algebra $\mathfrak{B}$ that is not superatomic or is of infinite ordinal type, there exists a continuum of countable Boolean algebras such that their quotient algebras by the atomless ideal is isomorphic to $\mathfrak{B}$ [146]. In the case of a superatomic Boolean algebra $\mathfrak{B}$ of finite ordinal type, any countable Boolean algebra $\mathfrak{A}$ such that $\mathfrak{A}/\text{Atomless}(\mathfrak{A}) \cong \mathfrak{B}$ is strongly constructivizable [146].

Proposition 3.7.6. *There exists a constructivizable Boolean algebra such that its quotient algebra by the atomless ideal is not constructivizable.*

PROOF. The reasoning is the same as in the proof of Theorem 3.5.1. We estimate the complexity of the invariant $\omega_{\mathfrak{B}}$ with respect to the Feiner hierarchy and construct a recursive linear order so that, for the algebra constructed by this order, the quotient algebra by the atomless ideal is not constructivizable. □

Using the quotient algebras by sequences of standard iterated ideals, it is possible to give a description of Boolean algebras that are constructive in different classes of the arithmetic hierarchy. The integrable characterization was obtained by Odintsov and Selivanov [145]. We present their method as well as a series of interesting results which clarify the algorithmic structure of recursive Boolean algebras.

Let $\langle \mathfrak{B}, \nu \rangle$ be an enumerated Boolean algebra and let $\{\nu[\sigma] \mid \sigma \in 2^{<\omega}\}$ be a set generating $\langle \mathfrak{B}, \nu \rangle$, where $\nu[\Lambda] = \mathbf{1}$, $\nu[r * 0] = \nu[r] \wedge \nu(|r|)$ and $\nu[r * 1] = \nu[r] \wedge C(\nu(|r|))$, The set $P = \{\sigma \mid \nu[\sigma] \neq \mathbf{0}\}$ is a segment since $\sigma \subset \tau$ implies $\nu[\tau] \subset \nu[\sigma]$. Furthermore, P satisfies the condition

$$\tau \in P \Rightarrow (\exists i \leqslant 1)(\tau * i \in P) \tag{3.7.1}$$

3.7. Algorithmic Properties

Indeed if $\nu[\sigma] \neq 0$, then $\nu[\sigma*1] = \nu[\sigma] \cap \nu i \neq \mathbf{0}$ or $\nu[\sigma*0] = \nu[\sigma] \cap C\nu i \neq \mathbf{0}$, $i = \mathrm{lh}(\sigma)$.

A segment satisfying the condition (3.7.1) is called *normal*. Normal segments allow us to introduce one more method of representation of Boolean algebras [145]. For $P, F \subseteq 2^{<\omega}$ we set $P[F] = \{\tau \in P \mid \exists \sigma \in F \, (\sigma \subseteq \tau)\}$. With each normal segment T we associate a Boolean algebra $\mathrm{al}(T)$ which is a subalgebra of the quotient algebra of the Boolean algebra of all subsets of T by the ideal of finite sets. It is generated by the congruence classes of the form $T[F]^*$, where F is a finite subset of $2^{<\omega}$. The Boolean algebra $\mathrm{al}(T)$ admits the natural numbering that is generated by the numbering of sequences.

REMARK 3.7.1. If $\langle \mathfrak{B}, \nu \rangle$ is an enumerated Boolean algebra and $P = \{\sigma \mid \nu[\sigma] \neq 0\}$, then $\mathfrak{B} \simeq \mathrm{al}(P)$ (the isomorphism is given by the mapping $T[F]^* \mapsto \bigcup_{\sigma \in F} \nu[\sigma]$).

We establish a connection between algorithmic properties of Boolean algebras and those of normal segments.

Proposition 3.7.7. *A Boolean algebra belongs to Σ_1^A (Π_1^A, Δ_1^A) if and only if it is isomorphic to $\mathrm{al}(T)$ for a suitable normal segment $T \in \Pi_1^A$ (Σ_1^A, Δ_1^A).*

PROOF. Let T be a normal segment of the class Π_1^A (Σ_1^A, Δ_1^A). It is clear that, with respect to the natural numbering, all the Boolean operations can be represented by recursive functions. It is only required to verify the fact that the equality relation on $\mathrm{al}(T)$ belongs to Σ_1^A (Π_1^A, Δ_1^A). To this end, it suffices to establish that the relation "$T[F]$ is finite" belongs to Σ_1^A (Π_1^A, Δ_1^A). Since T is normal, the above relation is equivalent to the relation $\forall \sigma \in F \, (\sigma \notin T)$ that belongs to Σ_1^A (Π_1^A, Δ_1^A). The converse assertion follows from Remark 3.7.1. □

We find a sufficient condition for isomorphism between the Boolean algebras $\mathrm{al}(S)$ and $\mathrm{al}(T)$.

Proposition 3.7.8. *Let S and T be segments and φ be an isomorphism between $\langle S; \subseteq \rangle$ and $\langle T; \subseteq \rangle$ such that for any $\tau \in T$ there is $\sigma \in S$ subject to the condition $\tau \subseteq \varphi(\sigma)$. Then the Boolean algebras $\mathrm{al}(S)$ and $\mathrm{al}(T)$ are isomorphic.*

PROOF. An isomorphism can be defined by the mapping $S[F]^* \to T[\varphi(F)]^*$, where $\varphi(F)$ is the image of a finite subset $F \subseteq 2^{<\omega}$ under the partial mapping φ. It is easy to verify that φ is an isomorphism.

For example, let us verify that φ is a mapping "onto" (the proof of the remaining properties is trivial). Since the algebra $\mathrm{al}(T)$ is generated by elements $T[\tau]^*$, $\tau \in T$, it suffices to check that for any $\tau \in T$ there is $\sigma \in S$ such that $T[\tau] =^* T[\varphi(\sigma)]$ ($=^*$ denotes the equality modulo finite sets). Let $\sigma \in S$ be a sequence of the least length such that $\tau \subseteq \varphi(\sigma)$. In the case $\sigma = \varnothing$, we have $T[\tau] =^* T[\varphi(\varnothing)]$. The inclusion $T[\varphi(\varnothing)] \subseteq^* T[\tau]$ is obvious in view of $\tau \subseteq \varphi(\varnothing)$. To prove $T[\tau] \subseteq^* T[\varphi(\varnothing)]$ it suffices to verify that $\tau \subseteq \rho$ and $\rho \in T$ imply that ρ and $\varphi(\varnothing)$ are comparable; moreover, $T[\tau] \setminus T[\varphi(\varnothing)] = \{\rho \mid \tau \subseteq \rho \subset \varphi(\varnothing)\}$. Let $\tau \subseteq \rho$, $\rho \in T$ and let $\sigma_1 \in S$ be such that $\rho \subseteq \varphi(\sigma_1)$. Since $\varphi(\varnothing) \subseteq \varphi(\sigma_1)$, we conclude that ρ and $\varphi(\varnothing)$ are comparable.

Let $\sigma \ne \varnothing$, $\sigma = \widetilde{\sigma} * k$ for some $\widetilde{\sigma}$ and $k \leqslant 1$. Since σ is minimal, we have $\varphi(\widetilde{\sigma}) \subseteq \tau \subseteq \varphi(\sigma)$. Let $l \leqslant 1$ and $l \ne k$. If $\tau \not\subseteq \varphi(\widetilde{\sigma} * l)$, then, as in the case $\sigma = \varnothing$, we can verify the equality $T[\tau] =^* T[\sigma]$. Otherwise, a similar reasoning shows that $T[\tau] =^* T[\widetilde{\sigma}]$. $\square$

We now establish a correlation between representations of Boolean algebras by two methods: by means of normal segments and by means of trees. We define an operation transforming a segment of the tree $2^{<\omega}$ into a tree. Let P be an initial segment of the tree $2^{<\omega}$ and $T(P) = \{\sigma \mid \sigma \in P \,\&\, \sigma = \tau * i \,\&\, \tau * (1-i) \in P\} \cup \{\Lambda\}$. It is clear that the set $T(P)$ itself is not a subtree of the complete tree $2^{<\omega}$; nevertheless, the order $\leqslant$ induces on $T(P)$ the structure of a binary tree. Let $\mathrm{BA}\,(P)$ denote the Boolean algebra generated by the tree $T(P)$.

Proposition 3.7.9. *For any normal segment $P \subseteq 2^{<\omega}$ the Boolean algebras $\mathrm{BA}\,(P)$ and $\mathrm{al}(P)$ are isomorphic.*

PROOF. The tree $T(P)$ is isomorphic to some subtree of $S \subseteq 2^{<\omega}$. We fix an isomorphism $\xi \colon S \to T(P)$ and define a mapping $\psi \colon S \to \mathrm{al}(P)$ by setting $\psi(\sigma) \rightleftharpoons P[\{\xi(\sigma)\}]^*$. Let us check that ψ is an admissible mapping of a tree into a Boolean algebra. To this end, we show that ψ satisfies all the conditions (1)–(5) of the definition (cf. Sec. 1.7). We have $\psi(\varnothing) = P[\{\xi(\varnothing)\}]^* = P[\{\varnothing\}]^* = \mathbf{1}_{\mathrm{al}(P)}$. Therefore, the condition (1) holds. Let $\sigma \in S$. We verify the condition (2). By the definition of $T(P)$, $\tau \supseteq \xi(\sigma)$ implies one of the following relations: $\tau \supseteq \xi(\sigma * 0)$, $\tau \supseteq \xi(\sigma * 1)$, and $\tau \subseteq \xi(\sigma * 0) \,\&\, \tau \subseteq \xi(\sigma * 1)$. Therefore, $P[\{\xi(\sigma)\}] =^* P[\{\xi(\sigma * 1)\}] \cup P[\{\xi(\sigma * 0)\}]$. If $\sigma | \tau$, then $\xi(\sigma) | \xi(\tau)$. Consequently, the condition (3) is fulfilled. Furthermore, for any $\sigma \in S$ we have $P[\{\xi(\sigma)\}] \ne^* \varnothing$, i.e., the condition (4) holds.

3.7. Algorithmic Properties

Any element of al(P) can be represented as a finite union of elements of the form $P[\{\sigma\}]^*$. If $\sigma \in P$ and $\tau = \bigcup\{\rho \mid \rho \subset \sigma \,\&\, \rho \in T(P)\}$, then $P[\{\sigma\}] =^* P[\{\tau\}]$. Therefore, (5) holds. Thus, BA $(P) \simeq$ al(P). □

We now establish connections between constructivizabilities for different arithmetic classes. Let Δ be an arithmetic class. Considering a Boolean Δ-algebra, we can assume that there is a numbering ν such that the operations on numbers are recursive and the equality relation η_ν belongs to Δ. This is based on the fact that any Δ-algebra is isomorphic to the quotient algebra of a free algebra by the Δ-congruence.

Theorem 3.7.3. *A Boolean Π_1^A-algebra is a Boolean Δ_1^A-algebra.*

Let $(\mathfrak{B}, \nu) \in \Pi_1^A$ and let $P^A(\sigma, x)$ be an A-recursive predicate such that $\nu[\sigma] = 0 \Leftrightarrow \forall x \, P^A(\sigma, x)$, where $\{\nu[\sigma] \mid \sigma \in 2^{<\omega}\}$ is the standard set generating $\mathfrak{B}$. It is clear that the normal segment $\mathcal{D} = \{\sigma \mid \nu[\sigma] \neq 0\} = \{\sigma \mid \exists x \, \neg P^A(\sigma, x)\}$ is recursively enumerated with the oracle A. We introduce the notation $\mathfrak{B}^s = \text{BA}(\widetilde{\mathcal{D}}^s)$, $s \in \omega$. We set $\widetilde{\mathcal{D}}^s = \{\sigma \mid \sigma \in \mathcal{D} \,\&\, \text{lh}(\sigma) \leqslant s\}$. Then $\mathcal{D} = \bigcup_{s \in \omega} \widetilde{\mathcal{D}}_s$. If $P^0 \subset P^1$, then $T(P^0) \subseteq T(P^1)$. Therefore, $\mathfrak{B}^s \subseteq \mathfrak{B}^{s+1}$, $s \in \omega$. Proposition 3.7.9 implies that the Boolean algebra $\mathfrak{B}$ is the limit of an increasing sequence of finite Boolean algebras $\mathfrak{B}^0 \subseteq \mathfrak{B}^1 \subseteq \ldots$.

A set $\mathcal{D}$ is recursively enumerable with the oracle A, hence it is possible to construct an A-effective approximation of $\mathcal{D}$ by finite initial segments of the tree $2^{<\omega}$. We fix such an approximation $\{\mathcal{D}^s \mid s \in \omega\}$, $\mathcal{D}^0 = \{\varnothing\}$, subject to the additional condition $|\mathcal{D}^{s+1} \setminus \mathcal{D}^s| \leqslant 1$. We define step-by-step a construction which turns out to be effective with the oracle A. At every step, a finite Boolean algebra $\mathfrak{A}^s$ and an isomorphism $\varphi^s \colon \text{BA}(\mathcal{D}^s) \to \mathfrak{A}^s$ must be constructed.

STEP 0. We set $\mathfrak{A}^0 = \{0, 1\}$. An isomorphism φ^0 is defined in a natural way.

STEP $s+1$. For $T(\mathcal{D}^s) = T(\mathcal{D}^{s+1})$ we set $\mathfrak{A}^{s+1} = \mathfrak{A}^s$ and $\varphi^{s+1} = \varphi^s$.

Let $T(\mathcal{D}^s) \neq T(\mathcal{D}^{s+1})$. At the step $s+1$, only one pair of elements can be added to the tree $T(\mathcal{D}^s)$ since $|\mathcal{D}^{s+1} \setminus \mathcal{D}^s| \leqslant 1$. Let $\sigma * 0$, $\sigma * 1$, be such a pair and $\tau = \bigcup\{\rho \mid \rho \leqslant \sigma \,\&\, \rho \in T(\mathcal{D}^s)\}$, $a = \varphi^t(\tau)$, $a \in \mathfrak{A}^s$. From now on, $\tau \in T(\mathcal{D}^s)$ is identified with the corresponding element of BA (D^s). We note that $|\text{Atom}((\widehat{\tau})_{\text{BA}(\mathcal{D}^{s+1})})| = |\text{Atom}((\widehat{a})_{\text{BA}(\mathcal{D}^{s+1})})| + 1$. Let the algebra $\mathfrak{A}^{s+1}$ be obtained from $\mathfrak{A}^s$ by dividing one of the elements of Atom $((\widehat{a})_{\mathfrak{A}^s})$ into two parts. Then

$$|\text{Atom}((a)_{\mathfrak{A}^{s+1}})| = |\text{Atom}((\widehat{\tau})_{\text{BA}(\mathcal{D}^{s+1})})|$$

Let $\xi^{s+1} \colon (\widehat{\tau})_{\text{BA}(\mathcal{D}^{s+1})} \to (\widehat{a})_{\mathfrak{A}^{s+1}}$ be an isomorphism. It can be extended to an isomorphism $\varphi^{s+1} \colon \text{Atom}(\mathcal{D}^{s+1}) \to \mathfrak{A}^{s+1}$ so that $\sigma \in T(\mathcal{D}^s) \,\&\, \neg(\sigma \supseteq \tau) \Rightarrow \varphi^{s+1}(\sigma) = \varphi^s(\sigma)$. The description of the construction is complete. It is effective with the oracle A.

We set $\mathfrak{A} = \bigcup_{s \in \omega} \mathfrak{A}^s$ and prove that $\mathfrak{B} \simeq \mathfrak{A}$. For any s_0 there is a step s_1 such that $\widetilde{\mathcal{D}}^{s_0} \subset \mathcal{D}^s$ for all $s \geqslant s_1$. Consequently, $\mathfrak{B}^{s_0} \subset \text{BA}(\mathcal{D}^s)$. Therefore, for all $s \geqslant s_1$ the restrictions of φ^s and φ^{s_1} to $\mathfrak{B}^{s_0}$ coincide. Thus, we can define a sequence of embeddings $\psi^s \colon \mathfrak{B}^s \to \mathfrak{A}$ such that $\psi^s \subset \psi^{s+1}$. The mapping $\psi = \bigcup_{s \in \omega} \psi^s$ is an isomorphism of $\mathfrak{B}$ into $\mathfrak{A}$.

Let $a \in \mathfrak{A}$. Then there is s such that $a \in \mathfrak{A}^s$. An element a can be represented as a union of atoms of the algebra $\mathfrak{A}^s$. Each of these atoms a_i has the form $a_i = \varphi^s(\tau_i)$, $\tau_i \in \text{BA}(\mathcal{D}^s)$. All the sequences τ_i are in $\widetilde{\mathcal{D}}^{s_1}$ for some sufficiently large s_1. We note that the preimage of a may be changed only at those steps s' for which $\widetilde{\mathcal{D}}^{s_1} \cap \mathcal{D}^{s'} \neq \widetilde{\mathcal{D}}^{s_1} \cap \mathcal{D}^{s'+1}$. However, there are only a finite number of such steps. Thus, for all $a \in \mathfrak{A}$ there exists $\lim_s (\varphi^s)^{-1}(a)$ and, consequently, $\lim_s (\psi^s)^{-1}(a)$. Therefore, ψ is an isomorphism.

Corollary 3.7.3. *If a Boolean algebra $\mathfrak{B}$ is constructivizable as an order, then it is constructivizable as a Boolean algebra.*

Let ν be a constructivization of $\langle |\mathfrak{B}|, \leqslant \rangle$ and let μ be the numbering of $\mathfrak{B}$ generated by ν with the help of the numbering of Boolean terms. It is easy to see that $\langle \mathfrak{B}, \mu \rangle \in \Pi_1$. By Theorem 3.7.3, $\mathfrak{B}$ is constructivizable.

Theorem 3.7.4. *A Boolean Δ_2^A-algebra is a Boolean Σ_1^A-algebra.*

PROOF. By Proposition 3.7.7, it suffices to verify that for any normal segment $S \in \Delta_2^A$ there is a normal segment $T \in \Pi_1^A$ such that $\text{al}(T) \simeq \text{al}(S)$. Let $S \in \Delta_2^A$. We assume $S \neq \varnothing$; otherwise, the assertion is obvious. We define a mapping $\varphi \colon S \to 2^{<\omega}$. Let $\varphi(\Lambda) = \Lambda$. Assume that $\sigma \in S$ and that $\varphi(\sigma)$ is defined. If $\sigma * 0$, $\sigma * 1 \in S$, then we set $\varphi(\sigma * i) = \varphi(\sigma) * i$ for $i \leqslant 1$. If $\sigma * k \in S$ for a single $k \leqslant 1$, then we set $\varphi(\sigma * k) = \varphi(\sigma) * 0$. It is clear that $\varphi(S) \in \Delta_2^A$, φ is an isomorphism of $\langle S; \subseteq \rangle$ onto $\langle \varphi(S); \subseteq \rangle$, and $\tau * 0 \in \varphi(S)$ in the case $\tau \in \varphi(S)$. Without loss of generality, we can assume that S is such that $\sigma \in S \Rightarrow \sigma * 0 \in S$ (otherwise, instead of S we can take $\varphi(S)$ since, in view of Proposition 3.7.8, $\text{al}(S) \simeq \text{al}(\varphi(S))$). Since $S \in \Delta_2^A$, there exists a function $f \colon 2^{<\omega} \times \omega \to \{0, 1\}$ that is recursive with respect

3.7. Algorithmic Properties

to A and for any $\sigma \in 2^{<\omega}$ there exists the limit $\lim_s f(\sigma, s)$; moreover, this limit is equal to 1 only if $\sigma \in S$. Let $g(e)$ be the least number s such that $f(\sigma, s) = f(\sigma, t)$ for all $t \geqslant s$ and $\sigma \in 2^{<\omega}$, $\mathrm{lh}(\sigma) = e$. We define $\psi \colon 2^{<\omega} \to 2^{<\omega}$ as follows: $\psi(\varnothing) = \varnothing$ and $\psi(i_1, \ldots, i_k) = 0^{g(1)} * i_1 * \ldots * 0^{g(k)} * i_k$, $k \geqslant 1$, $i_j \leqslant 1$. It is clear that ψ is an isomorphism of $\langle 2^{<\omega}, \subseteq \rangle$ onto itself. By Proposition 3.7.9, $\mathrm{al}(S) \cong \mathrm{al}(T)$, where $T = \{\tau \mid \exists \sigma \in S \, (\tau \subseteq \psi(\sigma))\}$. To complete the proof of the theorem it remains to verify that $T \in \Pi_1^A$. By the definition of ψ and T as well as the property of S, for all $\tau \in 2^{<\omega}$ we have

$$\text{if } \exists \sigma \, (\tau * 1 \subseteq \psi(\sigma)), \text{ then } \exists \sigma \, (\tau * 1 = \psi(\sigma)) \tag{3.7.2}$$

$$\text{if } \tau \in T, \text{ then } \tau * 0 \in T \tag{3.7.3}$$

It is required to verify that the relation "$\rho \in T$" belongs to Π_1^A. If $\rho = 0^k$ for some $k < \omega$, then $\rho \in T$ in view of (3.7.3). In the opposite case, we can effectively find $\tau \in 2^{<\omega}$ and $k \in \omega$ such that $\rho = \tau * 1 * 0^k$. By (3.7.3), $\rho \in T \Leftrightarrow \tau * 1 \in T$. Therefore, it remains to verify that the relation "$\tau * 1 \in T$" belongs to Π_1^A.

We define a function $g_s(e)$ as follows: $g_0(e) = 0$, $g_{s+1}(e) = g_s(e)$ for $f(\sigma, s+1) = f(\sigma, s)$ for all $\sigma \in 2^{<\omega}$, $\mathrm{lh}(\sigma) = e$ and $g_{s+1}(e) = s+1$ otherwise. It is clear that g is recursive with respect to A in both arguments, $g_s(e) \leqslant s$, $g_{s+1}(e) \in \{g_s(e), s+1\}$ and $\lim_s g_s(e) = g(e)$ for all e. Let $\psi_s \colon 2^{<\omega} \to 2^{<\omega}$ be defined in the same way as ψ with g_s instead of g. The properties of g imply that

$$\mathrm{lh}(\psi_s(\sigma)) \leqslant \mathrm{lh}(\psi_{s+1}(\sigma)) \text{ and } \psi_s(\sigma) \neq \psi_{s+1}(\sigma) \to \mathrm{lh}(\psi_{s+1}(\sigma)) \geqslant s+1 \tag{3.7.4}$$

We prove that the relation $\tau * 1 \in T$ is equivalent to the relation

$$(\exists \sigma)_{\mathrm{lh}(\sigma) \leqslant s} (\tau * 1 = \psi_s(\sigma) \wedge \forall i < \mathrm{lh}(\sigma) \\ \forall t \geqslant s \, (g_t(i+1) = g_s(i+1)) \wedge f(\sigma, s) = 1) \tag{3.7.5}$$

where $s = \mathrm{lh}(\tau * 1)$. Since (3.7.5) is a Π_1^A-relation, this completes the proof. If (3.7.5) is true, then from the second term of the conjunction we obtain $g_s(i+1) = g(i+1)$ for $i < \mathrm{lh}(\sigma)$, whence $\tau * 1 = \psi_s(\sigma) = \psi(\sigma)$ and $\lim_t f(\sigma, t) = f(\sigma, s) = 1$, i.e., $\tau * 1 \in T$. If (3.7.5) is false, then we first assume that $\tau * 1 \neq \psi_s(\sigma)$ for all σ. Then $\psi_t(\sigma) \neq \tau * 1$ for all σ and $t \geqslant s$; otherwise, $\mathrm{lh}(\psi_t(\sigma)) \geqslant s+1 > \mathrm{lh}(\tau * 1)$ in view of (3.7.4).

Therefore, $\psi(\sigma) \neq \tau * 1$ for all σ and $\tau * 1 \notin T$ in view of (3.7.2).

Let $\tau * 1 = \psi_s(\sigma)$ for some (single) σ. Since (3.7.5) is false, the second term or the third term of the conjunction is also false. For definiteness, let the second term be false, i.e., $g_t(i+1) \neq g_s(i+1)$ for some $t \geqslant s$ and $i < \mathrm{lh}\,\sigma$. Taking into account the properties of g_t and ψ_t, we obtain $\mathrm{lh}(\psi_t(\sigma)) \geqslant g_t(i+1) \geqslant s+1 > \mathrm{lh}(\tau*1)$. Therefore, $\psi_u(\rho) \neq \tau*1$ for all $u \geqslant t$ and ρ, $\mathrm{lh}(\rho) \geqslant \mathrm{lh}(\sigma)$, whence $\psi(\rho) \neq \tau*1$ for $\mathrm{lh}(\rho) \geqslant \mathrm{lh}(\sigma)$. Let $\rho \in 2^{<\omega}$, $\mathrm{lh}(\rho) < \mathrm{lh}(\sigma)$. Then $\mathrm{lh}(\psi_s(\rho)) < \mathrm{lh}(\psi_s(\sigma)) = s$. Therefore, if $\psi_t(\rho) = \tau*1$ for some $t \geqslant s$, then $\psi_t(\rho) \neq \psi_s(\rho)$ and we again have $\mathrm{lh}(\psi(\rho)) \geqslant \mathrm{lh}(\psi_t(\rho)) > s = \mathrm{lh}(\tau*1)$. Consequently, $\psi(\rho) \neq \tau*1$ for all $\rho \in 2^{<\omega}$. Hence $\tau*1 \notin T$.

Finally, let the first two terms of the conjunction in (3.7.5) be true and let $f(\sigma, s) = 0$. Since the second term is true, we have $\psi(\sigma) = \psi_s(\sigma) = \tau*1$, $f(\sigma, s) = \lim_t f(\sigma, t) = 0$, $\sigma \notin S$. Therefore, $\tau*1 \notin T$. □

Theorems 3.7.3 and 3.7.4 yield the following assertion.

Corollary 3.7.4. *A Boolean Π_2^A-algebra is a Boolean Σ_1^A-algebra.*

Let α be a recursive ordinal and let Σ_α, Π_α be some hyperarithmetic hierarchy classes. From Sec. 3.5 it follows that for any α there is a Boolean $\Sigma_{\alpha+1}$-algebra which is not a Σ_α-algebra. By Theorem 3.7.1 and Corollary 3.7.4, this result cannot be made more precise in terms of inclusions of the classes of Boolean Σ_α-algebras and Π_α-algebras. Any Boolean $\Pi_{\alpha+1}$-algebra is a Boolean $\Delta_{\alpha+1}$-algebra and every Boolean $\Pi_{\alpha+2}$-algebra is a Boolean $\Sigma_{\alpha+1}$-algebra. In particular, it is impossible to construct a hierarchy of Boolean algebras thinner than the hyperarithmetic one. For example, any Boolean Σ_n^{-1}-algebra (Σ_n^{-1} denote classes of the Ershov hierarchy) is a Boolean Σ_1-algebra.

We now proceed to the theorem on arithmetic representations of arithmetic Boolean algebras.

Theorem 3.7.5. *The following assertions hold*:

(a) *any Boolean Σ_2^A-algebra is isomorphic to the quotient algebra of a suitable atomic Boolean Δ_1^A-algebra by the Frechét ideal,*

(b) *any Boolean Σ_1^A-algebra is isomorphic to the quotient algebra by the Frechét ideal for a suitable atomic Boolean algebra that is strongly constructivizable with respect to A.*

PROOF. (a) Let ν be a numbering of a Boolean algebra $\mathfrak{A}$ such that $\langle \mathfrak{A}, \nu \rangle \in \Sigma_2^A$. The relation "$\nu[\sigma] = 0_{\mathfrak{A}}$" belongs to Σ_2^A. Hence there is a function $f\colon 2^{<\omega} \times \omega \to \omega$ that is recursive with respect to A and $f(\sigma, s) \leqslant f(\sigma, s+1)$. We also have $\nu[\sigma] = 0 \Leftrightarrow \lim_s f(\sigma, s) < \omega$. We construct a sequence $\{\mathfrak{B}_s\}$, computable with respect to A, of finite Boolean

3.7. Algorithmic Properties

algebras and mappings φ_s from $\mathrm{Atom}(\mathfrak{B}_s)$ into $2^{<\omega} \cup \{\alpha\}$, where α does not belong to $2^{<\omega}$. Let $\mathfrak{B}_0 = \{\mathbf{0}, \mathbf{1}\}$ and $\varphi_0(\mathbf{1}) = \varnothing$. We assume that $\mathfrak{B}_s$ and φ_s have been constructed. Let $C(\mathfrak{B}_s)$ be the set of all atoms $a \in \mathfrak{B}_s$ such that $\varphi_s(a) = \sigma \in 2^{<\omega}$ and $\forall \tau \subseteq \sigma \ (f(\tau, s) \geqslant \mathrm{lh}(\sigma))$. Let $\mathfrak{B}_{s+1}$ be the least Boolean algebra containing $\mathfrak{B}_s$. Let every element $a \in C(\mathfrak{B}_s)$ be divided into three atoms a_0, a_1, a_2. We set $\varphi_{s+1}(a_0) = \sigma * 0$, $\varphi_{s+1}(a_1) = \sigma * 1$, and $\varphi_{s+1}(a_2) = \alpha$ for $a \in C(\mathfrak{B}_s)$ and $\varphi_{s+1}(a) = \varphi_s(a)$ for $a \in \mathrm{Atom}(\mathfrak{B}_s) \setminus C(\mathfrak{B}_s)$. The construction is complete.

Let ψ_s be a homomorphism of $\mathfrak{B}_s$ into $\mathfrak{A}$ such that $\psi_s(a) = \nu[\sigma]$ for $a \in \mathrm{Atom}(\mathfrak{B}_s)$, $\varphi_s(a) = \sigma \in 2^{<\omega}$ and $\psi_s(a) = 0_\mathfrak{A}$ for $a \in \mathrm{Atom}(\mathfrak{B}_s)$, $\varphi_s(a) = \alpha$. It is easy to check that $\psi_0 \subseteq \psi_1 \subseteq \ldots$ and $\psi = \bigcup_s \psi_s$ is a homomorphism of $\mathfrak{B} = \bigcup_s \mathfrak{B}_s$ onto $\mathfrak{A}$ and its kernel is $F(\mathfrak{B})$. Therefore, $\mathfrak{B}/F(\mathfrak{B}) \simeq \mathfrak{A}$. It remains to verify that the algebra $\mathfrak{B}$ is atomic, i.e., for any nonzero element $b \in \mathfrak{B}$ there is an atom $c \leqslant b$. Let $b \in \mathfrak{B}_t$ and $a \leqslant b$, $a \in \mathrm{Atom}(\mathfrak{B}_t)$. If a is not divided at the steps $s \geqslant t$ (i.e., it does not get into $C(\mathfrak{B}_s)$), then a is an atom of $\mathfrak{B}$, and the required assertion is proved. If $a \in C(\mathfrak{B}_s)$ for some $s \geqslant t$, then, by construction, the element $a_2 \in \mathrm{Atom}(\mathfrak{B}_{s+1})$ is an atom of $\mathfrak{B}$ and $a_2 \leqslant b$. The assertion (a) is proved.

(b) Let $\langle \mathfrak{A}, \nu \rangle$ be a Boolean Σ_1^A-algebra. Consider an A-recursive predicate $P(\sigma, x)$ such that $\nu[\sigma] = 0 \Leftrightarrow \exists x \, P(\sigma, x)$. We define a new A-recursive predicate $Q(\sigma, x) \rightleftharpoons \forall \tau \, (\tau \subseteq \sigma \Rightarrow \neg P(\tau, x))$. As in the proof of (a), we construct a sequence $\{\mathfrak{B}_s\}$, computable with respect to A, of finite Boolean algebras and mappings φ_s from $\mathrm{Atom}(\mathfrak{B}_s)$ into $2^{<\omega} \cup \{\alpha\}$, where α does not belong to $2^{<\omega}$.

Let $\mathfrak{B}_0 = \{\mathbf{0}, \mathbf{1}\}$ and $\varphi_0 = \varnothing$. We assume that $\mathfrak{B}_s$ and φ_s are already constructed. Let $C_i(\mathfrak{B}_s)$, $i = 0, 1$, be the set of all atoms $a \in \mathfrak{B}_s$, such that $\varphi_s(a) = \sigma \in 2^{<\omega}$ and the predicate $Q(\sigma, s+1)$ is true for $i = 1$ and is false for $i = 0$. Let $\mathfrak{B}_{s+1}$ be the least Boolean algebra containing $\mathfrak{B}_s$ for which the following assertions hold. Every element $a \in C_1(\mathfrak{B}_s)$ is divided into three atoms: a_0, a_1, and a_2. If $\varphi_s(a) = \sigma$, then we set $\varphi_{s+1}(a_0) = \sigma * 0$, $\varphi_{s+1}(a_1) = \sigma * 1$, $\varphi_{s+1}(a_2) = \alpha$. Every element $a \in C_0(\mathfrak{B}_s)$ is divided into two parts: $a = a_1 \cup a_2$ and $\varphi_{s+1}(a_1) = \varphi_{s+1}(a_2) = \alpha$. For $a \in \mathrm{Atom}(\mathfrak{B}_s) \setminus (C_0(\mathfrak{B}_s) \cup C_1(\mathfrak{B}_s))$ we set $\varphi_{s+1}(a) = \varphi_s(a)$. The construction is complete. As in the proof of (a), we can prove that the Boolean algebra $\mathfrak{B}$ is atomic and $\mathfrak{B}/F(\mathfrak{B}) \simeq \mathfrak{A}$. It remains to verify that $\mathfrak{B}$ is strongly constructive with respect to A. An element $b \in \mathfrak{B}$ is an atom if and only if $b \in \mathfrak{B}^s$ for some s and $\varphi_s(a) = \alpha$. The last condition is effectively verified (with respect to A). As is known, an atomic Boolean algebra with decidable set of atoms is strongly constructivizable. $\square$

Let $F_n(\mathfrak{B})$ be the iterated Frechét ideal and let $\mathfrak{B}^{(n)} = \mathfrak{B}/F_n(\mathfrak{B})$. Applying Theorems 3.7.3 and 3.7.2 several times, we obtain the following claim.

Corollary 3.7.5. *Let $n < \omega$. Any Boolean Σ_{2n+2}-(Σ_{2n+1})-algebra is isomorphic to the quotient algebra of a suitable n-atomic constructive (strongly constructive) Boolean algebra by the ideal F_{n+1}.*

In accordance with the definition of the Frechét ideal, the fact that a Boolean algebra $\mathfrak{B}$ is constructive (strongly constructive) implies $\mathfrak{B}^{(n+1)} \in \Sigma_{2n+2}$ (Σ_{2n+1}). Therefore, $\mathfrak{B} \mapsto \mathfrak{B}^{(n+1)}$ is a mapping from the class of all constructive (strongly constructive) Boolean algebras onto the class of all Boolean Σ_{2n+2}-algebras (Σ_{2n+1}-algebras). This mapping is a one-to-one mapping of the class of n-atomic Boolean algebras.

Our next aim is to prove claims similar to Theorem 3.7.5 for the ideal $\mathrm{FAl}(\mathfrak{B})$ generated by atoms and atomless elements and the Ershov–Tarski ideal $\mathrm{I}(\mathfrak{B})$ generated by atomic elements and atomless elements of a Boolean algebra $\mathfrak{B}$. To this end, we need to define an operation on trees and to study its properties.

For $\sigma \in 2^{<\omega}$ and $T \subseteq 2^{<\omega}$ we set $\sigma * T = \{\sigma * \tau \mid \tau \in T\}$ and $T(\sigma) = \{\tau \mid \sigma * \tau \in T\}$. We introduce an operation such that it associates the segment $\prod_\tau T_\tau$ with a sequence of segments $\{T_\tau\}_{\tau \in 2^{<\omega}}$ as follows. Let W be the set of all sequences with zeros at "odd places." We set $\varphi(\Lambda) = \Lambda$ and $\widetilde{\sigma * i} = \widetilde{\sigma} * 0i$. The mapping $\sigma \mapsto \widetilde{\sigma}$ is an isomorphism from $(2^{<\omega}; \subseteq)$ into $(W; \subseteq)$, and for every $\tau \in W$ there is σ such that $\tau \subseteq \widetilde{\sigma}$. It is clear that $\{\widetilde{\sigma}*1 \mid \sigma \in 2^{<\omega}\}$ is the set of all minimal elements of $(2^{<\omega} \setminus W; \subseteq)$ and $W = \{\widetilde{\sigma}, \widetilde{\sigma}*0 \mid \sigma \in 2^{<\omega}\}$. If $\sigma \in W \setminus \{\widetilde{\tau} \mid \tau \in 2^{<\omega}\}$, then $\sigma = \widetilde{\tau}*0$ for some $\tau \in 2^{<\omega}$ and $\sigma * 0 = \widetilde{\tau * 0}$, $\sigma * 1 = \widetilde{\tau * 1}$. We set $\prod_\tau T_\tau = W \cup \left(\bigcup_\tau \widetilde{\tau}*1*T_\tau\right)$. The following assertion is a simple consequence of the definition.

Lemma 3.7.1. *Let $T = \prod_\tau T_\tau$.*

(a) *If $\{T_\tau\}$ is a Σ_1^A-sequence of trees, then T is a Σ_1^A-tree.*

(b) *For any $\sigma \in 2^{<\omega}$ the equalities $T(\widetilde{\sigma}*1) = T_\sigma$ and $T(\widetilde{\sigma}) = \prod_\tau T_{\sigma*\tau}$ hold.*

All the trees considered below are assumed to be nonempty.

3.7. Algorithmic Properties

Lemma 3.7.2. *Let $\{T_\tau\}$ be a sequence of trees, $\mathfrak{B}$ be the Boolean algebra defined by the tree $T = \prod_\tau T_\tau$, and b_τ be an element of $\mathfrak{B}$ corresponding to $\tau \in T$.*

(a) *If each of the elements $b_{\tilde{\sigma}*1}$ is atomic, then the Boolean algebra $\mathfrak{B}$ is atomic and $\mathfrak{B} \neq \mathrm{FAl}(\mathfrak{B})$.*

(b) *If each of the elements $b_{\tilde{\sigma}*1}$ belongs to $\mathrm{FAl}(\mathfrak{B})$, almost all of these elements are atomless, and at least one of them is atomic, then $1_\mathfrak{B}$ can be represented as the union of an atomless element and a nonempty finite set of atoms.*

(c) *If each of the elements $b_{\tilde{\sigma}*1}$ belongs to $\mathrm{I}(\mathfrak{B})$, almost all of these elements are atomic, and at least one of them is not atomic, then $1_\mathfrak{B}$ can be represented as the union of nonzero atomic and atomless elements.*

(d) *If an infinite number of elements of the form $b_{\tilde{\sigma}*1}$ are neither atomic nor atomless, then $1_\mathfrak{B} \notin \mathrm{I}(\mathfrak{B})$.*

PROOF. (a) It is clear that $\mathfrak{B}$ has an infinite number of atoms. Therefore, $1_\mathfrak{B} \notin \mathrm{FAl}(\mathfrak{B})$. It remains to verify that there is an atom under every b_τ ($\tau \in T$). This suffices because every element of $\mathfrak{B}$ is a finite union of elements of the form b_τ. If $\tau \notin W$, then $\tilde{\sigma} * 1 \subseteq \tau$ and $b_{\tilde{\sigma}*1} \geqslant b_\tau$ for some σ. Therefore, b_τ is atomic. Let $\tau \in W$. Then $\tau \subseteq \tilde{\sigma}*1$ and $b_\tau \geqslant b_{\tilde{\sigma}*1}$ for a suitable σ. There is an atom under $b_{\tilde{\sigma}*1}$. The assertion (a) is proved. The fact that $\mathfrak{B}$ is atomless if all $b_{\tilde{\sigma}*1}$ are atomless can be proved in the same way.

(b) Let k be such that $b_{\tilde{\sigma}*1} \in \mathrm{FAl}(\mathfrak{B})$ for $\mathrm{lh}(\sigma) \leqslant k$ and let $b_{\tilde{\sigma}*1}$ be atomless for $\mathrm{lh}(\sigma) > k$. It is easy to see that $1_\mathfrak{B} = (\bigvee_{\mathrm{lh}(\sigma) \leqslant k} b_{\tilde{\sigma}*1}) \vee (\bigvee_{\mathrm{lh}(\sigma)=k+1} b_{\tilde{\sigma}})$. By condition, $\bigvee_{\mathrm{lh}(\sigma) \leqslant k} b_{\tilde{\sigma}*1} \in \mathrm{FAl}(\mathfrak{B})$. By Lemma 3.7.1, $T(\tilde{\sigma}) = \prod_\tau T_{\sigma*\tau}$. The aforesaid means that the element $b_{\tilde{\sigma}}$ ($\mathrm{lh}(\sigma) = k+1$), being the unit of a Boolean algebra isomorphic to $\mathrm{BA}(T(\tilde{\sigma}))$, is atomless. Therefore, $1 \in \mathrm{FAl}(\mathfrak{B})$. The nontriviality of the representation of $1_\mathfrak{B}$ is obvious by condition.

(c) The assertion is established in the same way as the assertion (b), but $b_{\tilde{\sigma}}$ is atomic for $\mathrm{lh}(\sigma) = k+1$ in view of (a).

(d) We assume the contrary: $1_\mathfrak{B} = a \vee c$, a is an atomic element, and c is an atomless element of $\mathfrak{B}$. Let finite sets $F, G \subseteq T$ be such that

$a = \bigvee_{\tau \in F} b_\tau$ and $c = \bigvee_{\tau \in G} b_\tau$. Since $\mathbf{1}_\mathfrak{B} = \bigvee_{\tau \in F \cup G} b_\tau$, any sequence of T is comparable with some sequence of $F \cup G$. The set $\{\sigma \mid \exists \tau \in F \cup G \ (\tilde{\sigma} * 1 \subseteq \tau)\}$ is finite. Therefore, there are $\tau \in F \cup G$ and $\sigma \in 2^{<\omega}$ such that $\tau \subseteq \tilde{\sigma} * 1$ and $b_{\tilde{\sigma}*1}$ is neither atomic nor atomless. This contradicts the fact that the element $b_\tau \geqslant b_{\tilde{\sigma}*1}$ is either atomic or atomless. □

Theorem 3.7.6. *Any Boolean Σ_3^A-algebra is isomorphic to the quotient algebra of a suitable Boolean Δ_1^A-algebra by the ideal* FAl.

PROOF. Let $\langle \mathfrak{A}, \nu \rangle$ be a Boolean Σ_3^A-algebra. Then there exists a Σ_1^A-sequence of nonempty initial segments of the natural series $\{\psi_\tau \mid \tau \in 2^{<\omega}\}$ such that $\nu[\sigma] = \mathbf{0} \Leftrightarrow (\psi_\tau = \omega$ for almost all extensions $\tau \supset \sigma$), where $\{\nu[\sigma] \mid \sigma \in 2^{<\omega}\}$ is a standard set generating the enumerated Boolean algebra $\langle \mathfrak{A}, \nu \rangle$. A sequence of trees $\{R_\tau \mid \tau \in 2^{<\omega}\}$, where $R_\tau = \{\sigma \mid \mathrm{lh}(\sigma) \in \psi_\tau\}$, is a Σ_1^A-sequence. By Lemma 3.7.1, $S = \prod_\tau R_\tau$ is a Σ_1^A-tree.

We consider a mapping $\varphi \colon S \to \mathfrak{A}$ such that

$$\varphi(\tau) = \begin{cases} \nu[\sigma] & \text{if } \tau \in W \text{ and } \tau = \tilde{\sigma} \text{ or } \tau = \tilde{\sigma} * 0 \\ \mathbf{0} & \text{if } \tau \in S \setminus W \end{cases}$$

It is easy to verify that φ is an admissible mapping subject to the condition (5) (cf. Sec. 1.7).

Let $\mathfrak{B}$ be a Boolean algebra generated by the tree S, $\tau \mapsto b_\tau$ be the corresponding admissible mapping, and $\psi \colon \mathfrak{B} \to \mathfrak{A}$ be an epimorphism such that $\psi(b_\tau) = \varphi(\tau)$, $\tau \in S$. It remains to verify that $\ker \psi = \mathrm{FAl}(\mathfrak{B})$.

By construction, the tree $S(\tilde{\sigma} * 1) = R_\sigma$ is finite or complete. Then BA(R_σ) is a finite Boolean algebra in the first case and is atomless in the second case. It is easy to note that $(\widehat{b}_{\tilde{\sigma}*1})_\mathfrak{B} \simeq \mathrm{BA}(R_\sigma)$; therefore, $b_{\tilde{\sigma}*1} \in \mathrm{FAl}(\mathfrak{B})$ for all $\sigma \in 2^{<\omega}$. Each element of $\mathfrak{B}$ can be represented as a union of elements of the form $b_{\tilde{\sigma}}$ and b_τ, where $\tau \supseteq \tilde{\sigma}$, Hence it remains to verify that $b_{\tilde{\sigma}} \in \mathrm{FAl}(\mathfrak{B}) \Leftrightarrow \nu[\sigma] = \mathbf{0}$. Let $\nu[\sigma] = \mathbf{0}$. Then $\psi_\tau = \omega$ for almost all $\tau \supseteq \sigma$. Consequently, R_τ is a complete tree and $b_{\tilde{\tau}*1}$ is an atomless element. As was already mentioned, $b_{\tilde{\tau}*1} \in \mathrm{FAl}(\mathfrak{B})$ for all $\tau \in 2^{<\omega}$. $[b_{\tilde{\sigma}}]_\mathfrak{B} \simeq \mathrm{BA}(S(\tilde{\sigma}))$, $S(\tilde{\sigma}) = \prod_\tau R_{\sigma*\tau}$. Applying Lemma 3.7.2 to the tree $S(\tilde{\sigma})$, we obtain $b_{\tilde{\sigma}} \in \mathrm{FAl}(\mathfrak{B})$.

If $\nu[\sigma] \neq \mathbf{0}$, then there exist infinitely many extensions $\tau \supseteq \sigma$ such that R_τ is a finite tree, i.e., $b_{\tilde{\tau}*1} \in \mathrm{FAl}(\mathfrak{B})$. If $\tau \supseteq \sigma$, then $b_{\tilde{\tau}*1} \leqslant b_{\tilde{\sigma}}$. Therefore, under $b_{\tilde{\sigma}}$ there are infinitely many different atoms and $b_{\tilde{\sigma}} \notin \mathrm{FAl}(\mathfrak{B})$. □

3.7. Algorithmic Properties

Theorem 3.7.7. *Any Boolean Σ_4^A-algebra is isomorphic to the quotient algebra of a suitable Boolean Δ_2^A-algebra by the Ershov–Tarski ideal* I.

PROOF. Let ν be a numbering of a Boolean algebra $\mathfrak{A}$ such that $\langle \mathfrak{A}, \nu \rangle \in \Sigma_4^A$. The relation "$\nu[\sigma] = 0_{\mathfrak{A}}$" belongs to $\Sigma_4^A = \Sigma_3^{A'}$. Therefore, there is a $\Sigma_1^{A'}$-sequence $\{\psi_\sigma\}_{\sigma \in 2^{<\omega}}$ of nonempty initial segments ω such that $\nu[\sigma] = 0$ if and only if $\psi_\tau = \omega$ for almost all $\tau \supseteq \sigma$. The relation "$x \in \psi_\tau$" belongs to $\Sigma_1^{A'} = \Sigma_2^A$. Therefore, there is a double Σ_1^A-sequence $\{c_{\tau,x}\}_{\tau \in 2^{<\omega}, x \in \omega}$ of nonempty initial segments of ω such that $x \in \psi_\tau \Leftrightarrow c_{\tau,x} \neq \omega$. We define a sequence of nonempty trees $S_{\tau,x} = \{\rho \mid \text{lh}(\rho) \in c_{\tau,x}\}$ which is recursively enumerable with respect to A. We set $R_\tau = \prod_{x<\omega} S_{\tau,x}$ and $T = \prod_\tau R_\tau$ (the notation $\prod_x S_{\tau,x}$ means that the number x is identified with the corresponding sequence in the natural numbering of the sequences $0 \mapsto \Lambda, 1 \mapsto 0, 2 \mapsto 1, 3 \mapsto 00, 4 \mapsto 01, \ldots$). It is clear that the tree T is recursively enumerable with respect to A. Let $\mathfrak{B}$ be the corresponding Boolean Δ_1^A-algebra. We define a mapping $\varphi \colon T \to \mathfrak{A}$ as in Theorem 3.7.6. It again is admissible and satisfying the property (5) (cf. Sec. 1.7). Therefore, there exists an epimorphism $\xi \colon \mathfrak{B} \to \mathfrak{A}$ such that $\xi(b_\tau) = \varphi(\tau)$, where $\mathfrak{B}$ is the Boolean algebra generated by the tree T and $\tau \mapsto b_\tau$ is the corresponding admissible mapping. We show that $\ker \xi = \mathrm{I}(\mathfrak{B})$.

Let $\psi_\sigma = \omega$. Then the tree $S_{\sigma,x}$ is finite for any x and, by Lemma 3.7.2, $b_{\bar\sigma * 1}$ is an atomic element because $\widehat{(b_{\bar\sigma * 1})}_{\mathfrak{B}} \simeq \mathrm{BA}(R_\sigma)$.

If $\psi_\sigma \neq \omega$, then the tree $S_{\sigma,x}$ is complete for almost all x and is finite for the remaining x. By Lemma 3.7.2, $b_{\bar\sigma * 1} \in \mathrm{FAl}(\mathfrak{B})$. In any case, $b_{\bar\sigma * 1} \in \mathrm{I}(\mathfrak{B})$. It remains to verify that $b_{\bar\sigma} \in \mathrm{I}(\mathfrak{B}) \Leftrightarrow \nu[\sigma] = 0$. This can be done as in Theorem 3.7.6 with the help of assertions (c) and (d) of Lemma 3.7.2. $\square$

The iterated ideals I_n and FAl_n are defined in a standard way.

Corollary 3.7.6. *Any Boolean Σ_{3n+3}-algebra is isomorphic to the quotient algebra of a suitable constructive Boolean algebra by the ideal* FAl_{n+1}.

Corollary 3.7.7. *Any Boolean Σ_{4n+4}-algebra is isomorphic to the quotient algebra of a suitable constructive Boolean algebra by the ideal* I_{n+1}.

It is easy to see that the above-mentioned Boolean algebras are effectively constructed. Hence the following assertion is true: from any Boolean

Σ_{4n+1}-algebra $\mathfrak{A}$ one can effectively find a Boolean Σ_1-algebra $\mathfrak{B}$ such that $\mathfrak{B}/I_n(\mathfrak{B}) \simeq \mathfrak{A}$.

We consider a Boolean algebra B_{re} generated by the set of all recursively enumerable sets of the Boolean algebra $\langle \mathcal{P}(\mathbb{N}), \cup, \cap, C, \varnothing, \mathbb{N}\rangle$. In this case, any element of B_{re} can be represented as $b = \bigcup_{k=0}^{n} W_{i_k} \setminus W_{j_k}$, where W_{l_k} are recursively enumerable sets. We consider a natural number n as a number of the pair $\langle k, s \rangle$, where k and s denote the numbers of sequences $\langle i_1, \ldots, i_r \rangle$ and $\langle j_1, \ldots, j_r \rangle$ of the same length respectively. Defining $\mu(n) = \bigcup_{s=1}^{r} W_{i_s} \setminus W_{j_s}$, we obtain a numbering of the Boolean algebra B_{re}. By the properties of numberings $\{W_i\}_{i \in \mathbb{N}}$ of the family of all recursively enumerable sets, there are recursive functions f, g, and h such that $\mu(n) \cup \mu(m) = \mu f(n,m)$, $\mu(n) \cap \mu(m) = \mu g(n,m)$, and $C(\mu(n)) = \mu h(n)$. We estimate the complexity of the set of pairs $\eta_\nu = \{(n,m) \mid \mu n = \mu m\}$. The following equivalences are valid:

$$\mu(n) = \mu(m) \Leftrightarrow (\forall x)(x \in \mu(n) \Leftrightarrow x \in \mu(m))$$

On the other hand, the following equivalence holds:

$$x \in W_i \setminus W_j \Leftrightarrow (\exists t)(x \in W_i^t) \,\&\, (\forall t')(x \notin W_i^{t'})$$

Using the Tarski–Kuratowski algorithm, we find that the numbering μ is a Π_3^0-constructivization of this model. In view of the Lachlan theorem on characterization of hyperprime sets [101] and the existence of positive but not constructivizable systems, we conclude that B_{re} is not constructivizable. The Lachlan theorem [101] means that $\mathrm{ch}(B_{re}/F(B_{re})) = (\omega, 0, 0)$. However, the questions on describing the isomorphism type for this Boolean algebra and its automorphisms still remain open.

The question on the structure of subalgebras of constructive Boolean algebras has attracted the attention of many authors, in particular, the mathematicians of the Nerode school. There were many attempts to generalize the results on the structure of a lattice $\mathcal{E}$ of recursively enumerable subsets of $\mathbb{N}$ for the case of subalgebras. Odintsov obtained a general result (stated below) in accordance with which $\mathcal{E}$ is imbedded into the structure of recursively enumerable subalgebras of any constructive Boolean algebra.

We define a lattice $\mathcal{S}(\mathfrak{B}) = \langle S(\mathfrak{B}); \vee, \wedge \rangle$, where $S(\mathfrak{B})$ is the set of all subalgebras of the algebra $\mathfrak{B}$ that have recursively enumerable universe, the operation $\wedge$ is a standard set-theoretic intersection, and $\vee$ is defined as follows: $D_0 \vee D_1 = \mathrm{gr}(D_0 \cup D_1)$ for $D_0, D_1 \in S(\mathfrak{B})$. Let $D \in S(\mathfrak{B})$. Then $\mathcal{S}_\mathfrak{B}(D)$ denotes a lattice consisting of all recursively enumerable subalgebras of $\mathfrak{B}$ containing D.

3.7. Algorithmic Properties

If $D_0, D_1 \in \mathcal{S}(\mathfrak{B})$ and D_0 is finitely equivalent to D_1, then we write $D_0 =_F D_1$ if and only if there exist finite sets F_0 and F_1 such that $\mathrm{gr}(D_0 \cup F_0) = \mathrm{gr}(D_1 \cup F_1)$. In the lattice $\mathcal{S}(\mathfrak{B})$, the relation $=_F$ plays a role similar to that of the relation $\sim_f$ in a lattice $\mathcal{E}$ of recursively enumerable sets, where $\sim_f$ is defined as follows: $\alpha \sim_f \beta$, where $\alpha, \beta \in \mathbb{N}$, if and only if $\alpha \Delta \beta$ is a finite set. The relation $\sim_f$ is a congruence on the lattice $\mathcal{E}$, $\mathcal{E}^*$ denotes the quotient lattice of $\mathcal{E}$ by $\sim_f$, $\mathcal{E}(\eta)$ is the lattice of recursively enumerable supersets $\eta \in \mathcal{E}$, and $\mathcal{E}^*(\eta)$ is the quotient lattice of $\mathcal{E}(\eta)$ by $\sim_f$. It is obvious that $=_F$ is an equivalence relation. But, in contrast to the case of $\sim_f$, it is not a congruence on the lattice $\mathcal{S}(\mathfrak{B})$ [121]. Nevertheless, we define $\mathcal{S}_\mathfrak{B}^*(D) \rightleftharpoons \mathcal{S}_\mathfrak{B}(D)/_{=_F}$. Following [121], we define the maximal subalgebra and an elementary subalgebra of a recursive Boolean algebra $\mathfrak{B}$ which are natural analogous notions of maximal and elementary subsets of $\mathbb{N}$.

A subalgebra $M \in \mathcal{S}(\mathfrak{B})$ is called *maximal* if $M \neq_F \mathfrak{B}$ and $M \vee \mathfrak{A} =_F M$ or $M \vee \mathfrak{A} =_F \mathfrak{B}$ for any Boolean algebra $\mathfrak{A} \in \mathcal{S}(\mathfrak{B})$. A subalgebra $D \in \mathcal{S}(\mathfrak{B})$ is called *simple* if $D \cap \mathfrak{A} \neq \{\mathbf{0}, \mathbf{1}\}$ for any infinite $\mathfrak{A} \in \mathcal{S}(\mathfrak{B})$.

By a recursively enumerable sequence generating a recursive Boolean algebra $\mathfrak{B}$ we mean a recursively enumerable sequence of elements $b_0 \rightleftharpoons \mathbf{1}, b_1, b_2, \ldots$ such that $\mathfrak{B} = \bigcup_{s \geqslant 0} \mathfrak{B}^s$, where $\mathfrak{B}^s = \mathrm{gr}(\{b_0, \ldots, b_s\})$ and for any s there exists a unique atom $a \in \mathfrak{B}^s$ such that $b_{s+1} < a$. Every sequence defines some effective method of constructing the algebra $\mathfrak{B}$ such that, at the end of every step, a finite Boolean algebra must be constructed and, at the following step, one of the atoms of the constructed algebra is divided into two parts.

Theorem 3.7.8. *Let $\mathfrak{B}$ be an infinite recursive Boolean algebra. Then there exists a recursive subalgebra $D \in \mathcal{S}(\mathfrak{B})$ such that the lattice $\mathcal{S}_\mathfrak{B}(D)$ is isomorphic to the lattice of recursively enumerable sets.*

PROOF. Let $b_0 \rightleftharpoons \mathbf{1}, b_1, \ldots$ be an arbitrary recursively enumerable sequence generating the Boolean algebra $\mathfrak{B}$. The set $R = \bigcup_{s \geqslant 0} \mathrm{Atom}(\mathfrak{B}^s)$ is recursive. Indeed, any element $b \in |\mathfrak{B}|$ belongs to $\mathfrak{B}^s$ for some s. Let s_0 be the least s such that $b \in \mathfrak{B}^s$. The inclusion $b \in R$ holds if and only if $b \in \mathrm{Atom}(\mathfrak{B}^{s_0})$. This inclusion is effectively verified. If $b \in R$ and $b \neq \mathbf{1}$, then there is s such that $b \in \mathrm{Atom}(\mathfrak{B}^{s+1}) \setminus \mathfrak{B}^s$. This means that there exists $a \in \mathrm{Atom}(\mathfrak{B}^s)$ such that $b_{s+1} < a$ and $b \in \{b_{s+1}, a \setminus b_{s+1}\}$. In other words, every element $b \in R$ different from $\mathbf{1}$ is obtained as a result of dividing another element $a \in R$ into two parts $a = b \vee c$, where $b \wedge c = \mathbf{0}$, and $c \in R$. If $b < c$ in the sense of the natural order on the set of natural numbers, then b is said to be a *left element*; otherwise, it is called a *right element*. Thus, the set R is divided into two nonintersecting recursive sets R_0 of left elements and R_1 of right elements respectively (the unit is assumed to be a right element).

We give instructions for enumeration of the sets B, A_0, A_1, ... All these sets possess a number of interesting properties. In particular, they are subsets of R. At a finite step s, all elements of $\operatorname{Atom}(\mathfrak{B}^s)$ get into one of the enumerable sets. We denote by B^s, A_0^s, A_1^s, ..., the set of elements that are already enumerated in the sets B, A_0, A_1, ... to the end of the step s.

STEP 0. We set $\mathfrak{B}^0 = \{\mathbf{0}, \mathbf{1}\}$, $B^0 = \{\mathbf{1}\}$, $A_0^0 = \varnothing$, $A_1^0 = \varnothing$.

STEP $s+1$. Let an element $a \in \operatorname{Atom}(\mathfrak{B}^s)$ be such that $b_{s+1} < a$. Then $\operatorname{Atom}(\mathfrak{B}^{s+1}) = (\operatorname{Atom}(\mathfrak{B}^s) \setminus \{a\}) \cup \{b_{s+1}, a \setminus b_{s+1}\}$. Furthermore, in the set $\{b_{s+1}, a \setminus b_{s+1}\}$ there is exactly one left element and one right element.

If a belongs to B^s and i is the least natural number such that $A_i^s = \varnothing$, then we set $A_i^{s+1} \rightleftharpoons \{b_{s+1}, a \setminus b_{s+1}\}$ while the rest of the sets remain unchanged.

If a belongs to A_i^s for some i, then we set $A_i^{s+1} \rightleftharpoons A_i^s \cup \{c\}$ and $B^{s+1} \rightleftharpoons B^s \cup \{b\}$, where c is a left element and b is a right element of the set $\{b_{s+1}, a \setminus b_{s+1}\}$. The remaining sets are left unchanged. This completes the description of the construction. It is completely effective. Consequently, the sets B, A_0, A_1, ... are uniformly and recursively enumerable. Moreover, they are recursive, being mutually disjoint subsets of R, and each element of R gets into one of enumerable sets. This fact follows from the description of the construction. The following alternative holds: either $A_i \neq \varnothing$ for any $i \in \mathbb{N}$ (which is suitable for us) or there is i such that $A_j = \varnothing$ for all $j > i$ and $A_j \neq \varnothing$ for all $j \leqslant i$. The last possibility is not acceptable. However, in this case, the set of atoms of the algebra $\mathfrak{B}$ turns out to be recursive. Hence we can modify the construction in such a way that all A_i are nonempty. Indeed, let s be a step such that $A_j^s \neq \varnothing$ for all $j \leqslant i$, and let $\{a_0, \ldots, a_k\} \subseteq \operatorname{Atom}(\mathfrak{B}^s)$ be all those atoms of the subalgebra $\mathfrak{B}^s$ that do not belong to the ideal $F(\mathfrak{B})$. Then $\{a_0, \ldots, a_k\} \cap B = \varnothing$ since the next set A_i, $i \in \mathbb{N}$, becomes nonempty as soon as an element of the set B is divided into parts, which is impossible. If a_i, $i = 0, \ldots, k$, is divided, then all the right elements turn out to be atoms and all the left elements do not belong to $F(\mathfrak{B})$. Therefore, all atoms of the algebra $\mathfrak{B}$, except a finite number of them, are right elements and they are located under one of the elements $a_1, \ldots, a_k$.

3.7. Algorithmic Properties

If $\mathfrak{B}$ is a recursive Boolean algebra with recursive sets of atoms, then the set of left elements as well as the set of right elements can be redefined as follows. Let $a \in R$ and a be divided into two parts at the next step: $a = b \vee c$ and $b < c$. If $c \notin \text{Atom}(\mathfrak{B})$, then we regard b as a left element and c as a right element. In the case $c \in \text{Atom}(\mathfrak{B})$, conversely, we assume that c is a left element and b is a right element. The recursivity of the sets of left elements and right elements respectively is preserved, and in the above construction all the sets A_i, $i \in \mathbb{N}$, are nonempty.

The subalgebra $D = \text{gr}(B)$ satisfies the hypotheses of the theorem. Before the proof of this fact we make some remarks concerning the structure of the sets A_i, $i \in \mathbb{N}$. Let s be the largest step such that $A_i^s = \varnothing$. Then there is an element (denoted by $\bar{b}_i$) of $\text{Atom}(\mathfrak{B}^s)$, which is already related to B^s, such that, at the step $s + 1$, it is divided into two parts a_0^i and c_0^i. These parts form the set A_i^{s+1}. If one of these parts (say, a_0^i) is divided at one of the following steps into two parts (denote them by a_1^i and b_1^i), then one of them gets into A_i and the other part gets into B respectively and so on.

The structure of the set A_i can be represented as the tree depicted in Fig. 3.7.1.

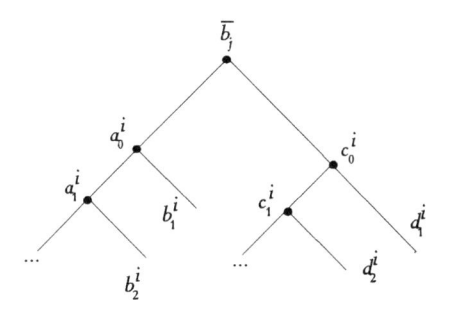

Fig. 3.7.1

Thus, acting in a natural way, with each set A_i we associate an element $\bar{b}_i \in B$ (a "vertex" of the set A_i) and two sequences of elements of the set B: $\{b_j^i \mid 0 < j < \alpha_i\}$ and $\{d_j^i \mid 0 < j < \beta_i\}$, where $\alpha_i, \beta_i \leqslant \omega$.

3. Constructive Boolean Algebras

The set A_i is the union of two decreasing sequences $\{a_j^i \mid j < \alpha_i\}$ and $\{c_j^i \mid j < \beta_i\}$; moreover, $a_0^i \vee c_0^i = \bar{b}_i$, $a_0^i \wedge c_0^i = \mathbf{0}$, $a_{j-1}^i = a_j^i \vee b_j^i$, and $c_{k-1}^i = c_k^i \vee d_k^i$, $0 < j < \alpha_i$, $0 < k < \beta_i$.

On the set $\mathbb{N}$, we introduce an order $<_A$ in the following way. Let $i <_A j$ if and only if $\bar{b}_i \leqslant \bar{b}_j$. We emphasize some properties of this construction.

⟨1⟩ *Let $\mathfrak{A}$ be a subalgebra of the algebra $\mathfrak{B}$ and $D \subset \mathfrak{A}$. If $A_i \cap \mathfrak{A} \neq \emptyset$, then $A_i \subset \mathfrak{A}$.*

We now introduce auxiliary notions. A set of elements $\{a_1, \ldots, a_n\} \subseteq |\mathfrak{B}|$ is called a *partition of an element* b if $b = \vee a_i$ and $a_i \wedge a_j = \mathbf{0}$, $i \neq j$. The set $\{a_1, \ldots, a_n\}$ is a *partition of an element* b *at the step* s if it is a partition of $b \in \mathfrak{B}^s$, and $a_1, \ldots, a_n \in \text{Atom}(\mathfrak{B}^s)$. It is clear that for any $b \in |\mathfrak{B}|$ and s a partition of b is uniquely defined at the step s provided that $b \in \mathfrak{B}^s$. If $b \notin \mathfrak{B}^s$, then the partition of b is not defined at the step s. This implies the correctness of the following definition. Let $\{a_1, \ldots, a_n\}$ be a partition of an element b at the step s. We set $\text{supp}_s(b) \rightleftharpoons \{i \mid |A_i \cap \{a_1, \ldots, a_n\}| = 1\}$. It is clear that $\text{supp}_s(b)$ is a finite set such that its canonical index can be effectively found from b and s. We note that $|A_i \cap \{a_1, \ldots, a_n\}| \leqslant 2$ for any i since A_i is the union of two decreasing sequences in R such that their first elements are nonintersecting.

⟨2⟩ *If partitions of an element $b \in R$ at the steps s and s' are defined, then $\text{supp}_s(b) = \text{supp}_{s'}(b)$.*

For any $b \in \mathfrak{B}$ there is a step s such that $\text{supp}_s(b)$ is defined and is independent of s according to ⟨2⟩. Therefore, we further use the notation $\text{supp}\, b$. We set $\text{supp}\, \mathbf{0} = \emptyset$.

⟨3⟩ *A subalgebra D is recursive and $D \cap A_i = \emptyset$ for any $i \in \mathbb{N}$.*

For $C \in \mathcal{S}(\mathfrak{B})$, $C \supset D$ we set $W_C = \{i \mid C \cap A_i \neq \emptyset\}$. Let W be an arbitrary recursively enumerable set. We set $D_W \rightleftharpoons \text{gr}(D \cup \bigcup\{A_i \mid i \in W\})$.

⟨4⟩ *Let C be a recursively enumerable subalgebra of $\mathfrak{B}$ and let $D \subset C$. Then $\text{supp}\, b \subset W_C$ for any $b \in C$.*

⟨5⟩ *For any $i \in W$ $D_W \cap A_i \neq \emptyset$ and $D_W = \{b \mid \text{supp}\, b \subset W\}$.*

3.7. Algorithmic Properties

⟨6⟩ *Let C be a recursively enumerable subalgebra of $\mathfrak{B}$ and let $C \supset D$. Then $C = D_{W_C}$.*

Indeed, the inclusion $D_{W_C} \subset C$ is an immediate consequence of ⟨1⟩. The converse inclusion follows from ⟨4⟩.

The property ⟨6⟩ means that the mapping $W \to D_W$ transforms the lattice $\mathcal{E}$ onto the lattice $\mathcal{S}_\mathfrak{B}(D)$. By ⟨5⟩, this mapping is one-to-one. The fact that $W \to D_W$ is an isomorphism can be immediately verified. □

REMARK 3.7.2. The isomorphism constructed in the proof of Theorem 3.7.1 preserves the Turing degrees.

It is obvious that $D_W \leqslant_T W$. In view of ⟨4⟩, we obtain $W \leqslant_T D_W$.

We state some corollaries of Theorem 3.7.8, which enables us to obtain a series of various claims by a common method.

Corollary 3.7.8. *Let $\mathfrak{B}$ be an infinite recursive Boolean algebra and let δ be a high recursively enumerable degree. Then there is a maximal subalgebra $\mu \in \mathcal{S}(\mathfrak{B})$ of degree δ.*

PROOF. In accordance with the classical results of Martin [115, 189], in any high recursively enumerable degree δ there exists a maximal set $M \equiv_T \delta$. We consider a subalgebra D_M. In view of Remark 3.7.2, $D_M \equiv_T M$. We show that the subalgebra D_M is maximal. Let $D_M \subset C$. We consider W_C, $M \subset W_C \subset \mathbb{N}$. From the maximality of M it follows that either $W_C \setminus M$ is finite (hence $D_M =_F C$ by ⟨1⟩) or $\mathbb{N} \setminus W_C$ is finite (in this case, $C =_F \mathfrak{B}$). Therefore, D_M is a maximal subalgebra. □

We now expand the Lachlan results about hypersimple sets to the lattice $\mathcal{S}(\mathfrak{B})$. As was established by Lachlan [101, 189], hypersimple sets are exactly recursively enumerable sets η such that $\mathcal{E}^*(\eta)$ is a Boolean algebra. Furthermore, starting from an arbitrary Σ_3^0-lattice $\mathfrak{A}$ that is a Boolean algebra, he constructed a set η such that $\mathcal{E}^*(\eta) \cong \mathfrak{A}$.

A subalgebra $\mathfrak{A} \in \mathcal{S}(\mathfrak{B})$ is called *hyperhypersimple* if, for any $D_0 \in \mathcal{S}(\mathfrak{B})$ such that $D_0 \supset \mathfrak{A}$, there is $D_1 \in \mathcal{S}(\mathfrak{B})$ such that $D_1 \supset \mathfrak{A}$, $D_0 \cap D_1 = \mathfrak{A}$, and $D_0 \vee D_1 = \mathfrak{B}$.

Corollary 3.7.9. *Let $\mathfrak{B}$ be an arbitrary recursive Boolean algebra and let $\mathfrak{A}$ be an arbitrary Σ_3^0-lattice and a Boolean algebra. Then there exists a hyperhypersimple subalgebra $C \in \mathcal{S}(\mathfrak{B})$ such that $\mathcal{S}_\mathfrak{B}^*(C) \cong \mathfrak{A}$.*

REMARK 3.7.3. As was mentioned, the relation $=_F$ is not, in general, a congruence on the lattice $\mathcal{S}(\mathfrak{B})$. Therefore, the notation $\mathcal{S}_\mathfrak{B}^*(C) \cong \mathfrak{A}$ is somewhat more significant than an isomorphism between lattices. Namely, the relation $=_F$ is a congruence on the lattice $\mathcal{S}_\mathfrak{B}(C)$ and the quotient lattice $\mathcal{S}^*_\mathfrak{B}(C)$ is isomorphic to the lattice $\mathfrak{A}$.

Indeed, for C we can take a subalgebra D_M in the notation of Theorem 3.7.8, where M is a hyperhypersimple set such that $\mathcal{E}^*(M) \cong \mathfrak{A}$. If $D_M \subset D_0$, $D_M \subset D_1$, and $D_0 =_F D_1$, then $W_{D_0} \sim_f W_{D_1}$. Indeed, if $\mathrm{gr}(D_0 \cup F_0) = \mathrm{gr}(D_1 \cup F_1)$, then

$$W_{\mathrm{gr}(D_0 \cup F_0)} = W_{D_0} \cup \bigcup_{b \in F_0} \mathrm{supp}\, b, \quad W_{\mathrm{gr}(D_1 \cup F_1)} = W_{D_1} \cup \bigcup_{b \in F_1} \mathrm{supp}\, b$$

Since the sets $\mathrm{supp}\, b$, F_0, and F_1 are finite, $W_{D_0} \sim_f W_{D_1}$. □

As was noted by Remmel [167, 171], not all maximal subalgebras are simple. However, all the maximal and hyperhypersimple subalgebras constructed here are simple. The corollary given below provides examples of simple but not hyperhypersimple subalgebras.

Let $\mathfrak{B}$ be a recursive Boolean algebra and let a subalgebra D be connected with $\mathfrak{B}$ in the same way as in Theorem 3.7.8. The following claim holds.

Corollary 3.7.10. *If M is a hypersimple set, then D_M is a simple subalgebra of the algebra $\mathfrak{B}$.*

PROOF. Assume the contrary. Let C be an infinite recursively enumerable subalgebra of the algebra $\mathfrak{B}$ such that $D_M \cap C = \{0, 1\}$. We try to construct an effective sequence of finite sets. Its existence contradicts the fact that the set M is hypersimple. Let $d_0 = 1, d_1, d_2, \ldots$ be an arbitrary recursively enumerable sequence generating the algebra C. We consider $d_1 \neq 0$. Since $d_1 \in D_M$, we have $\mathrm{supp}\, d_1 \cap \overline{M} \neq \varnothing$. We set $F_0 = \mathrm{supp}\, d_1$.

Let $|\mathrm{supp}\, d_1| = k$. From the definition of a recursively enumerable generating sequence it follows that $|\mathrm{Atom}(C^s)| = s + 1$. We consider $C^{2k} = \mathrm{gr}(\{d_1, \ldots, d_{2k}\})$. Then $|\mathrm{Atom}(C^{2k})| = 2k + 1$. It is clear that for any $i \in \mathrm{supp}\, d_1$ there exists at most two different elements $a \in \mathrm{Atom}(C^{2k})$ such that $i \in \mathrm{supp}\, a$. This follows from the fact that for any s and i at most two elements of A_i are included into the partition of unity at the step s. Therefore, among atoms of the algebra D^{2k}, there is at least one atom $a \in \mathrm{Atom}(C^{2k})$ such that $\mathrm{supp}\, a \cap \mathrm{supp}\, d_1 = \varnothing$. We set $F_1 = \mathrm{supp}\, a$. Then $F_1 \cap F_0 = \varnothing$ and $F_1 \cap \overline{M} \neq \varnothing$.

Assume that $F_0, \ldots, F_n$ and $F_i \cap \overline{M} \neq \varnothing$ are already constructed for any $i \leqslant n$; moreover, $F_i \cap F_j = \varnothing$, $i \neq j$. To define F_{n+1}, we proceed in the same way as at the first step. Let $\left|\bigcup_{i \leqslant n} F_i\right| = m$. We consider a finite

3.7. Algorithmic Properties

subalgebra of C^{2m}. Among $2m+1$ atoms of C^{2m}, there is at least one atom $a \in \text{Atom}(C^{2m})$ such that $\text{supp}\, a \cap \left(\bigcup_{i \leqslant n} F_n \right) = \varnothing$. We set $F_{n+1} \rightleftharpoons \text{supp}\, a$.
The existence of the sequence $\{F_i \mid i \in \mathbb{N}\}$ contradicts the fact that M is hypersimple. Consequently, the subalgebra D_M is simple. □

Corollary 3.7.11. *For any recursive Boolean algebra $\mathfrak{B}$ the elementary theory of the lattice $\mathcal{S}(\mathfrak{B})$ is undecidable.*

PROOF. Herrman [80] proved that the elementary theory of the lattice of recursively enumerable sets is undecidable. In particular, he proved that the class consisting of all recursive Boolean pairs is relatively elementarily definable in the lattice $\mathcal{E}$ of recursively enumerable sets. By a *Boolean pair* we mean a Boolean algebra and a unary predicate distinguishing its subalgebra.

Burris and McKenzie [14] constructed a class of Boolean pairs which have hereditarily undecidable theory. As was indicated by Herrman, this class consists of recursive Boolean pairs. Therefore, the undecidability of $\mathcal{E}$ follows from the result of [37] which can be stated as follows: *If the class $\mathcal{K}_0$ is relatively elementarily definable in the class $\mathcal{K}_1$ and the theory $\text{Th}(\mathcal{K}_0)$ is hereditarily undecidable, then the theory $\text{Th}(\mathcal{K}_1)$ is also hereditarily undecidable.* This theorem shows that Herrman proved not only the undecidability of the lattices $\mathcal{E}$ but also their hereditary undecidability. Theorem 3.7.1 implies that the lattice $\mathcal{E}$ is relatively elementarily definable in the lattice $\mathcal{S}(\mathfrak{B})$ for any recursive infinite Boolean algebra $\mathfrak{B}$. Using the same theorem, we can conclude that the lattice $\mathcal{S}(\mathfrak{B})$ also has hereditary undecidable theory. □

Exercises

1. Construct an example of a nonatomic Boolean algebra such that it is not constructivizable but its quotient algebra by the Frechét ideal is constructivizable [65].

2. Construct an enumerable ideal η of an atomless Boolean algebra $\mathfrak{B}$ such that $\mathfrak{B}/\eta$ is not constructivizable [34].

3. Prove that a constructive atomic Boolean algebra $\langle \mathfrak{B}, \nu \rangle$ with the Frechét ideal of complexity Δ_2^0 is strongly constructivizable [34].

4. Prove that for any constructivizable atomic Boolean algebra $\mathfrak{B}$ there exists an effectively atomic constructivization, i.e., a constructivization ν and a recursive function $f(n, x)$ such that for any n and

x the relations $\nu f(n,0) = \nu n$, $\nu f(n, x+1) \leqslant \nu f(n,x)$ hold and $\nu \lim\limits_{x\to\infty} f(n,x)$ is an atom of the Boolean algebra $\mathfrak{B}$ [64, 65].

5. Prove that any constructivization of the Boolean algebra $\mathfrak{B}_\omega$ is effectively atomic and the set of numbers of atoms is not hypersimple.

6. Prove that the algebras $\mathfrak{B}_{\omega^2}$, $\mathfrak{B}_{\omega\times(\omega\times(1+\eta))}$, and $\mathfrak{B}_{\omega\times(\omega\times\eta)}$ have constructivizations that are not effectively atomic.

7. Give a complete proof of Proposition 3.7.6. Prove that for any n there exists a constructivizable Boolean algebra $\mathfrak{B}$ such that $\mathfrak{B}/I_n(\mathfrak{B})$ is not constructivizable, where $I_n(\mathfrak{B})$ is the nth Ershov–Tarski ideal.

8. Prove that the quotient algebras $\mathfrak{B}/I_n(\mathfrak{B})$ and $\mathfrak{B}/\mathrm{Atomless}(\mathfrak{B})$ of a strongly constructivizable Boolean algebra $\mathfrak{B}$ are strongly constructivizable.

9. Construct an example of a strongly constructivizable atomic Boolean algebra such that its quotient algebra by the Frechét ideal is not constructivizable.

10. Prove that for any $A \in \Delta_1^1$ there exists a superatomic constructive Boolean algebra $\langle\mathfrak{B},\nu\rangle$ such that $\nu^{-1}(I) \equiv_T A$ for some maximal ideal I [53].

3.8. Automorphisms of Countable Boolean Algebras

In this section, we study the structure of the automorphism group of a countable Boolean algebra which characterizes the presence of symmetry in the algebra. All automorphisms of a Boolean algebra $\mathfrak{A}$ form a group which is denoted by $\mathrm{Aut}\,\mathfrak{A}$. For countable Boolean algebras the automorphism groups are close, in a sense, to the permutation groups. This fact allows us to establish some nontrivial relations between these groups and algebras. Automorphisms of Boolean algebras, in particular, recursive automorphisms of recursive Boolean algebras, have been extensively studied. The aim of this section is to demonstrate the technique relative to actions of automorphisms on atoms.

3.8. Automorphisms of Countable Boolean Algebras

Proposition 3.8.1. *Any automorphism of a countable Boolean algebra is completely defined by its actions on atoms and atomless elements.*

PROOF. It suffices to show that if an automorphism φ keeps all atoms and atomless elements fixed, then it is the identity mapping. Let $\varphi(a) \neq a$ for some a. Without loss of generality, we can assume that $\varphi(a) \wedge a = 0$. Conversely, let $a \backslash \varphi(a) \neq 0$ or $\varphi(a) \backslash a \neq 0$. In the first case, we consider $a \backslash \varphi(a)$ instead of a. In the second case we consider $\varphi(a) \backslash a$ instead of a and φ^{-1} instead of φ. If there exists an atom $\beta \leqslant \alpha$, then $\varphi(\beta) \leqslant \varphi(\alpha)$. From the equality $\alpha \wedge \varphi(\alpha) = 0$ we obtain $\varphi(\beta) \wedge \beta = 0$ and $\beta(\beta) \neq \beta$, which contradicts the hypothesis. Hence α is an atomless element. But, by condition, $\varphi(\alpha) = \alpha$, which contradicts the equality $\alpha \wedge \varphi(\alpha) = 0$. □

If a, b are two atoms of a Boolean algebra $\mathfrak{A}$, then (a, b) denotes the automorphism that transforms a into b and b into a and is identical on the rest of the atoms and atomless elements. Such an automorphism is unique. It is called an *atomic transposition* or simply a *transposition*.

It is easy to verify that all atomic transpositions are conjugate. Furthermore, they can be distinguished by a ∀-formula in the group.

Proposition 3.8.2. *An automorphism x of a countable Boolean algebra $\mathfrak{A}$ is a transposition if and only if* $\operatorname{Aut} \mathfrak{A} \models \operatorname{tr}(x)$, *where the denotation* $\operatorname{tr}(x) = (x \neq 1 \,\&\, x^2 = 1 \,\&\, \forall y([x,y]^6 = 1))$ *is introduced for brevity and* $[x, y] = x^{-1} y^{-1} xy$ *is the usual denotation of the group commutator.*

PROOF. We assume that x is a transposition. Then $x \neq 1$ and $x^2 = 1$. Furthermore, $y^{-1} xy$ is an atomic transposition under any automorphism y. Therefore, $[x, y] = x^{-1} y^{-1} xy = x \cdot y^{-1} xy$ either is the product of two transpositions moving disjoint set of atoms (hence $[x, y]^2 = 1$) or moves exactly three atoms (hence $[x, y]^3 = 1$); in any case, $[x, y]^6 = 1$.

We assume that $\operatorname{tr}(x)$ holds. We first prove that x identically acts on atomless elements. Let a be an atomless element such that $x(a) \neq a$. As in the proof of Proposition 3.8.1, we can assume that $x(a) \wedge a = 0$. It is obvious that the Boolean algebra $\widehat{a}$ possesses an automorphism y' of infinite order. Therefore, we can construct an automorphism y of infinite order such that $y \restriction \widehat{a} = y'$ and y identically acts on those elements that are less than $C(a)$. Hence $[x, y]$ acts on $\widehat{a}$ in the same way as y' and, consequently, has infinite order. The contradiction obtained shows that x is identical on atomless elements.

We consider two cases.

CASE 1: acting on atoms, x contains at least two 2-cycles (a,b) and (c,d), and leaves two atoms e and f fixed. We consider an automorphism defined by a permutation on atoms $y = (c,b,e)(d,f)(a)$. It is easy to check that $[x,y]$ contains the cycle (a,d,c,b). Consequently, the equality $[x,y]^6 = 1$ is impossible.

CASE 2: acting on atoms, x contains at least four cycles (a,b), (c,d), (e,f), and (g,h). We choose an automorphism y such that $y^{-1}xy$ includes the cycles (b,c), (d,f), (e,g), and (a,h). It remains to check that $[x,y] = x^{-1}(y^{-1}xy) = x(y^{-1}xy)$ contains the cycles (a,g,f,c). Thus, $[x,y]^6 = 1$ is impossible. From the conditions on $\mathfrak{A} \vDash \mathrm{tr}\,(x)$ and the cases considered it is follows that $x = (a,b)$ fo atoms a,b and x is their transposition. □

Theorem 3.8.1. *If $\mathfrak{A}$ and $\mathfrak{B}$ are countable Boolean algebras with isomorphic automorphism groups and $\mathfrak{A}$ is atomic, then $\mathfrak{A} \cong \mathfrak{B}$.*

PROOF. The two-sorted model $\langle \mathrm{Aut}\,\mathfrak{A}, \mathrm{Atom}, \circ, \mathrm{ap} \rangle$, where

$$\mathrm{ap} : \mathrm{Aut}\,\mathfrak{A} \times \mathrm{Atom}\,\mathfrak{A} \to \mathrm{Atom}\,\mathfrak{A}, \quad \mathrm{ap}(g,a) = g(a)$$

and the group operation $\circ$ on $\mathrm{Aut}\,\mathfrak{A}$ can be uniquely restored from the group $\mathrm{Aut}\,\mathfrak{A}$. With every atom α of the algebra $\mathfrak{A}$ we associate the set of pairs of transpositions $\langle x,y \rangle$ such that $x = (a,b)$, $y = (c,d)$, and $\{\alpha\} = \{a,b\} \cap \{c,d\}$. The elementary condition distinguishing such pairs in the group of all automorphisms can be written as follows: $\mathrm{tr}\,(x)\,\&\,\mathrm{tr}\,(y)\,\&\,[x,y] \ne 1$. A formula recognizing, among such pairs, the pair of transpositions $\langle x_0, x_1 \rangle$ and $\langle y_0, y_1 \rangle$ corresponding to the same atom [40] can be written in the form $E(x_0, x_1, y_0, y_1) = \underset{i,j=0,1}{\&} [x_i y_j]^3 = 1$. The proof of this fact consists in considering all possible variants of locations of moving atoms in transpositions. We now are able to interpret the set $\mathrm{Atom}\,\mathfrak{A}$ of the group $\mathrm{Aut}\,\mathfrak{A}$ as the quotient set of the set

$$\{\langle x,y \rangle \mid \mathrm{Aut}\,\mathfrak{A} \vDash \mathrm{tr}\,(x)\,\&\,\mathrm{tr}\,(y)\,\&\,[x,y] \ne 1\}$$

by the equivalence relation $\sim$ given by the formula E:

$$\langle x_0, x_1 \rangle \sim \langle y_0, y_1 \rangle \Leftrightarrow E(x_0, x_1, y_0, y_1)$$

If an atom β corresponds to a pair $\langle x,y \rangle$, then the atom $f(\beta)$ corresponds to the pair $\langle fxf^{-1}, fyf^{-1} \rangle$. This provides a way of interpreting the mapping ap. Consequently, for any countable Boolean algebras $\mathfrak{A}$ and $\mathfrak{B}$ such that $\mathrm{Aut}\,A \cong \mathrm{Aut}\,\mathfrak{B}$ we have $\langle \mathrm{Aut}\,\mathfrak{A}, \mathrm{Atom}\,\mathfrak{A}, \circ, \mathrm{ap} \rangle \cong \langle \mathrm{Aut}\,\mathfrak{B}, \mathrm{Atom}\,\mathfrak{B}, \circ, \mathrm{ap} \rangle$; more exactly, for any isomorphism $i : \mathrm{Aut}\,\mathfrak{A} \to \mathrm{Aut}\,\mathfrak{B}$ there exists a one-to-one correspondence h between $\mathrm{Atom}\,\mathfrak{A}$ and

3.8. Automorphisms of Countable Boolean Algebras

Atom $\mathfrak{B}$ such that $h(\mathrm{ap}(g,a)) = \mathrm{ap}(i(g), h(a))$ for any $g \in \mathrm{Aut}\,\mathfrak{A}$, $a \in$ Atom $\mathfrak{A}$. Thus, we have established an isomorphism between actions of automorphism groups on atoms. Let us show that $\mathfrak{B}$ is atomic. Otherwise, there exists a nontrivial automorphism Aut $\mathfrak{B}$ that leaves all atoms fixed, which is false for the group Aut $\mathfrak{A}$. This contradicts the isomorphism of actions of these groups on atoms. We convince ourselves that for any $X \subseteq \mathrm{Atom}\,\mathfrak{A}$ there exists $\sup X$ if and only if there exists $\sup h(X)$. Then we can continue the mapping h to an isomorphism $h^* : \mathfrak{A} \to \mathfrak{B}$ as follows: $h^*(a) = \sup\{h(\alpha) \mid \alpha \in \mathrm{Atom}\,\mathfrak{A}\,\&\,\alpha \leqslant a\}$ Thereby, the theorem is proved. We first construct three automorphisms φ, π, and τ. Without loss of generality, we can assume that the universe of the algebra $\mathfrak{A}$ is the set of natural numbers. We define a mapping $p : 2^{<\omega} \to \{0,1,2\}$ as follows: the value of p on the empty word Λ is 0 by definition. Further,

$$p(\varepsilon 0) = \begin{cases} 1 & \text{if } p(\varepsilon) = 0 \\ 2 & \text{if } p(\varepsilon) = 1 \\ 0 & \text{if } p(\varepsilon) = 2 \end{cases} \qquad p(\varepsilon 1) = \begin{cases} 2 & \text{if } p(\varepsilon) = 0 \\ 0 & \text{if } p(\varepsilon) = 1 \\ 1 & \text{if } p(\varepsilon) = 2 \end{cases}$$

We will perform the construction step-by-step.

STEP 0. Let an atom α regarded as a natural number be the least one among all atoms and let an atom c_Λ regarded as a natural number be the least one among all atoms except for α. Let g_Λ be equal to the transposition (α, c_Λ) and let $e_\Lambda = \mathbf{1}\setminus\alpha$, where $\mathbf{1}$ is the unit of the algebra $\mathfrak{A}$.

STEP $n+1$. For all $\varepsilon \in 2^n$ such that g_ε and c_ε are defined, we set $e_{\varepsilon 1} = (n \wedge e_\varepsilon)\setminus c_\varepsilon$, $e_{\varepsilon 0} = (C(n) \wedge e_\varepsilon)\setminus c_\varepsilon$. For $i = 0,1$, let $c_{\varepsilon i}$ be equal to an atom that does not exceed $e_{\varepsilon i}$, and (if such an atom exists) has the least number among all such atoms. If there is no atom with the above property, then $g_{\varepsilon i}$ and $c_{\varepsilon i}$ are assumed to be undefined. If $c_{\varepsilon i}$ is defined, then we assume that $g_{\varepsilon i}$ is equal to $(c_\varepsilon, c_{\varepsilon i})$. The description of constructing c_ε and g_ε is complete.

We note that Atom $\mathfrak{A} = \{\alpha\} \cup \{c_\varepsilon \mid \varepsilon \in 2^{<\omega}$ and g_ε is defined$\}$. We now put

$$\varphi = \prod_{p(\varepsilon)=0} g_\varepsilon, \quad \pi = \prod_{p(\varepsilon)=1} g_\varepsilon, \quad \tau = \prod_{p(\varepsilon)=2} g_\varepsilon$$

where infinite products are considered as pointwise limits. These definitions are well defined since, for any $a \in \mathfrak{A}$, the set $\{\varepsilon \mid g_\varepsilon(a) \neq a\}$ is finite and $g_\varepsilon g_\delta = g_\delta g_\varepsilon$ for all ε and δ such that g_ε and g_δ are defined and $p(\varepsilon) = p(\delta)$.

We note that for any function $f : 3^{<\omega} \to \{0,1\}$ the infinite products

$$\varphi^f = \prod_{p(\varepsilon)=0} g_\varepsilon^{f(\varepsilon)}, \quad \pi^f = \prod_{p(\varepsilon)=1} g_\varepsilon^{f(\varepsilon)}, \quad \prod_{p(\varepsilon)=2} g_\varepsilon^{f(\varepsilon)} \qquad (3.8.1)$$

regarded as pointwise limits are well defined and are automorphisms.

We show that for any $X \subseteq \operatorname{Atom} \mathfrak{A}$ there exists $\sup X$ if and only if the family $\{\lambda(X) \mid \lambda \in \operatorname{Aut} \mathfrak{A}\}$ is countable. It is clear that if $a = \sup X$ exists, then the family $\{\lambda(X) \mid \lambda \in \operatorname{Aut}(\mathfrak{A})\}$ consists of sets of the form $\{\alpha \in \operatorname{Atom} \mathfrak{A} \mid \alpha \leqslant b\}$, where $b = \lambda(a)$ for a suitable $\lambda \in \operatorname{Aut} \mathfrak{A}$, and consequently is countable. We now assume that $a = \sup X$ exists. Then at least one of the sets $\{\alpha \in X \mid \varphi(\alpha) \notin X\}$, $\{\alpha \in X \mid \pi(\alpha) \notin X\}$, $\{\alpha \in X \mid \tau(\alpha) \notin X\}$ is infinite. If all the sets are finite, then, beginning from some m, for any $\varepsilon \in 2^m$ we have $\{\alpha \in \operatorname{Atom} \mathfrak{A} \mid \alpha \leqslant e_\varepsilon\} \subseteq X$ or $\{\alpha \in \operatorname{Atom} \mathfrak{A} \mid \alpha \leqslant e_\varepsilon\} \cap X = \varnothing$, which implies the existence $\sup X$ in the form of the union of the corresponding e_ε and some finite family of atoms. Therefore, without loss of generality, we can assume that the set $\{\alpha \in X \mid \varphi(\alpha) \notin X\}$ is infinite. Since for any mapping $f : 3^{<\omega} \to \{0,1\}$ the automorphism φ^f is defined, the set X has an uncountable number of images of the form $\varphi^f(X)$. Thus, we have proved that for any $X \subseteq \operatorname{Atom} \mathfrak{A}$ there exists $\sup X$ if and only if there exists $\sup h(X)$. □

A stronger assertion is also valid.

Theorem 3.8.2 [117]. *If $\mathfrak{A}$ and $\mathfrak{B}$ are countable Boolean algebras with isomorphic automorphism groups and the union of all atoms exists in $\mathfrak{A}$, then $\mathfrak{A} \cong \mathfrak{B}$.*

In the proof of Theorem 3.8.2, we use the Anderson result stated below. Note that this result is of independent interest.

Theorem 3.8.3. *The group of all automorphisms of a countable atomless Boolean algebra is prime, i.e., it does not contain nontrivial subgroups.*

The following result of Rubin [179] looks more impressive.

Theorem 3.8.4. *Let $\mathfrak{A}$ and $\mathfrak{B}$ be arbitrary countable Boolean algebras such that the number of their atoms is not equal to 1. Then the elementary equivalence of the groups $\operatorname{Aut} \mathfrak{A}$ and $\operatorname{Aut} \mathfrak{B}$ implies the elementary equivalence of the algebras $\mathfrak{A}$ and $\mathfrak{B}$. Furthermore, there exists a sentence φ of the language of group theory such that $\operatorname{Aut} \mathfrak{A} \vDash \varphi$ and $\operatorname{Aut} \mathfrak{B} \vDash \varphi$ imply $\mathfrak{A} \equiv \mathfrak{B}$.*

3.8. Automorphisms of Countable Boolean Algebras

The description of atomic transpositions from Proposition 3.8.1 remains also valid for recursive automorphisms. Let $\text{Aut}_r(\mathfrak{A})$ denote the group of all recursive automorphisms of a system $\mathfrak{A}$. It suffices to note that all the automorphisms, whose existence is necessary in the proof of this proposition (in particular, those automorphisms that move a finite number of atoms and are identical on atomless elements), can be recursively chosen. It is useful to note that all transpositions are conjugate in the group of all recursive automorphisms of a Boolean algebra. We present only two results concerning nonconstructivizability for groups of recursive automorphisms of recursive Boolean algebras.

Theorem 3.8.5. *If, in an infinite recursive Boolean algebra $\mathfrak{A}$, the set of numbers of its atoms contains an infinite recursively enumerable subset, then the group $\text{Aut}_r \mathfrak{A}$ is not constructivizable.*

To prove this theorem we need the following auxiliary assertion.

Lemma 3.8.1. *Let $\mathfrak{A}$ be a recursive Boolean algebra and let the group $\text{Aut}_r \mathfrak{A}$ be constructivizable. Then any of its constructivizations can be extended to a constructivization of the two-sorted model $\langle \text{Aut}_r \mathfrak{A}, \text{Atom}\,\mathfrak{A}, \circ, \text{ap}\rangle$.*

PROOF. Let some constructivization of the group $\text{Aut}_r(\mathfrak{A})$ be given. Using the fact that all atomic transpositions are conjugate in this group, we can enumerate the set of their numbers as a set of elements of the form $g^{-1}\tau g$, where g run over the entire group $\text{Aut}_r(\mathfrak{A})$ and τ is a fixed transposition. With the help this numbering, we can effectively enumerate pairs of transpositions $\langle f, g\rangle$ such that $fg \neq gf$ (i.e., transpositions f and g have a unique common moving atom); moreover, we can act in such a way that it is possible to find effectively the numbers of elements of a pair from the number of the pair and the number of a pair constituted by non-commutative transpositions from the numbers of these transpositions. As in the case of usual automorphisms, the equality of common moving atoms of the pairs of atomic transpositions $\langle x_0, x_1\rangle$ and $\langle y_0, y_1\rangle$ is equivalent to the elementary condition $E(x_0, x_1, y_0, y_1)$. The proof is the same as in the case of usual automorphisms. Since the formula E expressing the equality of common moving atoms of pairs of atomic transposition is quantifier-free, there exists an algorithm which allows us to recognize whether common moving atoms of the pairs $\langle x_0, x_1\rangle$ and $\langle y_0, y_1\rangle$ coincide. Therefore, we can effectively enumerate classes of the form

$$A_{a_0, a_1} = \{\langle x_0, x_1\rangle \mid \text{common atoms of } \langle x_0, x_1\rangle \text{ and } \langle a_0, a_1\rangle \text{ coincide}\}$$

where a_0, a_1 are pairs of noncommutative transpositions, in such a way that, starting from the number of a class and the number of a pair of noncommutative transpositions, we can effectively define whether a given pair belongs to a given class. Each such class identified in a natural way with the corresponding moving atom which is common for all pairs of this class. Let the number of the class be the number of this atom, and let an atom α be the common moving atom of the pair $\langle x,y \rangle$. Furthermore, let $f \in \operatorname{Aut}_r \mathfrak{A}$. Then $f(\alpha)$ is the common moving atom of the pair $\langle fxf^{-1}, fyf^{-1} \rangle$. By this remark, from the number of an atom α (i.e., the number of a class) and the number of an automorphism f in $\operatorname{Aut}_r \mathfrak{A}$ we can effectively find the number of the atom $f(\alpha)$. Thus, we have constructed a constructivization of the two-sorted model $\langle \operatorname{Aut}_r \mathfrak{A}, \operatorname{Atom} \mathfrak{A}, \circ, \operatorname{ap} \rangle$. We note that the $\operatorname{Aut}_r \mathfrak{A}$ remains fixed, i.e., the constructivization of the two-sorted model extends the initial constructivization of this group.

We also note that the above numbering of the set of atoms is not necessarily connected with the set of numbers of this set in an initial algebra. □

PROOF OF THEOREM 3.8.5. We fix a recursive injective function g from $\mathbb{N}$ into $\operatorname{Atom} \mathfrak{A}$. We construct step-by-step a family of atoms $\{\alpha_i^j \mid j \leqslant \omega, \ i \leqslant j+1\}$.

STEP 0. We define α_0^0 and α_1^0 as $g(0)$ and $g(1)$ respectively. For the empty sequence Λ we set $r_\Lambda = C(\alpha_0^0 \vee \alpha_1^0)$.

STEP $s+1$. We set $r'_{\varepsilon 0} = r_\varepsilon \wedge s$, $r'_{\varepsilon 1} = r_\varepsilon \wedge C(s)$ for all $\varepsilon \in 2^s$. With the help of g, which enumerates atoms of the Boolean algebra B, we find $\varepsilon \in 2^{s+1}$ such that r'_ε contains $s+3$ different atoms $\alpha_0^{s+1}, \ldots, \alpha_{s+2}^{s+1}$ of the image of g. We set $r_\varepsilon \rightleftharpoons r'_\varepsilon \setminus (\alpha_0^{s+1} \vee \ldots \vee \alpha_{s+2}^{s+1})$ for any $\varepsilon \in 2^{s+1}$. Denote by $(c_0, \ldots, c_n)$ the automorphism of $\mathfrak{A}$ that realizes a cyclic permutation of the atoms $c_0, \ldots, c_n$ and leaves fixed the rest of the atoms and atomless elements.

Let $\lambda = \prod_{t \geqslant 0} (\alpha_0^{t+1}, \ldots, \alpha_{t+1}^{t+1})$. It is easy to see that it is a recursive automorphism of the algebra $\mathfrak{A}$. Moreover, for any computable function $f : \omega \to \{0,1\}$ the automorphism $\lambda^f = \prod_{t \geqslant 0} (\alpha_0^{t+1}, \ldots, \alpha_{t+1}^{t+1})^{f(t)}$ is well defined and is recursive. We now assume that ν is a constructivization of the group $\operatorname{Aut}_r \mathfrak{A}$. By Lemma 3.8.1, it can be extended to a constructivization of the two-sorted model $\langle \operatorname{Aut}_r \mathfrak{A}, \operatorname{Atom} \mathfrak{A}, \circ, \operatorname{ap} \rangle$.

We now define a recursive function h as follows: let α be a fixed atom and $n \in \mathbb{N}$. For any n we seek the least $k \in \omega$ such that $(\nu k)(\alpha) = \alpha_0 \overset{\lambda}{\to}$

3.8. Automorphisms of Countable Boolean Algebras

$\alpha_1 \overset{\lambda}{\to} \ldots \overset{\lambda}{\to} \alpha_{n+1} \overset{\lambda}{\to} \alpha_0$ and all $\alpha_0, \alpha_1, \ldots, \alpha_{n+1}$ are pairwise disjoint. Let

$$h(n) = \begin{cases} 0, & \nu n(\alpha_0) \neq \alpha_0 \\ 1 & \text{otherwise} \end{cases}$$

It is easy to see that h is a recursive function. Therefore, $\lambda^h \in \text{Aut}_r \mathfrak{A}$. We now prove that λ^h coincides with none of the automorphisms νn. We assume that $\lambda^h = \nu n$. In computing $h(n)$ in accordance with the above algorithm, the element $(\nu k)(\alpha) = \alpha_0$ goes to a cycle of length $n+2$. We note that λ has exactly one such cycle. By definition, in the case $\nu n(\alpha_0) \neq \alpha_0$, we must set $h(n) = 0$. Therefore, $\lambda^h(\alpha_0) = \alpha_0 \neq \nu(n)(\alpha_0)$. The fact that we arrive at an inequality in the case $\nu n(\alpha_0) = \alpha_0$ is proved in the same way. Thus, the equality $\lambda^h = \nu(n)$ is impossible. □

In the case $\varphi_1, \ldots, \varphi_k \in \text{Aut}_r \mathfrak{A}$, we call $\varphi_1, \ldots, \varphi_k$-*orbits* those orbits relative to the group that are generated by these elements.

Theorem 3.8.6 [130]. *Assume that for some finite family $\varphi_1, \ldots, \varphi_k \in \text{Aut}_r \mathfrak{A}$, in the set $\text{Atom} \mathfrak{A}$, there is an infinite $\varphi_1, \ldots, \varphi_k$-orbit. Then the group $\text{Aut}_r \mathfrak{A}$ is embedded in none of the constructivizable groups.*

PROOF. We assume that $\langle G, \nu \rangle$ is a constructive group and $\text{Aut}_r \mathfrak{A} \leqslant G$. We fix a pair of noncommutative transpositions $\langle \tau_0, \tau_1 \rangle$ and define $T = \{\langle \tau_0^f, \tau_1^f \rangle \mid f \in G\}$. Further, for $\langle \eta_0, \eta_1 \rangle, \langle \theta_0, \theta_1 \rangle \in T$ we define $\langle \eta_0, \eta_1 \rangle \sim \langle \theta_0, \theta_1 \rangle \Leftrightarrow E(\eta_0, \eta_1, \theta_0, \theta_1)$. Since the formula on the right-hand side of the equivalence is a quantifier-free one, the constructivization ν can be extended to a constructivization of the two-sorted model $\langle G, T, \text{ap} \rangle$ in the same way as in Theorem 3.8.5; here $\text{ap}(f, \langle \theta_0, \theta_1 \rangle) = \langle \theta_0^{f^{-1}}, \theta_1^{f^{-1}} \rangle$. We note that if $\langle \theta_0, \theta_1 \rangle \in T$ and $\theta_0 = \tau_0^{f^{-1}}, \theta_1 = \tau_1^{f^{-1}}$ for some automorphism $f \in \text{Aut}_r \mathfrak{A}$, then θ_0 and θ_1 are also transpositions. Moreover, τ_0 and τ_1 have a common moving atom α (such an atom exists and is unique since $[\tau_0, \tau_1] \neq 1$), and $f(\alpha)$ is such an atom for θ_0 and θ_1. If $\eta_i = \tau_i^g$, $i = 0, 1$, for some $g \in \text{Aut}_r \mathfrak{A}$, then $\langle \eta_0, \eta_1 \rangle \sim \langle \theta_0, \theta_1 \rangle$ holds if and only if common moving atoms of these pairs coincide. Enumerating the infinite $\varphi_1, \ldots, \varphi_k$-orbit by means of recursive automorphisms $\varphi_1, \ldots, \varphi_k$, we can construct a recursive automorphism λ, as in Theorem 3.8.5.

We now define a recursive function $g : \omega \to \{0, 1\}$ as follows. For a given $n \in \omega$, we seek a recursive automorphism t of the subgroup generated by $\varphi_1, \ldots, \varphi_k$ such that $\alpha_0 = \langle \tau_0^t, \tau_1^t \rangle$ implies

$$\text{ap}(\lambda, \alpha_0) = \alpha_1, \text{ap}(\lambda, \alpha_1) = \alpha_2, \ldots, \text{ap}(\lambda, \alpha_n) = \alpha_{n+1}, \text{ap}(\lambda, \alpha_{n+1}) \sim \alpha_0,$$

and $\neg(\alpha_i \sim \alpha_j)$ holds for all $i, j, 0 \leqslant i < j \leqslant n$. We define $g(n) = 1$ if $\nu n(\alpha_0) \sim \alpha_0$ and $g(n) = 0$ otherwise. It is easy to see that $\lambda^g \in \mathrm{Aut}\,_r\mathfrak{A}$. We show that λ^g coincides with none of the automorphisms $\nu n \in \mathrm{Aut}\,_r\mathfrak{A}$. We assume that $\lambda^g = \nu n(n)$. Since t appearing in the description of the computation of g is taken from $\mathrm{Aut}\,_r\mathfrak{A}$, we conclude that α_0 is a pair of noncommutative atomic transpositions. Moreover, a common moving atom of elements of this pair goes to a unique cycle of length $n + 2$ of the automorphism λ. If νn leaves this atom fixed (i.e., $\nu n(\alpha_0) \sim \alpha_0$), λ^g must move it (since $g(n) = 1$). Conversely, if νn does not leave it in place, ν^g keeps these atom fixed because $\neg(\nu n(\alpha_0) \sim \alpha_0)$. Therefore, $g(n) = 0$. Consequently, $\lambda^g \neq \nu n$, which yields a contradiction. □

Corollary 3.8.1. *The group* $\mathrm{Aut}\,_r\omega$ *of all recursive permutations of the set of natural numbers cannot be embedded in any consructivizable group.*

Corollary 3.8.2. *Let $\mathfrak{A}$ be an infinite recursive Boolean algebra with a recursive set of atoms. Then the group of its recursive automorphisms cannot be embedded in any constructivizable group. In particular, the group of recursive automorphisms of any decidable Boolean algebra cannot be embedded in any constructivizable group.*

We note that for many natural recursive Boolean algebras the groups of their automorphisms are not constructivizable. However, there exist recursive Boolean algebras with constructivizable groups of recursive automorphisms. Since the group of all finite permutations of $\mathbb{N}$ is constructivizable, such an example can be obtained on the basis of the following theorem.

Theorem 3.8.7 [40, 131]. *For any recursive Boolean algebra there exists an isomorphic recursive Boolean algebra each recursive automorphism of which moves only a finite family of atoms.*

In a number of cases, recursive Boolean algebras, as the usual countable Boolean algebras, can be completely restored from the group of their recursive automorphisms.

Theorem 3.8.8 [130]. *Let $\mathfrak{A}$ be a recursive atomic Boolean algebra with recursive set of atoms and let $\mathfrak{B}$ be an arbitrary recursive Boolean algebra. Then* $\mathrm{Aut}\,_r\mathfrak{A} \cong \mathrm{Aut}\,_r\mathfrak{B}$ *implies* $\mathfrak{A} \cong_r \mathfrak{B}$.

PROOF. We first construct recursive automorphisms φ, π, and τ of an atomic decidable Boolean algebra $\mathfrak{A}$ in the same way as in the proof of

3.8. Automorphisms of Countable Boolean Algebras 299

Theorem 3.8.1. We note that for any recursive function $f : 3^{<\omega} \to \{0,1\}$ the infinite products (3.8.1) are well defined and they are recursive automorphisms. We fix such automorphisms φ, π, and τ. Let v be an isomorphism between the groups of recursive automorphisms $\operatorname{Aut}_r \mathfrak{A}$ and $\operatorname{Aut}_r \mathfrak{B}$. As in the case of the usual automorphisms, the two-sorted models $\langle \operatorname{Aut}_r \mathfrak{A}, \operatorname{Atom} \mathfrak{A}, \circ, \operatorname{ap} \rangle$ and $\langle \operatorname{Aut}_r \mathfrak{B}, \operatorname{Atom} \mathfrak{B}, \circ, \operatorname{ap} \rangle$ are isomorphic, i.e., there exists a one-to-one mapping $h : \operatorname{Atom} \mathfrak{A} \to \operatorname{Atom} \mathfrak{B}$ such that $h(\operatorname{ap}(g,a)) = \operatorname{ap}(v(g), h(a))$ for all $g \in \operatorname{Aut}_r \mathfrak{A}$ and $\alpha \in \operatorname{Atom} \mathfrak{A}$. In other words, the actions of the group of $\operatorname{Aut}_r \mathfrak{A}$ on $\operatorname{Atom} \mathfrak{A}$ and of the group $\operatorname{Aut}_r \mathfrak{B}$ on $\operatorname{Atom} \mathfrak{B}$ are isomorphic. We fix a mapping with these properties. Since all atoms of the algebra $\mathfrak{A}$ form a single orbit under the actions of automorphisms φ, π, and τ, the same assertion is true for the algebra $\mathfrak{B}$. Therefore, the set of atoms of the algebra $\mathfrak{B}$ is enumerable. Since the set of atoms of any recursive Boolean algebra is co-enumerable (since atoms are distinguished by a $\forall$-formula), this set is recursive.

We note that the Boolean algebra $\mathfrak{B}$ atomic. Otherwise, $\mathfrak{B}$ has a nontrivial recursive automorphism acting identically on atoms, which is impossible since the group of recursive automorphisms of the algebras $\mathfrak{A}$ and $\mathfrak{B}$ acts isomorphically on atoms of these algebras.

Lemma 3.8.2. *For any $X \subseteq \operatorname{Atom} \mathfrak{A}$, in the algebra $\mathfrak{A}$, $\sup X$ exists if and only if $\sup h(X)$ exists in the algebra $\mathfrak{B}$.*

PROOF. Since the set of atoms is recursive for both algebras, the latter are in the same situation. Therefore, it suffices to show that if $\sup X$ does not exist in $\mathfrak{A}$, then $\sup h(X)$ does not exist in $\mathfrak{B}$ either. We assume that $\sup X$ does not exist in $\mathfrak{A}$; nevertheless, $\sup h(X)$ exists. It is clear that the set X is infinite in this case. By the construction of the automorphisms φ, π, and τ, for some $i \in \{0, 1, 2\}$, the set $\{vg_\varepsilon \mid vg_\varepsilon h(X) \neq h(X), p(\varepsilon) = i\}$ is infinite. Without loss of generality, we can assume that the set is infinite for $i = 0$. Since $\sup h(X)$ exists, the set $h(X)$ is recursive. By $\varphi = \prod_{p(\varepsilon)=0} g_\varepsilon$, there exists a computable sequence $\alpha_0, \alpha_1, \ldots$ of elements of the set $\operatorname{Atom} \mathfrak{B}$ such that $\alpha_i \notin h(X)$ for all $i \in \mathbb{N}$, $v\varphi(\alpha_i) \in h(X)$ for all $i \in \mathbb{N}$, and $\sup (h(X) \cup \{\alpha_i \mid f(i) = 1\})$ exists for any recursive function $f : \mathbb{N} \to \{0, 1\}$.

The function

$$p(n, i) = \begin{cases} 1, & \alpha_i \leqslant C(n) \\ 0, & \alpha_i \leqslant n \end{cases}$$

is recursive. (We recall that elements of the Boolean algebra under consideration are natural numbers.) We show that for any recursive function $f : \mathbb{N} \to \{0,1\}$ there exists n such that $p(n,i) = f(i)$ for all i. We take an arbitrary recursive function $f : \mathbb{N} \to \{0,1\}$. Let $n = \sup\left(h(X) \cup \{\alpha_i \mid f(i) = 1\}\right)$. Then n is the desired natural number, i.e., $p(n,i) = f(i)$ for all i. Thus, p is universal for all recursive functions from $\mathbb{N}$ into $\{0,1\}$. A simple diagonal construction shows that the construction of such a function is impossible. □

Lemma 3.8.2 implies that the algebras $\mathfrak{A}$ and $\mathfrak{B}$ are isomorphic; moreover, the isomorphism is defined by the mapping that transforms a into $\sup(\{\beta \mid \beta \in \operatorname{Atom}\mathfrak{A}, \beta \leqslant a\})$. It remains to prove the recursivity of this isomorphism. To this end, we use "topological" properties of the automorphisms φ, π, and τ.

Lemma 3.8.3. *For every n it is possible to write out effectively finite sets A_φ, A_π, and A_τ such that*

(a) $\varphi(n)\backslash n = \sup A_\varphi$, $\pi(n)\backslash n = \sup A_\pi$, $\tau(n)\backslash n = \sup A_\tau$,

(b) *if at least one of the sets A_φ, A_τ, and A_ρ is nonempty, n is a unique element of $\mathfrak{A}$ satisfying the above condition. If each of these sets is empty, then only $\mathbf{0}$ or $\mathbf{1}$ of the algebra $\mathfrak{A}$ satisfies the above condition.*

PROOF. The first condition follows from the construction of the automorphisms φ, π, and τ. It is required to prove that if $\varphi(n)\backslash n = \sup A_\varphi$, $\pi(n)\backslash n = \sup A_\pi$, $\tau(n)\backslash n = \sup A_\tau$, $\varphi(m)\backslash m = \sup A_\varphi$, $\pi(m)\backslash m = \sup A_\pi$, $\tau(m)\backslash m = \sup A_\tau$ and at least one of the sets A_φ, A_π, A_τ is nonempty, then $n = m$. To this end, it suffices to prove that for some fixed atom α and any finite sequence $\varphi_0, \varphi_1, \ldots, \varphi_k$ constituted by automorphisms φ, π, and τ, we have

$$\varphi_k \ldots \varphi_1 \varphi_0(\alpha) \leqslant m \Leftrightarrow \varphi_k \ldots \varphi_1 \varphi_0(\alpha) \leqslant n$$

As α we take an arbitrary atom of $A_\varphi \cup A_\pi \cup A_\tau$. If this set turns out to be empty, then, by the construction of n, only $\mathbf{0}$ or $\mathbf{1}$ may be such an element. If the sequence $\varphi_0, \varphi_1, \ldots, \varphi_k$ is empty, then the condition is satisfied because, in this case, $\neg(\alpha \leqslant m)$ and $\neg(\alpha \leqslant n)$ in view of $a \in A_\varphi \cup A_\pi \cup A_\tau$. We assume that the condition is satisfied for some subsequence $\varphi_0, \varphi_1, \ldots, \varphi_k$, and let ψ be one of the automorphisms φ, π, and τ. We show that

$$\psi\varphi_k \ldots \varphi_1 \varphi_0(\alpha) \leqslant m \Leftrightarrow \psi\varphi_k \ldots \varphi_1 \varphi_0(\alpha) \leqslant n$$

3.8. Automorphisms of Countable Boolean Algebras

We denote $\varphi_k \ldots \varphi_1 \varphi_0(\alpha)$ by β. Without loss of generality, we assume that $\psi = \varphi$ and $\varphi(\beta) \neq \beta$. We consider two cases.

CASE 1: $\beta \in A_\varphi$. Then $\beta \leqslant \varphi(n) \backslash n$ and $\beta \leqslant \varphi(m) \backslash m$. Consequently, $\varphi(\beta) \leqslant \varphi^2(n) \backslash \varphi(n) = n \backslash \varphi(n)$ and $\varphi(\beta) \leqslant \varphi^2(m)\varphi(m) = m \backslash \varphi(m)$. Therefore, $\varphi(\beta) \leqslant n$ and $\varphi(\beta) \leqslant m$, i.e., the assertion is valid.

CASE 2: $\beta \notin A_\varphi$. We assume that $\varphi(\beta) \leqslant n$, but $\varphi(\beta) \leqslant C(m)$. Then $\beta \leqslant \varphi(n)$ and $\beta \leqslant \varphi(C(n))$. Since $\beta \notin A_\varphi$, we have $\beta \leqslant n$. By the hypothesis, $\beta \leqslant \varphi(C(n))$. Further, $\beta \leqslant \varphi(C(m)) \wedge m = m \backslash \varphi(m)$. Consequently, $\varphi(\beta) \leqslant \varphi(m) \backslash m = \sup A_\varphi$ and $\varphi(\beta) \in A_\varphi$. From $\varphi(\beta) \leqslant n$ and $\beta \leqslant n$ it follows that $\varphi(\beta) \notin A_\varphi$, which yields a contradiction. □

This lemma gives an algorithm for finding, in an algebra $\mathfrak{B}$, the isomorphic image of an arbitrary element of an algebra $\mathfrak{A}$. Let an element $n \in \mathfrak{A}$ differ from 0 and 1. The procedure of constructing the automorphisms φ, π, and τ gives the possibility of writing out effectively the sets A_φ, A_π, and A_τ such that $\varphi(n) \backslash n = \sup A_\varphi$, $\pi(n) \backslash n = \sup A_\pi$, $\tau(n) \backslash n = \sup A_\tau$. Enumerating elements of the algebras $\mathfrak{B}$, we find an element k such that $v\varphi(k) \backslash k = \sup h(A_\varphi)$, $v\pi(k) \backslash k = \sup h(A_\pi)$, $v\tau(k) \backslash k = \sup h(A_\tau)$. This element is the isomorphic image of n.

Applying a more complex technique, it is possible to establish a more general result.

Theorem 3.8.9 [135]. *Let $\mathfrak{A}$ be a recursive atomic Boolean algebra with recursive set of atoms and let $\mathfrak{B}$ be an arbitrary recursive Boolean algebra. If the groups $\mathrm{Aut}_r \mathfrak{A}$ and $\mathrm{Aut}_r \mathfrak{B}$ are elementarily equivalent, then $\mathfrak{A}$ and $\mathfrak{B}$ are recursively isomorphic.*

Exercises

1. Let $\mathrm{Sym}\,(m)$ be the group of all permutations of the set formed by m elements. Prove that the group of automorphisms of any finite Boolean algebra with m atoms is isomorphic to $\mathrm{Sym}\,(m)$.

2. Prove that any countable atomless Boolean algebra has an automorphism of infinite order.

3. Prove that any countable Boolean algebra has a nontrivial automorphism.

4. Prove that any countable Boolean algebra has an automorphism of infinite order.

5. Prove that any permutation of a finite family of atoms of a Boolean algebra can be extended to some automorphism. If a Boolean algebra contains infinitely many atoms, then the group of all finite permutations of the natural numbers is embedded in the group of all automorphisms of this algebra.

6. Prove that the group of all recursive automorphisms of a recursive atomless Boolean algebra is prime.

7. Prove that for any two infinite recursive atomic Boolean algebras $\mathfrak{A}$ and $\mathfrak{B}$ there exist recursive Boolean algebras $\mathfrak{A}^*$ and $\mathfrak{B}^*$ such that $\mathfrak{A} \cong \mathfrak{A}^*$, $\mathfrak{B} \cong \mathfrak{B}^*$, and $\mathrm{Aut}_r \mathfrak{A}^* \cong \mathrm{Aut}_r \mathfrak{B}^*$.

References

1. ALTON, D. A., Iterated quotients of the lattice of recursively enumerable sets, *Proc. London Math. Soc.,* **28**, No. 3, 1–12, 1974.
2. ANDERSON, R., The algebraic simplicity of certain groups of homeomorphisms, *Amer. J. Math.,* **130**, No. 4, 955–963, 1958.
3. ARENS, R. E. and KAPLANSKY, I., Topological representation of algebras, *Trans. Am. Math. Soc.,* **63**, 457–481, 1948.
4. ASH, C. J. and NERODE, A., Intrinsically recursive relations, *Aspects of Effective Algebra (Clayton,* 1979), Upside Down a Book Co., Yarra Glen, Vic., 26–41, 1981.
5. ASH, C. J., Stability of recursive structures in arithmetical degrees, *Ann. Pure Appl. Logic,* **32**, No. 1, 113–135, 1986.
6. BERMAN, P., Complexity of the theory of atomless Boolean algebras, *Fundamentals of Computation Theory (Proc. Conf. Algebraic, Arith. and Categorical Methods in Comput. Theory, Berlin/Wendisch-Rietz,* 1979), Akademie-Verlag, Berlin, 1979, pp. 64–70.
7. BIRKHOFF, G. and BARTEE, T., *Modern Applied Algebra*, McGraw-Hill, New York, 1975.
8. BIRKHOFF, G., *Lattice Theory*, Am. Math. Soc., New York, 1948 (Am. Math. Soc. Colloquium Publ.; 25).

9. BLASZCZYK, A., A construction of a rigid Boolean algebra, *Bull. Polish Acad. Sci. Math.*, **35**, No. 7–8, 465–471, 1987.
10. BONNET, R. and RUBIN, M., Elementary embedding between countable Boolean algebras, *J. Symbolic Logic*, **56**, No. 4, 1212–1229, 1991.
11. BONNET, R., Rigid Boolean algebras, *Handbook of Boolean Algebras*, **2**, 637–678, North-Holland, Amsterdam–New York, 1989.
12. BOOLE, G., *The Mathematical Analysis of Logic*, Cambridge, 1847.
13. BUCUR, I. and DELEANU, A., *Introduction to the Theory of Categories and Functors*, John Wiley & Sons, London–New York, 1968.
14. BURRIS, S. and MCKENZIE, R., Decidability and Boolean representation, *Mem. Am. Math. Soc.*, **32**, No. 246, 1981.
15. BUSZKOWSKI, W., Embedding Boolean structures into atomic Boolean structures, *Z. Math. Logik Grundlag. Math.*, **32**, No. 3, 227–228, 1986.
16. CARROLL, J. S., The undecidability of the lattice of r.e. subalgebras of a recursive Boolean algebra, *Notices Am. Math. Soc.*, **30**, No. 3, 281, 1985.
17. CENZER, D. and REMMEL. J., Polynomial-time versus recursive models, *Ann. Pure Appl. Logic*, **54**, No. 1, 17–58, 1991.
18. CHANG, C. C. and KEISLER, H. J., *Model Theory*, Elsevier Science Publishers, New York, 1992.
19. CHOLAK, P., Boolean algebras and orbits of the lattice of r.e. sets modulo the finite sets, *J. Symbolic Logic*, **55**, No. 2, 744–760, 1990.
20. DAULETBAEV, B. K., Determining an atomic Boolean algebra from the action of the automorphism group, *Sib. Math. J.*, **34**, No. 6, 1041–1043, 1993.
21. DAY, C. W., Superatomic Boolean algebras, *Pac. J. Math.*, **23**, 479–489, 1967.
22. DEKKER, J. C. E., Isols and generalized Boolean algebras, *Rocky Mountain J. Math.*, **20**, No. 1, 107–115, 1990.
23. DOBBERTIN, H., On Vaught's criterion for isomorphism of countable Boolean algebras, *Algebra Universalis*, **15**, No. 1, 95–114, 1982.
24. DOWNEY, R. and JOCKUCH, C. G., Every low Boolean algebra is isomorphic to a recursive one, *Proc. Am. Math. Soc.*, **122**, No. 3, 871–880, 1994.
25. DOWNEY, R., Every recursive Boolean algebra is isomorphic to one with incomplete atoms, *Ann. Pure Appl. Logic*, **60**, No. 3, 193–206, 1993.
26. DRASHOVICHEVA, KH., KATRINÁK, T., and KOLIBIAR, M., Boolean algebras and lattices close to them, *Ordered Sets and Lattices*, pp. 7–77, Univ. Komenského, Bratislava, 1985.
27. DROBOTUN, B. N., Enumerations of prime models, *Sib. Math. J.*, **18**, No. 5, 707–715, 1977.
28. DULATOVA, Z. A., Boolean algebras with a distinguished subalgebra and a distinguished automorphism, *Some Problems in Differential Equations and Discrete Mathematics*, Novosibirsk State University, Novosibirsk, 1986, pp. 130–147.
29. DULATOVA, Z. A., Constructibility of Boolean algebras with a distin-

guished subalgebra, *Mat. Zametki*, **46**, No. 6, 53–56, 1989.
30. DULATOVA, Z. A., Extended theories of Boolean algebras, *Algebraic Systems. Algorithmic Problems and the Computer*, Irkutsk, 1986, pp. 31–39.
31. DULATOVA, Z. A., Extended theories of Boolean algebras, *Sib. Mat. Zh.*, **25**, No. 1, 201–204, 1984.
32. DZGOEV, V. D., Constructive Boolean algebras, *Mat. Notes*, **44**, No. 5–6, 896–907, 1988.
33. DZGOEV, V. D., Constructivization of Boolean lattices, *Algebra Logika*, **27**, No. 6, 641–648, 1988.
34. DZGOEV, V. D., On constructivizations of some structures, *Sib. Mat. Zh.*, **21**, No. 1, 231, 1980. Dep. VINITI, No. 1606-79, Moscow, 1979.
35. EISENBERG, E. F. and REMMEL J. B., Effective isomorphisms of algebraic structures, *Patras Logic Symposion (Patras, 1980)*, North-Holland, Amsterdam–New York, 1982, pp. 95–122.
36. ERSHOV, YU. L. and PALYUTIN, E. A., *Mathematical Logic*, Mir Publishers, Moscow, 1984.
37. ERSHOV, YU. L., LAVROV, I. A., TAĬMANOV, A. D., and TAĬTSLIN, M. A., Elementary theories, *Usp. Mat. Nauk*, **20**, 35–105, 1965.
38. ERSHOV, YU. L., Relatively complemented distributive lattices, *Algebra Logic*, **18**, No. 6, 431–459, 1979.
39. ERSHOV, YU. L., Decidability of the elementary theory of relatively complemented distributive lattices and the theory of filters, *Algebra Logika*, **3**, No. 3, 17–38, 1964.
40. ERSHOV, YU. L., *The Decidability Problems and Constructible Models*, Nauka, Moscow, 1980.
41. ERSHOV, YU. L., *Numbering Theory*, Nauka, Moscow, 1977.
42. ERSHOV, YU. L., *Numbering Theory. 3 (Constructive Models)*, Novosibirsk State University, Novosibirsk, 1974.
43. ERSHOV, YU. L., Constructive models, *Selected Questions of Algebra and Logic*, Novosibirsk, 1973, 111–130.
44. ESAKIA, L. L., Boolean algebras with subordinations, *Methods for Research in Logic*, Tbilisi, 1987, pp. 75–82.
45. FEINER, L., Hierarchies of Boolean algebras, *J. Symbolic Logic*, **35**, No. 3, 305–373, 1970.
46. FEINER, L., Orderings and Boolean algebras not isomorphic to recursive ones, Ph.D. dissertation, Mass. Inst. Tech., 1967.
47. FEINER, L., The strong homogeneity conjecture, *J. Symbolic Logic*, **35**, 375–377, 1970.
48. FEURSTEIN, S., Quantifier elimination for Stone algebras, *Arch. Math. Logic*, **28**, No. 2, 75–89, 1989.
49. FOURE, R. and HEURGONOVA, E., *Ordered Structures and Boolean Algebras*, Academia, Prague, 1984.
50. FREIDBURG, R. M., Three theorems on recursive enumeration, *J. Symbolic Logic*, **23**, 309–316, 1958.

51. FUCHINO, S., KOPPELBERG, S., and TAKAHASHI, M., On $L_{\infty k}$-free Boolean algebras, *Ann. Pure Appl. Logic*, **55**, No. 3, 265–284, 1992.
52. FUCHINO, S., Some remarks on openly generated Boolean algebras, *J. Symbolic Logic*, **59**, No. 1, 302–310, 1994.
53. GONCHAROV, S. S. and DROBOTUN B. N., Numeration of saturated and homogeneous models, *Sib. Math. J.*, **2**, No. 21, 164–175, 1980.
54. GONCHAROV, S. S. and DZGOEV, V. D., Autostability of models, *Algebra Logic*, **19**, No. 1, 28–37, 1980.
55. GONCHAROV, S. S. and NURTAZIN, A. T., Constructive models of complete sovable theories, *Algebra Logic*, **12**, No. 2, 67–77, 1973.
56. GONCHAROV, S. S., A recursively representable Boolean algebra, *Boundary-Value Problems for Differential Equations and Their Applications in Mechanics and Technology*, Nauka, Alma-Ata, 1983, pp. 43–46.
57. GONCHAROV, S. S., Autostability and computable families of constructivizations, *Algebra Logic*, **14**, No. 6, 647–680, 1975.
58. GONCHAROV, S. S., Autostability of models and Abelian groups, *Algebra Logic*, **19**, No. 1, 23–44, 1980.
59. GONCHAROV, S. S., The quantity of nonautoequivalent constructivizations, *Algebra Logic*, **16**, No. 3, 169–185, 1977.
60. GONCHAROV, S. S., Strong constructivizability of homogeneous models, *Algebra Logic*, **17**, No. 4, 247–262, 1978.
61. GONCHAROV, S. S., Universal recursively enumerable Boolean algebras, *Sib. Math. J.*, **24**, No. 6, 36–43, 1983.
62. GONCHAROV, S. S., *Countable Boolean Algebras*, Nauka, Novosibirsk, 1988.
63. GONCHAROV, S. S., Constructivizability of superatomic Boolean algebras, *Algebra Logic*, **12**, No. 1, 363–388, 1973.
64. GONCHAROV, S. S., Nonautoequivalent constructivizations of atomic Boolean algebras, *Mat. Zametki*, **19**, No. 6, 853–858, 1976.
65. GONCHAROV, S. S., Some propeties of constructivizations of Boolean algebras, *Sib. Math. J.*, **16**, No. 2, 264–278, 1975.
66. GONCHAROV, S. S., Restricted theories of constructive Boolean algebras, *Sib. Math. J.*, **17**, No. 4, 797–812, 1976.
67. GONCHAROV, S. S., *Constructive Models*, Clayton, Monash University, (Preprint No. 56), 1984.
68. GRYGIEL, J., Absolutely independent sets of generators of filters in Boolean algebras, *Rep. Math. Logic*, No. 24, 25–35, 1991.
69. GUICHARD, D., Automorphisms of substructure lattices in effective algebra, *Ann. Pure Appl. Logic*, **25**, 47–58, 1983.
70. GUICHARD, D., Automorphisms of substructure lattices in effective algebra, *Ann. Pure Appl. Logic*, **25**, 47–58, 1983.
71. HALMOS, P., *Lectures on Boolean algebras*, Van Nostrand, Toronto–New York–London, 1963.
72. HANDBOOK of Mathematical Logic, Nauka, Moscow, 1982.
73. HANF, W., On some fundamental problems concerning isomorphism of

Boolean algebras, *Math. Cand.*, **5**, 205–217, 1957.
74. HANF, W., Primitive Boolean algebras, *Proc. Sympos. Pure Math.*, 75–90, 1974.
75. HANSOUL, G. and WILLEMS, B., Recursive construction of some non-well-founded Boolean spaces, *Bull. Soc. Roy. Sci. Liege*, **63**, No. 5, 383–392, 1994.
76. HEINDORF, L. *Contribution to the Model Theory of Boolean Algebras*, Humboldt-Univerität, Sektion Math., Berlin, 1984.
77. HEINDORF, L., Alternative characterization of finitary and well-founded Boolean algebras, *Algebra Universalis*, **29**, No. 1, 109–135, 1992.
78. HEINDORF, L., Comparing the expressive power of some languages for Boolean algebras, *Z. Math. Logik Grundlag. Math.*, **27**, No. 5, 419–434, 1981.
79. HENDRY, H. E., Two remarks on the atomistic calculus of individuals, *Noûs*, **14**, No. 2, 235–237, 1980.
80. HERRMAN, E., Definable Boolean pairs in the lattice of recursively enumerable sets, *Proc. Conf. Model Theory,* Seminar berichte. N 49. Berlin: Humboldt-Univerität, Sektion Math., Berlin, 1984.
81. HORN, A. and TARSKI, A., Measures in Boolean algebras, *Trans. Am. Math. Soc.*, **64**, 467–497, 1948.
82. IVERSON, P., The number of countable isomophism types of complete extension of the theory of Boolean algebras, *Colloq. Math.*, **62**, No. 2, 181-187, 1991.
83. JECH, T. J., A note on countable Boolean algebras, *Algebra Universalis*, **14**, No. 2, 257–262, 1982.
84. JECH, T. J., *Lectures in Set Theory with Particular Emphasis on the Method of Forcing*, Springer Verlag, Berlin–Heidelberg–New York, 1971.
85. JOCKUSCH, C. G., JR. and SOARE, R.I., Boolean algebras, Stone spaces, and the iterated Turing jump, *J. Symbolic Logic*, **59**, No. 4, 1121–1138, 1994.
86. JONSSON, B., A survey of Boolean algebras with operators, *Algebras and Orders (Montreal, PQ, 1991)*, Kluwer Academic Publishers, Dordrecht, 1993, pp. 239–286.
87. JURIE, P.-F. and TOURAILLE, A., Elementary equivalent ideals in a Boolean algebra, *C. R. Acad. Sci. Paris, Ser. Math.*, **299**, No. 10, 415–418, 1984.
88. KELLEY, J. L., *General Topology*, Van Nostrand, Princeton, New Jersey, 1957.
89. KEMMERICH, S. and RICHTER, M. M., Remarks on the automorphism group of homogeneous Boolean algebras, *Ökonomie unde Mathematik*, Springer Verlag, Berlin–New York, 1987, pp.23–28.
90. KETONEN, J., The structure of countable Boolean algebras, *Ann. Math.*, **108**, 41–89, 1978.
91. KHISAMIEV, N. G., On strongly constructive models of a decidable theory, *Izv. Akad. Nauk Kaz.SSR, Ser. Fiz-mat.*, No. 1, 83–84, 1974.

92. KHISAMIEV, N. G., On strongly constructive models, *Izv. Akad. Nauk Kaz.SSR, Ser. Fiz-mat.*, No. 3, 59–63, 1971.
93. KOKORIN, A. I. and PINUS A. G., Questions of the decidability of extended theories, *Usp. Mat. Nauk,* **33**, No. 2(200), 49–84, 1978.
94. KOPPELBERG, S. and MONK, J. D., Homogeneous Boolean algebras with very nonsymmetric subalgebras, *Notre Dame J. Formal Logic*, **24**, No. 3, 353–356, 1983.
95. KOPPELBERG, S., A construction of Boolean algebras from first-order structures, *Ann. Pure Appl. Logic*, **59**, No. 3, 239–256, 1993.
96. KOPPELBERG, S., *Handbook of Boolean Algebras*, **1–3**, North-Holland, Amsterdam–New York, 1989.
97. KOSTRIKIN, A. I., *Introduction to Algebra*, Nauka, Moscow, 1977.
98. KURATOWSKI, K. and MOSTOWSKI A., *Set Theory*, North-Holland, Amsterdam–New York, 1967.
99. KURATOWSKI, K., *Topology*, Academic Press, London–New York, 1963.
100. LA ROCHE, P., Recursively presented Boolean algebras, *Notices Am. Math. Soc.*, **24**, 552, 1977.
101. LACHLAN, A. H., On the lattice of recursively enumerable sets, *Trans. Am. Math. Soc.*, **130**, No. 1, 1–36, 1974.
102. LIANG, P., The relative structure of the denumerable nonatomic Boolean algebras satisfying $\forall A_0 \in PA(VA_0, V(A - A_0) \notin B \to G_{A_0} = \varnothing)$ and research on the structure of the denumerable nonatomic Boolean algebras *J. Math./ Res. Exposition*, **7**, No. 1, 13–16, 1987.
103. MADISON, E. W. and NELSON, G. C., Some examples of constructive and nonconstructive extension of the countable atomless Boolean algebras, *J. London Math. Soc.*, **11**, 325–336, 1975.
104. MADISON, E. W., Cominatorial and recursive aspects of the automorphism group of the countable atomless Boolean algebra, *J. Symbolic Logic*, **51**, No. 2, 292–301, 1986.
105. MADISON, E. W., On Boolean algebras and their recursive completion, *Z. Math. Logik Grundlag. Math.*, **31**, No. 6, 481–486, 1985.
106. MADISON, E. W., The existence of countable totally nonconstructive extension of the countable atomless Boolean algebra, *J. Symbolic Logic*, **48**, No. 1, 167–170, 1983.
107. MAL'TSEV, A. I., Constructible algebras. I, *Usp. Mat. Nauk*, **16**, No. 3, 3–60, 1961.
108. MAL'TSEV, A. I., *Algebraic Systems*, Springer Verlag, Berlin, 1976.
109. MAL'TSEV, A. I., *Algorithms and Recursive Functions*, Wolters-Noordoff, Groningen, 1970.
110. MAL'TSEV, A. I., *Research in the Field of Mathematical Logic*, *Mat. Sb.*, **1**, No. 3, 323–335, 1963.
111. MAL'TSEV, A. I., Untersuchengen aus dem Gebiete der mathematischen Logik, *Rec. Math. N.S.*, **1**, 323–336, 1936.
112. MANASTER, A. B. and REMMEL, J. B., Co-simple higher-order indecom-

posable isols, *Z. Math. Logik Grundlag. Math.*, **26**, No. 3, 279–288, 1980.
113. MART'YANOV, V. I., Undecidability of the theory of Boolean algebras with automorphism. *Sib. Math. J.*, **23**, No. 3, 408–414, 1982.
114. MARTIN, D. A. and POUR-EL, M. B., Axiomatizable theories with few axiomatizable extensions, *J. Symbolic Logic*, **35**, No. 2, 205–209, 1970.
115. MARTIN, D. A., Classes of recursively enumerable sets and degrees of unsolvability, *Z. Math. Logik Grundlag. Math.*, **12**, No. 4, 295–310, 1966.
116. MAYER, R. D. and PIERCE, R. S., Boolean algebras with ordered bases, *Pac. J. Math.*, **10**, 925–942, 1960.
117. MCKENZIE, R., On the automorphism groups of denumerable Boolean algebras, *Can. J. Math.*, No. 3, 466–471, 1977.
118. MEAD, J. and NELSON, G. C., Model companions and K-model completeness for the complete theories of Boolean algebras, *J. Symbolic Logic*, **45**, No. 1, 47–55, 1980.
119. MEAD, J., Recursive prime models for Boolean algebras, *Colloq. Math.*, **41**, No. 1, 25–33, 1979.
120. METAKIDES, G. and NERODE, A., Recursively enumerable vector spaces, *Ann. Math. Logic*, **11**, 147–177, 1977.
121. METAKIDES, G. and NERODE, A., *Recursion Theory and Algebra*, Lecture Notes in Math., **450**, Springer Verlag, Berlin, 1975, pp. 209–219.
122. MIJAJLOVIC, Z., Saturated Boolean algebras with ultrafilters, *Publ. Inst. Math.*, Beograd, N. S., **26**, No. 40, 175–197, 1979.
123. MILDENBERGERM H., A rigid Boolean algebra that admits the elimination of Q_1^2, *Fundam. Math.*, **142**, No. 1, 1–18, 1993.
124. MOLZAN, B., On the number of different theories of Boolean algebras in several logics, *Workshop in Extended Model Theory (Berlin, 1980)*, Akad. Wiss. DDR, Berlin, 1981, pp. 102–113.
125. MOLZAN, B., The theory of superatomic Boolean algebras in the logic with the binary Ramsey quantifier, *Z. Math. Logik Grundlag. Math.*, **28**, No. 4, 365–376, 1982.
126. MONK, J. D., Automorphism groups, *Handbook of Boolean Algebras*, **2**, North-Holland, Amsterdam–New York, 1989, pp. 517–546.
127. MONK, J. D., Cardinal functions on Boolean algebras, *Lectures Math.* ETH Zürich. Birkhäuser Verlag, Basel, 1990.
128. MONK, J. D., On the automorphism groups of denumerable Boolean algebras, *Math. Ann.*, **216**, No. 1, 5–10, 1975.
129. MONK, J. D., *Mathematical Logic*, Springer Verlag, Berlin, 1976.
130. MOROZOV, A. S., Automorphisms of constructivization of Boolean algebras, *Sib. Math. J.*, **26**, No. 4, 555–565, 1985.
131. MOROZOV, A. S., Constructive Boolean algebras with almost identical automorphisms, *Mat. Zametki*, **37**, No. 4, 478–482, 1985.
132. MOROZOV, A. S., Countable homogeneous Boolean algebras, *Algebra Logic*, **21**, No. 3, 181–190, 1982.
133. MOROZOV, A. S., Decidability of theories of Boolean algebras with a dis-

tinguished ideal, *Sib. Mat. Zh.*, **23**, No. 1, 199-201, 1982.
134. MOROZOV, A. S., Groups of recursive automorphisms of constructive Boolean algebras, *Algebra Logic*, **22**, No. 2, 95–112, 1983.
135. MOROZOV, A. S., Recursive automorphisms of atomic Boolean algebras, *Algebra Logic*, **29**, No. 4, 310–330, 1990.
136. MOROZOV, A. S., Strong constructivizability of countable saturated Boolean algebras, *Algebra Logic*, **21**, No. 2, 130–137, 1982.
137. MOSES, M., Recursive linear orders with recursive successivities, *Ann. Pure Appl. Logic*, **27**, 253–364, 1984.
138. MOSES, M., Recursive properties of isomorphism types, *J. Aust. Math. Soc.*, **34**, 269–286, 1983.
139. MYERS, D., The Boolean algebra of the theory of linear orders, *Israel J. Math.*, **35**, No. 3, 234–256, 1980.
140. NAGAYAMA, M., On Boolean algebras and integrally closed commutative regular rings, *J. Symbolic Logic*, **57**, No. 4, 1305–1318, 1992.
141. NERODE, A. and REMMEL, J. B., A survey of the lattices of r. e. substructures, *Proc. Sympos. Pure Math.*, 323–376, 1985.
142. NERODE, A. and REMMEL, J. B., Generic objects in recursion theory, *Recursion Theory Week: Proc. (Oberwalfakh, 1984)*, Lecture Notes in Math., **1141**, Springer Verlag, Berlin, 1985.
143. NURTAZIN, A. T., Basis aggregates and two problems in the theory of Boolean algebras, *Izv. Acad. Nauk Kazakh. SSR, Ser. Fiz-Mat.*, No. 3, 33–36, 1986.
144. NURTAZIN, A. T., Strong and weak constructivizations and computable families, *Algebra Logic*, **13**, No. 3, 177–184, 1974.
145. ODINTSOV, S. P. and SELIVANOV, V. L., Arithmetical hierarchy and ideals of enumerated Boolean algebras, *Sib. Math. J.*, **30**, No. 6, 952–960, 1989.
146. ODINTSOV, S. P., Atom-free ideals of constructive Boolean algebras, *Algebra Logic*, **23**, No. 3, 190–202, 1984.
147. ODINTSOV, S. P., Intrinsically recursively enumerable subalgebras of a recursive Boolean algebra, *Algebra Logic*, **31**, No. 1, 24–29, 1992.
148. ODINTSOV, S. P., Lattice of recursively enumerable subalgebras of a recursive Boolean algebra, *Algebra Logic*, **25**, No. 6, 397–403, 1986.
149. ODINTSOV, S. P., Recursive Boolean algebras with a hyperhyperimmune set of atoms, *Mat. Zametki*, **44**, No. 4, 488–493, 1988.
150. ODINTSOV, S. P., *Restricted Theories of Constructive Booleans Algebras of Wide Layer*, Novosibirsk, Institute of Math. (Preprint No. 21), 1986.
151. PÜHRINGER, CH., A completeness proof for the theory of the atomless $\cap$-^{c}-partial Boolean algebras, *Conceptus*, **16**, No. 38, 81–88, 1982.
152. PADMANABHAN, R., A first-order proof of a theorem of Frink, *Algebra Universalis*, **13**, No. 3, 397–400, 1981.
153. PAL'CHUNOV, D. E., Countably categorical Boolean algebras with distinguished ideals, *Studia Logica*, **46**, No. 2, 121–135, 1987.
154. PAL'CHUNOV, D. E., Direct summands of Boolean algebras with distin-

guished ideals, *Algebra Logic*, **31**, No. 5, 295–315, 1992.
155. PAL'CHUNOV, D. E., Finitely axiomatizable Boolean algebras with distinguished ideals, *Algebra Logika*, **26**, No. 4, 435–455, 1987.
156. PAL'CHUNOV, D. E., Lindenbaum–Tarski algebra of a class of Boolean algebras with one distinguished ideal, *Algebra Logic*, **33**, No. 2, 179–210, 1994.
157. PAL'CHUNOV, D. E., Undecidability of theories of Boolean algebras with selected ideals, *Algebra Logic*, **25**, No. 3, 206–218, 1986.
158. PERETYAT'KIN, M. G., Strongly constructive models and numerations of the Boolean algebra of recursive sets, *Algebra Logic*, **10**, No. 5, 332–345, 1971.
159. PERETYAT'KIN, M. G., Turing machine computations in finitely axiomatizable theories, *Algebra Logic*, **21**, No. 4, 272–295, 1982.
160. PEROVIČ, Ž., Relatively complete 2-extension of Boolean algebras, *Math. Balkanica*, **6**, No. 2, 125–128, 1992.
161. PINUS, A. G., Applications of Boolean powers of algebraic systems, *Sib. Math. J.*, **26**, No. 3, 400–406, 1985.
162. PINUS, A. G., Constructivizations of Boolean algebras, *Sib. Math. J.*, **22**, 616–619, 1981.
163. PINUS, A. G., Theories of Boolean algebras in a calculus with the quantifier "infinitely many exist," *Sib. Math. J.*, **17**, No. 6, 1035–1038, 1976.
164. PINUS, A. G., *Boolean Constructions in Universal Algebras*, Kluwer Academic Publishers, Dordrecht, 1993.
165. POST, E. L., Recursively enumerable sets of positive algebras and their decision problems, *Bull. Am. Math. Soc.*, **50**, 284–316, 1944.
166. RASEVA, E. and SIKORSKI, P., *Mathematics of Metamathematics*, Nauka, Moscow, 1972.
167. REMMEL, J. B., R-maximal Boolean algebras, *J. Symbolic Logic*, **44**, No. 4, 533–548, 1979.
168. REMMEL, J. B., Complementation in the lattice of subalgebras of a Boolean algebra, *Algebra Universalis*, **10**, No. 1, 48–64, 1980.
169. REMMEL, J. B., Recursion theory on algebraic structure with an independent set, *Ann. Math. Logic*, **18**, No. 2, 153–191, 1980.
170. REMMEL, J. B., Recursive Boolean algebras with recursive atoms, *J. Symbolic Logic*, **46**, No. 3, 595–616, 1981.
171. REMMEL, J. B., Recursive Boolean algebras, *Handbook of Boolean Algebras*, **3**, North-Holland, Amsterdam–New York, 1989, pp. 1097–1166.
172. REMMEL, J. B., Recursive isomorphism types of recursive Boolean algebras, *J. Symbolic Logic*, **46**, No. 3, 572–594, 1981.
173. REMMEL, J. B., Recursively categorical linear orderings, *Proc. Am. Math. Soc.*, **83**, 387–391, 1981.
174. REMMEL, J. B., Recursively enumerable Boolean algebras, *Ann. Math. Logic*, **14**, No. 2, 74–107, 1978.
175. REMMEL, J. B., Recursively rigid recursive Boolean algebras, *Ann. Pure*

Appl. Logic, **36**, 39–52, 1987.
176. RICE, H. G., Classes of recursively enumerable sets and their decision problems, *Trans. Am. Math. Soc.*, **74**, 358–366, 1953.
177. ROBINSON, R. W., Simplicity of recursively enumerable sets, *J. Symbolic Logic*, **32**, 162–172, 1967.
178. ROGERS, H. J., *Theory of Recursive Functions and Effective Computability*, McGraw-Hill, New York, 1967.
179. RUBIN, M., On the automorphism groups of homogeneous and saturated Boolean algebras, *Algebra Universalis*, **9**, No. 1, 54–86, 1979.
180. RUBIN, M., The theory of Boolean algebras with a distinguished subalgebra is undecidable, *Ann. Sci. Univ. Clermont #60, Math.*, **13**, 129–134, 1976.
181. SACHS, D., The lattice of subalgebras of a Boolean algebra, *Can. J. Math.*, **14**, 451–460, 1962.
182. SARYMSAKOV, T. A., *Semifields and Probability Theory*, FAN, Uz. SSR, Tashkent, 1983.
183. SENUKOV, V. F., A sequential model for a Boolean algebra, *Dokl. Akad. Nauk Urk. SSR,* Ser. A, No. 2, 19–21, 1988.
184. SHEFFER, H. M., A set of five independent postulates for Boolean algebras with application to logical constants, *Trans. Am. Math. Soc.*, No. 14, 481–488, 1913.
185. SHI, N. D., Creative pairs of subalgebras of recursively enumerable Boolean algebras, *Acta Math. Sinica*, **25**, No. 6, 737–745, 1982.
186. SHI, N. D., Splitting recursively enumerable subalgebras in recursive Boolean algebras, *Acta Math. Sinica,* **4**, No. 1, 14–17, 1988.
187. SIKORSKI, P., *Boolean Algebras*, Berlin–Heidelberg–New York, 1964.
188. SOARE, R. I., Recursively enumerable sets and degrees, *Bull. Am. Math. Soc.*, **84**, 1149–1182, 1978.
189. SOARE, R. I., *Recursively Enumerable Sets and Degrees*, Springer Verlag, Berlin–Heidelberg–New York, 1986.
190. STONE, M. N., The theory of representations for Boolean algebras, *Trans. Am. Math. Soc.*, **40**, No. 1, 37–111, 1936.
191. ŠTEPHÁNEK, P., VAN DOVEN, E. K., MONK, J. D., and RUBIN, M., Embeddings and automorphisms, *Handbook of Boolean Algebras*, **2**, 607–636, Elsevier Science Publishers, 1989.
192. TAKAHASHI, M., Completeness of Boolean power of Boolean algebras, *J. Math. Soc. Japan*, **40**, No. 3, 445–456, 1988.
193. TARSKI, A., MOSTOWSKI, A., and ROBINSON, R., *Undecidable Theories*, North-Holland, Amsterdam, 1953.
194. TARSKI, A., Arithmetical classes and types of Boolean algebras, *Bull. Am. Math. Soc.*, **55**, 63, 1949.
195. THURBER, J. J., Recursive and r.e. quotient Boolean algebras, *Arch. Math. Logic*, **33**, No. 2, 121–129, 1994.
196. TOURAILLE, A., Theories of Boolean algebras equipped with distinguished

ideals. Part I, *J. Symbolic Logic*, **52**, No. 4, 1027–1043, 1987.
197. TOURAILLE, A., Theories of Boolean algebras equipped with distinguished ideals. Part II, *J. Symbolic Logic*, **55**, No. 3, 1192–1212, 1990.
198. TOURAILLE, A., Elimination of quantifiers in the elementary theory of Boolean algebras with a family of distinguished ideals, *C. R. Acad. Sci. Paris, Ser. Math.*, **300**, No. 5, 125–128, 1985.
199. VAN DOVEN, E. K., MONK, J. D., and RUBIN, M., Some questions about Boolean algebras, *Algebra Universalis*, **11**, No. 2, 220–243, 1980.
200. WARDEN B. L. VAN DER, *Algebra*. Springer Verlag, Berlin, 1971.
201. VAUGHT, R. L., *Topics in the Theory of Arithmetical Classes and Boolean Algebras*, Doctoral Thesis, University of California, Berkeley, 1954.
202. VINOKUROV, S. F., DULATOVA, Z. A., and PERYAZEV, N. A., Positive classification of Boolean algebras in an extended signature, *Algebraic Systems. Algorithmic Problems and the Computer*, Irkutsk, 1986, pp. 121–131.
203. VLADIMIROV, D. A., *Boolean Algebras*, Nauka, Moscow, 1962.
204. VLASOV, V. N. and GONCHAROV, S. S., Strong constructibility of Boolean algebras of elementary characteristic (1, 1, 0), *Algebra Logic*, **32**, No. 6, 334–341, 1993.
205. WEESE, M. and GOLTZ, H.-J., *Boolean Algebras*, Humboldt Universität, Sektion Math., Berlin, 1984.
206. WEESE, M., The theory of Boolean algebras with Q_0 and quantification over ideals, *Z. Math. Logik Grundlag. Math.*, **32**, No. 2, 189–191, 1986.
207. WEESE, M., *A New Product for Boolean Algebras and a Conjecture of Feiner*, Humboldt Universität, Sec. Math., Berlin, **29**, No. 4, 441–443, 1980.
208. WERNER, H., Boolean constructions and their role in universal algebra and model theory, *Universal Algebra and Its Links with Logic, Algebra, Combinatorics, and Computer Science*, Darmstadt, 1983, pp. 106–114.
209. WERNER, H., Sheaf constructions in universal algebra and model theory, *Universal Algebra and Applications*, Banach Center Publ., **9**, 1982, pp. 133–179.
210. WESOLOWSKI, T., Varieties of locally Boolean algebras, *Studia Sci. Math. Hungar*, **27**, No. 3-4, 339–347, 1992.
211. YABLONSKIĬ, S. V., GAVRIOLOV, G. P., and KUDRYAVTSEV, V. B., *Functions of Algebra and the Post Classes*, Nauka, Moscow, 1966.

Subject Index

Axiom, system 107
   enumerable 107
   recursive 107
Atom 71

Boolean lattice 19
Boolean pair 289
Boolean rank 82
Boolean ring 22
Boolean algebra 16
   atomic 29
   atomless 29
   dense 115
   superatomic 49
   universal 262
   α-atomic 29

Bound
   greatest lower 14
   least upper 14
   lower 14
   upper 14
Branch 66

Characteristic 109
Class
   axiomatizable 10
   effectively infinite 251
Class, congruence 3
Completion, ideal 32
Condition
   embedding 45
   epimorphism 45

isomorphism 45
Congruence 3
    strict 3
Constructivizations
    B-autoequivalent 164
    equivalent
        B-recursively 164

Dagram 97
    complete 97, 115
Dimension algorithmic 164

Element
    atomic 38, 52
    atomless 38
    complete 228
    greatest 14
    least 14
    orthogonal 75
    n-complete 228
    t-atomless 224
    t-principal 224
Embedding 3
    elementary 96
    isomorphic 2, 96
Enrichment 4
Epimorphism 2
Ershov algebra 30
    atomic 73
    atomless 75
    normal 82
    special 77
    superatomic 71
Ershov–Tarski ideal 108
Extension 32
    elementary 92

Feiner hierarchy 153

Filter 15, 26
    principal 26
    proper 15
Formula 7
    atomic 7
    complete 94
    consistent 9
    perfect 12
    T-equivalent 100
Fragment 161
Frechét ideal 28, 71
Function
    additive 76
    partial 150
    recursive 150
        partially 150

Gödel theorem 9

Henkin theory 94
Homomorphism 2
    canonical 28
    recursive 262
    C-recursive 162

Ideal 26
    dense 139
    maximal 39
    prime 39
    principal 26
        locally 35
    proper 15
Isomorphism 3
Interpretation 2

Kernel, homomorphism 28

Subject Index

Lattice 19
  distributive 15
Lindenbaum algebera 17

Mal'tsev theorem 9
Model 2
  almost n-complete 105
  almost complete 105
  branching 251
  decidable 190
  prime 98
  representation 250
  saturated countably 98
  universal 98
  ω-complete 104
  ω-saturated 98
  n-complete 104
  $\mathcal{B}$-decidable 161

Numbering 156
  Gödel 156
  positive 261
  principal 218

Order 19

Quotient set 3

Stone theorem 41
Segment 67
  initial 14
  normal 271
Sublattice
Sentence 10
Sequence

disjunctive 45
exact 31
principal 233
t-consistent 193
Set
  consistent 9
  cross-section 234
  decidable 158
  enumerable 158
  inconsistent 9
  recursive 150
  satisfiable 9
  unsatisfiable 9
Signature 1
Skolem enrichment 93
Skolem function 93
Subalgebra
  hyperhypersimple 287
  maximal 283
Subsystem 4
  elementary 92
Subtree 67
Symbol
  constant 2
  functional 1
  predicate 1
System 2
  algebraic 2
  atomic 98
  autostable 164
  enumerated 156
  $\mathcal{B}$-ω-constructive 161
  $\mathcal{B}$-constructive 156
  ω-homogeneous 96
  $\mathcal{B}$-n-constructive 161
  $\mathcal{B}$-$\mathfrak{F}$-constructive 161
  $\mathcal{B}$-recursive 161

Theory 10
  atomic 99

>   complete 10
>   countable categorical 99
>   decidable 107
>   definable 105
>   elementary 10
>   model-complete 100
>   superatomic 99
> Transposition 291
>   atomic 291
> Tree 60, 67
>   complete 67
>   generating 61

Ultrafilter 27
Universe 2

Vertex end 62
Vaught theorem 47

C-extension 162
C-homomorphism 162
E-rank 76
Π_n^B-set 151